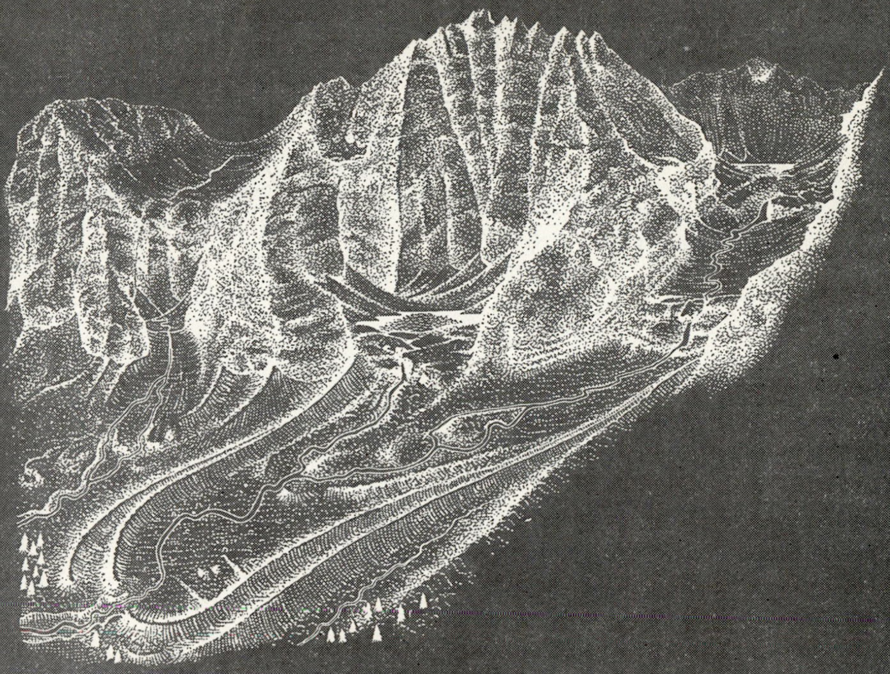

Science and Engineering

E. C. DAPPLES

Professor of Geology

Northwestern University

Library of Congress Catalog Card Number: 59-5880

Printed in the United States of America

Preface

The presentation of the subject matter of geology to the scientist and the engineer is beset with difficulties in the selection of what is to be considered the essential ingredients of the science. Fundamentally, this situation stems from the many facets that constitute its body of knowledge, which ranges between the taxonomy of fossil organisms and theoretical geophysics. Moreover, the very nature of the physical scale of the earth has dictated that the subject matter be largely descriptive, and only in limited fields has our information yielded to theoretical analysis. There exists, therefore, no clearly agreed upon selection of topics among professional geologists concerning where to begin and what to include in the root subjects to be taught the preprofessional student. Those students planning to become geologists, agronomists, mining, civil, and sanitary engineers, or other scientists whose future professions embody a basic understanding of physical geology require some special introduction presupposing an elementary knowledge of chemistry, physics, and mathematics. The profound interdependence which exists between the various specialties of geology poses problems of the organization of the material to be presented as well as the nature of the subjects to be included. For these reasons certain introductory courses have been developed in which the emphasis has been placed on the application of geology rather than on instruction in the fundamentals of the science. In this respect I differ from my colleagues who hold to the need of such an introduction, and I am firm in the belief that there are certain aspects of physical geology which must be considered basic to further study in the applied, descriptive, or theoretical form. With such a viewpoint in mind, I have organized the chapters of this book to introduce certain concepts in a particular order in which they may be employed in the chapters to follow. Such a presentation requires that from time to time the reader return to the same general subject in his progress through the book. The intent has been to introduce a concept where it appears to be most pertinent to the exposition. Return to the subject in later chapters is intentional and is considered appropriate because the reader has been instructed in the concepts concerned and because additional theoretical characterization can be revealed by graphs and tables.

The instructor who selects this book for his course will find that the pattern of presentation is organized as much as possible around empirical relationships; graphs are used wherever possible, and tables where description is preferred. Such an approach introduces an order to the presentation of the subject matter, and this order should not be altered from chapter to chapter. By using this scheme, the instructor will find increased freedom in the selection of his lectures and the liberty to expand certain subjects in the manner of his choosing. In this connection I have made an attempt to present each chapter in such form that he may prepare from data of his own choice certain homework problems to be utilized in further understanding of selected topics.

The logic of science demands transition from description into theory as the body of fact reaches certain propositions. Geological facts are by and large difficult to obtain even on the very skin of the earth, but with the progress of time much information is slowly yielding to classification and theoretical analysis. Tables and graphs have been utilized, therefore, wherever feasible as devices for reducing descriptive material and for organizing information as the level of precision dictates.

The broad outline of the book was conceived some years ago as the product of discussion with colleagues in the geological and engineering professions. In this respect I have been guided on innumerable occasions during preparation of the manuscript by the counsel of my associates at Northwestern University, principally L. L. Sloss, L. H. Nobles, W. C. Krumbein, and A. L. Howland. From time to time they have critically examined selected chapters, and their suggested revisions are most gratefully acknowledged. I am pleased to acknowledge also the significant contributions of Professors Don U. Deere and Joseph L. Rosenholtz, who by many helpful suggestions were instrumental in bringing about certain improvements in form and concept. Lastly, I wish to thank those who were so generous in supplying photographs to be used as illustrations. Principal among these contributors were Dr. M. M. Leighton, Dr. H. B. Willman, Dr. G. E. Ekblaw, Dr. L. H.

Nobles, Dr. T. R. Beveridge, Mr. Gerald Massie, The American Geographical Society, the Gulf Oil Corporation, and the U. S. Geological Survey.

E. C. Dapples

Evanston, Illinois
November, 1958

Contents

1 Introduction

THE SCOPE OF APPLIED GEOLOGY

In the years of evolution of science from its primitive beginnings to its current stages an aspect of this progress was the segregation of branches of learning into separate realms. Thus geology concerned with the nature and behavior of the earth was considered as related to other sciences only insofar as certain special aspects overlapped. With some exceptions most geologists had little in common with their colleagues in chemistry, engineering, agronomy, and the like. Now this stage of progress appears to be reaching an end as universal application of specialized knowledge has demanded the organization of the scientific team. Subjects once considered to be distinct and separate

Moving oil derrick. (Photograph by Bert Brandt, courtesy Gulf Oil Corporation.)

provinces have become nebulous in their interrelationship. Geology has become so intimately associated with certain aspects of chemistry, physics, and biology that certain research projects no longer can be catalogued as peculiar to geology any more than to such other fields of science.

Similarly with the progress in engineering construction and boldness in the projects undertaken an aspect of science has been developed known as engineering geology. In its broadest sense this is the specialized knowledge of earth processes and materials which establish the physical conditions restricting the design, erection, and operation of engineering works. Today virtually all major construction involves excavation of soil or rock, its transportation to a new site, and boring into the soil or the underlying rock. Under certain special conditions such as mining or tunneling an understanding of the properties of earth materials is a major aspect of successful practice. Likewise, thorough familiarity with geologic processes is of singular importance in engineering construction in localities subject to volcanic eruptions, earthquakes, storms, tidal waves, and other less spectacular phenomena such as shifting of beaches. In a parallel fashion as knowledge of the properties of soils is increased the soil scientist and agronomist find new demand for more precise understanding of the geologic processes which control the fundamental properties of soils.

As the name suggests engineering geology is a branch of earth science of particular interest to the engineer. In this connection certain aspects of geology assume greater importance than others. Also, as additional geologic information becomes available it must be embodied in engineering geology. Indeed, successful practice requires not only recognition of geologic features, but also an ability to translate such recognition into the mechanism by which they were produced. For this reason the soil scientist and the engineer must erect a framework of geologic knowledge. This must consist not merely of a long list of terms or isolated observations, but rather a closely knit relationship of principles many of which as yet can be expressed in qualitative terms only.

In its most general definition geology concerns the surface expression, rock composition, structure, internal activity, and history of the earth. As with other sciences its beginning is buried with civilizations long past and was centered in a philosophical interest in the shape of the earth and the origin of rocks. Later after the period of world exploration geology began to emerge as a clearly defined variant of natural history and to take its place as an individual field of science.

Modern geology is of recent origin, and all of its concepts have been expressed in the past one hundred and fifty years. It is a deductive science in which the geologist reconstructs events and processes of earth history on the basis of physical, chemical, and organic changes which he observes in a comparatively brief interval of time. Within such a framework the geologist records and interprets the significance of the composition of minerals, solid rocks, loose sediments, and fluids which rest upon and move through such rock materials. He is interested in how and when these materials formed, in what order they developed, and what properties they now possess. Interpretation of observed data controls reconstruction of the geology of an area, and as new data become available new interpretations may be required. Often these result in radical revisions in the former reconstruction of geologic conditions or in geologic thought. Through such constant revision of old concepts and the application of new knowledge from other sciences progress is made in understanding the behavior of the earth.

In its early period of growth geology was an academic science, and its many practical applications were scarcely apparent. As the industrial revolution progressed the demand for mineral raw materials placed emphasis on their discovery and exploitation, and transfer of the science into the applied field was accomplished. In this connection its most spectacular use has been the amazing pace of the continued discovery of new oil fields, but the location of deposits of metallic ores and nonmetallic materials (e.g., asbestos) is equally important though less publicized. Geology has contributed in an important way to knowledge in other sciences. Such contributions could be elaborated, but of concern to us herein is its application to the solution of problems arising from the physical conditions near the earth's surface.

Today applied geology is being recognized as a substantial subject in itself and not merely a field from which an engineer enlists the aid of a specialist. There is need therefore for the student interested in civil or mining engineering to recognize that a firm foundation in physical geology is an essential item in his education and that such knowledge must precede the application of geology to engineering design.

By virtue of custom more than for any other apparent reason the usual procedure of introducing the student to geology has been to present in more or less orderly succession a digest of the fields ordinarily considered to belong to that realm of science. In many instances this procedure has suffered from oversimplification and generalization, and the student has been left with the impression that he has explored all knowledge of the science. The presentation developed herein is concerned with fundamentals of physical geology with particular emphasis on the processes operating at or near the earth's surface. Discussion of the organisms buried in the rocks or of the events of geologic history have been omitted. It is not the intent to imply that physical geology overshadows these other branches of geology but that it is fundamental to their understanding. The interesting but complex other aspects of the science are left for further study.

From the beginning of modern geology, principles fundamental to the subject were recognized. At first these were proposed as obscure generalities. Later as hypotheses became increasingly supported by facts certain general laws of behavior were expressed. Actually very few such laws have been formulated because natural conditions are subject to so much variation and complexity that stipulation of rigid behavior has not been warranted. Nevertheless, with each passing decade new masses of data have become available, and there remains the ever-pressing desire to organize this knowledge. Of late the tendency is to assemble these data in such form that specific properties of earth materials can be expressed with some mathematical rigor. In this book this trend has been carried into analysis of geologic processes. Wherever desirable an attempt has been made to introduce the reader to concepts which can now be stated in terms of mathematical expression. It must be understood, however, that the general level of precision is low. Nevertheless, the intent is to present the reader with a mechanism for understanding the nature of the processes in question and a means of predicting the properties of geologic materials with greater accuracy than mere expression of concept.

GEOLOGIC PROCESSES

Primary among the basic precepts of geology is the *law of uniformitarianism* first proposed by Hutton near the end of the eighteenth century. In essence the law states that the geologic processes active today have been operative in the past and, hence, that past geologic events are to be interpreted on the basis of the physical, chemical, and biological laws recognized at present. Fundamentally, the earth is in a never-ending process of change. These changes occur slowly and in terms of what a human can observe may not appear to occur at all. We have no reason to believe that the geological events that occurred in the past were any more spectacular or proceeded at a different rate than those occurring today. Even earthquakes and volcanoes are not the result of spontaneous forces originating in the interior of the earth but are the result of the sudden release of energy following a long period of accumulation. Rock formations are not to be considered as inert layers that were formed by some process which occurred long ago and has since stopped but are to be recognized as undergoing constant change. All rocks are subjected to changing chemical and physical environments and to forces both vertically and laterally applied. These alter the rock aspect by rearrangement of chemical composition or physical appearance and replace the mineral constituents with new materials stable under the prevailing conditions.

If one could view the earth from a distant point for many thousands of years he would see rivers removing soils, eroding channels in the rocks, and depositing the transported sediment load in lakes or oceans. Glaciers could be seen moving forward and melting back to some central area. During their advance fragmental rock debris would be picked up, carried, and deposited while the ice was moving as well as during the melting stages. He would see great layers of sediments warping into folds, building into mountains, or being depressed downward into large basins. Some of the basins would become filled with fresh water; others would be filled with sediments carried by rivers eroding the mountains. Still others would permit ingress of the sea and become marine embayments within the continents. In each case he would see the basin eventually filled with sediments eroded from adjacent land areas. Evidence of the former existence of these basins would be indicated by the thickness and inherent characteristics of the filling sediments.

Throughout all the physical changes he would see the earth in constant adjustment to two opposing forces. One of these is internal and is derived from some energy source within the earth. Internal forces tend to distort and raise or lower the earth's surface. The other energy has its origin in the sun and is channeled into the forces of erosion by means of rainfall. Erosion tends to destroy the irregularities developed by the internal forces and to establish equilibrium. Our observer would see no tendency toward exhaustion of either of these primary forces nor the establishment of equilibrium. Everywhere he would see that the present represents a continuation of physical conditions long in existence and with no foreseeable end. For these reasons the geologist holds firmly to the law of uniformitarianism and that "the present is the key to the past."

Most geologic processes which are outgrowths of the earth's internal energy are directed toward the production of new rock types. In some instances this involves change in pre-existing rock materials. Deep-seated molten rock may be moved outward and emplaced near the earth's surface by contortion and lifting of the pre-existing rocks to make room for the new material. Upon cooling this is welded into a part of the crustal zone and remains as newly formed rock.

Rainfall as a product of solar energy is the primary destructive geologic agent. Wherever release of the earth's internal energy elevates the land surface the potential energy of a drop of water at the high point is greater than under previously existing conditions. As the water flows downhill this energy is transformed into kinetic energy, and work is done in eroding and transporting soil and fragmental rock material to some new site lower in elevation than its former position. Accumulation of this fragmental debris represents addition to the crustal thickness at that locality, whereas the region of erosion is losing crustal material and successively deeper rocks are exposed to destruction.

All geologic processes are both constructional and destructional although they are primarily one or the other depending upon the sources of their energy. Under natural conditions as well as inside the laboratory a physical or chemical process is driven in the direction of reduction in energy, and the process is gradually halted as the quantity of energy released approaches zero. There exists also a balance of materials as, for example, the loss of rock through erosion must be equaled by the total amount of sediment formed and carried away and the mineral matter dissolved. Frequently, natural conditions appear to reach an equilibrium or attain a steady state. Some lake levels maintain a nearly uniform elevation, and the discharge equals the inflow. Many great rivers of low gradient do not deepen their channels. Rather, sediment carried into a segment of the channel is deposited and removed periodically, and although the stream may alter its channel position the valley aspect remains the same and a steady-state condition prevails.

Such apparent equilibrium is temporary when considered in the light of geologic time. Nonetheless, the equilibrium condition may exist sufficiently long to permit its record to be preserved by certain geologic features. When the direction of energy drive is changed the equilibrium conditions are upset, and different processes become dominant. If these conditions are permitted to continue in operation eventually a new and different equilibrium condition will become established once more.

With such concepts in mind the reader should consider the subject matter of each chapter. Moreover, he should be mindful of the interrelationship of geologic processes, some of which appear to have little bearing upon one another. Oftentimes, this seeming lack of relationship is the result of restriction of the topic considered and the need to subdivide the subject matter for ease of treatment. Nevertheless, no single geologic process should be considered as completely independent. The general circumstance is that the release of some form of energy initiates a geologic process which in turn creates a series of new conditions leading to the appearance of other geologic processes. So long as the fundamental sources of earth internal energy and solar energy continue in operation there remains a continuum of never-ending geologic processes which alter the surficial aspect of the earth.

SEQUENCE OF PRESENTATION OF SUBJECT MATTER

Precisely what constitutes a logical progression through the basic tenets of physical geology is a matter of considerable debate. The philosophy herein adopted is to consider first those materials most familiar to the beginning student. Such materials are soils. From a brief consideration of soils the reader is led successively through a series of chapters each of which considers a separate process. The order of presentation is designed to pyramid the introduction of information in a manner leading to the consideration of some of the most difficult material in the final chapter.

The progression of chapters is arranged around three general units, the subjects of which are respectively: earth materials, processes of erosion and deposition, and processes of crustal deformation and rock metamorphism. Each unit purports to present descriptions of processes and organization of concepts in such succession as to complete more or less a cycle of reasoning and return the reader to the beginning subject matter of the unit. In like manner each of the general units is bound to the other in the overlap of the processes discussed and the concepts considered. Thus each unit consists of a series of individual chapters considering special geologic processes, yet each is in some manner connected to the others.

The intent herein is the presentation, under the title of basic geology, of a series of interrelated subjects of principal concern to the student of science and engineering. The subject matter has been selected as that most desirable to provide him with a background necessary to the application of his own specialized field of interest. This may be geology, mining, hydraulics, soil mechanics, foundation design, highway construction, agriculture, geophysics, geochemistry, forestry, and the like. Whatever his objective it is hoped that this initial invasion into the concepts of physical geology will kindle a desire to read with increasingly wider interest.

Selected Supplementary Readings

Chamberlin, T. C., "The Method of Multiple Working Hypotheses," *Journal of Geology,* Vol. 5, 1897, pp. 837–847. A classic paper which outlines the most powerful method of logical reasoning which has been developed toward the solution of geologic problems.

Croneis, Carey, and Krumbein, W. C., *Down to Earth,* University of Chicago Press, 1936, Chapters 1, 2, 3, and 4. An elementary treatment of the framework of geology.

Daly, R. A., *The Architecture of the Earth,* D. Appleton-Century Co., 1938, Chapter I. An excellent organization of modern thinking concerning the large features of the earth.

Emmons, W. H., Thiel, G. A., Stauffer, C. R., and Allison, I. S., *Geology Principles and Processes,* 4th ed., McGraw-Hill Book Co., 1955, Chapters 1 and 2. An introduction to the framework of the earth.

Gilluly, James, Waters, A. C., and Woodford, A. O., *Principles of Geology,* W. H. Freeman and Co., 1952, Chapter 2. Reviews the broad framework of the earth.

Leggett, R. F., *Geology and Engineering,* McGraw-Hill Book Co., 1939, Chapters I, V. An outline of the history and methods of geologic investigation. The relationship between civil engineering and geology.

Longwell, C. R., and Flint, R. F., *Introduction to Physical Geology,* John Wiley and Sons, 1955, Chapters 1 and 2. An introduction to the subject matter of geology, the tools of the geologist, and a general aspect of the earth.

Umbgrove, J. H. F., *The Pulse of the Earth,* Martinus Nijhoff, 1947, Chapter I. Describes the cycle of development of rocks of the earth's crust.

2

Soil materials

THE SURFACE VENEER

Definition and Geographic Distribution of Soils

Doubtless the most familiar feature of the earth's surface is a thin veneer of fragmental rock material known as soil. So widely distributed is this surficial coating that only in locally scattered areas is the underlying rock exposed to view. Elsewhere rock is seen in stream valleys or where excavations extend below the base of the soil and expose underlying rock. It is customary to refer to soil with respect to very general properties which it possesses. Often color is its most obvious property, but equally common is the reference to its texture,

Prairie soil.

hence descriptions such as sandy, clayey, or gravelly soils. Soil also carries a certain connotation with respect to its agricultural value, and poor or good soils are known to concern the capacity to grow plants. None of the above properties actually clarifies the meaning of the term and are completely ineffective in explaining the source of the material. When the question arises as to whether a sandy beach or an extensive gravel deposit is a soil no satisfactory answer is obtained without a more suitable definition.

Definition Used in Engineering

One all-inclusive definition is to consider all loose mineral or small rock fragments as soil. Under this definition beach sands, broken rock, and soils growing crops are included. Each material will display differences in specific physical properties such as the amount of change in volume when placed under a load, the tendency to swell and increase in volume when wet, or the rate at which water will move through a unit cross section. By means of such an approach a list of physical properties has been prepared which is considered suitable to appraise, from an engineering viewpoint, the qualities of loose-rock debris irrespective of its content of organic matter. Thus, a loose rock or mineral aggregate may be distinguished on the basis of its strength properties or hydraulic properties, and each may be recognized as a different soil. Justification for this concept exists inasmuch as this is the procedure applied in the science of soil mechanics. This subject concerns itself with the stability of loose earth materials in all engineering works, particularly when the material is to be placed under loads. The thesis applied in soil mechanics stipulates that ideally all granular aggregates consist of a solid fraction of discrete particles of mineral matter and a pore fraction of irregularly shaped voids which form an interconnecting network between the grains. Under most circumstances the solid particles are in direct contact and rest upon one another, whereas the void space is filled with air or water or both. Despite the usefulness of this concept as a generalization applicable to silts, sands, and gravels the profession is aware that the simple relationship does not hold for very fine-grained clays. In such material the individual particles may remain as a suspension in a water medium, and as the particles are not in contact such clays display special properties.

Definition Used in Agronomy

A second definition of soil stems from the knowledge that at the earth-air interface certain chemical reactions occur which tend to change the composition of the mineral particles. With progressive depth below the surface the composition of the mineral matter may change rather abruptly and impart special properties such as change in color, permeability to water, or even particle size. Some of these changes are not necessarily important in engineering aspects, but they hold strong interest to the agronomist inasmuch as they affect soil fertility. Such a scientist is not concerned with the same problems as the engineer, and his basis of defining a soil is necessarily different. Thus, in the understanding of the scientist interested in agriculture a soil is formed only at the earth's surface through the operation of climatic processes upon a parent earth material. In some cases this parent may be a loose aggregate of alluvium, or it may be solid rock.

The two definitions are somewhat incompatible, but as the knowledge of soil mechanics continues to increase there is a growing tendency to apply the dual usage. The engineer refers to glacial soils, dune soils, or alluvial soils as indicating loose material of specified origin. He also recognizes that a soil in the agricultural sense may be developed from these same materials at the earth's surface. Actually, the latter is somewhat more in keeping with the proper geologic understanding of the origin of earth materials as suggested in the following paragraphs.

Soil Varieties

Organization of the properties of soils reveals strong similarities between certain widely, or closely, spaced samples and equally pronounced dissimilarities between others. The logical inference from such relationships is that soils of similar properties have had reasonably similar origins, whereas soils of dissimilar properties are genetically unrelated. The origin of the soil therefore may be used as a framework of classification and distinction.

From the earth's surface downward the typical soil displays zones of vertical variability becoming less conspicuous with depth. Other loose fragmental mineral materials lack this vertical zonation and, hence, appear to have originated in somewhat different ways. Each of the vertical zones merges into the underlying one indicating an intimate relationship and mutual development. The zones roughly parallel the contour of the earth-air boundary, are present only where the boundary exists, and appear in all ways to be a response to some physical-chemical environment being impressed downward from the base of the atmosphere. From this we are to conclude that the material distinguished as soil by the agronomist is a product of the conditions at the earth's surface operating upon some pre-existing rock fragments. On such a basis soil is defined as a surficial coating of dominantly incoherent debris which with the progress of time passes through a series of physical and chemical changes characterizing the stages of development from its parent material.

The true soil forms in the site where it occurs and progresses through a series of physical, chemical, and biological changes which are controlled by the prevailing environment. Environments of soil formation are local in distribution and depend upon the nature of the soil-forming material and upon the immediate climate. The latter appears to be a major item in controlling the progress of soil development, and the response is the appearance of characteristic properties. Areas which straddle temperature-moisture belts show corresponding differences in soil types, whereas areas of uniform climate tend to have uniform soil. This characteristic is well shown by the map of the world distribution of some general soil types as illustrated by Fig. 2.1. In this illustration the over-all belted aspect of the different soils and their parallelism to climatic zones is readily apparent. For example the *red and yellow* soils are characteristic of the humid tropics, and *podzols*[1] occur in regions of cold wet climates.

Each soil group is recognized by name, oftentimes with a Russian derivation in recognition of the pioneer work on soils done in that country. Somewhat more detailed but nevertheless still highly generalized is the soil distribution map of the United States, Fig. 2.2. Observe in particular the geographic restriction of the types indicated as *podzol, red and yellow, prairie, chernozem, chestnut, brown* (semiarid) and *desert*.[2] In a region as large as the United States the climatic control of soil varieties is demonstrated readily. Appearance of podzol

[1] A podzol is a soil with a thin surface layer of organic rich mineral matter lying upon a leached zone of gray color essentially free from organic matter. Podzols form typically in areas of northern pine forests. The origin of the name is from the Russian "resembling ash." See Table 2.1 for other characteristics.

[2] Prairie soils have a very dark brown, or gray brown, surface color and are developed under tall grasses in a temperate rather humid climate. The term is restricted to soils in which carbonates have not been concentrated in the lower part of the soil.

Chernozem (from the Russian for "black earth") is a soil of nearly black surface color developed under tall grasses in a cool subhumid climate. They are distinguished from prairie types by the presence of a zone of carbonates in the lower part of the soil.

Chestnut soils are gradational with chernozems but form under more semi-arid conditions.

Brown (semi-arid) soils have a brown colored surface layer resting upon lighter colored material. They are typical in semi-arid climates where shrubs and bunch grasses are the vegetation.

Desert soils have a light-colored surface layer often resting upon a zone rich in carbonates or a layer of hardpan. The vegetation is scanty scrub.

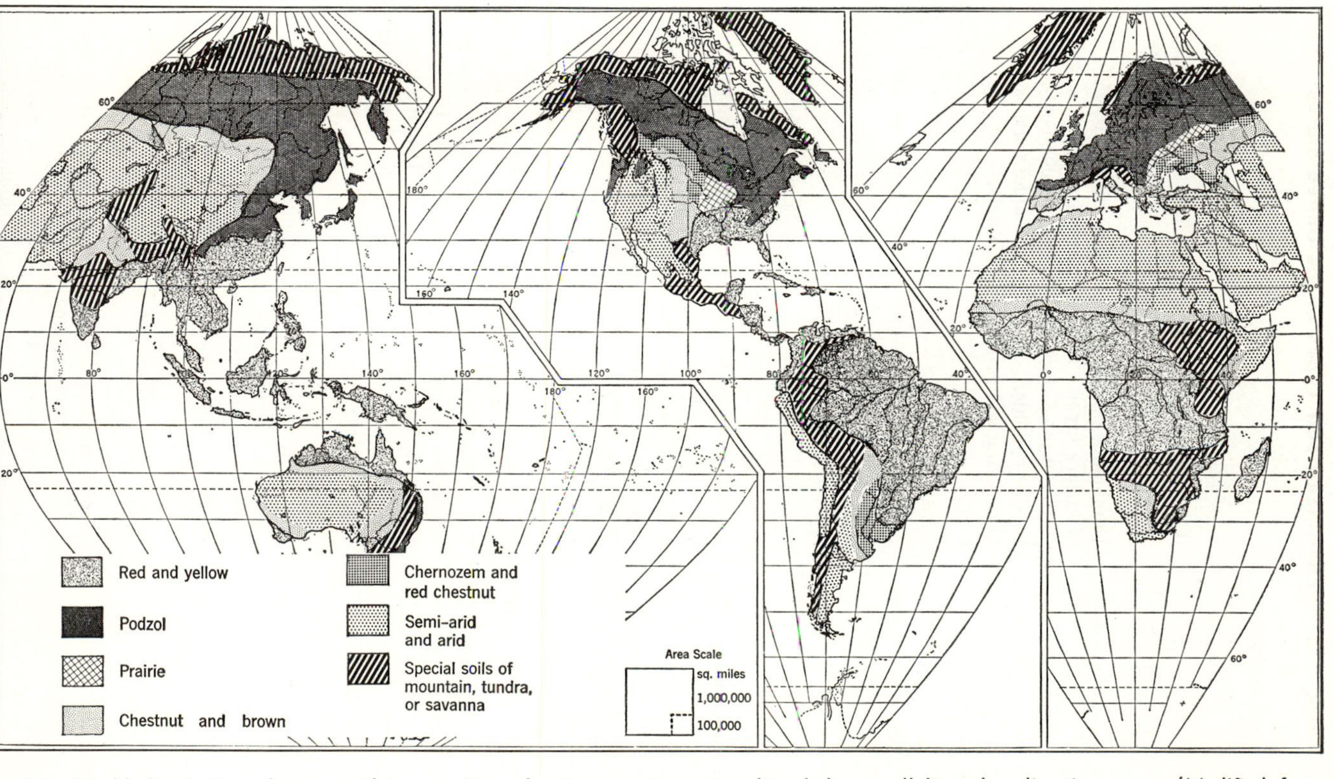

Fig. 2.1 World distribution of major soil types. Note the strong east-west trending belts paralleling the climatic zones. (Modified from Kellogg, *Scientific American*, 1953. Base map courtesy Rand McNally and Co., copyright R.L. 58S53.)

Table 2.1. Generalized Soil Types of the United States

Soil Type	Color of Surface Layer (upper 6″ to 1′)	Constituents of Surface and Underlying Layers	Climate in Localities Where Soil Occurs
Podzol	Black to dark brown	Dominance of raw humus in surface layer. Mineral matter stained dark by organic colloids. Underlying soil is light gray.	Rainfall 30″ to 45″; frost-free days less than 150; climate cold-humid. Example: northeastern Canada.
Gray brown podzol	Gray brown	Abundant humus in surface layer. Mineral matter stained brown by organic colloids. Underlying soil gray brown.	Rainfall 30″ to 45″; frost-free days 150 to 210; climate cool-humid. Examples: New York, northern Kentucky.
Northern and southern brown prairie soils	Dark brown to black	Abundant finely divided humic matter, chiefly decayed grass. Mineral matter stained dark brown by organic colloids. Underlying soil brown.	Rainfall 30″ to 45″; frost-free days 120 to 240; climate cool but warm summers, humid. Examples: Illinois, Iowa.
Chernozem	Very dark brown to black	Abundant humus in excess of that in brown prairie soils. Mineral matter stained brown to yellow brown by organic colloids. Underlying soil contains calcium carbonate.	Rainfall 20″ to 30″; frost-free days 120 to 240; climate cool but warm summers, dry to subhumid. Examples: eastern North and South Dakota.
Chestnut	Dark brown, gray brown, or red brown.	Humus in moderate amounts. Underlying soil contains light gray zones of carbonates.	Rainfall 12″ to 20″; frost-free days 65 to 290; climate cold winters, hot summers, subhumid. Examples: western Texas, western North Dakota.
Brown	Brown	Humus in small amounts. Underlying soil contains gray zones of carbonates.	Rainfall 12″ to 20″; frost-free days 200+; climate mild winter, long hot summer. Example: northeast New Mexico.

Table 2.1. Generalized Soil Types of the United States (*continued*)

Soil Type	Color of Surface Layer (upper 6″ to 1′)	Constituents of Surface and Underlying Layers	Climate in Localities Where Soil Occurs
Desert gray	Buff to gray buff	Humic matter absent, mineral matter gray buff, unstained. Underlying soil high in calcium carbonate.	Rainfall ±10″; frost-free days 90 to 240; cool to hot; arid. Examples: Arizona, eastern Oregon.
Red and yellow	Red brown to yellow red, locally light gray	Humic matter absent; mineral matter stained by iron oxides to red and yellow red colors; light gray soils lack iron oxides but rich in aluminum oxides.	Rainfall ±45″; frost-free days 210+, hot humid. Examples: southern Georgia, Alabama.

soils in the cool wet area of the northwestern states immediately adjacent to the localities of the desert soils is strikingly displayed. Also, gradation of soil types between prairie and chernozem and

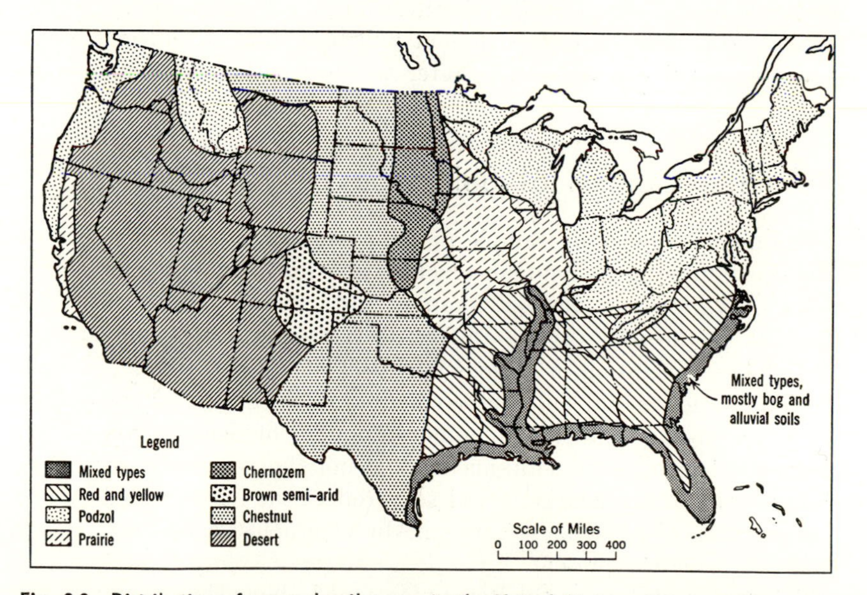

Fig. 2.2 Distribution of general soil types in the United States. Compare the position of the belts with respect to those of the world (Fig. 2.1). (Modified from *Soils and Man*, U. S. Dept. Agriculture, 1938.)

chestnut brown and desert is coincident generally with the belts of decreasing rainfall. (See Fig. 7.4.)

Sediment and Soil

If the term "soil" is to be restricted in use to the definition employed by the soil scientist the question naturally arises in the mind of the engineer as to the proper term to be applied to describe the materials of a sand-bar, dune, or gravel deposit. To designate such material the term "sediment" is suggested. Thus a glacial sediment or dune sediment or beach sediment is used as an all-encompassing term to indicate the loose mineral matter of a specific origin. The term soil however would be used to describe that particular part of a sediment which has been altered in the manner previously defined by the conditions at the earth-air boundary. The reader should bear in mind, however, that the term soil is engrained in engineering literature to describe all loose aggregates of mineral matter smaller than boulders, and by virtue of custom it will continue to be used by the profession. For the beginning student, however, use of the term sediment is recommended to serve as a reminder of the distinction between the parent unaltered loose aggregate and the altered product or true soil. Henceforth, in this text the term sediment will be applied to all loose rock debris, and soil will be used to refer to the weathered product only.

Climatic Control of Soil Types

Examination of a soil sample under a lens reveals the presence of two substances in predominance. (1) An assortment of mineral particles of various dimensions, shapes, and colors. These constitute the bulk of the soil. (2) Intermixed with the mineral fragments are particles of vegetation in various stages of decay. A product of such an examination is that soils whose colors are light gray, red, and yellow definitely are associated with minor amounts of decayed vegetation (humus); hence, the mineral matter and not the vegetation is responsible for the color. Dark-colored soils reflect the incorporated humus and the degree of blackness varies in direct proportion to the quantity of organic matter present.

Tabulation of data of the items mentioned above shows some of the relationships between climate, percentage of humus, and mineral matter which constitute a soil of certain type. Table 2.1 is such a compilation. Note that soils occurring in cool humid areas contain abundant humus, and soils of the warm or arid climates show a general lack of included decayed vegetation. At this juncture it is not the intent to investigate the causes for such color changes. Suffice it to state that the absence of humus depends to a large degree on the ability of soil bacteria to consume the humus as rapidly as it accumulates. Under conditions of optimum bacterial growth the humus is destroyed, but where bacterial development is inhibited by low temperatures rate of consumption of organic matter declines, and partially destroyed vegetation becomes an integral part of the soil.

PARTICLE SIZE DISTRIBUTION IN SEDIMENTS

One principal difference in soils and sediments in general is the variation in the dimension of mineral particles observed laterally and vertically. Some sediments are sandy, others clayey, and still others contain pebbles. Many of the properties of sediments are physical responses to the ranges in size of the particles and frequency distribution of such sizes. Thus, the appearance which characterizes a gravel is due to the presence of a conspicuous percentage of pebbles embedded in a matrix primarily of sand. If the pebbles are surrounded by a matrix of clay the sediment is a pebbly clay, whereas if it consists of a mixture of sand and clay it is called a sandy clay. Distinctions between sediments, therefore, are made partly upon the dimensions of the individual particles of which the material is composed. The sizes of the particles, however, represent a continuous series from diameters ranging in meters to fractions of a micron. Within this great size range particles can be recognized as boulders, pebbles, sand grains, or clay, but no uniformity can exist in defining such size grades without resort to established diameter limits.

Size Grade Scales

A rather large number of grade scales has been proposed by individuals and organizations throughout the world. Each is an attempt to permit subdivisions into size grades which will best reveal the properties of a sediment from the viewpoint of the profession concerned and oftentimes in the light of special materials being handled. For example, regions where the sediments largely are of glacial origin display types whose properties are quite distinct from those of a region such as northern Africa where the sediments are products of desert and stream environments. The primary concern, however, of each size grade system is to subdivide particles into groups possessing similar properties, and the segregation of particles which by virtue of size contribute certain special characteristics to a sediment. Four systems of particle size subdivision in common use in the United States are listed in Table 2.2 and Fig. 2.3. The reader should understand,

Table 2.2. Typical Systems of Classification of Particle Size*

System	Grade Limits (Diameters)	Designation of Particles or Size Grade
Wentworth	Above 256 mm	boulder
	256–64 mm	cobble
	64–4 mm	pebble
	4–2 mm	granule
	2–1 mm	very coarse sand
	1–$\frac{1}{2}$ mm	coarse sand
	$\frac{1}{2}$–$\frac{1}{4}$ mm	medium sand
	$\frac{1}{4}$–$\frac{1}{8}$ mm	fine sand
	$\frac{1}{8}$–$\frac{1}{16}$ mm	very fine sand
	$\frac{1}{16}$–$\frac{1}{256}$ mm	silt
	less than $\frac{1}{256}$ mm	clay
Bureaus of Plant Industry, Chemistry & Soils, and Agri. Engineering; U.S. Dept. of Agriculture	>2 mm	gravel
	2–1 mm	very coarse sand
	1–$\frac{1}{2}$ mm (1–0.5)	coarse sand
	$\frac{1}{2}$–$\frac{1}{4}$ mm (0.50–0.25)	medium sand
	$\frac{1}{4}$–$\frac{1}{10}$ mm (0.25–0.10)	fine sand
	$\frac{1}{10}$–$\frac{1}{20}$ mm (0.10–0.05)	very fine sand
	$\frac{1}{20}$–$\frac{1}{500}$ mm (0.05–0.002)	silt
	below $\frac{1}{500}$ mm (<0.002)	clay
Massachusetts Institute of Technology	Above 2000 mm	very large boulder
	2000–600 mm	large boulder
	600–200 mm	medium boulder
	200 – 60 mm	small boulder or cobble
	60 – 20 mm	coarse gravel
	20 – 6 mm	medium gravel
	6 – 2 mm	fine gravel
	2 – 0.6 mm	coarse sand
	0.6 – 0.2 mm	medium sand
	0.2 – 0.06 mm	fine sand

* See also P. E. Truesdell and D. J. Varnes, *Chart Correlating Various Grain Size Definitions of Sedimentary Materials*, U. S. Geological Survey, 1950.

Table 2.2. Typical Systems of Classification of Particle Size (*continued*)

System	Grade Limits (Diameters)	Designation of Particles or Size Grade
	0.06 – 0.02 mm	coarse silt
	0.02 – 0.006 mm	medium silt
	0.006 – 0.002 mm	fine silt
	0.002 – 0.0006 mm	coarse clay
	0.0006– 0.0002 mm	medium clay
	<0.0002 mm	fine clay (colloidal)

System	Passing† Sieve Number	Retained† on Sieve Number	Designation of Particles or Size Grade
Unified Soil Classification (Basic types)	3″	$\frac{3}{4}''$	coarse gravel
	$\frac{3}{4}''$	#4 (4.76 mm)	fine gravel
	#4	#10 (2 mm)	coarse sand
	#10	#40 (0.42 mm)	medium sand
	#40	#200 (0.074 mm)	fine sand
	#200	Liquid limit ‡ 28 or less. Plastic limit 6 or less.	silt
	#200	Liquid limit ‡ more than 28. Plastic limit more than 6.	clay

† Standard sieving screens are used to separate a sediment into its respective size grades. This technique is employed irrespective of the system of classification.

‡ Liquid limit and plastic limit are measures of shear strength of clays and are in effect measures also of the moisture content of the clay inasmuch as the strength decreases with increase in liquidity.

however, that the classifications listed are only representative and that others exist equally in common use. Of principal concern herein is the understanding of the significance of size grade scales and that the chief difficulty of classification by size is in the clay group. For this reason, systems recently proposed employ certain other physical properties such as degree of plasticity to distinguish between silt and clay sizes. (See Unified Soil Classification System, Table 2.2.) The

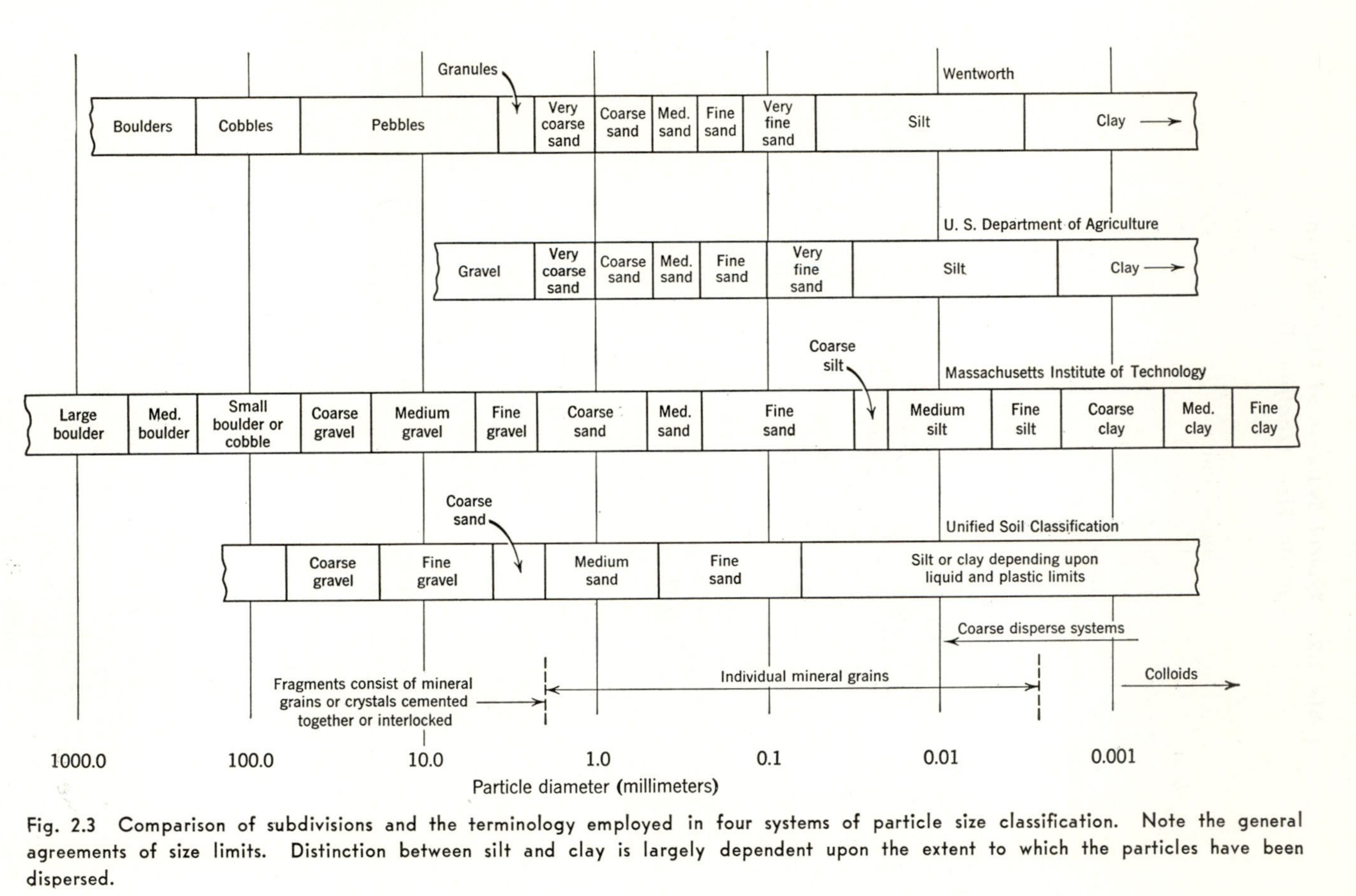

Fig. 2.3 Comparison of subdivisions and the terminology employed in four systems of particle size classification. Note the general agreements of size limits. Distinction between silt and clay is largely dependent upon the extent to which the particles have been dispersed.

student interested in soil mechanics will be introduced to special techniques in this connection.[3]

Limiting Boundaries

Pebble. Coincidence of limits characteristic of a grade size is only roughly the same between grade scales and is attributed to the necessity of establishing boundaries in a continuous system of changing properties. For example, the upper limiting diameter of a pebble is arbitrarily placed inasmuch as pebbles and cobbles consist of aggregates of cemented mineral grains or interlocked crystals and differ only in size. Sand grains, however, are not diminutive pebbles but are specific mineral grains whose composition and density differ widely from pebbles. The size boundary where individual mineral grains are released from rock fragments is the upper limit of sand grain size and represents a physical and chemical change in the nature of the material. Generally the change from bonded aggregates of minerals to individual mineral grains occurs in fragments ranging between 4 and 2 mm in diameter. Very nearly all systems in current use recognize this natural boundary. (Fig. 2.3.) Among geologists the Wentworth scale has met with greatest favor, whereas in the engineering profession other scales, some of which are listed in Table 2.2, are preferred.

An assemblage of particles of one grade size can be diluted by admixture of material of another grade size. Pebbles diluted with sand result in material which is called gravel. If the percentage of pebbles is very large the sand occupies only the openings between the closely packed pebbles. As the percentage of sand is increased it becomes a matrix in which the pebbles are embedded, and individual pebbles are no longer in contact with one another. With additional increase in sand the bulk properties of the sediment become progressively those of sand and less like gravel.

Silt. Distinction between silt and very fine sand is difficult to make inasmuch as silt represents individual mineral grains whose greatest diameter does not exceed .06 mm (Wentworth scale) and very fine sand is slightly larger in grain size. The size limits of silt, however, are established where some cohesion between individual grains is observed. Dry sand displays no tendency in this direction, but cohesion between silt particles is noted by its ability to stand in vertical slopes without collapse.

[3] For a thorough description of grade scales refer to W. C. Krumbein, and F. J. Pettijohn, *Manual of Sedimentary Petrography,* pp. 76–90. D. Appleton Century Co., New York, 1938.

Clay. Clay is a term used to connote a grade size whose physical properties are distinct from those of silt. Most of the properties of clay can be attributed to the properties of colloids inasmuch as the solid fragments constituting clay are in a state of dispersion. Clay often swells on moistening and shrinks and cracks on drying; individual particles suspended in water coagulate on discharge of electrical charges, and chemical reaction occurs between particles and the medium in which they are suspended. The upper size limit where such properties are no longer observed marks the boundary between silt and clay.

Sand. Sand can be diluted with silt or clay. If present in minor amounts silt or clay particles are small enough to occupy only the interstitial space between individual sand grains in relationships similar to the pebbles and sand already mentioned. As fine material is increased it constitutes the matrix in which the sand grains are incorporated, and ultimately the bulk properties of sand are no longer even remotely present.

Size Mixtures

With very few exceptions sediments are mixtures of fragments of varying size ranges, and their bulk properties are to a large degree controlled by the degree of predominance of certain size groups. In the broadest sense these can be classified into pebbles, sand, silt, and clay, and soils can be described as admixtures of these four major grade sizes. Each of these grade sizes can be visualized as consisting of particles of a single average size, and the percentage of occurrence of that size in the composite represents its addition to the properties of the sediment.

A convenient and extremely useful system of representing the percentage occurrence of such individual major grade sizes is to plot their occurrence on the faces of a tetrahedron (Fig. 2.4). The tetrahedron can be opened and each equilateral triangle face treated individually. An example is illustrated by Fig. 2.5 in which has been plotted the relationship between mixtures of the sizes sand, silt, and clay. For ease of handling, each of these "end-member" sizes is considered as representing one grade, and the spread in sizes is indicated by the relative proportion of admixture. The illustration has been selected to outline the proportional mixtures of the end-member sizes to constitute certain special types of sediments such as loam and "lean" clay. From the positions of the fields on the triangle, the significance of such terms appears. *Loam* is seen to represent a soil in which the admixture of sand and silt is the important control and the percentage of clay is of secondary significance. Note that as sand is increased in amount the properties of the loam change materially from a silty loam to a sandy loam and as clay is added in major amounts loams

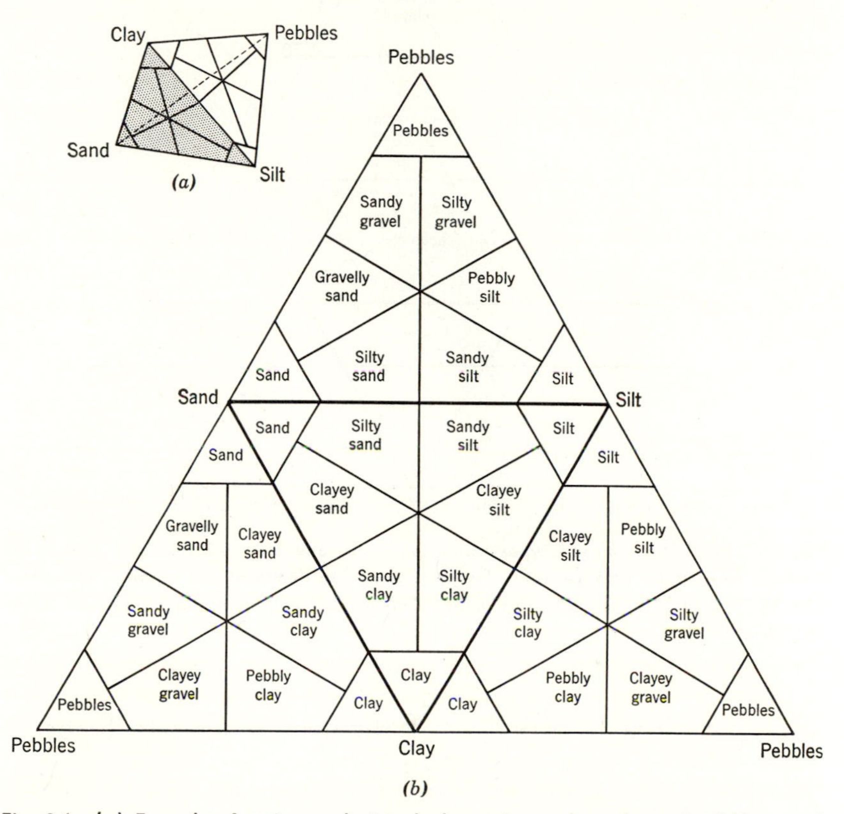

Fig. 2.4 (*a*) Example of a size grade tetrahedron using end members of pebbles, sand, silt, and clay. Subdivision into sediment types is based upon the division of each triangular face into approximately equal parts. (*b*) Tetrahedron opened at the pebble end to show the names assigned to the sediment types.

are transitional into clays. Terms such as "fat" and "lean" clay occur in engineering literature. Fat clay consists of extremely small particles generally associated with a high water content. This combination produces a highly plastic characteristic and a greasy feeling when the clay is molded between the fingers. With increase in percentage of silt and sand the greasy feeling is replaced by a grittiness, and the degree of plasticity is reduced as well.

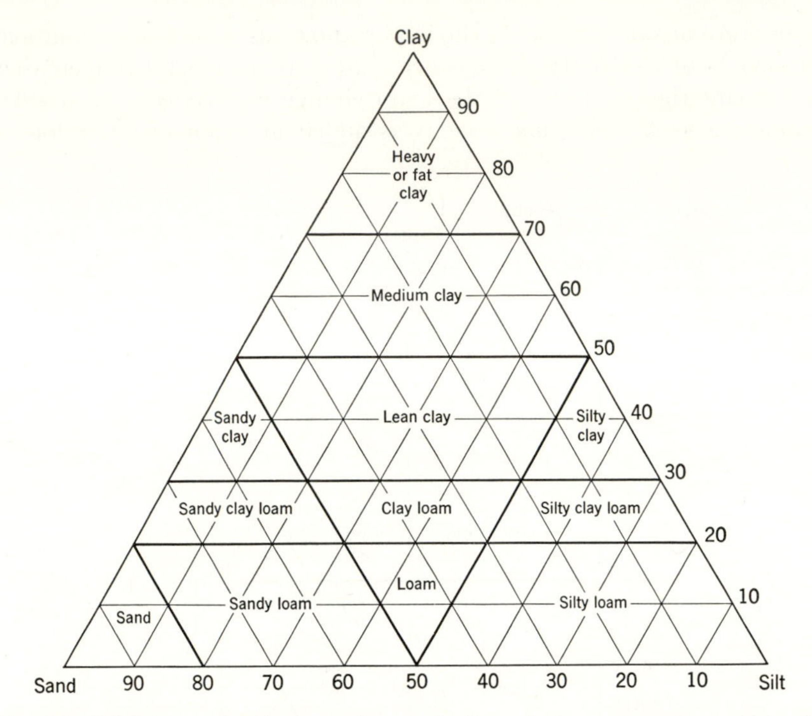

Fig. 2.5 Subdivisions of soil types using sand, silt, and clay end members. Note the relationship of terms such as "fat" clay and "loam" to dominant size grades. (Modified from Millar and Turk, *Fundamentals of Soil Science*, 1st ed., copyright 1943, John Wiley and Sons, Inc.)

Bulk Physical Properties of the Major Grade Sizes

Each major size grade, pebble, sand, silt, and clay, possesses certain properties in bulk which are products of the diameter of individual particles. Special techniques are available to establish numerical values for these properties and are introduced to the student of soil mechanics. Much value, however, is to be gained by a qualitative appraisal of the contribution of such properties, and the reader's attention is directed to this approach. Certain properties pertain to changes in volume between the wet and dry sample. Sediments which change in volume in direct proportion to their water content generally are undesirable engineering materials, and their presence should be recognized in any locality where construction is to occur. Clays have a high tensile strength in comparison to sand, whereas under compression sands change in volume very little and clays a great deal.

Some sediments such as gravel or sand permit the passage of large amounts of water through a unit cross section. Clays display an opposite tendency and inhibit the flow of water. The particular property in question is called *permeability* and in qualitative terms is expressed as the capacity which a sediment or rock possesses to transmit a fluid. Sometimes this property is confused with *porosity* which is the percentage of the total volume of a sediment not composed of solid particles. Porosity and permeability are related but not always in the same direction. Large pore openings are associated with high permeability, whereas very small pore openings are associated with low permeability. Gravels and sands with large pore openings have high porosity and high permeability. Clays tend to have high porosity, and some have exceptionally high porosity (80 to 90 per cent), but the surface effects of the walls of capillary openings tend to inhibit the passage of water, and hence clays display low permeability.

Employing similar lines of reasoning comparison may be drawn between physical properties of the major size grades as illustrated by Table 2.3. Each property listed serves to distinguish the primary attributes of each of the sediments in bulk and provides a basis for predicting the general properties of a sediment consisting of a mixture of sizes.

Under most circumstances unconsolidated earth material consists of mixtures of major grade sizes, and the bulk physical properties of the debris are composites of the properties of each major group. This is illustrated by selecting three ideal samples representing mixtures of pebbles, sand, and clay, Fig. 2.6.

Sample 1 consists of 60 per cent pebbles, 10 per cent sand, and 30 per cent clay. Each pebble is partially embedded in a matrix of clay diluted with small amounts of sand. The high percentage of pebbles suggests that they are locally in contact with one another and the clay and sand matrix is moderate in amount, much of it filling interstitial spaces between pebbles. Such a sediment would show a distinct volume change between wet (swelling) and dry (shrinkage) conditions due to the clay matrix. Its tensile strength when dry would be high and would be lower when wet, but in either case much higher than pebbles alone. This results from the high tensile strength of the clay which bonds the pebbles together. The change in volume under compression should range between low and very low, particularly in the zones where the pebbles are in contact. Under such conditions the stress is transmitted through the pebbles and does not involve the clay. Filling of the interstitial openings with

Table 2.3. Qualitative Physical Properties of Major Size Grades in Aggregates

Property	Pebbles	Sand	Silt	Clay
Volume change from dry to wet condition	None	None	Slight	Large (marked swelling)
Strength under tension:				
When wet	Low	Low	Intermediate	High
When dry	Low	Lower than when wet	Higher than when wet	Very high
Volume change under compression:				
When wet	Very low	Very low unless particles separated by water when change is high	Intermediate unless saturated with water when change is high	Very high
When dry	Very low	Very low	Low	Intermediate to low when very dry
Plasticity:				
When wet	None	Slight	Intermediate	Very high
When dry	None	None	None	None, partial cementation
Porosity	Very high	High	High	Very high
Permeability	Very high	High	Intermediate to low	Very low
Size of pore openings	Large	Intermediate	Capillary	Subcapillary
Shape of fragments	Rounded	Round to Angular	Angular	Thin tabular
Water retention	Very low	Low	High	Very high
Base exchange capacity	Absent	Absent	Low	High

clay and sand reduces both the porosity and the permeability; hence, water filtration will be slow and little water can be held within the sediment.

Sample 2 contains 5 per cent pebbles, 80 per cent sand, and 15 per cent clay. This sediment is chiefly sand with some clay acting as a matrix about infrequently occurring pebbles. The sample should have the bulk properties of sand but with higher tensile strength and lower porosity and permeability (reduced by volume occupied by pebble lacking pore space, and clay filling interstices between sand grains).

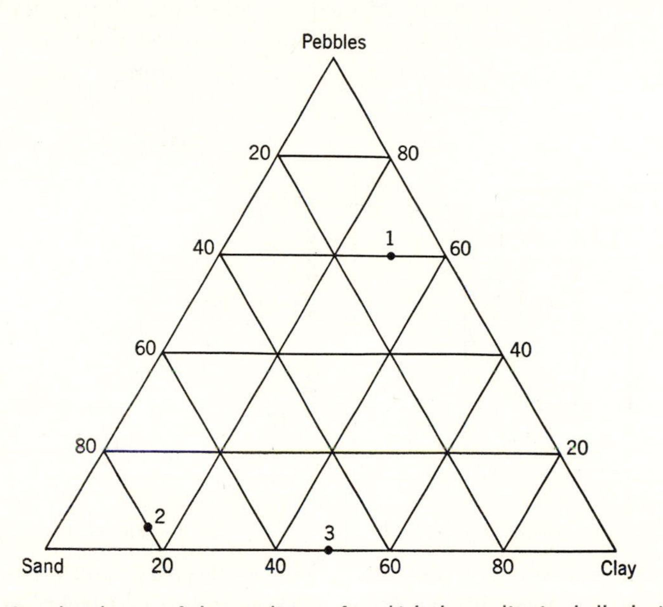

Fig. 2.6 Size distribution of three sediments for which the qualitative bulk physical properties may be determined using Table 2.3. Sample 1, clayey gravel; sample 2, sand; sample 3, sandy clay.

Sample 3 consists of 50 per cent sand and 50 per cent clay. Such a sediment would possess many of the bulk properties typical of silt but would differ in grain size and in appearance. Tensile strength and change in volume under compression would show wide ranges between wet and dry material due to the high percentage of clay, whereas the high percentage of sand would tend to decrease the reduction in volume under compression to less than that expected from most silts.

Manipulation of Particle Size Distribution Data

Some sediments consist of particles which are prevailingly of one major size (sand, silt, or clay) and as such are homogeneous and show little change in properties both laterally and vertically. More commonplace are those sediments which are composites as already described. Recognition of this composite nature is obtained by segregation of selected size classes. Often this is done by sieving the sediment through screens of known diameter openings and weighing the amounts retained on each sieve. Or when the material is silt or clay separation can be accomplished by collecting aliquot parts of a sample settling through a water column at selected intervals of time. From the known rate of settling of particles through water the per

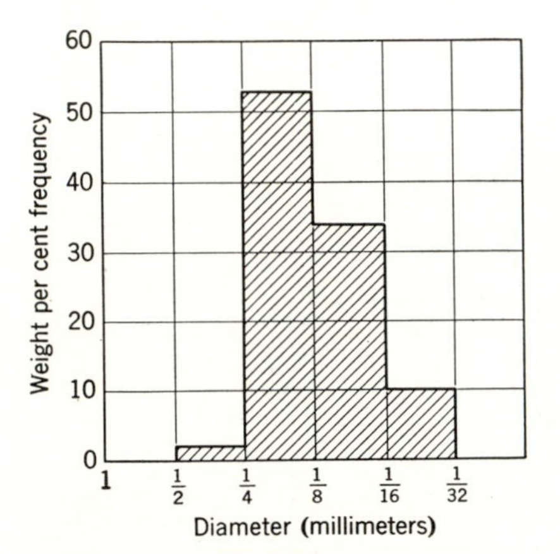

Fig. 2.7 Histogram of a dune sand. The weight per cent occurring within the limits of each size is plotted as a series of columns of equal width. Note that the diameter sizes decrease geometrically and that the concentration of sizes is readily visualized.

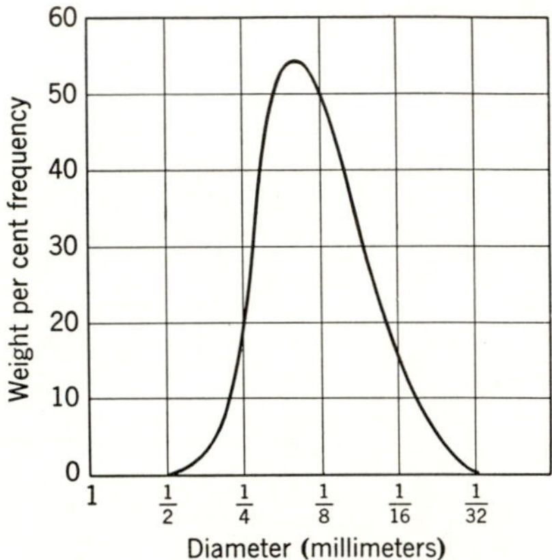

Fig. 2.8 Frequency distribution of particle sizes in the dune sand plotted in Fig. 2.7. The smooth curve of the frequency distribution is in effect drawn through the points representing the heights of infinitely closely spaced columns of the histogram. The frequency curve reveals the symmetry of the distribution about the modal (most abundant) size. In symmetrical distributions the mode, median, and mean sizes are identical.

cent size distribution of the original samples is determined. Either of these methods furnishes data on the percentage distribution of various size classes constituting the sample in question.

Size data are assuming greater significance in characterization of the properties of a sediment as more precise techniques of comparison between different particle size distributions are developed. In this connection the methods of statistical manipulation are proving to be most appropriate. For this reason the following sections are presented to introduce the reader to certain procedures of comparison of sediments in common use among geologists.

The Histogram and Frequency Curves

Perhaps the most simple representation of the distribution of sizes of particles in a sediment is a bar graph on which is drawn a series of columns whose height is a measure of the percentage of relative

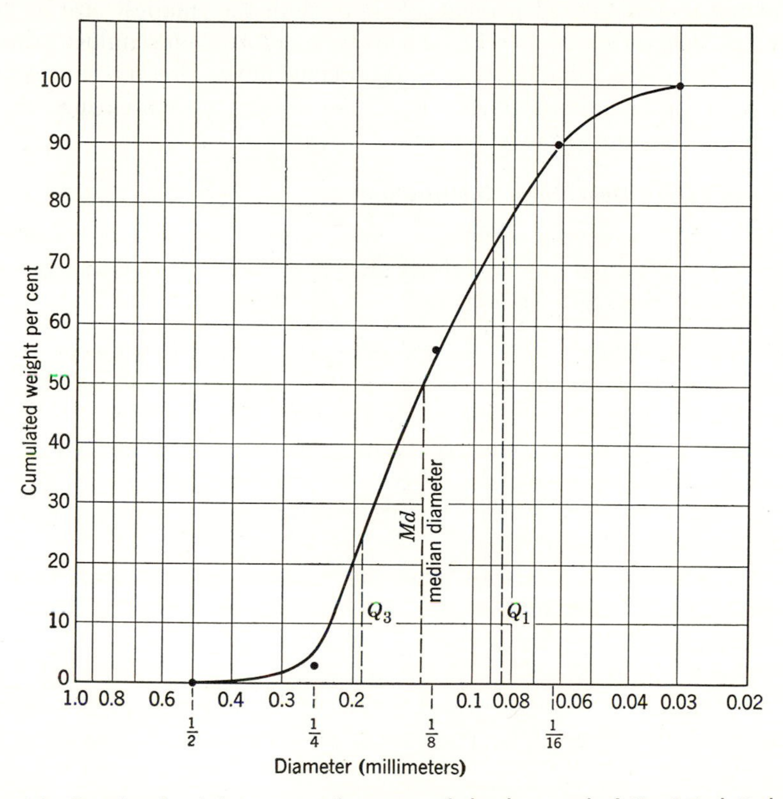

Fig. 2.9 Cumulated weight per cent frequency of the dune sand of Fig. 2.7 plotted on a semilogarithmic base. Points along the curve are percentage values cumulated by adding the successive heights of the histogram columns. Note that in this system cumulating the weight per cent begins by setting 100 per cent of the distribution as coarser than $\frac{1}{32}$ (0.031) mm or 0 per cent coarser than $\frac{1}{2}$ mm. Q_1 = the quartile diameter, of which 75 per cent of the sample is coarser; whereas Q_3 = the quartile diameter, of which 25 per cent is coarser. *Md* = the median diameter or the size of which 50 per cent of the sample is larger. A steep slope of the central part of the curve indicates well-sorted deposits.

abundance of the sieved sizes arranged in order of decreasing size. Such representation called a *histogram* is shown in Fig. 2.7. A smooth curve which in effect connects the mid-point of the top of infinitely closely spaced columns of the histogram is a *per cent frequency curve,*

Fig. 2.8. The frequency curve shows at a glance the size which occurs in greatest percentage and the symmetry of the distribution about that size. Frequency distribution curves are somewhat laborious to prepare, and instead use is made of the *cumulative-size frequency curve.* This involves cumulating the values of each column of the histogram and plotting the cumulated percentage more than the smallest size or less than the largest size. (See Fig. 2.9.) When plotted on semilogarithmic graph paper the cumulative-size frequency curve presents an aspect whose slope is often diagnostic of sediments accumulating under specific depositional environments.

Parameters Describing Distribution

Manipulation of values obtained from the cumulative-size frequency curve results in parameters which are useful in characterizing certain features of the size distribution.

1. The *median diameter* is that size which has 50 per cent larger and 50 per cent smaller in the distribution. This is read directly by erecting an ordinate from the x axis (representing diameter grain size) to the point where the 50 per cent line intersects the cumulative curve.
2. *Quartiles* represent sizes of which three-fourths of the sample is larger and one-fourth smaller or three-fourths of the sample is smaller and one-fourth larger.
3. The point, or points, of inflection of the cumulative curve represent the most abundant size, or sizes, and would appear as the peak, or peaks, of the frequency curve. Such sizes are known as the *mode,* or *modes.*
4. A measure of the degree to which all the grains approach one size is the sorting. One parameter of sorting is obtained by the ratio[4]

$$\frac{Q_3}{Q_1} = \text{sorting coefficient}$$

where, Q_3 = diameter of quartile of larger grain size than Q_1, diameter of quartile of smaller grain size.

5. Size frequency distribution curves range from symmetrical about the modal size to a high degree of asymmetry. The relationship between the size of the respective quartiles compared to the median size will show asymmetry of the frequency distribution.

$$\frac{Q_1 \times Q_3}{M_d{}^2} = \text{coefficient of skewness}^4$$

where Q_1 and Q_3 are diameter values of the respective quartiles, and M_d is diameter of the median.

Data on the size distribution between individual soils or parts of the same soil aid in the distinction between types and in the changes occurring within the soil as it develops from its parent material.

The particle-size frequency distribution of a dune sand as illustrated in Fig. 2.8 has been plotted on a scale which does not show a true arithmetic progression of sizes. Rather, spacing of millimeter sizes is in the form of a geometric progression such that equal increments represent a change in millimeter size by one-half. Should the same frequency distribution be plotted on a scale in which the millimeter ruling is arithmetic a much longer scale would be necessary. Of more significance, however, would be the change in the shape of the frequency distribution curve. Instead of the symmetry which it displays as plotted in Fig. 2.8 it would be a highly asymmetric curve with a long tail extending into the small millimeter sizes. In contrast the curve as plotted in Fig. 2.8 approaches very closely that of the *"normal curve of error"* and, hence, is subject to easy statistical manipulation.

The Phi Scale and the Symmetrical Curve

Most all sediments show a highly asymmetric size distribution in terms of relative particle size. This distribution may be symmetrized by a transformation in which the particle diameters are plotted in some progression such as illustrated in Fig. 2.8. Perhaps the most satisfactory of these transformations is the *phi scale*[5] which is the negative logarithm to the base 2 of the millimeter size of a particle. The result is an arithmetic scale beginning with a value of 0ϕ equal to 1 mm and progressing as $+1\phi$, $+2\phi$, $+3\phi$, $+4\phi$ for values of $\frac{1}{2}$, $\frac{1}{4}$, $\frac{1}{8}$, and $\frac{1}{16}$ mm, respectively. Negtive phi values of -1ϕ, -2ϕ, -3ϕ are equal to 2, 4, and 8 mm, respectively. (See Fig. 2.10.) Plotting size-frequency data using the phi transformation permits easy computation of the arithmetic mean size which is a measure

[4] The coefficients of sorting and skewness parameters developed by P. D. Trask. These parameters are compared with other similar ones in Krumbein and Pettijohn, *op. cit.*

[5] The phi scale devised by W. C. Krumbein has enabled ease in statistical manipulation of particle size data in a manner unequaled by any other system. Students interested in this aspect of characterization of sediments should read W. C. Krumbein, "Size Frequency Distribution of Sediments," *Jour. Sedimentary Petrology,* Vol. 4, 1934, pp. 65–77; and Krumbein and Pettijohn, *op. cit.*

of central tendency of size distribution. In a symmetrical distribution the *arithmetic mean* and the median size are the same.

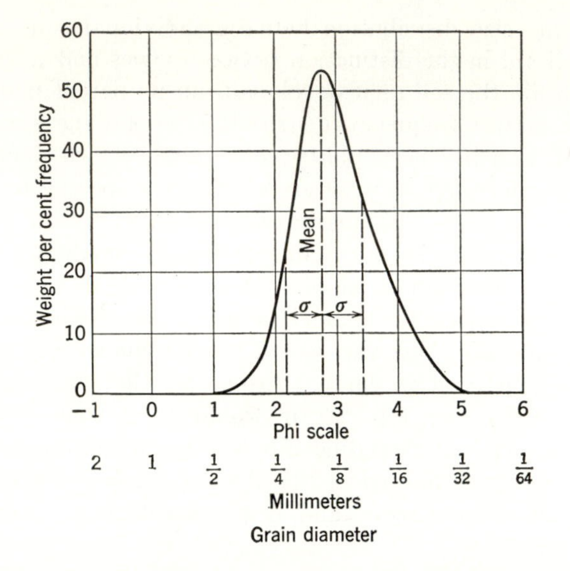

Fig. 2.10 Weight per cent frequency distribution of the dune sand of Fig. 2.7 plotted with the phi scale size transformation (phi size = $-\log_2$ of millimeter diameter). The phi scale is an arithmetic progression positive and negative about the zero phi value (1 mm). The effect of using the phi transformation is to produce a "log normal" distribution in well-sorted sediments and to permit use of statistical parameters of the normal probability distribution, namely the mean and standard deviation (σ) of the mean.

The Mean and Standard Deviation

Most statistical manipulations involve the arithmetic mean and the *standard deviation* (σ) of the mean. The standard deviation is a measure of the spread of the distribution and for the general case is defined as the square root of $1/N$ of the sum of the squares of the deviations from the arithmetic mean of N items as indicated below (see also Table 2.4):

$$\frac{\Sigma m_i}{N} = M \tag{1}$$

$$\sigma^2 = \frac{1}{N}\Sigma f_i(m_i - M)^2 \tag{2}$$

$$\sigma = \sqrt{\sigma^2}$$

where m_i = mean diameter of each size class.

N = number of items; when per cent is used $N = 100$.

M = arithmetic mean diameter of particle size.
σ = standard deviation.

Table 2.4. Size Distribution of a Beach Sand Illustrating Calculation of Mean and Standard Deviation

Grade Size		Mean ϕ	Weight Per Cent Frequency				
(mm)	ϕ	(m)	(f)	fm	$m-M$	$(m-M)^2$	$f(m-M)^2$
4–2	−2–−1	−1.5	0.5	−.75	−3.06	9.1	3.64
2–1	−1–0	−0.5	5.6	−2.80	−2.06	4.3	24.07
1–$\frac{1}{2}$	0–1	0.5	11.7	5.85	−1.06	1.12	13.10
$\frac{1}{2}$–$\frac{1}{4}$	1–2	1.5	53.7	80.55	−0.06	.00	.00
$\frac{1}{4}$–$\frac{1}{8}$	2–3	2.5	26.4	66.00	.94	.88	23.23
$\frac{1}{8}$–$\frac{1}{16}$	3–4	3.5	2.1	7.35	1.94	3.75	7.87
Totals			100	156.20			71.91

$$\text{Arithmetic mean } (M) = \frac{156.20}{100} = 1.56\phi$$

$$\text{Standard deviation } (\sigma) = \sqrt{\frac{71.91}{100}} = 0.85\phi \text{ units}$$

In a symmetrical curve (i.e., the "normal curve of error") the standard deviation on either side of the mean lies approximately between the 16 per cent and 84 per cent sizes as shown by the cumulative size distribution; and two standard deviations include 68 per cent of the frequency distribution. Four standard deviations (i.e., two on either side of the mean) include 95 per cent of the distribution, and three standard deviations on each side of the mean include 99 per cent of the distribution, as shown in Fig. 2.11. By symmetrizing the size distribution using the phi scale the shape of the curve of distribution can be described by the arithmetic mean and the standard deviation. In this manner sediments of different environments of deposition can be compared and characterized.

In a general way the shape of the cumulative size frequency distribution has been used to characterize or distinguish sediments of various types as, for example, the dune sand (Fig. 2.9) and the wind-blown silt (Fig. 2.11).

Such cumulative distributions normally are plotted on semilogarithmic paper on which the cumulated weight per cent is plotted along the arithmetic ruling and particle size along the logarithmic rulings. This has the effect of spacing, at uniform intervals, sizes

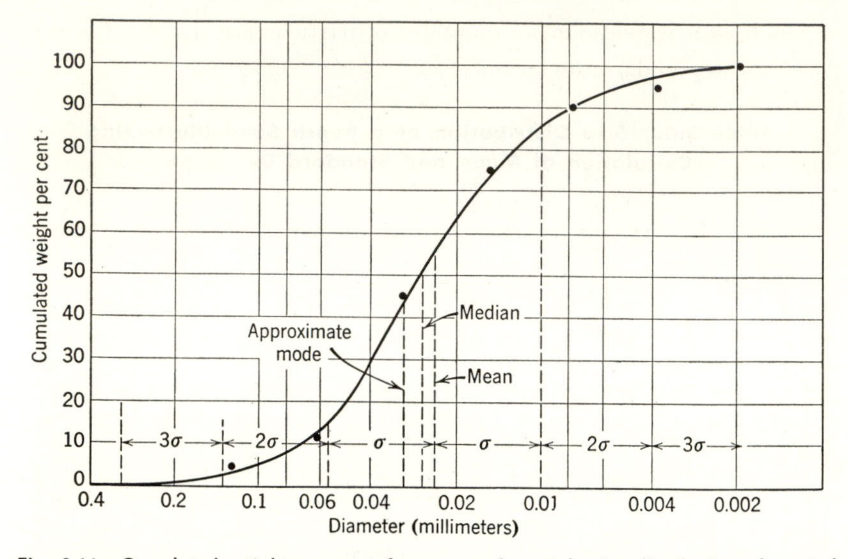

Fig. 2.11 Cumulated weight per cent frequency of particle size distribution of a wind-blown silt (loess) plotted on semilogarithmic grid. The distribution is skewed slightly toward the fine sizes, as indicated by the relative positions of the mean and median, which in symmetrical distributions are coincident. In slightly skewed distributions the modal size can be approximated, since the distance between the median and mean is one-third that between mode and mean. Note the spread of one standard deviation on either side of the mean, i.e., 2σ includes 68 per cent of the distribution, 4σ includes 95 per cent, and 6σ, 99 per cent.

such as 1, $\frac{1}{2}$, $\frac{1}{4}$, $\frac{1}{8}$, and $\frac{1}{16}$ mm and establishes the progression similar to that illustrated in Fig. 2.9.

The Otto Graph

An improvement on the semilogarithmic ruling is that prepared as shown in Fig. 2.12. The base consists of equally spaced horizontal rulings in phi units and vertical rulings spaced according to the equation of a probability distribution.[6] On such a base a symmetrical curve (i.e., the "normal curve of error") plots as a straight line when cumulated. This is illustrated by plotting the cumulative weight per cent of the size distribution of the wind-blown silt on a semilogarithmic base (Fig. 2.11) and on the normal probability distribution base (Fig. 2.12).

The Otto paper is ruled in such a fashion as to show the position

[6] This technique was developed by G. H. Otto, "A Modified Logarithmic Probability Graph for the Interpretation of Mechanical Analysis of Sediments," *Jour. Sedimentary Petrology,* Vol. 9, 1939, pp. 62–76.

of the mean (50 per cent) and standard deviation limits (see 15.7 per cent and 84.3 per cent); a straight line connecting the diameters between the limits of one standard deviation on either side of the mean describes the slope of the normal curve. A parallel to this slope, extended from the center of the small circle (at the intersection of 6ϕ

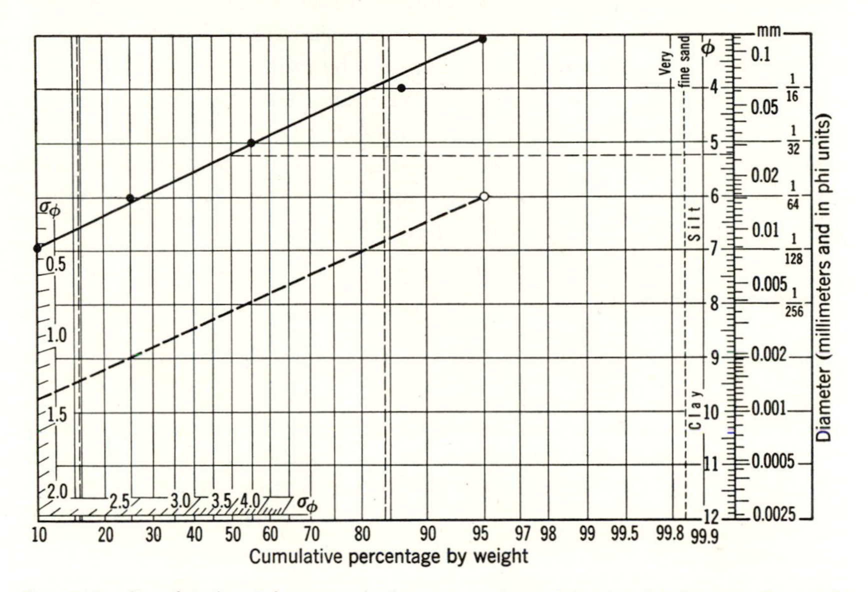

Fig. 2.12 Cumulated weight per cent frequency of particle size distribution of a wind-blown silt (loess), Fig. 2.11, plotted on Otto base. Note that on this base the ordinates are particle sizes and that the cumulated weight per cent is plotted as abscissae. Spacing of the ordinates is determined by a slight modification of the equation of the "normal curve of error." Symmetrical distributions plot as straight lines on the Otto graph. The mean size on the phi scale is determined by the height of the ordinate of the 50 per cent size (5.2ϕ). The standard deviation is determined by constructing a parallel to the line of the idealized distribution through the plotted points. The parallel line begins at the circle at the junction of the 95 per cent and 6ϕ lines and extends into the σ scale along the left-hand and bottom margins (in this case 1.3 phi units).

and 95 per cent into the left-hand margin, permits direct reading of the value of the standard deviation (σ) in phi units.

In illustrations scattered throughout this book, the technique of using size-distribution data plotted on Otto paper has been employed as a basis of comparison between sediments deposited under various environmental conditions. The mean size, the standard deviation, and the range in sizes are important indicators of the environments of deposition. Sediments of small standard deviation in size have been well sorted during their transportation. This property developed by current flow indicates that the environment in which the sediment was moving was that of wind or water. Conversely, sediments of large size range and large standard deviation are products of dumping such as by landslides or by glaciers. A size distribution of average size in the pebble grade is known not to be a wind-blown deposit because of the inability of the wind to move fragments as large as pebbles. However, sediments of such characteristics may represent a residue of the winnowing action of the wind. Similar lines of reasoning can be employed to advantage in recognizing deposits accumulated by gently flowing streams or clays deposited in glacial lakes. In fact entire fields of investigations are being based upon standard statistical procedures as controls in the attempt to reduce the degree of qualitative reasoning.

THE SOIL PROFILE

Soil was stated to be a product of alteration of a sediment or rock at the earth's surface and to represent material having special characteristics. The actual transformation proceeds through certain recognizable stages which in descending order from the earth's surface appear as a series of zones collectively described as the *soil profile.*

The soil profile is exposed in road cuts or in borings where in either case may be observed the physical and chemical changes which have occurred as the parent material has progressed through the soil-forming stages. A typical profile, Fig. 2.13, beginning from the surface downward is described in brief as follows:

Upper or "A" Horizon

The unit designated the "A" horizon ranges from a few inches to as much as 2 feet thick. It is the zone in which the grass roots and other forms of life are embedded or living and hence in moist regions contains much humus. Humus stains the mineral fragments dark making individual mineral species indistinguishable. Particles are generally finely divided and range between clay and sand in decreasing proportion respectively.

The "B" Horizon

Below and often sharply defined from the "A" zone is the lighter, or sometimes deeper, colored "B" horizon, ranging between a few

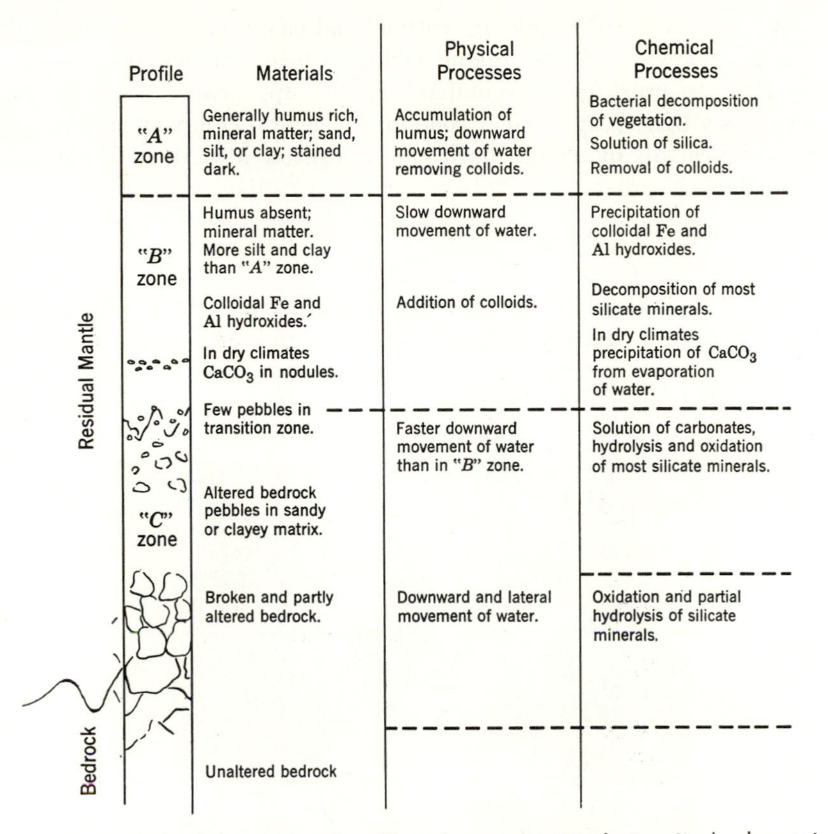

Fig. 2.13 Generalized composite soil profile and processes attendant on its development from the underlying bedrock. Each zone is characterized by the attributes indicated.

inches and about 3 feet in thickness. Colors range between light gray and light buff and dark red brown and are marked by the absence of staining by humus. Ordinarily the material in this zone is more plastic than the "A" horizon due to the presence of colloidal iron and aluminum hydroxides and development of clays. With increasing aridity nodules of precipitated $CaCO_3$ also make their appearance. Except for the secondarily deposited nodules the particle size distribution of the "B" zone contains somewhat more clay-sized particles than that of the "A" horizon but increases in coarseness downward and is gradational with the underlying "C" zone. Within the gradational zone a few rock fragments are observed. These differ from the individual mineral grains, clay aggregates, or nodules of precipitated $CaCO_3$ higher up in the "B" zone inasmuch as they are aggregates of differing mineral species. When such "pebbles" are first noted they are friable or "rotten" and easily crushed between the fingers. Downward the fragments increase in number and are more resistant to crushing. Accompanying the appearance of rock fragments is a lightening of color tone and a general compact and massive appearance, all of which mark the transition into the "C" horizon.

The "C" Horizon

The "C" horizon marks a gradual transition from a few pebbles embedded in a matrix ranging between sand and clay, downward through a zone where the pebbles become dominant, to a zone of solid rock broken only by fissures. The zone, therefore, marks the

Fig. 2.14 A soil profile being developed on glacial moraine (central Indiana). The "C" horizon (about middle of soil zone) is transitional into the underlying unaltered bouldery-clay sediment overlying the bedrock. Note the sharp boundary between the fresh limestone bedrock and the overlying moraine. The bedrock is not involved in the soil-forming processes as in Fig. 2.13.

change from disintegrated or broken rock to solid rock and is of variable thickness ranging from several feet to more than 100 feet in the hot moist climates.

Zones "A," "B," and "C" are often called the *mantle* from the tendency of the soil to obscure exposures of solid rock. Local reference may also be made to the "A" and "B" zones as representing the true soil and to zone "C" as the subsoil. Despite such minor differences in terminology there is no doubt that the gradation from soil into rock demonstrates the latter to be the parent material. In fact, as will be described later the soil has been derived from the bedrock by a process known as weathering. Leaching by downward-moving surface water is one of the prime factors in the weathering process, and the soil represents the products of the leaching action. The "A" horizon is the most thoroughly leached of all, whereas the "B" and "C" are progressively less altered.

Some profiles in road cuts or borings do not fulfill the requirements of gradation into the underlying bedrock. In such cases the relationships of the zones "A," and "B," and "C" are normal, but the "C" horizon is gradational with an underlying sediment. (See Fig. 2.14.) This may be a beach sand, a wind-blown deposit, alluvium in a river valley, or glacial debris, each carried to its present site by some geologic agent of transportation. The uncemented sediment blankets the bedrock upon which it rests, but no genetic relationship exists between the two. The sediment and its soil constitute a unit to be contrasted with the exposed solid rock and its soil, and differences in the properties of the two soils are to be expected.

Unconsolidated transported debris and its derived soil covering the bedrock are known also as *mantle.* Normally they are called *transported mantle* to distinguish them from *residual mantle* representing weathered and partially weathered material being developed from the bedrock.

ROCKS CONSTITUTING THE BEDROCK

"Bedrock" is a useful term signifying the uppermost unit of rock below which no more unconsolidated debris occurs. In some localities the boundary between the mantle (transported mantle) and the bedrock is sharply defined, and the underlying rocks are solid and unweathered. Elsewhere the mantle (residual mantle) is gradational with strongly coherent rock and the integrity of the rock increases with depth. There are, however, many areas where the bedrock despite its unweathered character is weakly coherent, and little physical distinction appears between the bedrock and its overlying soil. This situation is commonplace in much of the Coastal Plain of the United States, and confusion arises from the reference to certain "bedrock" deposits as partly consolidated "soils." Such difficulties can be avoided by noting the positions and relationships of the soil horizons and by certain other attributes to be described later.

Bedrock is not of uniform type but shows considerable variation

Table 2.5. Summary of Significant Attributes of Ideal Bedrock Types

Color of Rock	Size of Grains or Crystals	Shape of Grains or Crystals	Mineral Types	Mass Appearance	Major Rock Group
Light gray, dark gray, black, or red	2 cm to .01 mm, average 1 mm; crystalline, crystal size uniform or large crystals surrounded by smaller ones	Regular, often tabular; crystals interlock	Commonly three different varieties	Massive, homogeneous; often cut by fissures of arcuate or irregular pattern	Igneous
Light gray, green gray, blue gray, dark gray, red, or brown	Range from 2 mm to colloidal dimensions, also pebbles in sand matrix	Grains round and cemented together; rarely interlocking	Generally three or less varieties	Generally in parallel layers ranging from 1 mm to 1 m thick	Sedimentary
Green gray, mottled gray, dark gray, brown gray	Range 2 cm to .01 mm; crystals of uniform size or large crystals interlocked with smaller	Thin, tabular or rodlike, crystals interlocked and distinctly oriented in one plane. Rock often shows banding due to concentration of minerals in zones	Generally less than three varieties	Generally shows pronounced parallel cleavage or banding	Metamorphic

(*a*) (*b*) (*c*)

Fig. 2.15 Outcrops of (*a*) igneous (granite), (*b*) sedimentary (limestone) and (*c*) metamorphic (schist) rocks illustrating characteristics in bulk. Note the typical aspects, massive igneous, layered sedimentary, and banded or cleaved units in metamorphic rocks. (Photograph of granite by Dale, U. S. Geological Survey.)

in aspect. This results from inherent properties of the rocks summarized in Table 2.5. (See Fig. 2.15.) Three great groups of rocks (igneous, sedimentary, and metamorphic) have long been recognized, not only on the basis of the differences in their properties but primarily on their mode of origin.

The processes leading to the development of the major rock groups perform certain functions of constantly forming new rock varieties from pre-existent material. Igneous rocks are generated at depth, are forced toward the earth's surface where they raise the temperature of the rocks which they intrude, and remelt and assimilate much of the rock material which is intruded. Metamorphism alters the sedimentary and igneous rocks in the zones of compression. Erosion destroys the exposed rocks, transports the debris to sites of deposition, and leads to the generation of sedimentary rocks. These processes establish a flow of material from one major rock group into another in a route identified as the *metamorphic cycle.* An understanding of the energy, materials, and processes involved is the framework of geology.

Selected Supplementary Readings

Jenny, Hans, *Factors of Soil Formation,* McGraw-Hill Book Co., 1941, Chapters I and II. Presents definitions and concepts of the nature of soil and methods of presentation of soil data.

Krumbein, W. C., and Pettijohn, F. J., *Manual of Sedimentary Petrography,* D. Appleton-Century Co., 1938, Chapters 4, 5, 6, 7, 8, and 9. An excellent presentation of the fundamental concepts of particle size scales and manipulation of size distribution.

Krynine, D. P., and Judd, W. R., *Principles of Engineering Geology and Geotechnics,* McGraw-Hill Book Co., 1957, Chapter IV. Introduces the concepts of soil mechanics.

Lyon, T. L., and Buckman, H. O., *The Nature and Property of Soils,* The Macmillan Co., 1943, Chapters I and III. An introduction to the nature of soil and physical properties of mineral soils.

Millar, C. E., and Turk, L. M., *Fundamentals of Soil Science,* 2nd ed., John Wiley and Sons, 1951, Chapters I, II, and III. Soil materials; classification and physical and chemical properties.

"Soils and Men," *Yearbook of Agriculture 1938,* Parts IV and V, U. S. Department of Agriculture, Government Printing Office, 1938. Contains articles dealing with soil properties, classification, and maps.

Twenhofel, W. H., and Tyler, S. A., *Methods of Study of Sediments,* McGraw-Hill Book Co., 1941, Chapters IV and VII. Describes methods of preparing data from sediments.

Weir, W. W., *Soil Science,* J. B. Lippincott Co., 1936, Chapters 2, 3, and 4. Concerns the physical and chemical properties of soils.

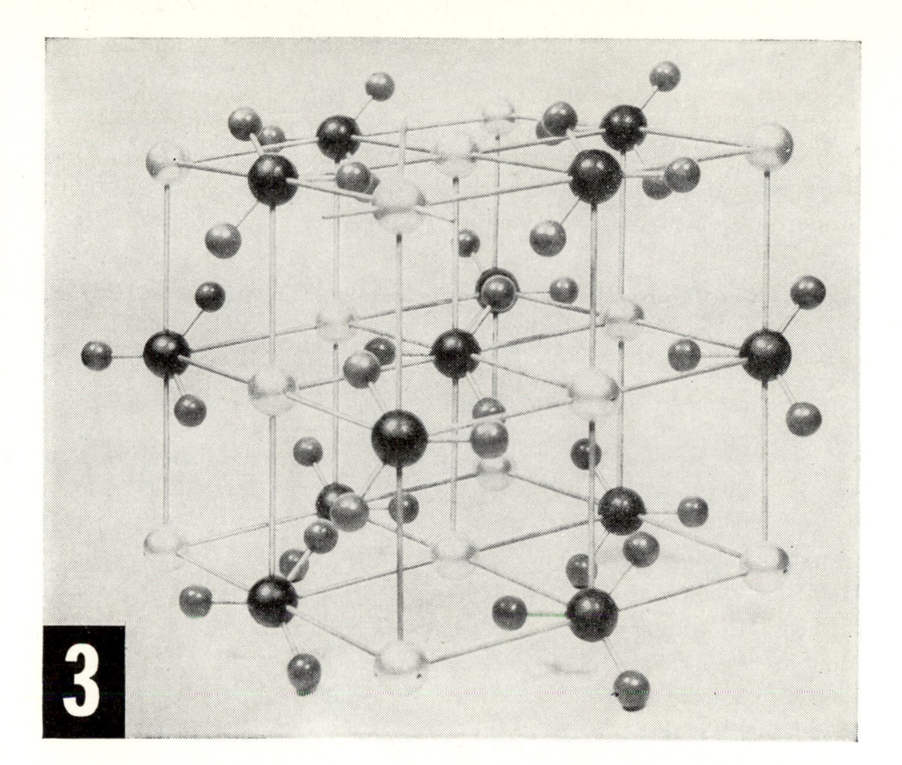

3

Physical and chemical properties of rock materials

PHYSICAL AND CHEMICAL ENVIRONMENTAL ZONES IN THE EARTH

The Geothermal Gradient

Available data indicate that with increase in depth from the earth's surface downward a progressive rise in temperature is recorded. This condition is known as the *geothermal gradient*. Most of the information concerning the geothermal gradient is gathered from vertical borings the deepest of which exceeds 20,000 feet. Although drilling

Lattice structure of calcite.

to such a depth is a spectacular achievement by man the rock column represented is only a thin surface skin when considered in terms of the radius of the earth (20.8×10^6 feet). Within the zone of observations there exists, between temperature and depth, an over-all direct linear proportion which approximates 30°C for every kilometer of depth.

Table 3.1. Determined Geothermal Gradients in the United States*

Locality	Depth to Last Observation in Meters	Mean Thermal Gradient in Degrees Centigrade per Kilometer from 30.5m to Deepest Observation	Oil Field Area	Area outside Oil Field
Georgia, La Grange (gneiss)	188	14.6		x
Louisiana, Pine Island	1067	40.4	x	
Michigan, Houghton	504	14.3		x
Montana, Anaconda	488	38.5		x
Nevada, Virginia City	701	56.6		x
New Jersey, Franklin Furnace	762	10.7		x
North Dakota, Lonetree	1143	36.4	x	
Oklahoma, Oklahoma City	1829	20.6	x	
Oregon, Astoria	1152	30.4		x
Texas, Edwards	2012	30.4	x	
Wyoming, Lance Creek	1067	49.6	x	

Mean value (22 observations)	30.1
Standard deviation of mean	13.2
Mean value—oil fields (16 observations)	33.3
Mean value—outside oil fields (8 observations)	24.6

* H. C. Spicer, "Observed Temperatures in the Earth's Crust," *Handbook of Physical Constants*, Geological Soc. Amer., Spec. Pap. 36, section 19, 1942.

This value is not to be interpreted as signifying that everywhere the same geothermal gradient is observed, i.e., that at a selected depth below the earth's surface the same temperature will obtain. Rather the figure given is an average of many whose spread is large. Table 3.1 indicates the ranges of geothermal gradients in localities within the United States. Most of these observations are from oil-field borings and represent higher gradients than those obtained in areas distant from oil fields. The higher values obtained in oil fields represent the addition of heat from two different sources. One is the radiation from the high-temperature zones deep within the earth (i.e., from "normal" heat diffusion); the other is heat produced in exothermic reactions between hydrocarbons within the oil-producing strata.

If the mean value of 30°C per km is extrapolated to depths beyond the limits where data are now available the inferred temperatures are far in excess of those necessary to melt rocks under atmospheric conditions. Figure 3.1 illustrates the extrapolation of the average thermal

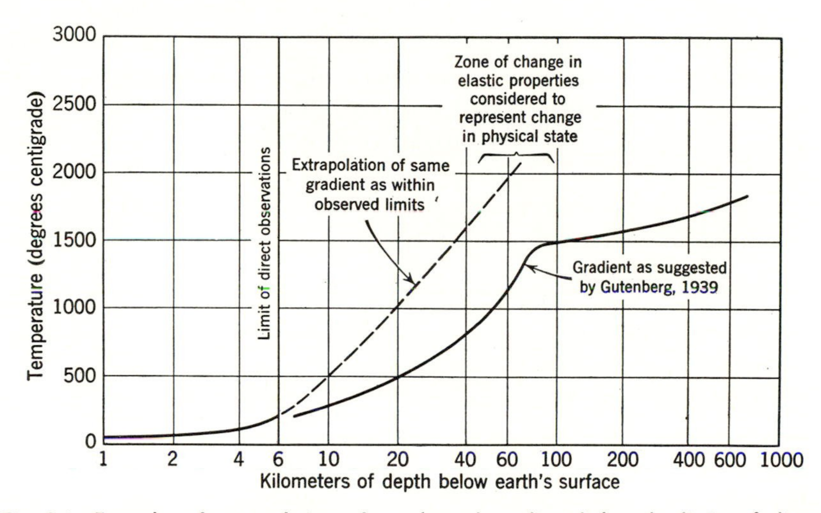

Fig. 3.1 Examples of extrapolation of geothermal gradient below the limits of direct measurements.

gradient to depths of 60 km where temperatures of 2000°C should prevail. If one accepts the existence of such temperatures rocks below the critical zone where melting occurs could be considered to be fluid to the center of the earth. Whether such is the case is dependent upon other factors, namely the validity of the linear relationship between depth and temperature when extrapolated to great depths and change of rock-melting temperature with increased pressure. Evidence which has been obtained from the study of earthquakes indicates that sudden fracture of the rock material under shear stress still occurs at depths approaching 1000 km below the earth's surface and, hence, that rock materials behave as solid substances to great depths. These data are subject to two interpretations: namely, (1) the melting temperatures are rising with increased depth as fast or faster than the geothermal gradient, or (2) the material is above the crystallization temperature but behaves as a solid despite its noncrystalline state.

Not uncommonly lava escapes from volcanoes to solidify and cool as great flows on the earth's surface. Where such volcanoes occur there is considerable doubt in the minds of geologists that the liquid material has come from hundreds of kilometers of depth. Yet in the same geographic region the deep-seated earthquakes indicate that rupture of rocks occurs at nearly 1000 km of depth. This apparent paradoxical occurrence of liquid lava at shallower depths than the solid is explained by the behavior of rock material under the enormous hydrostatic pressures which prevail at great depths. As a rough approximation rocks at depths exceeding 200 km can be considered to be under hydrostatic pressures sufficient to cause the material to behave as a solid and to fail by rupture when subjected to a great unbalanced stress. However, at depths less than 200 km where the earth's internal-stress conditions permit local unbalanced forces to exist the solid rock can move under plastic flow toward the surface. Eventually this solid will enter a realm where the rock melting point at the prevailing pressure is less than the temperature of the intruding rock and a true liquid may form. Given sufficient time the liquid cools below the crystallization temperature and a generation of igneous-rock species will develop.

Elastic waves spreading outward from earthquake centers indicate that at a depth of 2900 km a pronounced change in rigidity of the medium of transmission prevails, and there is a strong suggestion that such material has the properties of the liquid state. This condition extant to the earth's center outlines the "core," the constitution of which appears to have much to do with the earth's bulk properties. What temperatures obtain within the core are purely speculative, but it is doubtful that the geothermal gradient continues with the same average values as exist within the surface layers, and there is much support from the characteristics of the elastic properties that the slope of the geothermal gradient flattens very appreciably beyond 70 km of depth (see Fig. 3.1).

Pressure Zones

A pressure gradient similar to the geothermal gradient exists from the surface downward. This is due to the weight of the overburden of rock exerting a force toward the earth's center. The pressure gradient is important not only in affecting the melting temperature of the rocks, but also in controlling the physical behavior of the plastic flow of solids and reorientation of elements into compounds of higher density, and in facilitating crystallization of minerals of bladed or rodlike shape. The pressure gradient, however, is merely an average value, and deviations are recognized less than and greater than the average. In localities of recently elevated mountains the pressure gradient appears to exceed the average value, whereas in regions such as deltas the gradient appears to be below the mean. These data are difficult to obtain and to interpret, and more reliance can be placed upon evidence supplied by the structure of the rocks. Localities where strong differential stress was once applied are identified by the presence of strongly folded and fractured strata. Some localized areas have been under tension as recognized by systems of planar open fissures (*joints*) spaced at regular intervals across the rock. Where such fissuring is observed application of tensional stress has been greatest in a horizontal direction and least in the vertical plane. Elsewhere, strong folding in sedimentary rocks indicates that compressional stress was applied in the same directions. Nearly everywhere that rocks are exposed to view similar evidence of the former existence of shear stresses is to be observed, and the geologist must conclude such forces dominate pressure conditions in the upper part of the crust.

Abundant evidence of compression is to be found in subsurface zones particularly in the upper 10 miles (17 km). Near the surface, folds tend to be open, but with increased depth the application of differential stress has been so severe that individual mineral grains have become oriented along parallel planes and rocks are metamorphosed. In exceptional areas where erosion permits examination of the deepest part of the crust the effects of shear stress appear to have resulted in intense plastic flow. Metamorphic rocks are contorted in such an extreme fashion that the failure by rupture which is so characteristic of the uppermost part of the crust is subordinate to yield by plastic flow.

Below the depth of metamorphism prevailing temperatures cause localized melting, and differential stress drives the fluid to the surface to erupt as a volcano. Also, under the shear-stress conditions molten rock intrudes rock in the plastic state, and a complex rock is developed whose characteristics are both igneous and metamorphic. Still deeper, at depths of the order of magnitude of between 100 and 200 km, the load of rock overburden is sufficiently great to exert a pressure approaching that generated by the primary internal forces. Here, the magnitude of the applied stress is approaching the same value in all directions. Indeed, at increasingly greater depths shear no longer prevails, and hydrostatic stress dominates (see Fig. 3.2). Thus, the earth can be imagined as consisting of three concentric shells of pressure conditions: the lowermost and thickest of which is the shell of hydrostatic stress, the middle zone is one of transition to shear

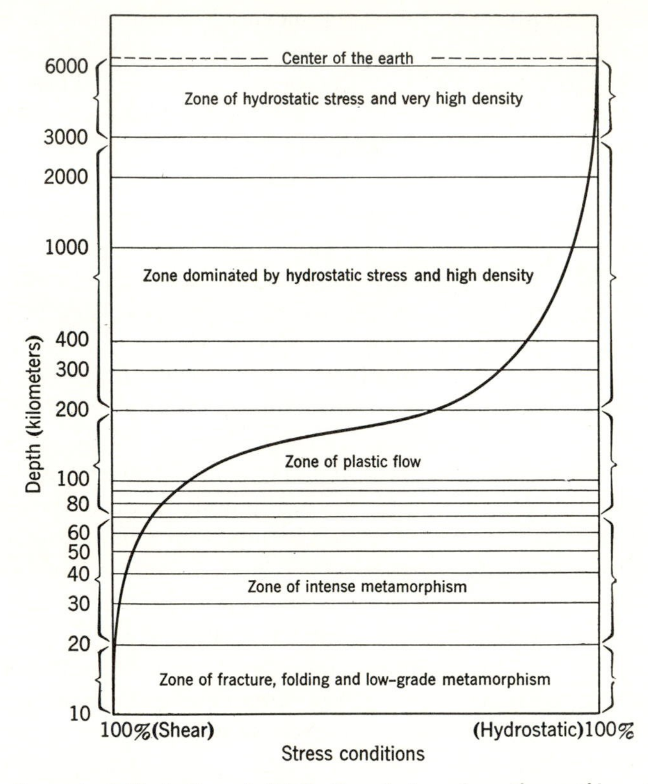

Fig. 3.2 Conceptual illustration of distribution of stress from the earth's surface downward. The limits of zones should be considered as extremely uncertain.

stress, and the upper thin shell is overwhelmingly one of shear conditions. Insofar as the human observer is concerned shear stresses dominate the rocks which he examines; an important field of geology is devoted to this aspect of the science.

Zones of Predominant Chemical Units

A concept of an earth consisting of shells of different pressure conditions has been presented, but a similar concept of concentric layers of rock of different density is supported by much more factual information. With the exception of the core which is very large and sharply defined the remaining shells are of variable thickness and separated

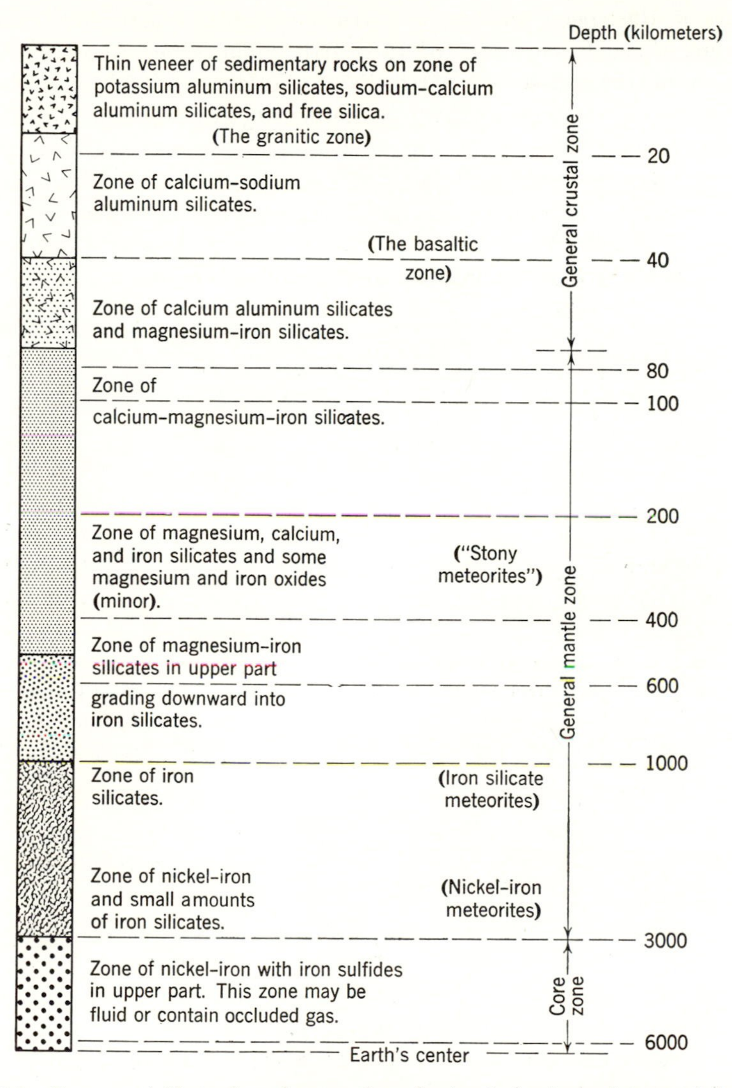

Fig. 3.3 Conceptual illustration of zones of predominant chemical substances below the earth's surface. (Modified from Buddington, *Am. Mineralogist*, 1943.)

by gradational or indefinite boundaries. (See Fig. 3.3.) Difference in the elastic properties characteristic of each of the shells is considered in part to result from stratification of different chemical units.

Although understanding of the nature of the interior of the earth is fragmentary in this respect a generalized framework can be erected on the basis of information derived from various sources. One of these sources is the speed and path of transmission of earthquake waves traveling through the body of the earth. The speed of travel is a function of the elastic properties and the density of the material through which the waves travel. The density distribution within the earth therefore can be determined. This distribution must satisfy the moment of inertia, the average density, and the shape of the earth as established from astronomical data. Table 3.2 illustrates this

Table 3.2. Distribution of Density within the Earth

Depth in Kilometers		Range in Density from Top to Base of Zone	Pressure (Bars) at Base of Zone*
0–70	crustal zone	2.67– 3.05	21×10^3
70– 450		3.18– 3.50	15×10^4
450– 950		3.80– 4.65	35×10^4
950–2700		4.65– 5.60	12×10^5
2700–2900		5.60– 5.68	13×10^5
2900–5100	core	9.7 –11.9	31×10^5
5100–6371	(center)	11.9 –12.2	35×10^5

*A bar is 0.987 atmosphere, or 14.504 lbs per sq. in., or 1.000×10^6 dynes per sq cm. Data are interpolated from Francis Birch, "Geodetic Constants," *Handbook of Physical Constants,* Sec 8, Geological Soc. Amer., Spec. Pap. 36, 1942; and R. A. Daly, *Strength and Structure of the Earth,* Prentice-Hall Inc., 1940, p. 123.

change in density with increasing depth below the surface. It will be noted that zones of more or less similar density are distributed beginning with the lighter materials at the surface and increasing in density downward. Note also that there are two conditions of density change: (1) a gradual transition from one value to a higher one over an interval of 10 or more kilometers of depth, and (2) a sudden change in density value across a sharp boundary. Together these two varieties of density change produce concentric shells each of which has its characteristic average density.

The Core

Some of the density distribution characteristic of the individual shells can be attributed to a bulk difference in chemical units; for example, the electrical field of the earth appears to be controlled largely by the nature of the core (see Fig. 3.3). Its high density together with its magnetic properties suggest that its bulk composition is dominated by iron alloyed with some nickel. Some support to this assumption is furnished by the occurrence of certain meteorites constituted principally of iron, but nickel may be present in amounts which may range as high as 20 per cent. Other meteorites contain iron in a silicate molecule and have a much lower density. Assuming that iron silicate meteorites have been derived from the same parent body as those of nickel-iron composition the position of the latter material by virtue of its higher density should be stratified beneath the iron silicates and, hence, be representative of core material in a former astronomical body of planetary proportions and properties.

The Intermediate Zone

Rare occurrences of igneous rocks known to come from deep within the earth contain oxides of iron and magnesium and sulfides of nickel and iron. These are of lower density than nickel-iron and are believed to represent material from a zone some distance above the core. The gradual density drop toward the surface is attributed in part to a proportional increase in magnesium, iron, and calcium silicates, the compositions of which are similar to some of the "stony" meteorites falling to the earth's surface. Silicates of this type in contrast to other igneous rocks are exceptionally low in silica, approximately 25 to 35 per cent.

With the exception of the upper 40 km the remaining composition of the rocks appears to be dominated by a gradual increase in silica of 35 to approximately 50 per cent, and a sharp increase in aluminum of low values (a few per cent) to about 15 per cent. Material of this composition issues from numerous volcanoes particularly in the region of the Pacific Ocean. Such lava known as *basalt* is widespread on the earth's surface and is characteristically of uniform composition, irrespective of the geologic age at which such lavas were erupted.

The Crustal Zone

Resting upon the basalt zone with sharp discordance is the crustal zone. Its thickness is variable and normally ranges between 25 and 70 km. This boundary is established on the basis of the sudden increase in the velocity of seismic waves as they enter the basalt layer. Crustal rocks are highest in silica ($\pm$70 per cent), aluminum, sodium, potassium, rare earths, and the other metallic elements not typical of the rocks of the deeper zones. *Granite* is the dominating rock of the crustal zone; hence, the latter is frequently termed the granitic zone although it should be understood that in its upper portion, at least, metamorphic rocks constitute an important percentage.

Locally, on the continents, the granitic zone is exposed at the surface, but the greatest proportion of the continental surface is coated with a veneer of sedimentary rocks, the maximum thickness of which does not exceed 15 km. The sedimentary veneer is characteristically high in silica, hydrous aluminum silicates, hydrated iron oxides, carbonates, sulfates, phosphates, water, restricted organic compounds (coal and oil are examples), and many other inorganic minerals stable under the physicochemical conditions of the atmosphere-lithosphere (rock sphere) interface.

PROPERTIES OF SOLIDS

General Properties

Attention has been directed to minerals as the fundamental units of which rocks are constituted and to the variation in rock types as primary responses to the kinds, and percentage distribution, of minerals within the rock. Except for one noteworthy example minerals are solids under the temperature and pressure regimes which prevail at or near the earth's surface,[1] and must respond to physical laws which govern the behavior of solids. The exception is the mineral water, which exists on the earth's surface in gas, liquid, and solid phases.

Among other physical properties which distinguish the different states of matter the solid state is characterized by its high values of rigidity or resistance to shear. This property is important in controlling contortion, folding, or fracturing of minerals and rocks as well as certain other properties within the elastic limit such as speed of transmission of sound waves. Certain substances such as water change at precise temperatures, or over a very narrowly defined temperature range, from the fluid state with extremely low values of rigidity to the solid state with high values. This is the crystallization temperature. Other liquids gradually increase in rigidity over a wide temperature range, and there is no precise temperature at which they are acknowledged to become solids. Such liquids have entered the solid state by a gradual increase in viscosity and should be considered fluids of extremely high viscosity at the prevailing temperatures. These supercooled liquids have finite strength when a shear stress is applied

[1] Temperature ranges between approximately −50°C and 50°C and pressure ranges between slightly less than 1 atmosphere and as much as 10 atmospheres.

rapidly, but when the stress is applied slowly deformation occurs at much lower pressure values. A case in point is tar or sealing wax which will fail by fracture when a stress is applied rapidly but which will deform by internal flow under low forces applied slowly.

Amorphous and Crystalline Solids

One of the most significant differences between the two types of solids is their appearance upon solidification. In the case of the crystalline solids when the temperature of the liquid reaches the crystallization

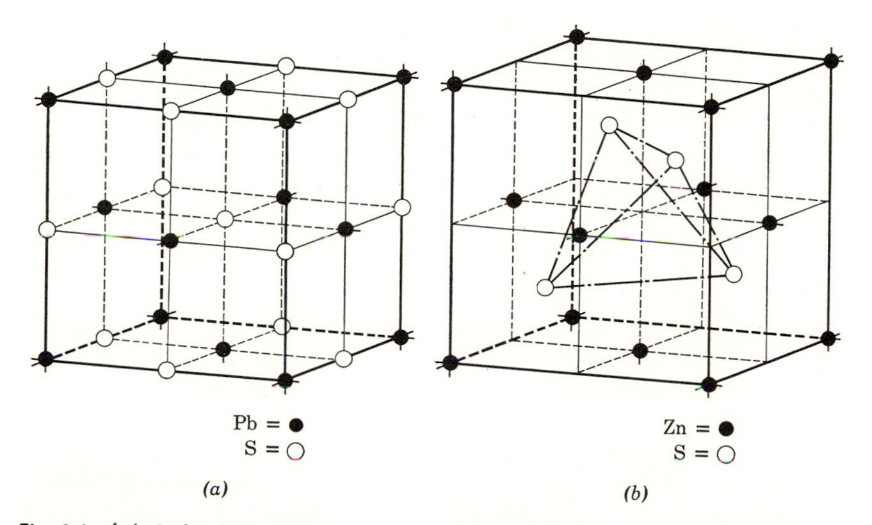

Fig. 3.4 (*a*) Stylized illustration of the space lattice of galena with atoms of Pb and S alternating in positions and arranged at the corners of the cubes. (*b*) Space lattice of sphalerite showing the positions of atoms arranged within the lattice cube. Note that the atoms of Zn and S do not alternate as above but S atoms are centered within the lattice. (By permission from Kraus, Hunt, and Ramsdell, *Mineralogy*, copyright 1951, McGraw-Hill Book Co., Inc.)

point the resulting solid (crystal) assumes a precisely definable geometric pattern. This is accompanied by a release of heat, the latent heat of fusion, and the acceptance of a lower free-energy state. For the amorphous or shapeless solids no definite geometric pattern results when the fluid solidifies, and no release of latent heat is observed. Rather, the gradual increase in viscosity permits the solid to assume the characteristics of gels or supercooled liquids such as glass.

The differences between the two solids are even more convincingly illustrated when the framework of molecular structure is examined by means of X-ray bombardment. In the case of the crystalline solids a definite geometric pattern of arrangement of atoms into a unit form is observed. The position of the atoms with respect to the geometry of the unit form delimits the *unit cell* and is important in defining the type of unit cell. Examples are: (1) the unit cell described as a cube with an atom occupying each corner, or (2) a cube similar to that of (1) but in which additional atoms are arranged internally (see Fig. 3.4). In the amorphous substance no such geometric orientation of atoms prevails, and the only arrangement is that necessary to establish the individual molecule.

The unit cell is the fundamental order of arrangement of atoms in the crystal and is repeated over and over again to construct the large external geometric pattern of the crystal. Absence of such a unit cell or of an orderly arrangement of cells in amorphous substances demonstrates that they cannot be represented as crystals. Truly amorphous substances are very exceptional among the minerals of the earth's crust. Except for certain lavas which are cooled before crystallization can occur amorphous rocks are virtually unknown.

PROPERTIES OF MINERALS

Definition

Use of the word "mineral" has not met with uniform application. Economists have generally applied it to include any nonliving material which is extracted from the earth. Until recently mineralogists have restricted the term to designate a naturally occurring, inorganic, crystalline substance. A more flexible definition such as the following is suggested: Minerals are both inorganic and organic substances which normally occur in the crystalline state, which for specific species have well-defined physical properties varying between narrowly restricted limits, a chemical composition that can be represented by either a precise formula or variation within specifically defined ranges, and are both naturally occurring and synthetically produced.

Under natural conditions inorganic minerals are formed far more commonly than organic ones because of the extremely low percentage of carbon in the earth's elemental composition. Inorganic minerals are the principal rock-forming group and to these our attention will be restricted.

Physical Properties

Crystal Systems

Addition of one unit cell to another establishes the geometric form of a crystal bounded by plane surfaces known as crystal faces. These may enclose space in such a manner as to produce simple or complex geometric solids depending on the number of faces developed in the individual crystal. If the angles between faces of a single crystal are

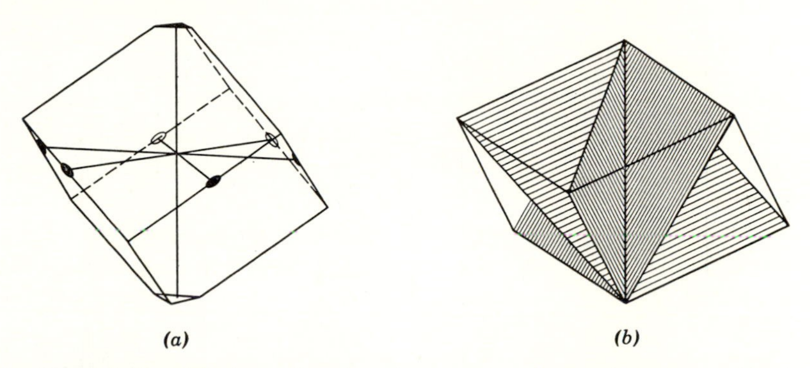

Fig. 3.5 (*a*) Rhombohedron oriented to show position of axes of symmetry. This form has a center of symmetry, three axes of two-fold, and one axis of three-fold symmetry. (*b*) Three planes of symmetry cut the rhombohedron. (Reprinted with permission from Hurlbut, *Dana's Manual of Mineralogy*, copyright 1952, John Wiley and Sons, Inc.)

measured with respect to a plane of reference it is noted that there are repetitions of certain faces having the same angular relationships. Such faces may be repeated one after the other, or they may follow some symmetrical arrangement in which certain other faces lie between two corresponding ones. This orderly repetition of certain selected plane surfaces endows the crystal with symmetry.

Symmetry may be axial in which case faces are symmetrical about an axis of reference, or it may be planar and an imaginary plane will divide a crystal into two parts each of which is a mirror image of the other. A third and lowest form of symmetry is a center about which crystal faces may be symmetrically arranged (see Fig. 3.5). Degree of symmetry ranges between forms which possess only a center and forms with as many as seven planes and seven axes of symmetry in an individual crystal. This range permits grouping those with common degrees of symmetry into six major systems, customarily arranged in order beginning with that one having the highest order and ending with that possessing a center only.

Although the degrees of symmetry displayed by a crystal are fundamental in classifying it into its proper system more use is made of intercept distances along imaginary axes of reference intersecting at the crystal center. These crystallographic axes, one of which by custom is oriented vertically, serve as coordinates along which distances are measured to the edges of a crystal face. The position of the crystal face can, therefore, be uniquely represented by ratios of intercept distances along the axes. (See Fig. 3.6.) With the exception of

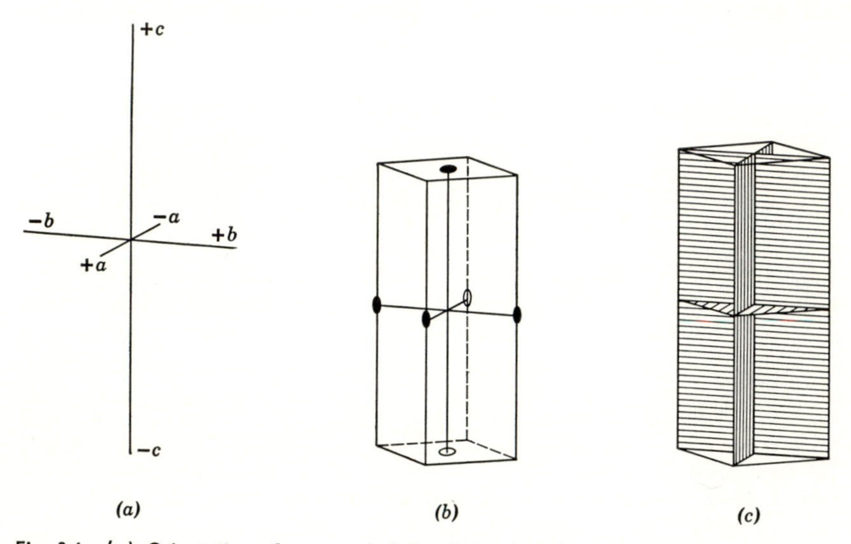

Fig. 3.6 (*a*) Orientation of axes and their relative lengths in the orthorhombic system. (*b*) Three two-fold axes of symmetry in a crystal of the orthorhombic system. (c) Crystal such as (*b*) showing positions of planes of symmetry. (Reprinted with permission from Hurlbut, *Dana's Manual of Mineralogy*, copyright 1952, John Wiley and Sons, Inc.)

one system the angular relationships and intercept lengths of three axes are used to characterize the geometric forms of each system. The exception is the hexagonal system whose symmetry requires three horizontal axes and one vertical axis of reference.

For purposes of illustration the angular relationships of the crystallographic axes characteristic of each system are given below. Reference to the intercept ratios listed are for the most simple and typical geometric form, but it should be understood that other geometric forms having different axial ratios are associated with each system. The order of listing below is also that of decreasing degrees of symmetry.

1. *Isometric* or *cubic*. The three crystallographic axes are mutually perpendicular, and intercept ratios are of equal length.

2. *Tetragonal.* The three axes are mutually perpendicular as in the isometric system, but the intercept ratios are not equal. When properly oriented the intercepts along the two horizontal axes are of equal length, but that along the vertical or *c* axis is either longer or shorter than the other two.

3. *Hexagonal.* Four axes characterize this system. Proper orientation places three of these in the horizontal plane, and each is separated from its neighbor by an angle of 60 degrees; the fourth is the vertical or *c* axis. Intercept lengths along the horizontal axes are equal, but that along the *c* axis may be longer or shorter than the others.

4. *Orthorhombic.* The three axes all intersect at right angles, but the intercept length is different along each axis, and this serves to distinguish this system from the tetragonal. As intercept lengths along two axes approach one another the symmetry of the tetragonal system is reproduced.

5. *Monoclinic.* The three axes lie within two planes which intersect perpendicularly. Proper orientation requires that the plane which contains only one axis be placed in a horizontal position. The vertical plane contains one axis, *c*, which is oriented vertically, and a second axis, *a*, which does not form a right angle with *c*. By custom the *a* axis is held inclined toward the observer, but the degree of inclination from the *c* axis varies considerably for minerals included in the system.

6. *Triclinic.* The symmetry of the triclinic system requires three axes, all of which form variable angles with one another. Axial intercept lengths are also variable. In consequence, the crystals that belong to this system have a very low order of symmetry which in some cases is restricted to a center.

Hardness

Hardness is a measure of the ease with which a mineral is scratched and doubtless reflects the composition and arrangement of atoms in the space lattice. Degrees of hardness vary widely between very soft and extremely hard, but the variation in hardness is small for any selected mineral species. In short the hardness of a single species such as quartz is consistent irrespective of its form of occurrence. Some mineral groups show a range in hardness, but the range is small, and, hence, minerals can be classified according to their hardness value.

Hardness scales or hardness values are more or less arbitrarily selected. The earliest approach to this subject was the establishment of Mohs' scale, based upon a qualitative measure of the varying hardness between mineral species. Certain minerals are selected as standards of reference and to these the hardness of other minerals is compared. Standard minerals of the Mohs' scale are: (1) talc, (2) gypsum, (3) calcite, (4) fluorite, (5) apatite, (6) orthoclase, (7) quartz, (8) topaz, (9) corundum, (10) diamond. It is to be understood that this is purely a qualitative scale and that its significance lies in comparing one hardness value with another. It is not to be interpreted

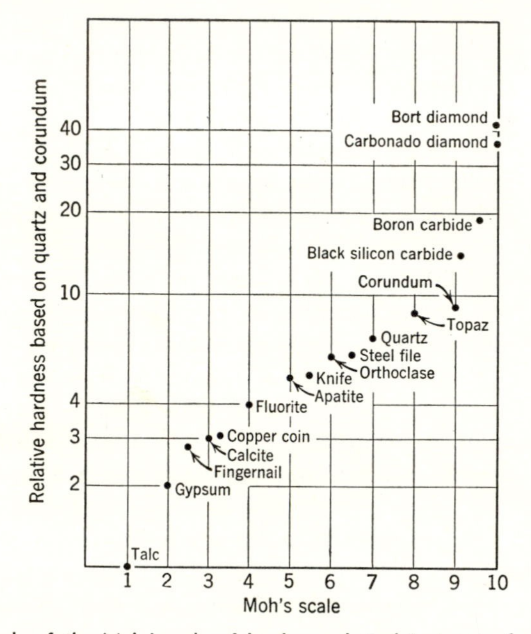

Fig. 3.7 Minerals of the Mohs' scale of hardness plotted in terms of relative hardness.

that the span of hardness value from 4 to 5 represents an equivalent increase as from 3 to 4, and it is to be noted that in some instances the real range in hardness is very much greater than the value indicates. As a result of this qualitative scale and the inadequacies involved, the American Society of Testing Materials has established a system wherein the range in hardness above 9 is more properly represented (see Fig. 3.7). Although the ASTM scale is more satisfactory for precise measurements Mohs' scale is standard in elementary mineralogy. The virtue of Mohs' scale lies in the fact that most minerals have been catalogued according to that scale, and the hardness of common minerals can be determined by simple qualitative tests. For example, the fingernail has a hardness of 2.5, a pen knife has a value of about 5, and ordinary window glass a value of about 5.5. These common items along with the common minerals in Mohs' scale can be used as standards of reference and as such they are useful both in the field and laboratory.

Color and Luster

Color is the most obvious of all of the optical properties and is useful in distinguishing certain mineral species. Normally, color is not a diagnostic characteristic inasmuch as most common minerals display varieties of colors. Quartz, for example, will range from colorless through reds and yellow to smoky brown or black.

Table 3.3. Common Varieties of Nonmetallic Luster

Adamantine	Exceedingly brilliant, or flashy, clear; luster exhibited by minerals of very high index of light refraction such as diamond.
Dull (sometimes called earthy)	High light absorption by small, granular particles to produce the luster of chalk or dry clay.
Greasy	The smooth, unctuous appearance developed by grease, wax, or soap.
Pearly, or opalescent	The appearance of mother of pearl, often displayed by mica along a broken surface.
Resinous	A clear, somewhat glassy, luster typical of clear resins.
Silky	The dull gloss reflected from a surface of woven silk fibers. This is commonly developed only in minerals whose habit is that of closely spaced parallel fibers.
Vitreous (glassy)	The clear luster developed from light reflection deep within the mineral. This luster is typified by that of clear glass.

A much more useful property is luster which represents the degree of reflected light returned from the specimen to the observer. The degree of reflected light is controlled by the roughness of the reflecting surface and the depth of light penetration or absorption by the specimen. In the latter respect it is also a measure of opacity. Normally, two major varieties of luster are recognized. One is characterized by polished metal or metallic minerals and is called metallic, whereas nonmetallic luster is representative of minerals which transmit light through thin edges or sections. Between the two extremes lies a poorly defined zone known as submetallic luster. (See Table 3.3.)

Cleavage

Cleavage along certain planes is controlled by the arrangement of atoms in the space lattice of the crystal. Certain orientations of the space lattice result in low tensile strengths along selected planes. Under stress, failure will occur in a series of planes all of which are parallel to existing or possible crystal faces. Cleavage is characteristic of certain mineral species, whereas in others no cleavage is recognized whatsoever. Quartz, for example, will not cleave, but calcite always shows a well-developed cleavage in three directions. Inasmuch as cleavage planes are parallel to possible crystal faces the orientation of cleavage is defined by intercept distances along the crystallographic axes of the system in which the mineral occurs. Hence, reference is made to the cleavage of calcite (hexagonal) as rhombohedral because the intercept ratios are those of the rhombohedron of the hexagonal system.

Cleavage is useful in orienting a known mineral crystallographically or often in determining the mineral species. The rhombohedral cleavage of calcite is so diagnostic that the student will recognize this mineral with ease on the basis of cleavage and luster. Micas have cleavage which is so perfectly developed that laminas less than 10^{-3} mm thick can be peeled from a crystal fragment.

Twinning

When a mineral species crystallizes repetition of the same compositional units of the space lattice occurs until the exterior shape of the crystal is completed. Although the composition of the units is identical their orientation in the space lattice is not always the same but may be arranged in repetitive order bearing relationships such as the right and left hands or mirror images of one another from a common plane of reflection. This feature often observed in minerals is called *twinning*. Individual twins appear to be joined along a plane (*twinning plane*) having specific crystallographic orientation or rotated about an axis (*twinning axis*) and interpenetrated. Twinning planes and axes are, therefore, described in relation to the crystallographic axes of the system in which the mineral species occurs. In certain minerals or allied mineral groups a distinct preference to twin along certain crystallographic planes or axes is noted. Such twinning is described as following certain laws generally named after mineral species which characteristically exhibit the law. Figure 3.8 illustrates several common varieties of twins.

Repetition of parallel grouping of successive crystal laminas alternating in obverse and reverse orders is defined as *multiple* or *polysynthetic twinning*. Normally the individual laminas are extremely thin, and their projection on a smooth surface approximately normal to the twinning plane gives rise to a series of parallel lines or striae on the crystal face.

Twinning serves as an additional means of mineral identification, for the occurrence of certain twinning laws is extremely common among specific minerals. In identifying fragments of minerals large enough

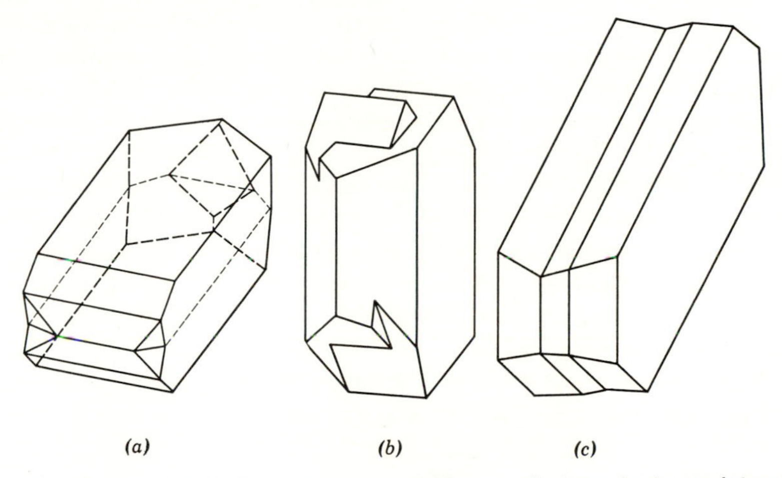

Fig. 3.8 Varieties of twinning typical among feldspars. The Mannebach twin (*a*) represents parts of crystals which are mirror images along a common plane (the twinning plane). In the Carlsbad twin (*b*) the right side of the crystal appears to have been rotated in position 180 degrees about an horizontal axis (the twinning axis) and interpenetrated. In (c), albite twinning, is illustrated the type example of multiple or polysynthetic twinning. Individual parts of the crystal are arranged in left-hand, right-hand, left-hand, right-hand, etc., relationship in repetitive order separated by parallel twinning planes. (Reprinted with permission from Hurlbut, *Dana's Manual of Mineralogy*, copyright 1952, John Wiley and Sons, Inc.)

to be studied with the unaided eye twinning is recognized largely on the basis of the degree of light being reflected from a smooth surface normal to the twinning plane. When the specimen is oriented so that light is reflected back to the observer it will be noted that parts of the twinned crystal will reflect more light than others. Moreover, there will be a sharp line of demarkation between the strongly reflecting and mildly reflecting surfaces. This arises from the slight difference in inclination of the reflecting face of one twin with respect to its neighbor. The projection line of the twinning plane bounds the reflecting and nonreflecting parts of the crystal surface. Single or multiple twins are distinguished by comparing the number of reflecting surfaces with the outline of the crystal fragment. If as the crystal is slowly rotated light is reflected first from one half and then from the other half of the face of the fragment a single twin is recognized. In multiple twins the reflection of light is not so easily recognized. Rather, the demarkation between the individual twins appears as a series of scratches or striations appearing as parallel rulings on the face of the crystal. The potash feldspar, orthoclase, is representative of the variety typically displaying single twins, whereas the sodium-calcium feldspars, plagioclase, are representative of polysynthetic twinning. Inasmuch as other properties of the two types of feldspars are similar distinction is made, in the field, between these two substances primarily on the basis of the type of twinning observed.

Special Features

Some minerals are recognized by certain special features or characteristics which enable them to be distinguished from others by simple tests. Among these are the color of the *streak* or powder, *specific gravity,* a *test for carbonates* (dilute hydrochloric acid applied to the specimen produces an effervescence of carbon dioxide), and elastic cleavage laminas of micas. Oftentimes such crude features suffice to identify a common mineral, but elaborate techniques are required to identify others.

General Chemical Properties

Solution

Many minerals are precipitates from aqueous solutions. Such solutions are rarely of high concentration, but when ionic solubility products are exceeded crystals will appear and grow as ions are removed from solution. There is, therefore, an equilibrium relationship between the solid crystals, the undissociated molecule of similar composition in solution, and the individual ions. The balance between ions, molecules in solution, and solid crystals will vary according to the pressure, temperature, and concentration relationships which exist in the solution. If the physical controls are altered from an equilibrium condition so that certain ions are removed from solution such ions will appear as a precipitated crystal. Should the physical conditions be altered in the reverse order, material which was solid will become molecules in solution, and in turn an equilibrium will be established between these and the equivalent ions. Under natural conditions all three exist at any instant of time.

When water from streams, lakes, or oceans is chemically analyzed an extensive list of ions is found to be present. If the complex ionic equilibria are unbalanced some of the ions combine and are converted into minerals of varying composition. In addition certain complex and simple molecules exist in a finely divided suspension defined as colloidal.

Colloids

Colloids are particles of minerals or other molecules of 0.2 micron (one micron = 0.001 mm) or smaller in diameter suspended in a surrounding medium. The particles remain in suspension indefinitely unless they are allowed to coalesce into large enough masses to settle out of the suspension. The presence of an electrolyte such as a dilute acid or NaCl will neutralize a colloidal solution by a combination of the ions of the electrolyte and the negative charge normally carried by all colloidal particles. Neutralization of the charges causes the particles to adhere when they collide, forming larger groups of particles or "flocs" which attain sufficient weight per unit area to sink through the suspended medium. This process is called flocculation. Colloids are abundant in nature and may be precipitated by certain organic or inorganic salts as flocculates or aggregates of individual colloidal particles, often amorphous in character. Other colloids are finely crystalline but maintain an over-all amorphous appearance.

As solutions and colloidal suspensions make their way below the earth's surface through fissures or the intergranular pore space of the rocks precipitation of minerals may result. Conversely, should the solutions be undersaturated some of the already existing crystalline material will pass into solution. Minerals are, therefore, constantly being precipitated or dissolved as water moves downward, laterally, and upward through openings in existing rocks.

Oxidation and Reduction Reactions

Near the lithosphere-atmosphere interface oxidation and reduction are common reactions. For any element involved in the reaction oxidation is defined as a gain in positive valence or loss of electrons, and reduction as a loss in positive valence, or gain in electrons. When such reactions occur in the earth's surface layers they generally involve actual addition or loss of oxygen. For example, one of the most abundant elements in the earth's crust is iron. In many minerals ferrous iron is a part of the composition. This is particularly true of minerals which are associated with deep-seated igneous or metamorphic rocks. When such rocks as a result of profound erosion are exposed to the atmosphere the iron is oxidized to the ferric form and appears as one of the iron oxide minerals.

Ferric iron should be regarded as characteristic of sediments and sedimentary rocks (formed at or near the surface of the earth). Sedimentary rocks often contain carbonaceous material derived through the decay of organisms, particularly plants. Ferric iron is unstable when buried with carbonaceous matter and is reduced to the ferrous state generally with a corresponding loss of oxygen. Ordinarily the difference between the ferrous and ferric state is readily recognized by the differences in color of the two ions. Ferrous minerals are colorless, green, or greenish black, whereas the ferric minerals are yellow, brown, maroon, or red.

Substitution in the Space Lattice

The unit cell has been described as an arrangement of a series of atoms in a geometric pattern to constitute the fundamental unit of a mineral. Repetition of such units in specific order establishes the space lattice of the crystal and determines its properties. The relative size of atoms arranged within a unit cell is important in interatomic linkage and is fundamental to the framework of the lattice. Atoms which make up the units of the space lattice are not all of equal radii. In short, a unit cell can be made up of several different atoms, each of which may vary substantially in diameter.

Relatively few crystal susbtances exist whose composition is completely represented by a single formula. Analysis of most minerals indicates the presence of aberrant atoms occupying positions in the space lattice. Such stray atoms formerly were considered to represent absorbed matter on the surface of the crystal. It is now known that they are an actual part of the crystal lattice and are occupying the space normally held by other atoms of similar atomic radii.

Certain groups of minerals exhibit substitution in the lattice more than others; for example, the plagioclase group ranges in composition between pure sodium aluminum silicate ($NaAlSi_3O_8$) and calcium aluminum silicate ($CaAl_2Si_2O_8$). In this mineral group there is substitution of Ca atoms for Na atoms and of Al atoms for Si atoms. Such substitution is not haphazard. Calcium and sodium commonly replace one another, whereas there is little substitution between sodium and potassium, which chemically show more affinity than sodium and calcium. The apparent control of this substitution within the space lattice is not the chemical similarity between two elements such as sodium or potassium but the approach to common atomic radii or the respective sizes of the atoms or ions. In short, atoms or ions of similar dimensions may substitute one for the other, whereas those whose radii are considerably different will not substitute. There are certain limits in which substitution is possible and others in which it is impossible. In the minerals of common occurrence it will be noted that the atomic or ionic substitutions are sodium and calcium, oxygen and hydroxyl, silicon and aluminum, and ferric iron, ferrous iron, and magnesium. As a result certain minerals have complex formulas because of these substitutions, and chemical properties will vary as the substitutions increase. The reader should bear in mind, therefore, that there are pure or ideal compositions for each mineral group and these are varied in accordance with substitutions which are made within the space lattice. Thus there appears a series of minerals all related under one group but whose specific formulas and physical properties will vary as the degree of substitution increases.

Base Exchange

Base or cation exchange connotes the exchange or substitution of certain elements in solution for ions occupying unimportant positions in the space lattice of a crystal. Normally, the effect is to replace one metallic ion with another on the outer surface or internal channels of the crystal. The mechanism is in part controlled by mass action in which if sodium is to be exchanged for calcium the concentration of Ca ions must be greater in solution than the concentration of Na ions in the mineral.

Base exchange often results in very radical alterations in the physical properties of minerals; this is particularly true of the group known as the clay minerals which are so widespread in sediments. When certain clay minerals undergo base exchange their permeability to water, shrinkage or *consolidation,*[2] and plasticity may undergo enormous change. For example, a ceramist may increase or decrease the plasticity of his brick clay by artificially causing base exchange to occur.

The principal exchangeable bases of concern to the geologist and

[2] In soil mechanics the term "consolidation" is used to refer to the process of reduction in volume of a soil due to the pressing out of pore water. Consolidation begins when the initially applied load causes water to be driven from the pore space. At first the reduction in volume is large, but with the progress of time there is gradual diminution until scarcely any change occurs. If after this stage the load is increased consolidation is initiated once more.

Geologists use the term to refer to the aggregation of a sediment into a rock. Usually this is accomplished by the addition of a cementing agent, but the implication is that loose material has been transformed into a rock by any process.

the engineer are sodium and calcium. When calcium in the clays is exchanged for sodium a marked decrease in permeability to water and pronounced increase in plasticity are observed. So important is this change in physical properties that special attention must be given to the subject. For this reason further consideration is reserved for a later section following a discussion of the minerals in question.

Selected Supplementary Readings

Daly, R. A., *Our Mobile Earth,* Chas. Scribner and Sons, 1926, Chapter III. A generalized concept of the earth's interior.

Gilluly, James, Waters, A. C., and Woodford, A. O., *Principles of Geology,* W. H. Freeman and Co., 1950, Chapter 3. A summary of the concept of density balance within the earth's crust and the relative strength of rocks.

Gutenberg, B., and Richter, C. F., *Seismicity of the Earth and Associated Phenomena: Structure of the Earth,* Princeton University Press, 1954.

Hurlbut, C. S., *Minerals and How to Study Them,* John Wiley and Sons, 3rd ed., 1949. An elementary introduction to minerals. A revision of the book by the late Edward S. Dana.

Kraus, E. H., Hunt, W. F., and Ramsdell, L. S., *Mineralogy,* 4th ed., McGraw-Hill Book Co., 1951, Chapters I–X. An elementary text on the properties of minerals.

Leet, L. Don, and Judson, Sheldon, *Physical Geology,* 2nd ed., Prentice-Hall, Inc., 1958, Chapter 2. A good elementary description of atomic structure.

Mason, Brian, *Principles of Geochemistry,* John Wiley and Sons, 1952, Chapters 3 and 4. These chapters concern the structure and composition of the earth as a whole and the atomic structure of minerals.

4 Rock-forming minerals

MINERALS PRIMARILY OF IGNEOUS ROCKS

Molten Rock Material

Minerals of the igneous rocks are known to crystallize from melts containing small amounts of water. This can be demonstrated by observing and sampling lava escaping from volcanoes. Under con-

Calcite crystals.

trolled laboratory conditions such fluids will solidify into rock containing more or less interlocked crystals whose mineralogic composition and texture are identical to crystalline rocks obtained in areas where volcanic activity is no longer present.

Observed temperatures of the melts are, without exception, above the critical temperature of water. Maximum temperatures are known to exceed 1000°C, and the minimum temperatures do not fall below about 500°C. Lava from Vesuvius ranges between 1100 and 1200°C and from Kilauea between 1185 and 1200°C. Since the temperature of such melts is above the critical point of water, water does not exist in the same state as the molten material. Such essentially "dry" melts differ considerably from solutions which are overwhelmingly aqueous in nature and which are commonplace in the development of minerals associated with the sedimentary rocks.

Despite an impressive list of minerals which crystallize from melts the bulk of all igneous rocks is derived from less than a dozen minerals occurring in association and under most circumstances readily identifiable in the field. Many of these minerals are the principal constituents of certain metamorphic and sedimentary rocks as well, indicating that the major rock-forming minerals are few in number.

Knowledge of the properties of the important mineral species is fundamental to the identification and classification of rocks as well as to furnish the observer with a qualitative analysis of the rock specimen. In the following section some of the more significant characteristics are mentioned, but the primary features used in identification are listed in the Appendix. The reader should become familiar with the determinative table and the listed properties of the common minerals inasmuch as they are not repeated below.

Quartz

Composition: SiO_2 (oxygen 53.3%, silicon 46.7%). See Appendix, p. 592.

With the exception of water quartz is the most widely distributed mineral substance. This is not entirely unexpected as the two most common elements in the upper 10 miles of the earth's crust are oxygen and silicon (see Table 4.1). Moreover, quartz is stable under environments of high temperature and high pressure, high temperature and low pressure, and low temperature and low pressure prevailing at the earth's surface. This stability is in part due to the existence of SiO_2 in several varieties and to its rigid lattice structure. The lattice con-

Table 4.1. Per Cent Occurrence of Most Abundant Elements in the Earth's Crust*

Element	*Per Cent*
O	46.60
Si	27.72
Al	8.13
Fe	5.00
Ca	3.63
Na	2.83
K	2.59
Mg	2.09
Ti	0.44
H	0.14
P	0.12
Mn	0.10
S	0.05
C	0.03
	99.47

* Part of the list in Brian Mason, *Principles of Geochemistry,* John Wiley and Sons, 1952, 1st ed., p. 41.

sists of a framework of units of silica in which four ions of oxygen are arranged one in each corner of a tetrahedron and surround the small Si ion which occupies the interstitial space at the center. The framework of tetrahedra is arranged so that each oxygen is shared by an adjacent tetrahedron and the charge of each Si ion $(4+)$ is balanced by the total of one-half charge of each of the four O ions, $(4 \times \frac{1}{2} \times 2^{-} = 4^{-})$. (See Fig. 4.1.)

Varieties of Quartz

Alpha-quartz (see Table 4.2) exists and is stable at temperatures below 573°C. This is the variety of quartz that is associated with veins, that forms well-terminated crystals lining cavities, and that is found in certain very coarse igneous rocks (pegmatites) which crystallize from highly aqueous igneous solutions below 573°C and in quartz cement in sedimentary rocks.

Above 573°C alpha-quartz is no longer stable and inverts to *beta-quartz.* Beta-quartz differs from alpha-quartz in symmetry. Of the two the alpha form has the higher symmetry, and, when beta-quartz inverts, strains in the crystal appear due to distortion of the space lattice as a higher form of symmetry is developed. Beta-quartz is stable within the temperature range of 573 to 870°C and is the variety to crystallize in igneous and high-temperature metamorphic rocks. The partly rounded grains of gray-blue greasy-luster quartz observed in granite and related igneous rocks crystallized as the beta variety.

Fig. 4.1 Schematic framework of the space-lattice structure of quartz. Each tetrahedron consists of a grouping of four O ions, one at each corner of the tetrahedron, and an Si ion in the interstitial space between the large oxygens. Such is the fundamental building block of all silicates. In quartz individual tetrahedra are connected to others by sharing oxygens at each corner. A similar framework is typical of the feldspars. Note that the structure is extended by growth in three dimensions and that each Si ion is balanced by the $\frac{1}{2}$ charge of four O ions, i.e., SiO_2 is electrically neutral. (Model by L. H. Nobles.)

Beta-quartz inverts to another form (*tridymite*) at temperatures higher than 870°C. Tridymite is stable between 870° and 1470°C. When temperatures exceed 1470°C, the variety *cristobalite* forms and is stable to the melting point. Tridymite and cristobalite are mineral curiosities as most igneous rocks remain fluid to temperatures below the stability range of tridymite.

Table 4.2. Common Varieties of Quartz

Alpha-quartz—Coarsely crystalline varieties:

Vitreous luster—(commonly gem varieties).

Rock crystal	Well terminated with crystal faces, occurs lining cavities or in veins, crystals often radially arranged or as crusts of small (1 mm) crystals lining cavities (drusy quartz).
Amethyst	Clear, purple or violet, other properties as rock crystal.
Smoky	Smoky yellow to smoky brown, other properties as rock crystal.
Milky	Milky white color, translucent, luster often greasy, crystalline outline rare. Occurs in veins in igneous or metamorphic rocks. Specimens are splintery or irregular fragments often showing conchoidal fracture.

Alpha-quartz—Cryptocrystalline varieties (crypto-, very finely):

Greasy to waxy luster.

Chalcedony	Translucent, colors white, gray, blue gray, brown, or black. Generally as irregular nodules and masses which show well-developed conchoidal fracture. Chert (flint) is typical.
Jasper	Opaque red or brown color, in veins or as massive nodules or beds in sedimentary rocks.

Alpha-quartz and beta-quartz—granular varieties:

Clear or translucent grains of quartz or chert, forming the major constituent of sandstones. Such quartz is now in the alpha form either as a primary deposit or having inverted from the higher temperature variety.

Irregularly shaped crystals of dull or greasy luster in igneous or metamorphic rocks. Much of this quartz is in the alpha form having inverted from the beta variety.

Feldspar Group

The feldspars represent the most abundant group of minerals although more restricted in occurrence in sedimentary rocks than quartz. This important group crystallizes in two major systems, but all species are closely related by many properties in common. Among these are their manner of occurrence, composition, crystal habit, cleavage, hardness, and color. (See Table 4.3.)

The lattice of feldspars consists of a three-dimensional framework of silica tetrahedra (SiO_4) and alumina tetrahedra (AlO_4) in which all O ions are shared with those of neighboring tetrahedra. For each alumina tetrahedron present the structure is electrically unbalanced by having an excess negative charge, and this condition requires introduction into some interstitial space of a metal ion such as K^+ or Na^+ to create electrical neutrality. (See Fig. 4.1.)

Table 4.3. General Properties of the Feldspar Group

Crystal systems	Monoclinic variety—Orthoclase Triclinic varieties—Microcline and plagioclase group
Cleavage	Well developed in three directions approaching 90°
Hardness	6 to 6.5
Specific gravity	2.55 to 2.75
Color	White, pale yellow, red, green, and less commonly dark gray
Composition	Orthoclase and microcline—$KAlSi_3O_8$ Plagioclase—infinite series from $NaAlSi_3O_8$ to $CaAl_2Si_2O_8$
Twinning	Single—orthoclase Polysynthetic in parallel direction—plagioclase Polysynthetic in perpendicular directions—microcline

Orthoclase and Microcline (The Potash Feldspars)

Composition: $KAlSi_3O_8$; oxide formula $K_2O \cdot Al_2O_3 \cdot 6SiO_2$; silica 64.7%, alumina 18.4%, and potassa 16.9%. See Appendix, p. 591.

Orthoclase and microcline represent two minerals of identical composition but which crystallize in two different systems (polymorphous), due to space-lattice relationships. Normally there is little reason to distinguish one from the other inasmuch as their bulk properties are identical. They occur in crystals of prismatic or thin tabular habit and as massive, coarsely cleaved, granular or compact aggregates. In some coarsely crystalline igneous rocks well-developed single twins of orthoclase varying from flesh color to white are common. In the more finely crystalline igneous rocks othoclase is a compact mass usually of flesh color constituting the major part of the rock, enclosing individual grains of quartz more or less rounded. Microcline occurs in the same habits and in the same rocks and is distinguished from orthoclase by the presence in the former of two sets of polysynthetic twinning which intersect at nearly right angles. With the unaided eye microcline is generally indistinguishable from orthoclase. Either orthoclase or microcline is an essential constituent in granitic (high silica) rocks, but they occur also in very large crystals in pegmatites.

Ordinary orthoclase crystallizes below 900°C, but a glassy variety, *sanidine,* is a higher temperature form. Sanidine is rare in contrast to ordinary orthoclase and is usually observed in fine-grained high-silica lavas (rhyolites).

Plagioclase (Sodium-Calcium Feldspars)

The plagioclase feldspars differ from the potash feldspars in representing a continuous series of minerals between two compositional limits. The limiting compounds or end members are $NaAlSi_3O_8$ and $CaAl_2Si_2O_8$. When fluid the two end members are completely miscible and crystallize in all proportions to form an isomorphous (same crystal form) series. It will be noted that the sodium end-member molecule contains one more Al atom and one less Si atom than the calcium end-member molecule. This is an example of identical lattice structure in which an Al atom proxies for an Si atom. The exchange is accomplished without any major alteration in properties, but there is corresponding reduction in the proportion of silica.

Inasmuch as the possible number of individual minerals representing mixtures of $NaAlSi_3O_8$ and $CaAl_2Si_2O_8$ is extremely large certain bounding limits have been established as mineral species, and to these composition ranges have been assigned arbitrarily as indicated in Table 4.4. One plagioclase may be distinguished from another by

Table 4.4. Compositional Ranges of the Plagioclase Group

Mineral Name		Per Cent Sodium End-Member Molecule ($NaAlSi_3O_8$)	Per Cent Calcium End-Member Molecule ($CaAl_2Si_2O_8$)
Albite		100–90	0–10
Oligoclase	Intermediate plagioclases	90–70	10–30
Andesine	Intermediate plagioclases	70–50	30–50
Labradorite	Intermediate plagioclases	50–30	50–70
Bytownite (rare in occurrence)		30–10	70–90
Anorthite (rare in occurrence)		10–0	90–100

the indices of refraction of plane polarized light, but in the hand specimen this is not possible. Light-colored varieties tend to be rich in the albite molecule and dark-gray varieties are about the composition of labradorite.

Mica Group

Minerals classified as micas show certain typical properties enumerated below which permit their recognition as a distinct group.

General characteristics:

1. All cleave perfectly into extremely thin *elastic* laminas. This is mica structure.
2. All yield water on ignition, often with some corresponding swelling of laminas.
3. All are monoclinic, but they closely approximate orthorhombic and hexagonal (rhombohedral) symmetry.
4. All are closely related to talc, serpentine, and the clay minerals.

Mica Structure

In mica the theoretical arrangement of the space lattice consists of a linkage of tetrahedral SiO_4 units, sharing three of the four oxygens, into individual groups in an hexagonal ring. Such rings are arranged in sheets which occur one upon the other (see Fig. 4.2). Unshared oxygens are directed toward one another and held by Al ions each of which is surrounded by four such oxygens (from the silica sheets) and by two OH^- ions. Substitution is important in this structure as for example some Fe^{++} for Al^{+++} (which results in two mica varieties, muscovite and biotite) and charges are balanced by other metal ions in spaces in the rings.

Structure of the individual mica sheets appears responsible for the cleavage, whereas the ringed linkage provides an explanation of the pseudohexagonal symmetry. Linkage of Si and O ions in tetrahedra results in the radical Si_2O_5, but since Al atoms occupy one out of every four Si positions, $AlSi_3O_{10}$ units result. Sheets of Si_2O_5 units are considered to be the framework of talc and the clay minerals; hence, the presence of certain common properties between these mineral groups. (See Fig. 4.3.)

Muscovite (Common or Potash Mica)

Composition: $(OH)_2KAl_2(AlSi_3O_{10})$; oxide formula $2H_2O{\cdot}K_2O{\cdot}3Al_2O_3{\cdot}6SiO_2$; silica 45.2%, alumina 38.5%, potassa 11.8%, water 4.5%. See Appendix, p. 587.

Muscovite is the most common mica and is almost universally found in granites and granite pegmatites but is rare in volcanic rocks. It is an essential constituent of certain metamorphic rocks (mica

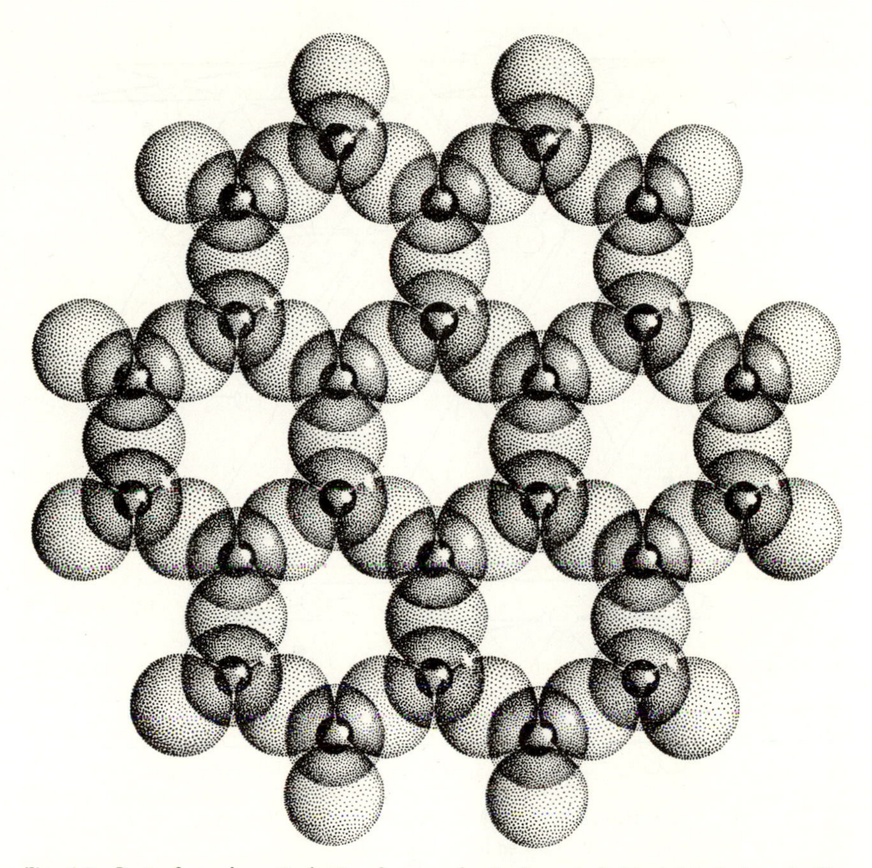

Fig. 4.2 Part of a schematic lattice framework of mica. Individual tetrahedra of silica are shown with oxygens surrounding their enclosed Si ion (small sphere). Tetrahedra are linked together in one plane such that three of the four oxygens are shared by neighboring tetrahedra to form an hexagonal ring which in turn is linked to other similar rings to develop a continuous sheet structure. Unshared oxygens (facing upward in the diagram) are directed toward others from a similar sheet (not shown) piled on top in a mirror-image relationship. Note that an individual ring consists of six silicons, six unshared oxygens, and eighteen shared oxygens (a total of fifteen complete oxygens) as Si_6O_{15} or Si_2O_5 units. Also note that the central opening in the ring structure is large enough to accommodate cations such as potassium. (Reproduced by permission from *Physical Geology*, by L. Don Leet and Sheldon Judson, copyright 1954, Prentice-Hall, Inc., Englewood Cliffs, N.J.)

schists) and is abundant in many gneisses (metamorphic) and sandstones.

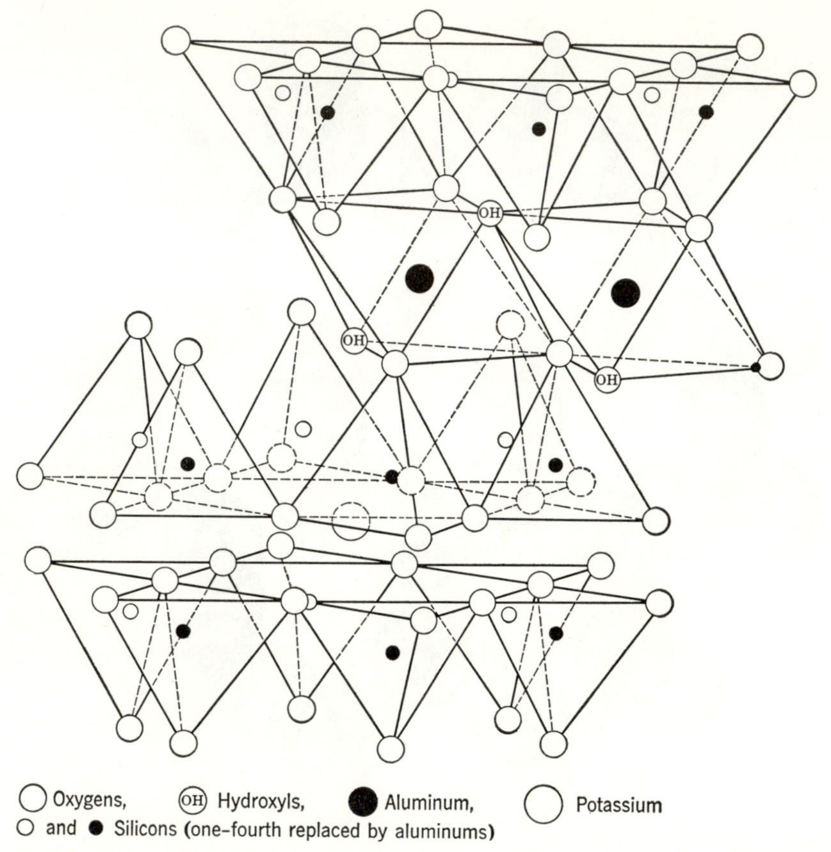

Fig. 4.3 Arrangement of ions in an ideal structure of muscovite. Note in the upper part of the diagram that the structure consists of three layers, an upper sheet of silica tetrahedra as in Fig. 4.2 with unshared O ions directed downward, a lower layer of ringed silica tetrahedra with unshared O ions directed upward, and a middle layer of alumina octahedra joining the two silica sheets. Each alumina octahedron consists of an Al ion surrounded by two oxygens from the upper sheet, two oxygens from the lower sheet, and two OH ions. Such three-layer units are partly held to one another by a K ion which occupies an interstice in the ring and balances the electrical charges. Illite clay displays a similar structure. (By permission from *Clay Mineralogy*, by Grim, copyright 1953, McGraw-Hill Book Co., Inc.)

Biotite (Black Mica)

Composition:[1] $(OH)_2K(Mg,Fe)_3AlSi_3O_{10}$; oxide formula $2H_2O \cdot K_2O \cdot 6(MgO,FeO) \cdot Al_2O_3 \cdot 6SiO_2$. See Appendix, p. 587.

[1] The composition listed is for common biotite; other varieties have formulas which differ appreciably from the one given above due to cation substitution.

Biotite is almost as widespread as muscovite but is much less resistant to weathering and, hence, is more rare in sandstones. Coarsely grained igneous rocks high in potash or magnesia generally contain biotite as do certain metamorphic rocks (schists and gneisses).

Analyses of some biotites indicate the presence of ferrous iron or magnesium, whereas others have large amounts depending upon the number of ions for which the ferrous iron or magnesium has been substituted. The common proxy relationship between ferrous iron and magnesium has resulted in a general descriptive term *ferromagnesian* for such minerals that prominently display this substitution.

Chlorite

Composition: Selected end members $(OH)_8Mg_5Al_2Si_3O_{10}$ and $(OH)_8Fe_5(Fe^{+++})_2Si_3O_{10}$; oxide formulas $4H_2O{\cdot}5MgO{\cdot}Al_2O_3{\cdot}3SiO_2$ and $4H_2O{\cdot}5FeO{\cdot}Fe_2O_3{\cdot}3SiO_2$. See Appendix, p. 588.

Chlorite is distinguished from other minerals of this group by its characteristic green color. It is so abundant in certain metamorphic rocks such as chlorite schists and greenstones that the green color is imparted to the rocks. In igneous rocks and sediments its occurrence is sporadic, and only in metamorphic rocks does it constitute a major or essential mineral. Chlorite is a group name representative of a number of individual mineral species whose composition is best illustrated in terms of percentage occurrence of certain end-member chlorite minerals. The complexity of its composition arises from the numerous substitutions of ions in the lattice. Thus, Mg, Fe^{++}, and Mn proxy for one another, and Al, Fe^{+++}, and Cr are found in many combinations. The noteworthy presence of considerable amounts of Mg and Fe^{++} in certain end members catalog these chlorites as ferromagnesian minerals. All chlorites, however, show an absence of Ca, Na, and K atoms in the lattice.

Pyroxene Group

Pyroxene like feldspar is a term which embraces a number of mineral species crystallizing in different systems (orthorhombic, monoclinic, and triclinic) but which are bound into a single group by common characteristics. They cleave parallel to prism faces whose angular relationships to one another range between 87 and 93 degrees. The arrangement of atoms in their unit cells is in infinite single chains of SiO_4 tetrahedra each of which shares two O ions with its neighbors. (See Fig. 4.4.) Chains are linked by various cations, but this bond is weaker than between the Si and O ions.

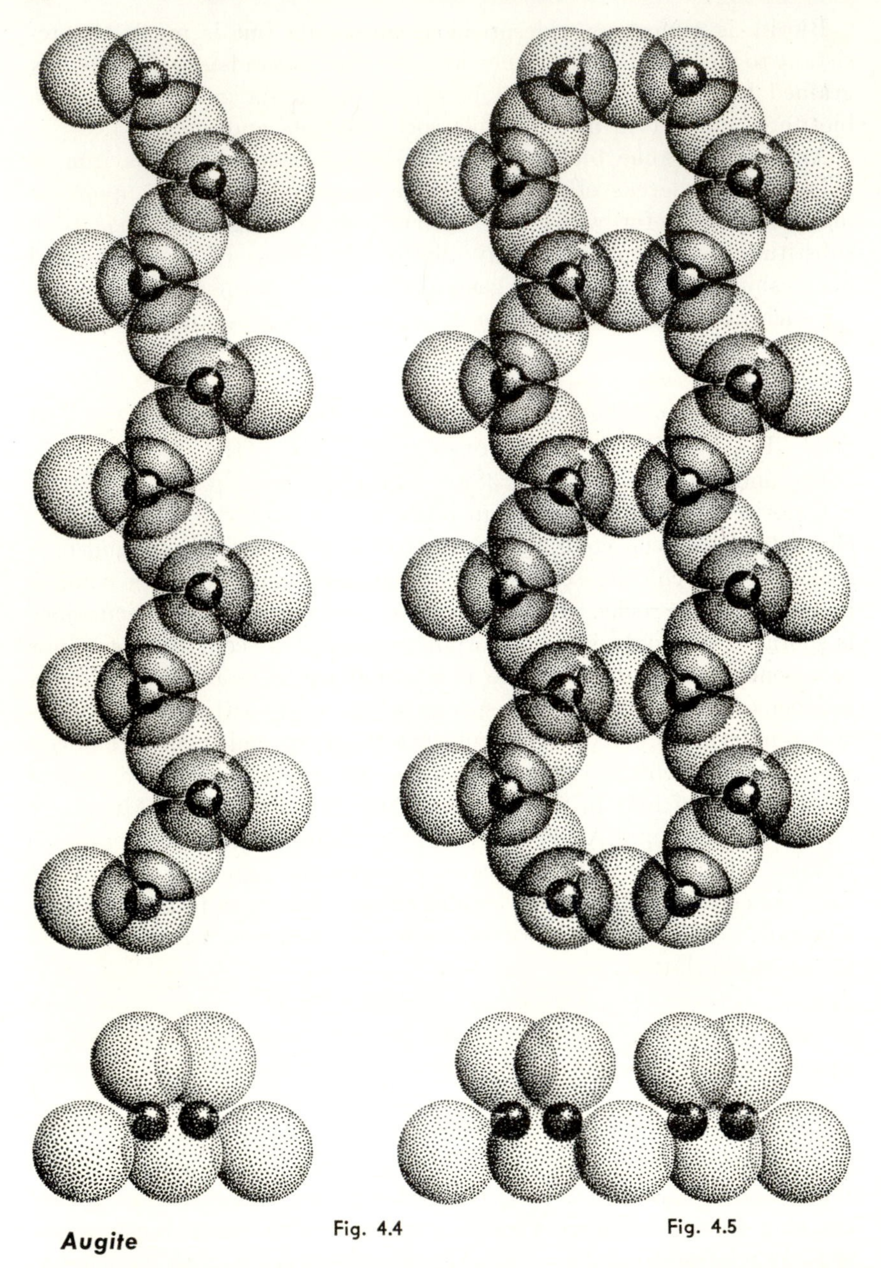

Fig. 4.4 Fig. 4.5

Augite

Composition: Ranges between two end members $CaMgSi_2O_6$ and $(Mg,Fe)(Al,Fe^{+++})_2SiO_6$; oxide formulas $CaO \cdot MgO \cdot 2SiO_2$ and $(MgO,FeO) \cdot (Al_2O_3 \cdot Fe_2O_3) \cdot SiO_2$. See Appendix, p. 589.

◀ Fig. 4.4 Chainlike major framework of the pyroxene lattice structure. Individual silica tetrahedra are linked together, each sharing two oxygens. Note two tetrahedra consist of two silicons (small spheres) and eight oxygens, four of which are unshared and four of which are shared, giving a total of six oxygens for every two silicons (Si_2O_6 units). The lower part of the figure is an end view showing the packing of a silicon ion surrounded by large oxygens. (Reproduced by permission from *Physical Geology* by L. Don Leet and Sheldon Judson, copyright 1954, Prentice-Hall, Inc., Englewood Cliffs, N.J.)

◀ Fig. 4.5 Basic double-chain structure of amphibole. Two individual chains are linked together not as sheets but as blades. Individual silica tetrahedra in each chain share two and three oxygens alternately, and a count of one side of the chain shows a total of eleven oxygens for every four silicons (Si_4O_{11}). (Reproduced by permission from *Physical Geology* by L. Don Leet and Sheldon Judson, copyright 1954, Prentice-Hall, Inc., Englewood Cliffs, N.J.)

Pyroxenes are very important ferromagnesian minerals. Some igneous rocks very low in silica consist almost entirely of pyroxenes, and in low-silica volcanic rocks they are along with labradorite (plagioclase) an essential constituent. Pyroxenes are prominent in crystalline metamorphic rocks also, but they are very rare in sandstones.

Amphibole Group

Amphibole like pyroxene represents an extensive list of minerals of complex composition representing admixtures of certain end-member molecules crystallizing in orthorhombic, monoclinic, and triclinic systems. They are similar to pyroxenes in many respects, and the two frequently are associated, but amphiboles differ in their common bladed-crystal habit and smaller cleavage angle (54 to 56 degrees). In pyroxenes the lattice structure consists of single chains of tetrahedra. Amphiboles are double chains in which each silica tetrahedron shares two O ions with its neighbor, and chains are linked together by alternate tetrahedra (in which case three oxygens are shared). (See Fig. 4.5.) The large openings between the linked chains are the sites of the OH^- ions, and various cations are attracted to the unshared oxygens in order to balance the charges. By such an arrangement a fundamental framework of Si_4O_{11} units characterizes amphibole to contrast with Si_2O_6 units in the single chain structure of pyroxene.

Amphiboles are very important in metamorphic rocks (gneisses and schists), are common in igneous rocks whose silica content is high or intermediate, but rare in sediments. Whereas pyroxenes crystallize from anhydrous melts, amphiboles require hydroxyl as a part of their structure and tend to crystallize from melts of lower temperature in which aqueous vapors are present.

Hornblende

Composition: Selected end members $(OH)_2Ca_2Na(Mg,Al)_5[(Al,Si)_4O_{11}]_2$ and $(OH)_2Ca_2(Fe^{++})_5[(Fe^{+++},Si)_4O_{11}]_2$. See Appendix, p. 589.

Olivine Group

Olivines are ferromagnesian minerals crystallizing from melts in which the percentage of silica is low. They are also products of metamorphism of high-magnesia and high-silica carbonate rocks and for this reason are associated with the metamorphic mineral serpentine.

Common Olivine

Composition: $(Mg,Fe)_2SiO_4$; oxide formula $MgO{\cdot}FeO{\cdot}SiO_2$ representing mixtures of two end members Mg_2SiO_4 and Fe_2SiO_4 in all proportions. See Appendix, p. 593.

Olivine is a silicate of very simple structure and consists of independent tetrahedra of silica in which the oxygens are unshared. Electrical charges are balanced by Fe^{++} and Mg^{++} which serve to link the independent tetrahedra.

A common habit of olivine is as a granular aggregate of small crystals of olive green color and found in metamorphic rocks. Among igneous rocks the crystals are small and isolated. The presence of olivine, however, connotes the low silica content of the rock.

MINERALS PRIMARILY OF METAMORPHIC ROCKS

Garnet Group

Garnet is an important group of minerals crystallizing in the isometric system and with similar habit. Individual species range widely in elemental composition although they can be grouped into aluminum-bearing and iron-bearing varieties. Their general formulas are basically similar and can be represented by $R_3^{++}R_2^{+++}(SiO_4)_3$ in which R^{++} is Ca, Mg, Mn, or Fe^{++} and R^{+++} is Al, Fe^{+++}, or Cr. The basic lattice structure is like that of olivine, i.e., individual tetrahedra of silica held together by cations which establish electrical neutrality of the compound.

Common Garnet (Andradite)

Composition: $Ca_3Fe_2(SiO_4)_3$; (oxide formula $3CaO{\cdot}Fe_2O_3{\cdot}3SiO_2$); Aluminum will proxy for Fe^{+++}; and Fe^{++}, Mg, and Mn will replace some but not all of the Ca atoms. See Appendix, p. 594.

Common garnets are widely distributed in a large variety of metamorphic rock types such as mica and hornblende schists, gneisses, and carbonate rocks that have been thermally metamorphosed. They occur, but are not commonplace, in granites and also in low-silica igneous rocks. In schists the surface of garnet is commonly altered to chlorite; hence, such crystals show the characteristic green color of chlorite and are readily scratched.

Serpentine and Talc

Serpentine

Composition: $(OH)_4Mg_3Si_2O_5$; oxide formula $2H_2O{\cdot}3MgO{\cdot}2SiO_2$. See Appendix, p. 589.

Talc

Composition: $(OH)_2Mg_3Si_4O_{10}$; oxide formula $H_2O{\cdot}3MgO{\cdot}4SiO_2$. See Appendix, p. 588.

Serpentine and talc are closely related minerals allied to the micas (particularly chlorite) and the clay minerals in appearance and lattice structure. Sheets of SiO_4 tetrahedra each sharing three O ions are assembled in layers with unshared oxygens directed toward each other and held by Mg ions. Either mineral occurs as a fibrous mass which is not an intermeshed network of individual crystals. This generally fibrous habit and soapy "feel" to the touch distinguish serpentine or talc from mica. Asbestos and soapstone are noteworthy examples.

Serpentine and talc are always metamorphic minerals and frequently occur in veins or irregular masses inclosed in the surrounding rock. They are alteration products of such minerals as olivine, garnet, pyroxene, or amphibole. Occurrence with dolomite in coarse marbles is common.

MINERALS PRIMARILY OF SEDIMENTARY ROCKS

Calcite

Composition: $CaCO_3$; oxide formula $CaO \cdot CO_2$ (lime 56%, carbon dioxide 44%). See Appendix, p. 588.

Calcite is the most common mineral of a group of related carbonates (dolomite and siderite) whose occurrence is in sedimentary rocks typically.

Calcite is the stable form of calcium carbonate at low pressures prevailing at the earth's surface and at temperatures up to 970°C. It crystallizes from aqueous solution; hence, it is characteristic of sedimentary and metamorphic rocks which have not been subjected to very high temperature and pressure conditions. Varieties of calcite are numerous and diverse in appearance but are of two general types: (1) coarse crystals lining cavities or fissures in rocks, and (2) massive aggregates of crystals constituting widespread sedimentary rock layers (limestone, chalk, marble).

Dolomite

Composition: $CaMg(CO_3)_2$; oxide formula $CaO \cdot MgO \cdot 2CO_2$ (lime 20.4%, magnesia 21.7%, carbon dioxide 47.9%). See Appendix p. 589.

Dolomite is structurally related to calcite but differs in composition being a double salt of calcium and magnesium carbonate. It occurs with calcite under the same conditions of development and in similar habits. The compositional range however is restricted, and no complete series of crystals from pure calcite to pure dolomite occur. Ferrous iron and manganese often proxy for some of the magnesium.

With an increase in ferrous iron replacing magnesium, dolomite gradually grades into the iron carbonate *siderite* ($FeCO_3$). Siderite is a common mineral in sedimentary rocks and is the dominant constituent of *clay ironstones.* These are nodular masses (concretions) distributed in shales particularly those associated with coal beds.

Massive strata of dolomite, often very fine grained, are widespread in area. Such beds are distinguished from limestone (calcite rich) on the basis of the prompt reaction of the latter to dilute HCl.

Hematite

Composition: Fe_2O_3 (iron 70%, oxygen 30%). See Appendix, p. 586.

Hematite is the principal ore of iron and is widespread both in small amounts and as extensive layers. Its characteristic brown or cherry-red color provides the most important coloring agent in nature. In small quantities hematite occurs in virtually all rocks; the extensive deposits are nearly always associated with sediments or where intensive weathering of rocks high in iron has occurred. *Specularite,* the crystalline variety, has a steel gray color, a pronounced metallic luster, and is generally granular and massive with directionally oriented individual grains. The occurrence of specularite is in association with metamorphic rocks.

Limonite

Composition: Approximately $2Fe_2O_3 \cdot 3H_2O$ (iron 59.8%, oxygen 25.7%; water 14.5% ranges widely). See Appendix p. 586.

Limonite is an amorphous, colloidal variety of hydrated iron oxide which like hematite is extremely widespread as a coloring agent in rocks and soils. Often it is finely disseminated, but common occurrences are granular, porous masses of brown or yellow-brown color. In extensive deposits it is an ore of iron and in such localities the common habit is in radiating fibers or rounded (mammillary) forms. Often it is precipitated around impurities of rock or soils in marsh areas under atmospheric temperatures and pressures; elsewhere it is a weathering alteration product of minerals in which iron occurs in the lattice.

Gypsum and Anhydrite

Gypsum

Composition: $CaSO_4 \cdot 2H_2O$; oxide formula $CaO \cdot SO_3 \cdot 2H_2O$; (lime 32%, sulfur trioxide 48%, water 20%). See Appendix, p. 588.

Anhydrite

Composition: $CaSO_4$; oxide formula $CaO \cdot SO_3$; (lime 41.2%; sulfur trioxide 58.8%). See Appendix, p. 588.

Gypsum is typically a sedimentary mineral and often is widespread and associated with sandstone, shales, and carbonate rocks. The common habit is as veins and individual crystals (called *selenite*) in shales and also in massive beds interstratified with other sedimentary rocks. Gypsum is deposited from evaporating sea waters and saline lakes at temperatures below 90°C and under such conditions is often associated with an anhydrous variety *anhydrite* and with common salt, *halite*.

Pyrite

Composition: FeS_2 (iron 46.6%, sulfur 53.4%). See Appendix, p. 586.

Iron sulfide is extremely widespread in all rocks and occurs as *pyrite* or its dimorphic form (i.e., crystallizes in unlike crystallographic systems) *marcasite*. Well-formed crystals (cubes) and irregular or fibrous masses either disseminated throughout the rock or in veins are characteristic. The mineral is typical of sedimentary and metamorphic rocks but is found in igneous rock particularly in areas where ores are present. In dark shales pyrite is frequently finely disseminated throughout or in other sedimentary rocks is found as nodules and replacements of the calcareous or organic parts of fossils. Pyrite oxidizes readily to iron sulfate and to limonite, and large masses of pyrite exposed at the earth's surface generally are altered to a porous mass of limonite.

Clay Mineral Group[2]

Clay minerals represent groups of mineral species consisting of aggregates of extremely minute crystal particles. These range in dimensions between diameters of 100 microns and less than 0.1 micron, but the most common sizes are of colloidal dimensions. Many of the physical properties of colloids such as swelling when moist, shrinking on drying, and flocculation in a water suspension are typical of the clay minerals. Other substances particularly silica, iron oxide, aluminum oxide, and organic material also occur in the colloidal state and under such conditions display some of the same physical properties

[2] An excellent summary of the properties of clay minerals has been presented by R. E. Grim, "Properties of Clay," *Recent Marine Sediments*, Amer. Assoc. Petroleum Geologists, 1939, pp. 466–495.

mentioned. Distinction should be made, however, between clay as a dimensional term signifying mineral particles less than 4 microns in diameter and clay minerals which are hydrous aluminum silicates.

Almost all clay minerals are related to three major groups, as follows:

Kaolinite

Composition: $(OH)_8Al_4Si_4O_{10}$; oxide formula $4H_2O \cdot 2Al_2O_3 \cdot 4SiO_2$.

Illite

Composition: $(OH)_4\ K_y(Al_4,Fe_4,Mg_{4\text{-}6})\ (Si_{8\text{-}y}\ Al_y)\ O_{20}$; where y varies from 1–1.5; oxide formula $4H_2O \cdot K_2O \cdot 4Al_2O_3 \cdot 16SiO_2$ approximately.

Montmorillonite

Composition: $(OH)_4Al_4Si_8O_{20} \cdot XH_2O$; oxide formula $3H_2O \cdot 2Al_2O_3 \cdot 8SiO_2$ approximately.

Each of the clay mineral groups probably is monoclinic in habit and characterized by pronounced cleavage (basal) which controls its occurrence in extremely thin laminas. The lattice structure is best understood by considering it to be composed of sheets of silica tetrahedra in units of Si_4O_{10}, i.e., as in the micas, separated by aluminum-hydroxyl units arranged so that each aluminum is surrounded by a total of six oxygens and hydroxyls. (See Fig. 4.6.) Kaolinite consists of two "layers," one of the silica tetrahedra and one of aluminum-hydroxyl. Montmorillonite (three "layer") structure has two basic sheets of Si_4O_{10} tetrahedra arranged as layers and with the unshared oxygens directed toward one another. These sheets are held together by the aluminum-hydroxyl units fitted into position so that each aluminum is surrounded by four O and two OH ions. (See Fig. 4.7.) Composite layers of the type described are bound loosely to each other by water in layers of variable thickness. Thus differing amounts of water are introduced as the lattice expands, and water is ejected as the lattice contracts. Swelling which is characteristic of montmorillonite is explained by the expanding lattice. Illite has a basic lattice much like that of montmorillonite with the exception that K ions occupy the position of water between the lattice layers.

Substitution in the kaolin lattice apparently does not occur, but in montmorillonite magnesium is nearly always present and with ferric iron proxies for aluminum. Illite differs from kaolinite and montmorillonite in the complexity of its formula because of substitution, the presence of potash, and its structural affinity to muscovite. For these reasons the illite group is frequently called *hydromica*. (See Fig. 4.3.) Despite some variation in the ratios of alumina to silica uniformity prevails within each group; in kaolinite the ratio Al_2O_3/SiO_2 approaches one, in illite and montmorillonite one-half.

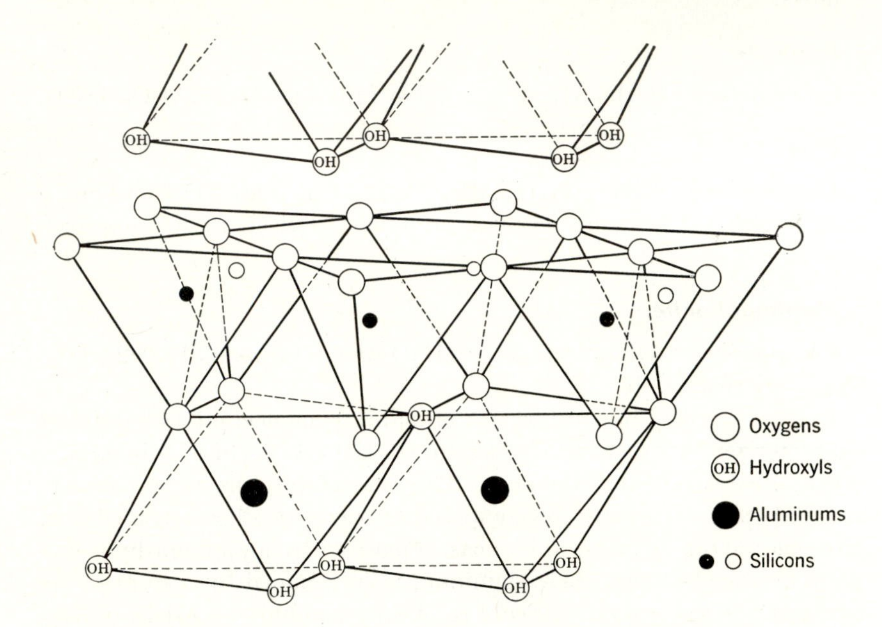

Fig. 4.6 Basic layer structure of kaolinite which consists of one layer of silica tetrahedra sheets such as Fig. 4.2 and a layer of Al ions in octahedra with oxygens and hydroxyls. Oxygens are shared with those in the silica tetrahedra and hydroxyls occupy other corners of the octahedra. Note similarity to upper part of muscovite structure in Fig. 4.3. (By permission from *Clay Mineralogy* by Grim, copyright 1953, McGraw-Hill Book Co., Inc.)

Base or Ion Exchange

The three clay mineral groups can be arranged in order of their increasing base-exchange capacities. This tendency to exchange both cations and anions is characteristic of clay minerals, and there is a preferential degree of exchange among the major groups. Kaolinite shows the weakest tendency to exchange cations and montmorillonite a very strong one. Ion-exchange capacity can be indicated in terms of the theoretical milliequivalents of an oxide which can be replaced at pH-7. For kaolinite the exchange capacity is small and ranges from 3 to 15 milliequivalents of oxides per 100 g of clay, depending upon the oxide. The exchange capacity for illite (10 to 40 milliequivalents) somewhat overlaps the range for kaolinite, indicating

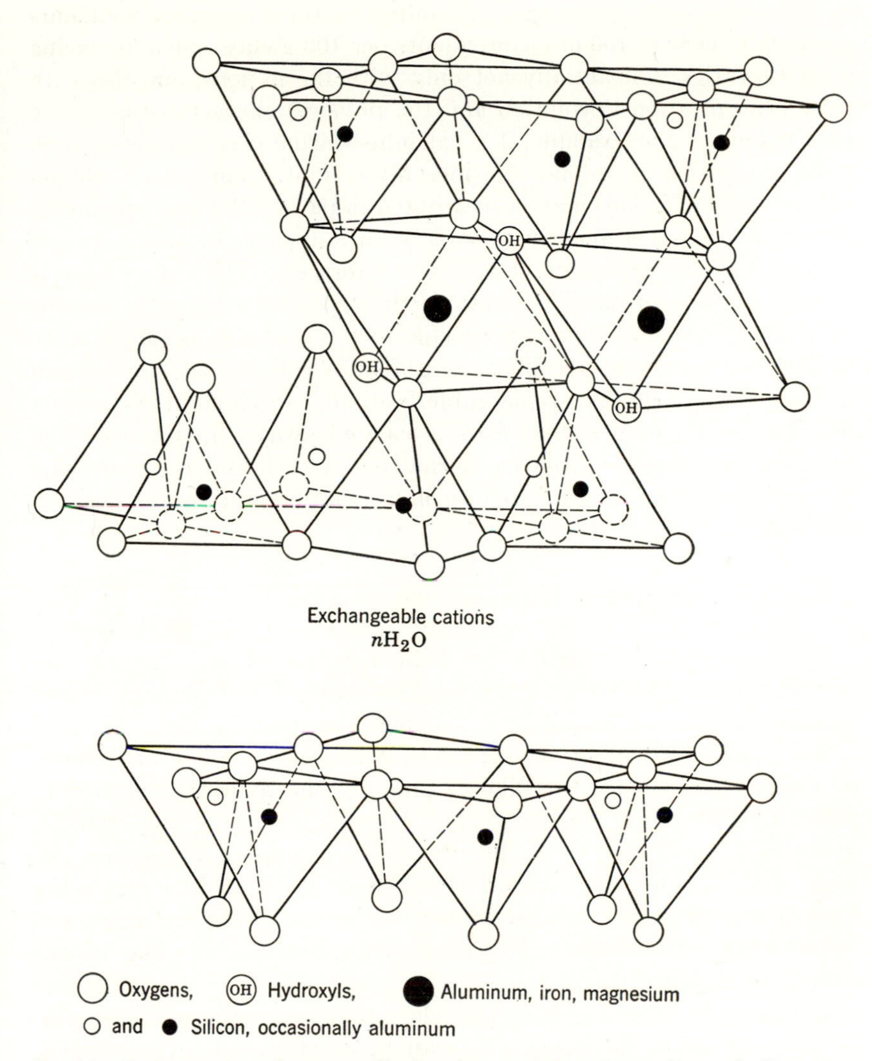

Fig. 4.7 Basic structure of montmorillonite clay consisting of two sheets of silica tetrahedra (similar to muscovite Fig. 4.2) oriented with their unshared oxygens toward one another and separated by an octahedral layer of Al ions sharing the oxygens from the silica sheets and holding OH ions in the other corners of the octahedra. Each such three-layer unit is separated from its underlying, or overlying, sheet by a layer of water of variable thickness causing the lattice to expand or contract as water is added or removed. (By permission from *Clay Mineralogy* by Grim, copyright 1953, McGraw-Hill Book Co., Inc.)

that for some ions the tendency to exchange is less than the most exchangeable ones in the case of kaolinite. Base exchange for montmorillonite is 80 to 150 milliequivalents per 100 g clay. Such replacing activity varies substantially not only with the cations, but also with their concentration in solution and the physical characteristics of the clay mineral. For example, if a calcium-bearing clay is leached with a dilute solution of sodium chloride only a limited amount of calcium will be replaced. This may be attributed in part to the lower replacing power of Na, but a more important feature appears to be the impervious nature of the sodium clay which is formed. The sodium clay is so impermeable to aqueous solutions that the replacement is brought to a virtual halt. However, if a sodium-saturated clay is subjected to prolonged leaching with a dilute solution of a calcium salt the calcium will replace all of the replaceable sodium, providing the solution bearing the replaced sodium ions is carried away. In this case the calcium clay which is formed is permeable, and the solutions carrying the replacing Ca ions can reach an individual crystal and replace the sodium.

Concept of the Water Structure

Crystals of clay minerals are held together in an aggregate by bonding forces which attract and hold the individual flakes. As water is added to such an aggregate it occupies the space between the individual grains by separating them with a water film. The mechanism of the development of this film is possibly the hydration of certain cations inasmuch as the variety of cations present is important in controlling the thickness of the film. For example, Na cations will build a thicker water film than H cations. With an increase in the thickness of the water film the attractive forces holding the individual flakes must operate through greater distances and swelling of the aggregate results.

Certain physical properties of clay minerals such as plasticity, bonding strength, shrinkage, etc., can be attributed to the structure and composition of the clay mineral and the type of exchangeable base. For example, two montmorillonites which differ only in the amount of Mg^{++} replacing Al^{+++} will have different physical properties. The water film between two clay mineral flakes will vary not only depending upon the particular clay mineral group, but also upon the type of exchangeable base. Plasticity is dependent upon the thickness of the water film between the flakes and the lubricating property of the water.

Dehydration

Water in clays exists in intergranular openings and within the space lattice. Much of the water which is held in the pores is removed by ordinary air drying, and in the removal of this water bulk properties such as plasticity are altered appreciably. Some clays tend to retain more of their pore water after air drying than others, and comparison of the bulk properties is made difficult by the difference in moisture content. The remaining pore water can be removed by heating the sample to 105°C until a constant weight is attained. When such is done kaolinite has lost about $2\frac{1}{2}$ per cent of its air dry weight, whereas montmorillonite has lost more than 15 per cent (see Fig. 4.8).

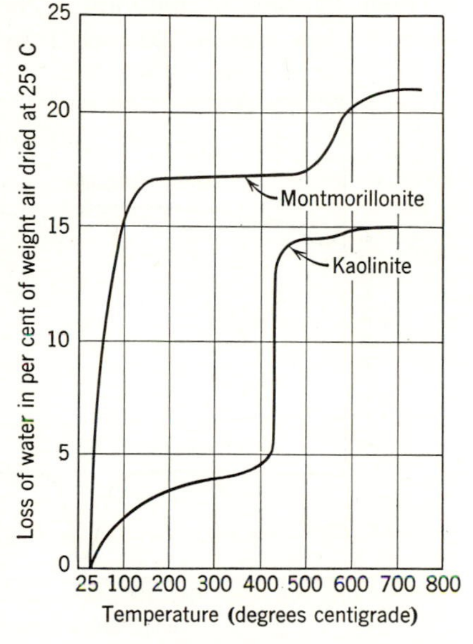

Fig. 4.8 Dehydration curves of montmorillonite and kaolinite clays. Note the differences in the early stage of pore water loss below 105°C. Note also that the release of water from the kaolinite lattice starts sharply at 430°C, whereas the release of water from the montmorillonite lattice is over a range of temperature from about 450 to over 600°C.

If heating is continued beyond 105°C there is a strong tendency for montmorillonite to maintain a constant weight, whereas kaolinite continues to lose moisture but at a much lower rate than the loss between 25° and 105°C. This water loss represents water loosely tied to the space lattice and is released without radical change in structure of the unit cell. For kaolinite the slow moisture loss is continued until approximately 430°C when with increased heating, but without temperature rise a moisture loss of about 9 per cent occurs. Sudden release of water at high temperatures represents emission from the space lattice and is accompanied by reorganization of the Al_2O_3 and Si_2O_5 layers into the new lattice of an anhydrous aluminum silicate. The new mineral which results from the "baking" of clays is a ceramic product and as such has none of the physical properties of clays. Additional heating of the ceramic product results in no further change of weight until the fusion temperature is reached. Montmorillonite displays a similar behavior. Above 105°C a very small amount of moisture is lost up to temperatures of 450°C when approximately $3\frac{1}{2}$ per cent of water is again released. Above 650°C no further moisture loss is observed.

The temperature at which rearrangement of the space lattice occurs with accompanying emission of water is unique for the important clay-mineral groups, and, hence, dehydration curves are useful as a means of identification of individual clay mineral species (see Fig. 4.8). Dehydration curves illustrated by Fig. 4.8 represent conditions of

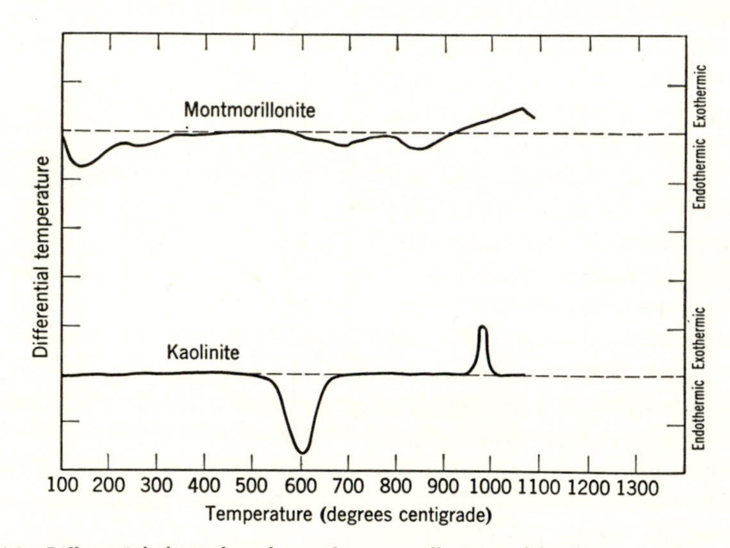

Fig. 4.9 Differential thermal analyses of montmorillonite and kaolinite. Kaolinite shows a well-developed endothermic reaction, i.e., release of water between 500° and 650°C; and an exothermic reaction, i.e., rearrangement of lattice at 950°C. Note that montmorillonite shows release of water (endothermic reaction) distributed over an extensive temperature range and an exothermic reaction above 900°C; neither reaction is well marked.

equilibrium established at selected temperatures and represent the total water to be released if the temperature is maintained indefinitely at a given point. Gathering data for the construction of such curves is laborious in that much time is consumed in waiting for equilibrium conditions to be established at successively higher temperatures.

In order to circumvent this loss in time a radical departure in theory and in the method of analysis has been developed. The method known as differential thermal analysis involves heating the mineral at a constant rate and recording the temperature change without resort to establishment of equilibrium conditions. The rate of temperature rise remains constant so long as no water is released from the mineral.

When water is released additional heat is required for its vaporization, and the rate of temperature rise is reduced, a condition which is typical of an endothermic chemical reaction.

At the end of the period of moisture release the rise in temperature is again resumed at the same or somewhat different rate. With further heating another rearrangement of the space lattice may occur to produce a crystalline state of a lower energy level and concommitant release of heat. This exothermic reaction supplies extra heat and the heating-rate curve must then rise faster than normal until the reaction is completed. Endothermic and exothermic reactions are illustrated in Fig. 4.9 which are typical differential thermal curves for kaolinite and montmorillonite. Note for kaolinite the endothermic reaction beginning at 500°C and ending about 650°C followed by an exothermic reaction at 950°C.

Comparison of the two types of thermal curves for kaolinite, Figs. 4.8 and 4.9, shows a disparity between the temperatures at which the reactions begin. Figure 4.8 which represents the heating curve for equilibrium conditions indicates more precisely the temperatures at which reaction is initiated. The differential thermal curves showing the continuous change of rate of heating must necessarily record the temperature change after the reaction is well underway and, hence, the lag in recording the temperature at which the reaction occurs. Nevertheless, the reproducibility of the curves and the position of the breaks in the heating rate are important, and these serve to identify the minerals in question by comparison with heating curves of known minerals.

Selected Supplementary Readings

Clarke, F. W., "The Data of Geochemistry," *U. S. Geological Survey Bull. 770,* Government Printing Office, 1924, Chapter X. One of the earliest attempts to assemble the chemical properties of rock-forming minerals.

Evans, R. C., *An Introduction to Crystal Chemistry,* Cambridge University Press, 1952, Chapters I, II, and III. An advanced treatment of a general theory of the solid state.

Grim, R. E., *Clay Mineralogy,* McGraw-Hill Book Co., 1953, Chapter 9. A detailed description of the changes in clay minerals during dehydration; also describes the methods of differential thermal analysis.

Hurlbut, C. S., *Dana's Manual of Mineralogy,* 16th ed., John Wiley and Sons, 1952, Chapter 2. Chiefly concerns crystallographic notation.

Kraus, E. H., Hunt, W. F., and Ramsdell, L. S., *Mineralogy,* 4th ed., McGraw-Hill Book Co., 1951, Chapter II. Describes the concepts of the symmetry of crystals and their characterization using axial ratios.

Leet, L. Don, and Judson, Sheldon, *Physical Geology,* 2nd ed., Prentice-Hall, Inc., 1958, Chapter 3. Illustrates the space-lattice structure of minerals.

Mason, Brian, *Principles of Geochemistry,* John Wiley and Sons, 1958, Chapter V. The development of common igneous minerals from the chemical viewpoint.

Pirsson, L. V., and Knopf, Adolph, *Rocks and Rock Minerals,* 3rd ed., John Wiley and Sons, 1952, Chapters 3 and 4. A systematic description of the properties of minerals and the characteristics of those which are rock forming.

Tutton, A. E. H., *The Natural History of Crystals,* Kegan Paul, Trench, Trubner & Co. Ltd., 1924, Chapter V. A description of the technique of describing crystal faces using axial ratios.

5

Igneous rocks

GENERAL FEATURES

Intrusions and Extrusions

That igneous rocks are products of melts generated some unknown number of miles below the earth's surface is knowledge of long standing. Such material called *magma* rises from the zone of origin and may escape through conduits to the surface and spread outward as a

Eruption of Vesuvius, 1822. (From Credner, *Elemente der Geologie*.)

lava flow until cooling solidifies the entire mass. At depth much magma is injected into the surrounding rock where it stagnates and crystallizes in homogeneous masses of varying shapes and dimensions. Rocks developing under each of these conditions possess distinctive textural and structural properties and from these has stemmed the subdivision into two major groups. *Extrusions* are representative of emissions of lava poured upon the earth's surface, whereas *intrusions* are magmas injected into the surrounding host rocks at depth. Intrusions are visible only as the result of subsequent erosion and exposure of the igneous rock mass. Extrusions typically occur in regions of volcanoes whose growth at the surface is the response to periodic eruptions of lava and *ejecta* (fragmental material) propelled through the air under the explosive force of escaping gases. Extrusions represent a heterogeneous accumulation of one lava flow superposed upon another, often separated by a layer of ejecta which has settled out of the atmosphere following a period of explosive violence of the volcano. Intrusions lack such physical heterogeneity and are characterized by general uniformity of texture and mineralogy throughout the individual mass.

Chemical Variation

Igneous rocks should be considered as being roughly chemically homogeneous throughout an individual body. For example, analyses of a single flow of lava will show small ranges both laterally and vertically, and ejecta from a single volcanic vent is approximately uniform in composition. However, wide chemical variation does exist between flows from different vents and between separate intrusions. Generally intrusive rocks show wider ranges of chemical composition than extrusive rocks which in some instances display a marked uniformity despite a great geographical separation.

In some respects variation in chemical composition in igneous rocks represents change in the proportions of the commonly occurring oxides (see Table 5.1). Of the ten most abundant oxides (i.e., exceeding 1 per cent) silica is by far the most preponderant (59 per cent). Alumina (15 per cent) is about one-fourth as abundant as silica and is followed in importance by the oxides of iron. In the average analysis ferric and ferrous iron oxides occur in about equal proportions (3.1 and 3.8 per cent), and the total constitutes about one-third of the quantity of alumina. Magnesium reported as an oxide has about the same percentage of occurrence (3.5 per cent) as each of the iron oxides, and lime (5.1 per cent) is slightly more abundant than magnesia. Sodium and potassium oxides are next in importance and occur in about equal proportion ranging in quantity somewhat more than 3 per cent. Water and titania occur in proportions of approximately 1 per cent each. The remaining substances in the average igneous rock constitute an imposing list but are present in low percentage values. Obviously, when the total bulk of the igneous rocks is considered there are large quantities in terms of tons of minerals rich in vanadium, chromium, phosphorous, and barium oxides, but these are not important as rock-making constituents.

Table 5.1. Compositions of Average Rock of the Upper 10 Miles of the Earth's Crust and the Average Igneous Rock*

	Average of 10 Mile Crust	Average Igneous Rock
SiO_2	59.08	59.14
Al_2O_3	15.23	15.34
Fe_2O_3	3.10	3.08
FeO	3.72	3.80
MgO	3.45	3.49
CaO	5.10	5.08
Na_2O	3.71	3.84
K_2O	3.11	3.13
H_2O	1.30	1.15
TiO_2	1.03	1.05
ZrO_2	.037	.039
CO_2	.350	.101
Cl	.045	.048
F	.027	.030
Totals	99.65	99.32

* Data from F. W. Clarke, "Data of Geochemistry," *U. S. Geological Survey Bull.* 770, 1924, pp. 29 and 34.

The similarity between the analysis of the average 10-mile thickness of the earth's crust (Table 5.1) and that of the average igneous rock is dependent upon the very large percentage of igneous rocks which make up the bulk of the earth. In consequence, the analysis of the 10-mile crustal zone represents the dilution of the analysis of the average igneous rock by sedimentary and metamorphic rocks.

Oxide Variation

If the analysis of the average igneous rock is selected as a norm from which departures are to be measured distinction can be drawn between those whose percentage of SiO_2 is higher and those whose percentage is lower than the average (see Table 5.2). This has led to

Table 5.2. Analyses of Selected Igneous Rocks*

(weight per cent)

Column number	1	2	3	4	5	6	7	8	9	10
Number of analyses (average of)	546	20	40	55	50	70	27	24	17	10
SiO_2	70.18	66.64	65.01	61.59	60.19	56.77	56.12	49.50	46.49	40.49
TiO_2	0.39	0.50	0.57	0.66	0.67	0.84	1.10	0.84	1.17	0.02
Al_2O_3	14.47	15.57	15.94	16.21	16.28	16.67	16.96	18.00	17.73	0.86
Fe_2O_3	1.57	1.91	1.74	2.54	2.74	3.16	2.93	2.80	3.66	2.84
FeO	1.78	1.94	2.65	3.77	3.28	4.40	4.01	5.80	6.17	5.54
MgO	0.88	1.41	1.91	2.80	2.49	4.17	3.27	6.62	8.86	46.32
CaO	1.99	3.50	4.42	5.38	4.30	6.74	6.50	10.64	11.48	0.70
Na_2O	3.48	3.41	3.70	3.37	3.98	3.39	3.67	2.82	2.16	0.10
K_2O	4.11	3.72	2.75	2.10	4.49	2.12	3.76	0.98	0.78	0.04
H_2O	0.84	1.15	1.04	1.22	1.16	1.36	1.05	1.60	1.04	2.88

Rock type (column number):

1. Granite
2. Quartz monzonite
3. Granodiorite
4. Quartz diorite
5. Syenite
6. Diorite
7. Monzonite
8. Gabbro
9. Olivine gabbro
10. Dunite

* Data from R. A. Daly, *Igneous Rocks and the Depths of the Earth,* McGraw-Hill Book Co., 1933.

an organization of groups of rocks based upon the silica content. When the percentage of silica shown in the analysis of each of a selected group of igneous rocks of wide compositional range is plotted against the percentage of each of the other oxides present, certain relationships are to be noted as follows (see Figs. 5.1 and 5.2):

1. Rocks which contain the lowest amount of silica (± 40 per cent) are rich in magnesia and iron oxide. Such rocks are also characterized by an extremely low percentage of alumina (± 1 per cent), whereas those containing 45 per cent or more of silica contain from 14 to 18

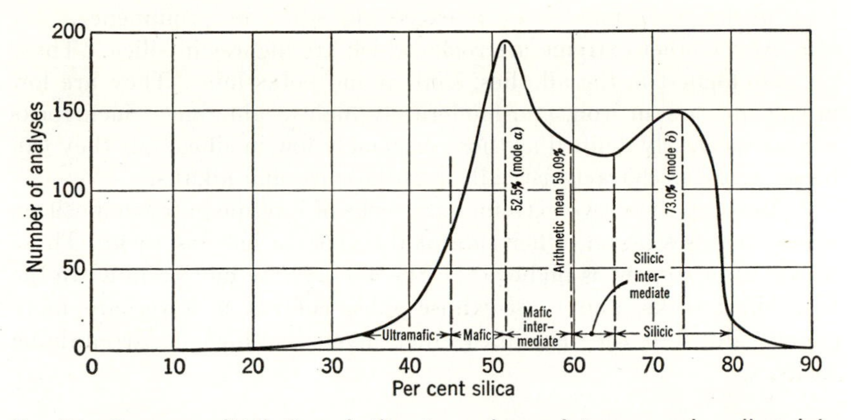

Fig. 5.1 Frequency distribution of silica in analyses of igneous rocks collected by Washington (*U. S. Geol. Survey Prof. Paper* 99). Note the modal positions of the silicic and mafic-intermediate groups. Boundaries separating the recognized igneous-rock groups are approximately as indicated. (Modified from Richardson and Sneesby, *Mineralogical Magazine*, 1922.)

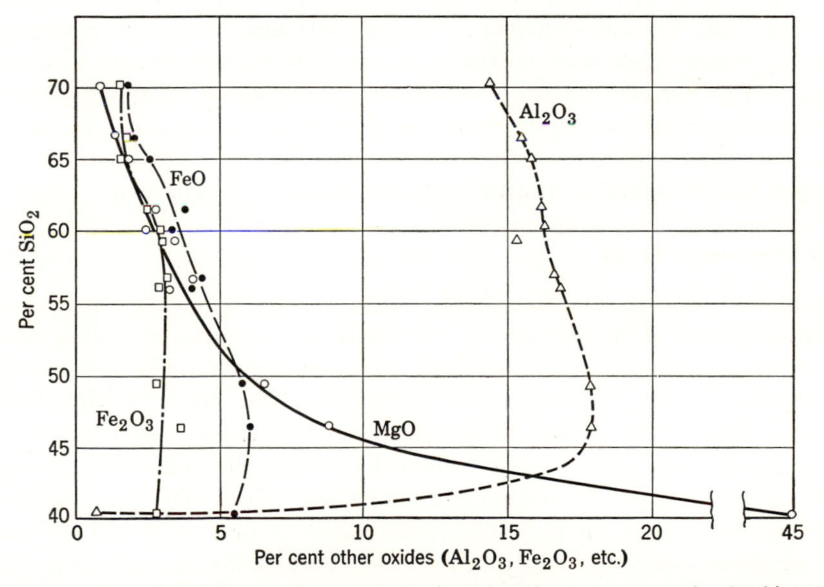

Fig. 5.2 General distribution of per cent of selected oxides in igneous rocks of Table 5.2 plotted against the per cent silica. Note the rapid decline in alumina and increase in magnesia in rocks where the silica content falls below 45 per cent.

per cent of alumina. Magnesia or iron may become the most important oxides, occasionally even exceeding silica in prominence.[1]

2. At the other extreme are rocks which are highest in silica. These are also highest in the alkalies, sodium and potassium. They are low in calcium, low in iron, and moderately high in alumina. Such rocks are not as clearly defined as those extremely low in silica, but they can be set apart as characteristically high in silica and alkalis.

3. Between these two extremes are rocks of intermediate composition whose analyses lie on either side of the average igneous rock. Those whose silica content is higher than the average are more related to the high silica rocks, and those whose silica content is lower are more closely allied to the low-silica, high-magnesia, high-iron-containing rocks.

Igneous Rock Groups

The major groups of igneous rocks recognized on the basis of the above described variation in oxide composition form the natural framework of igneous rock classification. With the exception of the silica- and alumina-deficient group a continuous series of changing composition exists between rocks whose silica content ranges between 45 and 70 per cent. Rocks whose silica composition falls within this range are, therefore, by necessity subdivided into groups on the basis of somewhat arbitrary distinctions although natural boundaries based upon the difference in mineral composition are used wherever possible.

Ultramafic Rocks

These consist of uncommonly occurring rocks constituting that group which is unique in its extremely low content of silica and alumina and its high magnesium and iron. Such rocks are called *ultramafic* and *ultrabasic* interchangeably. The term "mafic" is a contraction of the words "magnesium" and "ferrum" and is used to denote minerals or rocks in which such elements dominate the composition. Often the term ferromagnesian is used in the same connotation. Reference to these rocks as ultrabasic is from long established custom and stems from the early interpretation that in the igneous melt the silica was behaving as an acid radical in contrast to the other oxides which behaved as basic radicals. Hence, low-silica, high-magnesia

[1] Ferrous iron in some analyses is more abundant than magnesia and under such circumstances dominates the composition.

minerals or rocks were considered as extraordinarily basic when compared with others of higher silica content.

Mafic Rocks

Mafic or *basic rocks* are those whose chemical analyses show more silica (45 to 50 per cent) and alumina (± 18 per cent) but less magnesia and iron than the analysis of the ultramafic rocks. Like the ultramafic rocks they are dark in color, but here the similarity ends because of

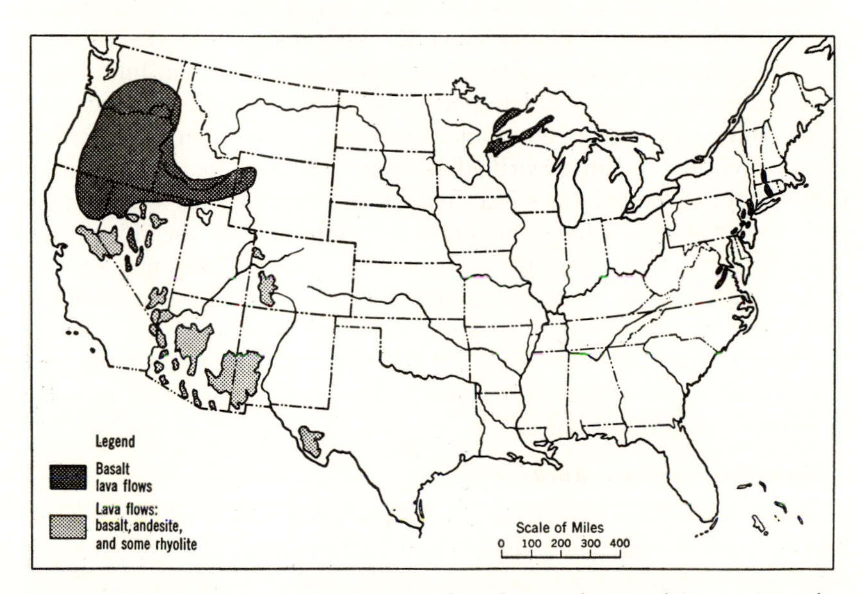

Fig. 5.3 Distribution of generally continuous lava flows in the United States. Note that basalt (mafic) flows dominate areally.

the difference in mineralogy. In composition mafic rocks are closely related to and grade imperceptibly into the mafic-intermediate rocks, and they are closely associated with these in the field. Mafic rocks are extremely widespread geographically particularly as great thicknesses of lava flows. (See Fig. 5.3.)

Silicic Rocks

Silicic rocks are recognized as containing more than 65 per cent of silica. They are also called *acidic rocks* inasmuch as the silica content of the magma is so high that uncombined quartz will crystallize, and the melt was formerly considered to have been dominated by the acid-behaving radical SiO_3. Rocks of this group are also characterized by the highest content of the alkalis, potassium and sodium, and low amounts of iron. Silicic rocks are widely distributed geographically and are typically represented by the familiar rock, granite. Silicic rocks occur both as intrusions and extrusions although intrusions are probably more commonplace, particularly in the very oldest rocks.

Silicic-Intermediate Rocks

A group of rocks exists with a silica content ranging approximately between that of the average igneous rock and the silicic rocks. Such rocks also are high in alkalis and often occur with the silicic rocks. As their composition grades towards the analysis of the average igneous rock they contain increasing amounts of alumina, lime, magnesia, and iron oxides which place them intermediate between the silicic and mafic groups but somewhat more silicic than the average. For this reason they are identified as *silicic intermediate.* Occasionally, rocks of this type have a silica content as low as 55 per cent, but in such cases the crystallizing minerals are more akin to the silicic than to mafic rocks and are properly placed in among silicic-intermediate rocks. Based on the many hundreds of available analyses it has been found that the silicic-intermediate group together with those of silicic composition are the most commonly occurring of all igneous rocks and are typical of enormous intrusions in the core of the more recently developed mountain ranges.

Mafic-Intermediate Rocks

These rocks lie on the mafic side of the composition of the average igneous rock, ranging in silica content as low as 50 per cent, but the most commonly occurring varieties are somewhat more silicic (55 per cent). *Mafic-intermediate* species are consistently lower than mafic rocks in lime, alumina, magnesia, and iron oxides and bridge the compositional gap between the mafic and silicic-intermediate groups. Normally mafic-intermediate varieties are associated with mafic rocks, particularly as extensive lava flows. They are less common as discrete intrusive bodies, rarely making very large masses, and are more common as lens-shaped pods connected with silicic-intermediate intrusions. The number of widespread extrusive masses is sufficient to place this group as the most common of all igneous rocks (see Fig. 5.1).

Orders of Silicate Mineral Crystallization

Crystallization of silicate melts of diverse chemical composition produces minerals of unrelated types. This selection is primarily a response to the composition of the melt, but secondary controls are the liquid-solid phase relationships during crystallization. Together these result in an over-all general order in which the crystals make their appearance. The basic tenets behind this order of crystallization are beyond the scope of this text, but the order in which minerals crystallize is significant to an understanding of the variability in the mineralogy of igneous rocks. If the premise is accepted that the prog-

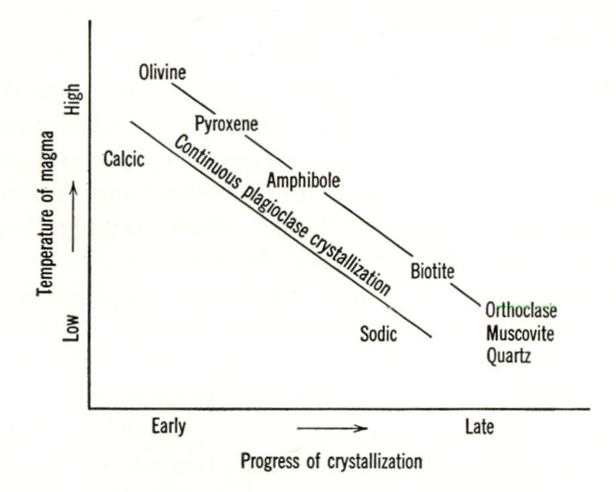

Fig. 5.4 Diagram of the progress of crystallization of minerals from the igneous-rock melt. (Based upon Bowen, *Jour. Geology,* 1922.)

ress of silicate crystallization proceeds with more or less regular order beginning with those minerals in which the percentage of silica is extremely low to those in which the percentage of silica is highest, high-silica minerals cannot form in the early stages of crystallization of a melt. In ultramafic rocks the only minerals which can crystallize from such low-silica-content melts are minerals whose oxide composition has the lowest ratio of silica to other oxides. Olivines and pyroxenes are therefore the first and only common silicate minerals to crystallize from such melts.

During the course of crystallization of a magma withdrawal of minerals such as pyroxene increases the per cent silica in the melt by removing other oxides in greater proportion, and the new minerals to crystallize are of higher silica content. The order of crystallization, therefore, proceeds from olivine and pyroxene to amphiboles, plagioclases, and finally such minerals as potash-rich feldspar, muscovite, and quartz (see Fig. 5.4). Inasmuch as quartz is the last mineral to crystallize from silicate melts it occurs only in those rocks where the residual liquid was left with an excess of silica beyond that necessary to crystallize such minerals as the high-silica feldspars. Primarily this condition prevails in those rocks in which the silica content is above 65 per cent. Melts which contain a lower silica percentage rarely produce quartz as the silica has been entirely utilized in the crystallization of other minerals.

Mineralogic Variation

The differences in oxide composition of igneous rocks reflected by each of the groups established above are more strongly emphasized by the variation in observed mineral composition. Among some rocks oxide composition may show only small distinctions, but the crystallizing minerals are markedly different. For example, among silicic rocks in which the composition of the melt is 70 per cent silica, crystallization of minerals in the order listed above does not require all the silica which is available, and the latter crystallizes as quartz. In melts of silica composition between 65 and 70 per cent there is only sufficient silica available to crystallize as potassa and soda feldspars and mica, and no quartz is present. Among melts of silicic-intermediate composition all of the available silica tends to combine preferably with sodium and calcium to form plagioclases, and quartz is rare or absent. Thus small differences in the oxide composition may result in rather different mineral associations, but the chemical analysis of the rock can be approximated by the identification of the incorporated minerals. Certain of these minerals are more important in this respect than others, and particular emphasis is to be placed on the association in specific rock groups.

Minerals Crystallizing in Ultramafic Melts

Analyses of ultramafic rocks (Table 5.2, column 10, dunites) shows that the ratio (in molecular proportions) of SiO_2 to the total of the other oxides exceeding 1 per cent is approximately 1 to 2. From a melt of such composition the minerals which crystallize are correspondingly low in silica. Reference to the list of common igneous-rock-forming minerals indicates that olivine $(Mg, Fe)SiO_4$ and pyroxene $CaMgSi_2O_6$ have ratios of silica to other oxides approximating that of such a melt. Moreover, such compounds become insoluble at temperatures at which the melt is still too hot for other minerals to make their appearance, and crystallization of olivine and pyroxene is initiated. When the melt is rich in iron, magnetite often crystallizes along with these two silicates, and crystallization proceeds until the liquid is entirely transformed into an intergrowth of crystals of olivine, pyroxene, and magnetite. During the progress of the crystal growth the developing minerals remove the oxides from the liquid in approximately the same proportions as present in the magma, and the composition of the liquid remains substantially the same throughout the entire span of crystallization. The final crystallized product consists of olivine, pyroxene, and magnetite which characterize the ultramafic group.

Minerals Crystallizing in Mafic Melts

Type examples of mafic rocks are *gabbros* (Table 5.2, columns 8 and 9). Melts of the compositions represented by such analyses are characteristically high in alumina and lime and richer in silica than ultramafic liquids. The ratio of SiO_2 to the total of other oxides exceeding 1 per cent is approximately 1:1, and minerals requiring a somewhat higher silica percentage than olivine may crystallize. Of even greater importance is the presence of much alumina in the melt. This permits crystallization of minerals in which aluminum occupies an essential position in the space lattice. Reference to the list of common igneous rock minerals shows that the most important rock-forming minerals, the feldspars, are included in this group. Calcic plagioclases along with olivine and pyroxene will crystallize at high temperatures, but the aluminum-silicate lattice apparently represents a lower free-energy level than those of olivine and pyroxenes. Calcium and sodium move to positions in the plagioclase lattice in preference to higher free-energy forms. As crystallization of a mafic melt proceeds an initial, but brief, period of olivine and pyroxene crystallization is superseded by a long stage of plagioclase growth. Mafic rocks are, therefore, intergrowths of plagioclase (calcic) which is dominant and pyroxene and olivine which are secondary in abundance.

Minerals Crystallizing in Mafic-Intermediate Melts

Mafic-intermediate rocks represented by *diorite* and *monzonite* (Table 5.2, columns 6 and 7) differ in oxide composition from mafic rocks in reduced amount of lime, magnesia, ferrous iron and increases in the alkalis and silica. In such analyses the ratio of silica to the total of the other oxides (greater than 1 per cent) is increased to approximately 1.5 to 1, and minerals requiring high silica content can crystallize from the melt. The progress of crystallization follows much the same pattern established in the mafic rocks, but there is an absence of those minerals whose ratio of silica to other oxides is low.

With addition of hydroxyl, amphiboles rather than pyroxenes crystallize first, and intermediate plagioclases rather than calcic ones. The increase in alkalis permits crystallization of minor amounts of sodic plagioclase, orthoclase, and mica, particularly biotite. Rocks representative of the complete crystallization of mafic-intermediate melts are dominated by intermediate plagioclase and amphibole but contain minor amounts of more sodic plagioclase, orthoclase, and mica.

Minerals Crystallizing in Silicic-Intermediate Melts

Strict adherence to the limiting boundary between silicic and silicic-intermediate rocks, as stated in earlier paragraphs, permits recognition of *quartz diorite* and *syenite* (Table 5.2, columns 4 and 5) as the only representatives of the latter group among the selected analyses. Experience has shown, however, that *quartz monzonites* and *granodiorites* (columns 2 and 3) more properly are to be included in this group rather than among the silicic rocks. This decision is based upon their inherent mineralogy and their association with typical silicic-intermediate rocks. The analyses show higher silica percentages than the average igneous rock and show decreases in the amounts of iron oxides, magnesia, and lime compared to mafic-intermediate rocks. Variability is also to be noted in the percentage of potassa. Crystallization of melts of *quartz-diorite* composition is initiated by the appearance of amphiboles and is soon followed by sodic-intermediate plagioclase and some mica (chiefly biotite). If the potassa is not already exhausted by biotite and muscovite crystallization small amounts of orthoclase are formed.

Quartz diorites have a ratio of silica to total of other oxides of 2 to 1, which is considerably higher than the ratio in the mafic-intermediate rocks. Moreover, the demand for silica in some of the early crystallizing minerals is less than this ratio with the result that as crystallization proceeds the liquid tends to become enriched in silica. This enrichment permits the residual liquid to have an excess of silica over the other oxides, and quartz is the final crystallization product.

Syenites (column 5) are characterized by their high potassium content. Crystallization of syenite melts proceeds with crystallization of amphiboles which may consume much of the available lime and soda so that very little sodic plagioclase may crystallize. Rather, the alumina and potassa form orthoclase or microcline which dominates the composition. Crystallization of the potash feldspar consumes silica at a higher rate than its percentage occurrence in the crystallizing liquid, and, hence, quartz is not a product. Syenites are often somewhat deficient in silica in their final stages of crystallization, and potassium aluminum silicates and sodium aluminum silicates closely related to feldspars, but lower in ratio of silica to other oxides, may be contained in the crystallized rock.

Quartz monzonites (column 2) and granodiorites (column 3) are closely related. They contain some amphibole, soda plagioclase, and biotite which utilize the iron oxide, magnesia, lime, and soda; but orthoclase and microcline may be more prominent than soda plagioclase as practically all of the potassa forms potash feldspar. Near the end of such crystallization the ratio of silica to the total of other oxides rises above 2 : 1, and uncombined silica is left as crystallization is completed. This appears as quartz scattered throughout the rock.

Minerals Crystallizing in Silicic Melts

Granites (column 1) constitute type silicic rocks. They are characterized by a dominance of potassa over soda and a ratio of silica to the total of other oxides of approximately 3 to 1. Normally the amount of amphibole crystallizing from such melts is small, and biotite is far more typical. In part this is due to the high silica content required by minerals such as micas. Orthoclase and microcline overshadow other minerals, but muscovite also crystallizes abundantly until the potassa is entirely consumed. The remaining silica appears as quartz, generally very conspicuous among the potash feldspar crystals. If the soda content is high some sodic plagioclase (albite) will appear as a crystallization product and may be observed in many granites.

Textures

Texture is defined as the size of, and boundary relationship between, adjacent crystals in the rock mass. In typical igneous rocks this textural relationship has the over-all aspect of an interlocking crystal meshwork which is particularly prominent when the crystal dimensions are large. Reference to igneous rock textures implies this interlocking relationship, and varieties of textures are established to indicate the size of the individual crystals, their degree of uniformity, or their absence. Under most circumstances textural description is implicit that the classification is based upon the appearance to the unaided eye or under lenses of low magnification.

In igneous rocks textural development is primarily the response to the rate of crystallization. When the magma has been allowed to remain in position and where the progress of crystallization is slow the sizes of the developing crystals are more or less uniform in dimension and are large enough to be readily recognizable to the unaided eye. Under such conditions crystallization proceeds unhindered, and sharp, clearly defined, limiting boundaries develop between individual minerals. Slow cooling impedes the tendency of the melt to increase

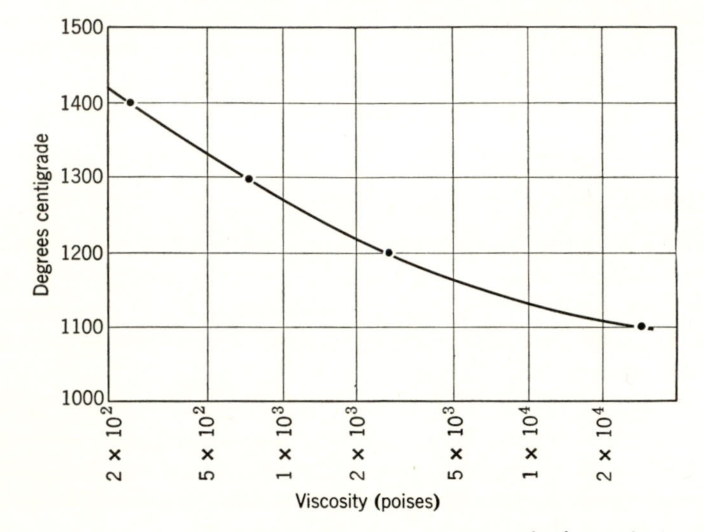

Fig. 5.5 Change in viscosity of lava from Mt. Vesuvius with change in temperature. (Data from Birch and Dane, *Geological* Soc. *Am. Spec. Paper,* 36, 1942.)

its viscosity, and even the late crystallizing minerals may thus grow to equal dimensions with the earlier crystallizing ones. When rapid cooling prevails the increase in viscosity with decrease in temperature may be sufficient to retard the crystallization rate, and only small crystals can develop during the interval of cooling. (See Figs. 5.5 and 5.6.) Such rocks present a dull stony appearance. The effect is to produce a mass of rock in which individual crystals are imbedded in a supercooled solution or glass. Progress across the crystallization-temperature range may be so rapid that no crystals develop at all and the viscosity rises to that of a solid. This is true of all glasses whether naturally developed or manufactured for commercial purposes.

Texture is also modified by the rate at which gases will escape from the melt. Of these, water is the most important and occurs to the extent of approximately 1 per cent in the analysis of the average crystallized igneous rock. There is good reason to consider that in the uncrystallized magma water may be as high as 10 per cent, and as the temperatures are above the critical point of water, the vapor pressure of the magma must be high. If the magma begins to cool below a thick mantle of overlying rocks and cooling proceeds very slowly the low viscosity of the fluid will permit the gradual emission of water vapor as the individual crystals grow, and the interlocking texture develops. On the other hand those rocks which have been

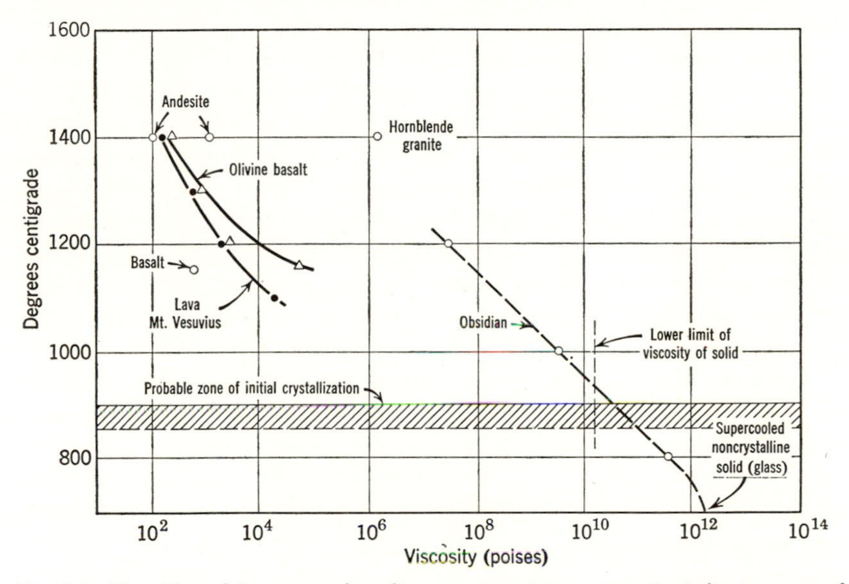

Fig. 5.6 Viscosities of igneous-rock melts at various temperatures to indicate zones of crystallization and appearance of glass. Note that during the cooling of obsidian the viscosity of the melt reaches that of a solid before the temperature of the zone of initial crystallization is reached. (Data in part from Birch and Dane, *Geological Soc. Am. Spec. Paper*, 36, 1942.)

brought to the earth's surface rapidly become extremely viscous while the vapor pressure is still high. Escape of gases is impeded, and they leave often with explosive violence. Ejection of the gases from solution occurs as the fluid becomes increasingly viscous or as crystallization proceeds. In the highly viscous solutions a froth is produced. This froth is preserved as a textural characteristic. Such rocks are called *vesicular* and are characteristic of lava flows and ejecta. *Pumice* is a classic example of vesicular rock, and often the pore space is sufficiently large to reduce the bulk density of the rock to less than that of water.

Some magmas are believed to be essentially anhydrous or may have lost a large percentage of their water on their upward rise to the locality of final emplacement. This appears true of some silicic varieties, and such lavas poured out on the earth's surface cool to a glass without developing a vesicular texture.

Textural Varieties

Recognition of the major textural units is based upon the following general properties, a knowledge of which is essential to the satisfactory classification of rocks.

Glassy texture. This texture is characterized by pronounced homogeneity throughout. It has the vitreous luster and conchoidal fracture of glass and a color which is dark gray, black, or dark red. Light is transmitted through the thin edges of fragments, and through such edges the glassy character is recognized. With the exception of a few nuclei of incipient crystallization all typical glasses show an absence of crystals. Some lava flows had already partly crystallized before the residual liquid solidified to glass and, hence, show individual crystals embedded in a matrix of glass. Vesicular and nonvesicular textures are distinguished.

Aphanitic texture. Aphanitic is the name applied to textures previously described as stony. Under the microscope most aphanitic rocks show the presence of variable amounts of glass as matrix, but this is not apparent to the unaided eye. Also scattered throughout the rock are larger crystals, some of which may be recognized with the unaided eye, but they occur as isolated individuals. Aphanitic texture is often also vesicular and associated, or intergradational, with glass. In *aphanites* (i.e., rocks of aphanitic texture) vesicles are not closely spaced but appear as scattered cavities of varying dimensions throughout individual fragments.

In some rocks vesicles have been filled with sedimentary minerals such as calcite deposited from cold waters percolating through the openings long after the rock had cooled. The resulting rock texture is a variety known as *amygdaloidal.* Individual filled cavities are called *amygdules.*

Phaneritic texture. The third major textural group is visibly granular. This is characterized by uniformity of crystal size in the individual specimen, but the range of dimensions is between 1 and 10 or more millimeters with an average of about 2 or 3. This texture is called *phaneritic.* Phaneritic texture is typically developed in rocks in which the cooling process has been so slow that the entire liquid has been consumed in growing crystals. The interlocking position of crystals is clearly visible, particularly if plane surfaces have been ground on the rock specimen.

Certain bodies of phaneritic rocks are noted for their exceptionally large crystal dimensions. These rocks, known as *pegmatites,* are widely scattered geographically, are not especially abundant, and show a particular preference to be associated with silicic or silicic-intermediate compositions. The development of such unusually large crystals which may reach proportions of 10 or more feet for the individual is associated with the last stages of magmatic crystallization. Pegmatites occur as irregular masses within large bodies of the finer grained rock and are associated with deep-seated intrusions. An important vapor phase appears necessary to the development of *pegmatitic* texture which in addition to being coarse grained shows individual minerals to interpenetrate and displays complex interrupted growth stages. Pegmatites are rich in rare elements such as boron, beryllium, lithium, tungsten, and phosphorous, and they are important sources of gem stones. The large dimensions of the individual crystals make them the only sources of sheet mica (muscovite), an essential item in the electrical industry.

Porphyritic texture. Sometimes the progress of cooling is not uniform and begins with the melt remaining at the crystallization temperature of certain minerals for a considerable span of time. Then the magma may be squeezed upward into a much cooler zone where the rate of temperature drop is accelerated. The cooling magma has been subject to two different temperature regimes. In the early stages of crystallization, conditions for the development of phaneritic texture prevail and large crystals of uniform dimensions grow. Later the fluid and its contained crystals are emplaced in the new position where the environment is one of accelerated cooling. Under this steep temperature gradient development of phaneritic texture is no longer possible, and either an aphanitic or glassy texture must result from solidification of the remaining fluid. The texture of the solidified rock is now a mixture of two different varieties. In part it is characterized by the presence of crystals of uniform dimensions readily recognizable by the unaided eye and in part by a groundmass of much finer crystals. This composite texture is called *porphyritic.* It is defined as a texture in which larger crystals are surrounded by a matrix or groundmass of smaller crystals. Obviously, the dimensions of the groundmass crystals can range considerably in magnitude. For example, the groundmass can be glassy, aphanitic, or phaneritic, so long as the crystals which it surrounds are larger in size. Rocks showing porphyritic texture are called *porphyries,* and much difference in appearance can be expected between the extreme textural aspects. Porphyries with a phaneritic groundmass are called *porphyritic phanerites* and are recognized on the basis of the coarse, uniformly granular groundmass. *Porphyritic aphanites* are those in which the groundmass is stony, and the larger crystals (*phenocrysts*) are clearly defined. *Porphyritic glasses* are also commonplace particularly in lava flows.

Colors of Igneous Rocks

Inasmuch as most igneous rocks are constituted of crystallized products of the melt the colors of these rocks are dependent entirely upon the percentage and composition of the contained minerals. The exception is glass in which no individual minerals are to be observed. In hand specimens glasses are dark, ranging between dark grays, often with a green shade, and dark red hues irrespective of whether they are high or low in silica content. In thin slices these colors lighten markedly and become transparent, but chemical analysis appears to be the only satisfactory means of recognizing compositional differences.

Aphanitic rocks represent the composite color of finely divided individual minerals. Colors of aphanites, therefore, reflect the bulk mineralogy of the rock. Those which are light in color are representative of silicic or silicic-intermediate rocks, whereas dark colors are typical of mafic-intermediate and mafic rocks. The light or dark color is a response to the colors of the minerals present. For example, silicic and silicic-intermediate rocks contain such light-colored minerals as sodic plagioclase, potash feldspar, quartz, and muscovite. The corresponding aphanites are light colored. Aphanites whose chemical composition is that of the mafic-intermediate or mafic rocks will show colors which are increasingly dark with decrease in the percentage of silica in the melt. This is the response to the inverse relationship between darkness of color and the ratio of silica to total other oxides. Colors of olivine, pyroxene, amphibole, and calcium plagioclase are dark; hence, rocks containing such minerals will be dark in color.

Porphyries and phanerites show colors corresponding to their mineralogically equivalent aphanites although in porphyries colors are spotted due to the isolation of individual minerals. Light-colored phanerites are dominated by the presence of sodium- or potassium-rich feldspars and muscovite. Dark-colored phanerites are characterized by the presence of pyroxene, amphiboles, and dark calcium-rich plagioclases. The color of the phaneritic rock is somewhat diagnostic of the mineralogy although it must be remembered that certain dark minerals such as biotite are high in silica and occur in the more silicic rocks. Thus, coarse grained porphyries vary in color although primarily the lighter colored ones are more siliceous than the darker colored ones. Porphyries which have a glassy or aphanitic groundmass will normally be light or dark colored depending upon the general composition of the groundmass.

CLASSIFICATION OF IGNEOUS ROCKS

Classification Based upon Texture and Mineralogy

Numerous attempts to explore methods of classifying igneous rocks have met with varying degrees of success depending upon the approach selected. The most successful systems from elementary and field viewpoints are based upon the rock texture as well as the mineralogic composition. Table 5.3 is representative of a systematic tabular arrangement modified from that first prepared by German and French geologists over seventy years ago. This table is primarily for use in establishing the broad grouping of a rock. It is not suitable to identify the special varieties of which there are many but which fortunately are uncommon in occurrence. Identification of the special varieties is accomplished only by combined mineralogic and chemical analyses. Nevertheless, the bulk of all common rocks can be assembled under the rock groups enumerated in Table 5.3 without committing violence to proper rock classification. In particular, the table has much use in the field where the available equipment for rock identification is generally restricted to a knife and a hand lens. Best results in identification are obtained by following the systematic procedure outlined below.

Procedure of Igneous Rock Identification Using Table 5.3

Color grouping. Classify rock by color.

Textural grouping. Classify by texture.

Identification of quartz. If quartz is abundantly present the rock is provisionally identified as a granite. If quartz is sparingly present the rock is placed in the quartz monzonite or granodiorite group. If quartz is not present the rock is known to be either intermediate, mafic, or ultramafic in composition.

Identification of feldspar. Importance is stressed in distinguishing between the potash and plagioclase feldspars. It will be recalled that one of the characterizing features of the potash feldspar is its pink color. It must also be remembered that not all orthoclase or microcline is pink, but if the pink color of the feldspar is present the mineral can be assumed to be either orthoclase or microcline. On the other hand if the feldspar is light gray or white in color it may be either

Table 5.3. Generalized Igneous Rock Classification for Use with Hand Specimens*

Observed Textures	Composition					
	Colors: ← Increasingly Light / Increasingly Dark →					
	Mineralogy					
	Quartz Orthoclase Biotite Muscovite Hornblende	Orthoclase Biotite Muscovite Hornblende	Orthoclase Na plagioclase Quartz Biotite Hornblende Muscovite	Na-Ca plagioclase Hornblende Biotite	Ca-Na plagioclase Augite Olivine	Olivine Augite Magnetite Chromite
	Approximate Percentage of Silica					
	>65+	±60	65–60	55+	50–45	±40
Phaneritic	Granite	Syenite	Quartz mon-zonite or granodiorite	Diorite	Gabbro	Dunite
Phaneritic-porphyritic	Granite porphyry	Syenite porphyry	Quartz mon-zonite or granodiorite porphyry	Diorite porphyry	Gabbro porphyry	
Porphyritic-aphanitic or porphyritic-glass	Rhyolite porphyry	Trachyte porphyry	Dacite porphyry	Andesite porphyry	Basalt porphyry	Not recognized
Aphanitic	← Felsite →			← Basalt		
Glass: Nonvesicular Vesicular Filled vesicles	← Obsidian → Pumice Amygdaloidal pumice				Scoria Amygdaloidal basalt	

* This table is designed for use in the field and is intended for hand specimens primarily. For advanced or specialized studies more precise identification is required.

plagioclase, orthoclase, or microcline. In this connection, search is made for the characteristic plagioclase twinning which is identified by the presence of multiple parallel striations extending the length or across the entire surface of selected cleavage surfaces. To the unaided eye orthoclase and microcline show a development of single twins or an entire absence of twinning.

Identification of accessory minerals. In silicic rocks micas are the most common accessory minerals and in such cases are little more than of incidental importance in rock classification. In some of the intermediate rocks distinction between either biotite or hornblende is useful in designation of the rock variety (e.g., hornblende syenite or biotite diorite).

Distinction between mafic-intermediate and mafic rocks is difficult because it is dependent upon distinction between hornblende and augite but is aided somewhat by reliance upon the general rule that as the plagioclase becomes more calcic it correspondingly darkens in color. Light-gray plagioclase is likely to be sodic or intermediate in composition.

Quantitative Classification Based upon Mineralogy and Texture

Use of Table 5.3 imposes certain limitations by rigidly requiring the rock identification to conform with one of the boxes of the table. Erection of artificial boundaries between each species effectively destroys the natural intergradations which exist between these rocks. Also there is no means of determining the variations in percentage of minerals between groups. Still another objection appears in that certain common rock names are omitted from the table, and their mineralogic composition and texture must be known before the approximate synonyms are recognized. In order to offset partially some of these objections a supplementary system of classification is presented below. This system has advantages in clarifying the compositional ranges and provides a more satisfactory means of comparing the mineralogic composition between rock species, but is restricted to those rocks whose silica content is greater than 45 per cent. Ultrabasic rocks are, therefore, eliminated from consideration. This is not as unfortunate as might appear at first inasmuch as the ultramafic rocks are extremely uncommon in occurrence, and most varieties are so unique in their mineralogy that they do not lend themselves to classification with other more silicic rocks.

The list of diagnostic minerals in the classification of the common igneous rocks can be restricted to quartz, orthoclase, microcline, and the plagioclases ranging between albite and anorthite. If each of these is considered an end member plotted at an apex of a compositional tetrahedron the proportionate percentage occurrence of each can be

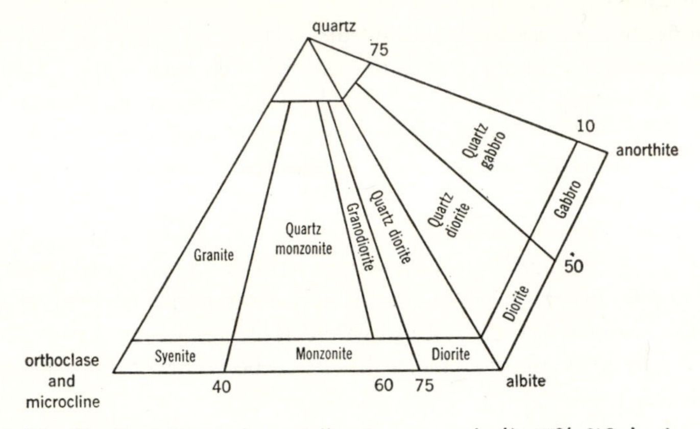

Fig. 5.7 Classification of coarsely crystalline igneous rocks (>45% SiO_2) using selected feldspars and quartz as end members.

indicated by the position on a face of the tetrahedron. (See Fig. 5.7.) The percentage variation between three end members is represented on each triangular face, and the boundaries of individual rock species established on the percentage relationships between feldspar and quartz. Such relationships are better shown in the "exploded" tetrahedron, Fig. 5.8, in which the fields represented by each common rock species are represented. Note that certain positions in the triangles have been left blank. This has been done to indicate those rocks which are extremely rare in occurrence and are representative of very special varieties.

Figures 5.9 and 5.10 are similar representations of porphyries and aphanites showing the compositional boundaries of each. Among the aphanites the term "trap" (a common commercial term) has been added.

Use of the Classification Tetrahedron

For elementary work the chief use of the classification tetrahedron is in determining the mineral composition of rock names which are used in technical reports. If, for example, a rock is described as a monzonite the field limiting its compositional range can be found on the tetrahedron and the per cent distribution of the major mineral constituents determined. In this case quartz is present to the extent of

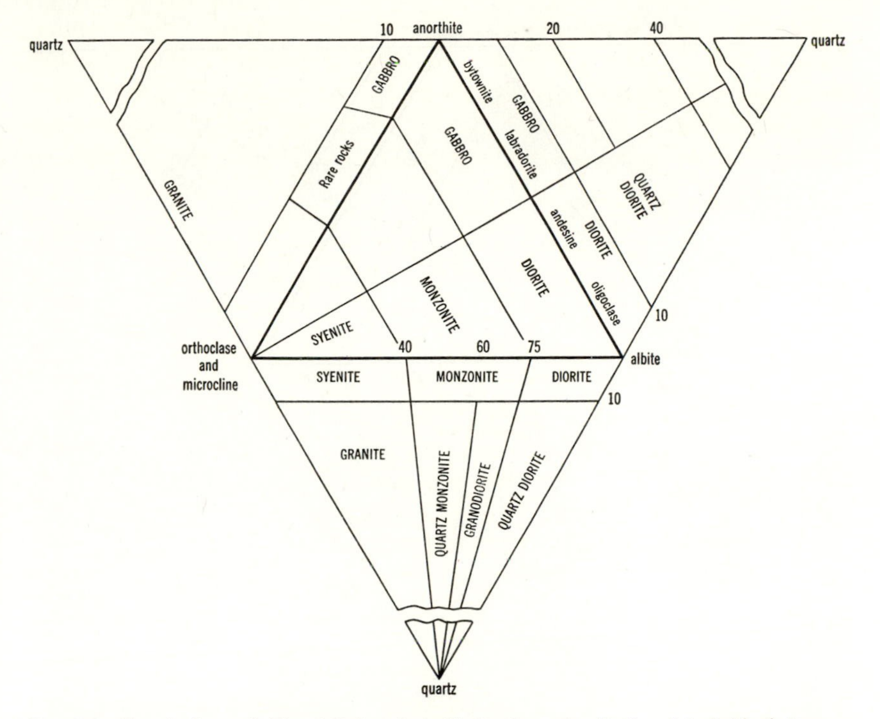

Fig. 5.8 Tetrahedron of Fig. 5.7 "exploded" to show the limits of individual igneous rocks in terms of the selected feldspar and quartz end members.

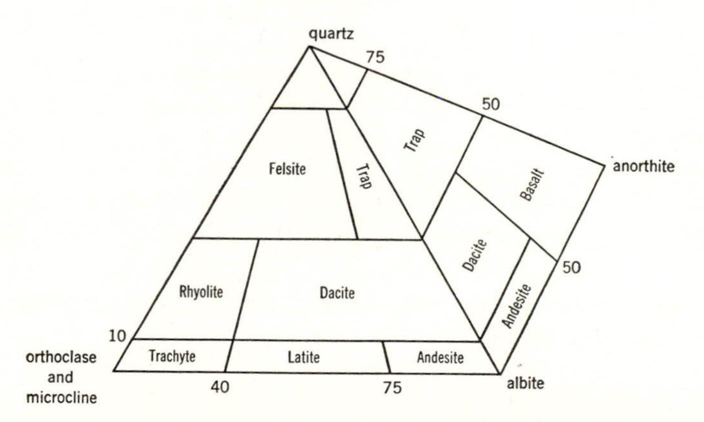

Fig. 5.9 Classification of finely crystalline and glassy igneous rocks (>45% SiO_2) using the end members of feldspar and quartz. Compare with Fig. 5.7.

less than 10 per cent, the albite molecule 40 to 75 per cent, and orthoclase and microcline together can range between 25 and 60 per cent of the total feldspar.

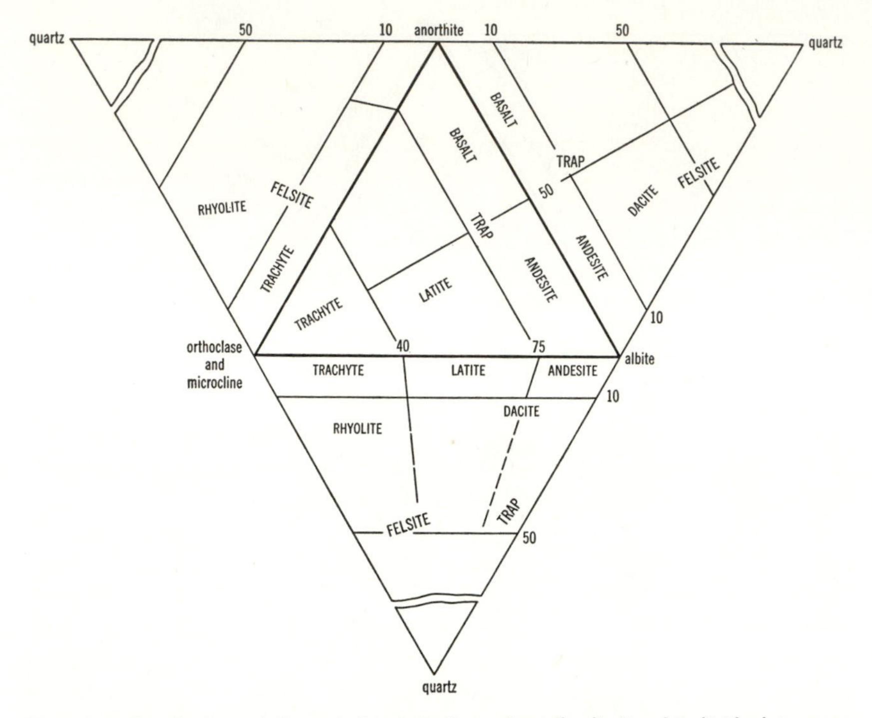

Fig. 5.10 Tetrahedron of Fig. 5.9 "exploded" to show the limits of individual igneous-rock species in terms of selected feldspar and quartz end members.

Primarily the classification tetrahedron is to be used after the composition and per cent occurrence of the feldspars is known. This requires the use of a petrographic microscope and knowledge of mineralogy beyond the scope of this text. Introduction of this system of classification is inserted herein primarily to provide an understanding of the field identification table and also to illustrate a somewhat more rational method of classification than that illustrated by Table 5.3.

SHAPES, DIMENSIONS, AND STRUCTURAL CHARACTERISTICS OF IGNEOUS BODIES

Extrusions

Lava Flows

Extrusions are typically represented by the lava flow and the volcanic cone. With few exceptions the two are closely associated and the lava flow is a fundamental part of the volcanic cone. More commonly in the ancient rocks no record of the volcanic cone is present, and the entire evidence of extruded rock is in the form of lava flows. Basaltic lavas are not necessarily associated with volcanic cones. They are known to cover areas of thousands of square miles and have risen to the surface along numerous fissures from which the lavas quietly flowed without any associated explosive cone-building activity. Such mafic lava issues from fissure vents and spreads as thin sheets encircling the fissure zone. The areal extent of each flow is controlled by the distance it has moved before its viscosity has increased to the degree where the flow can no longer advance. Few individual flows exceed 300 feet in thickness, and the average is of the order of 50 feet. Inasmuch as the flow is poured out on the former land surface (erosion surface) the thickness of an individual flow may vary substantially when the flow has taken the course of a pre-existing valley. One such case at Williams Canyon, Arizona has a width of 14 kilometers, a length of 22 kilometers, and its greatest thickness is 240 meters.

Thick flows often have a very irregular surface due to the tendency for the interior of the flow to remain fluid when the upper surface has solidified. This condition tears the frozen surface crust into fragments which pile upon one another in a manner identical with ice blocks up-ended along pressure ridges in a frozen stream. Lava surfaces also develop corded, ropelike surfaces and large bulbous masses called "*pillows*" lying one upon the other.[2] (See Fig. 13.3.) Far less spectacular but more commonplace is flow structure which shows the lines of fluid flow of the partially consolidated lava as it moved in forward motion. (See Figs. 13.1 and 13.2.)

Differential cooling rates between the bounding surfaces of the flow and its central portion result in textural differences. The upper and lower surfaces tend to be glassy, whereas toward the center an increasing number of crystals normally occurs. Vesicular texture is also commonplace particularly on the upper surface.

Flows are piled one upon the other frequently to enormous thicknesses and areas (in excess of 10,000 feet in total thickness and 100,000 sq. km in surface area as for example in Greenland and Iceland). In some cases the time interval between successive outpourings is short so that the individual flows are welded together without any prominent zone of separation between them. In such cases distinction between the tops and bottoms of individuals is accomplished by certain structural features as vesicle zones or concentration of zones of oxidized iron minerals formed during exposure to the atmosphere before burial

[2] Pillow structure is considered to form typically when basaltic lava flows into water and is suddenly chilled.

by the succeeding flow. When the interval of lava emission is punctuated by periods of explosive violence individual flows are separated by zones of fragmental debris such as scoria, pumice, or ash. If long periods of inactivity alternate with the intervals of lava emission the individual flows are often separated by soil zones which have developed on the older lava. Also, streams may cut local valleys in the flows and deposit sand and gravels in such depressions. These exist as zones of pronounced lithologic discordance in an otherwise monotonous rock type. (See Fig. 13.5.)

Volcanic Cones

Because of their spectacular qualities much interest is shown in volcanoes, and most texts generally devote more space to their discussion than to other igneous bodies. Volcanic cones, however, are subject to rapid destruction by agencies of erosion, and when buried by later rocks evidence of their former existence is restricted to their feeding conduits. All cones contain some ejecta accumulated during periods of intense volcanism. Some cones are built almost entirely from fragmental debris, whereas others are almost exclusively lava. (See Fig. 5.11.) Still others contain alternations of lava flows and beds of ejecta.

Correlation exists between the degree of slope inclination of the cone and the nature of the constructional material. Cones dominated by lava flows have low inclinations from the horizontal, generally less than 10 degrees, and cover extensive areas. These are also associated with the lavas of low viscosity and are typical of volcanoes erupting basalt. Lavas of greater viscosities obstruct the normal escape of gas, and eruptions terminate in periods of explosive violence and the addition of much fragmental material to the cone. Symmetry of profile as well as steepness of slope (30 degrees) characterize the cone overwhelmingly constructed from fragmental debris. They are called *cinder cones,* are of variable size, but never attain the areal extent of the low sloping lava cones called *shield volcanoes.* Most cones are intermediate between these two extremes and are constructed of alternating flows and fragmental ejecta.

Fragmental material of the cone. Ejecta includes a heterogenous group of volcanic fragments of varying shapes and characteristics. Largest among these are the *blocks.* These are extremely angular fragments of dimensions ranging greater than one foot, torn from the throat of the volcano during intermittent explosions. The blocks are cast upward and fall on the slope of the cone. Obviously the force of such an explosion creates a particle size range in distribution from blocks

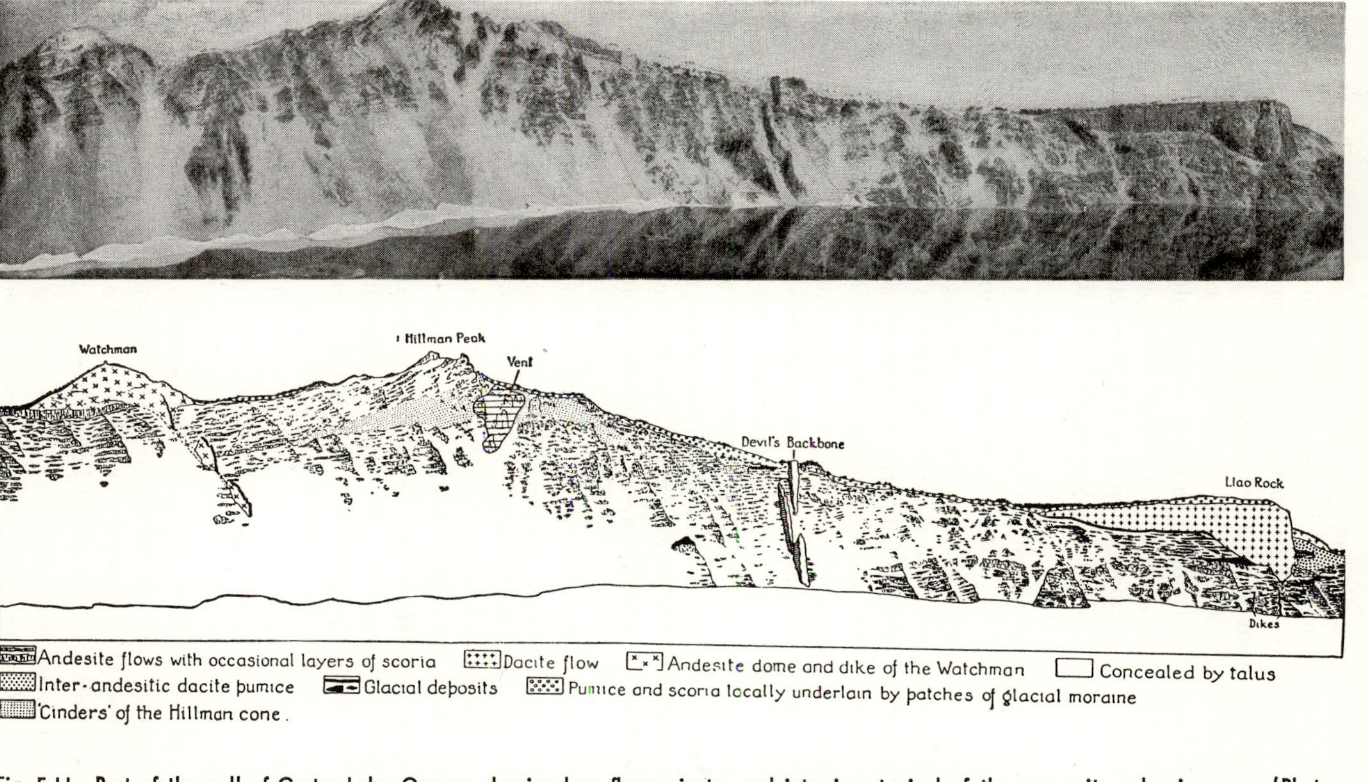

Fig. 5.11 Part of the wall of Crater Lake, Oregon, showing lava flows, ejecta, and intrusions typical of the composite volcanic cone. (Photograph by Elmer Aldrich and drawing from Williams, *Carnegie Inst. Wash. Publ.*, **540**, 1940.)

one hundred or more feet on a side to particles in the clay range of dimensions. Much fluid is also hurled into space, solidifies in the air, and falls as solid fragments. These are often characteristically spheroids or pear shaped becoming stream lined during their flight as liquid projectiles. Such spheroidal-shaped fragments are called *bombs, lapilli,* and *cinders* in order of their decrease in diameter (see Table 12.1). The very fine material cast upward constitutes the ash and dust. This ranges from granule through sand and clay in size and is carried upward many thousands of feet before beginning its downward descent.

During an episode of explosive activity there is accomplished a gradual segregation of the coarse and very fine ejecta. The fine ash and dust which is carried upward to great heights is held in suspension by the turbulence of the atmosphere and is eventually deposited some distance from the site of its origin.[3] Other material broken into fragments by the force of the explosion is propelled only a short distance and adds to the structure of the cone. The bulk of the material of the cone is, therefore, an accumulation of particles of continuous size range between cinders and blocks, whereas ash and dust are typical of volcanic debris deposited primarily outside of the periphery of the cone. The ash is representative of a deposit which is segregated from other material with which it was simultaneously generated. Debris on the cone tends to be locally unsegregated and consists of a wide range in particle size resulting from the crushing force of the explosion. Inasmuch as ash is representative of particles of a restricted diameter range the size frequency distribution of such material should show a strong concentration in certain sizes. This is due to the factor of selective sorting, accomplished during transportation through the atmosphere.

Little segregation of debris has occurred in the cone, the material of which has been thrown to its site of accumulation by the explosion, and large and small particles rest together in a random assortment. Randomness is typical of the size distribution of material whose deposition has not been controlled by laws of selective transport. Movement of debris from one site to another is controlled by a different set of laws than the laws typically attributed to deposition from air and water.

The size distribution of ejecta is a satisfactory means of illustrating fundamental laws governing the textural aspect of loose rock debris. One of these, *the law of crushing,* describes the size distribution of rock passed through a mechanical jaw crusher, or fragmented by explosive

[3] Dust from the 1883 explosion of Krakatoa off the coast of Java is reported to have been blown upward to a height exceeding 20 miles.

violence. The other laws govern particle settling velocities when debris sinks through a suspending medium. In such cases a tendency exists to concentrate certain size ranges.

Rosin and Remmler's law of crushing. This law is expressed in terms of the weight percentage of fragments which will remain on a sieve of selected size when the entire crushed sample is passed through a nest of sieves of graduated size openings. The mathematical expression is a function of a negative exponential type and is represented as follows:[4]

$$W = 100\, e^{-\left(\frac{x}{k}\right)^n} \tag{1}$$

where W = weight per cent of crushed material retained on a sieve whose mesh diameter is x millimeters.

e = the base of the natural logarithms = 2.719.

k = an average size in millimeters, i.e., size of x for which the cumulated weight per cent coarser than x = 36.8.

n = a measure of the per cent of material which is concentrated in a narrow size range. For material of uniform size the value of $n = \infty$, and with infinite range in size $n = 0$. For most crushed material the value of n ranges between 0.6 and 1.3.

The size distribution of material in accordance with Rosin and Remmler's law is not symmetrically distributed about the average size (k), but a larger weight per cent is finer, rather than coarser, than the average size. There is also a greater percentage of material in a selected size class which is coarser than the average than there is in a corresponding size class which is finer than the average. Such a distribution is considered skewed when compared with the normal size frequency distribution which is symmetrically distributed about a central value (see Fig. 5.12).

On a specially ruled paper with weight frequency per cent as the ordinate and particle size as the abscissa points in accordance with Rosin and Remmler's law will plot as a straight line. Similarly on such paper connection of points representative of a normal size frequency distribution will make a curved line. The curvature of the line increases as the value of n approaches infinity. Size distribution of ejecta from Crater Lake, Oregon, and Hawaii plotted on the special ruled paper are shown in Fig. 5.13 (see Table 5.4). Note except for the finest material the particle size distribution of the younger pumice

[4] J. G. Bennett, "Broken Coal," *Jour. Inst. of Fuel,* Vol. 10, 1936, pp. 22–39.

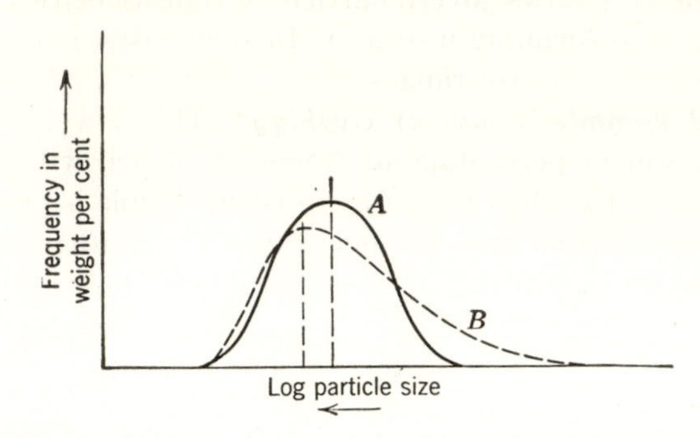

Fig. 5.12 Comparison of the size frequency distribution of the log of particle diameter of the sediment accumulated by settling from water or air (*A*) and the distribution of particle diameters according to the law of crushing (*B*). Note the tendency for the distribution developed by crushing to be skewed toward the small diameters.

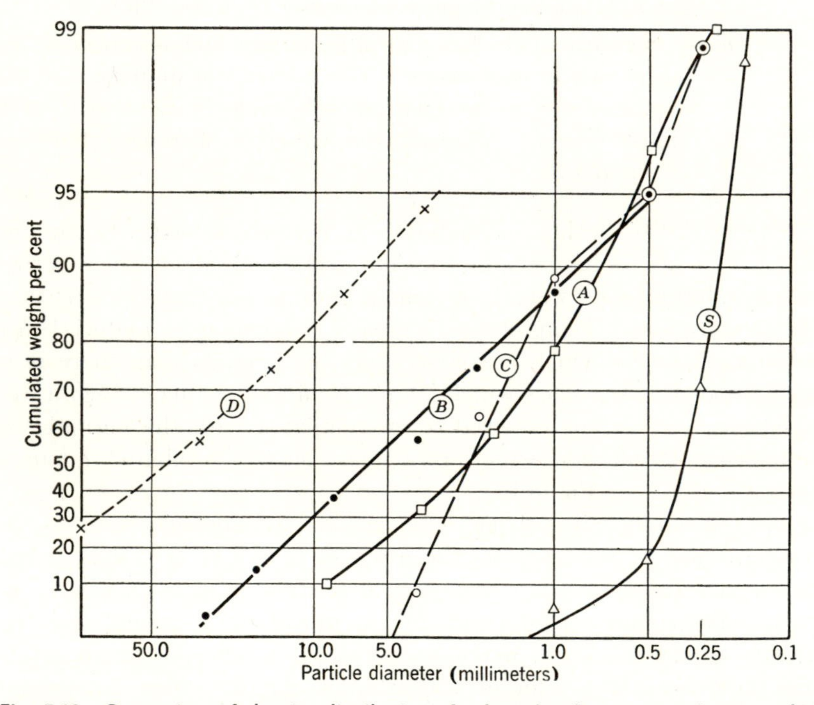

Fig. 5.13 Comparison of the size distribution of selected sediments on a base on which a distribution following the crushing law plots as a straight line. (*A*) Tuff, Crater Lake, Oregon; (*B*) younger pumice, Crater Lake, Oregon; (*C*) ideal Hawaiian ash; (*D*) crushed coal, LaSalle Co., Illinois; (*S*) beach sand, Lake Erie. (For data see Tables 2.5 and 5.7.)

Table 5.4. Size Distribution of Volcanic Ejecta

Screen Diameter (mm)	Cumulative Weight Per Cent Retained on Screen			
	*A**	*B*	*C*	*D*
64	0	1	0	28
32	0	5	0	58
16	0	15	0	75
8	10	38	1	88
4	34	59	8	94
2	60	76	63	
1	79	87	89	
$\frac{1}{2}$	96	95	95	
$\frac{1}{4}$	99	98	98	
$\frac{1}{8}$	100	100	100	

* *A*. Tuff, Crater Lake, Oregon; data from B. N. Moore, "Deposits of Possible Nuée Ardente Origin in the Crater Lake Region," *Jour. Geology,* Vol. 42, 1934, p. 367. See also Fig. 5.13.

B. Younger pumice, Crater Lake, Oregon; Sample 86 data from B. N. Moore, *op. cit.*

C. Ideal size distribution of Hawaiian ash from C. K. Wentworth, "Pyroclastic Geology of Oahu," *Bernice P. Bishop Mus., Bull.* 30, 1926, Fig. 25.

D. Coal from #2 bed, LaSalle Co., Illinois, crushed by jaw crusher; data from M. R. Geer and H. F. Yancey, "Expression and Interpretation of the Size Composition of Coal," *Amer. Inst. Mining & Metall. Eng., Trans. Coal Div.*, Vol. 130, 1938, p. 253.

from Crater Lake is in accordance with that dictated by the crushing law as shown by the straight line relationship of the plotted points and the similarity to the distribution of crushed coal. The pumice at the site of sampling must represent ejecta deposited directly without modification following the explosive violence of the eruption. To a lesser degree the distribution of particle sizes of the Hawaiian ash conforms to a straight line in the coarser sizes, but particles smaller than 1 mm in diameter do not conform to the expected distribution. Substantially greater departure from a straight line is indicated by the plotted points of the Crater Lake tuff, the distribution of particle size of which tends to simulate that of the beach sand described in Table 2.5. Such relationship establishes the basis for the inference that some ejecta is not a product of simple crushing alone but represents material which has been segregated in part by the rate of settling through the atmosphere. (See Fig. 5.14.)

Laws of settling. During the transportation of a rock particle from one resting point to another one of the most fundamental aspects of its behavior is the tendency to sink through the suspending medium. This tendency can be translated into terms of the settling velocity which is controlled by the difference in density between the particle and that of the suspending medium, the shape of the particle, its size, and the inertial impact of the medium upon the particle. Primarily the laws of settling have been established on the relationships of particles sinking through water, but the general concepts are applicable to air as

Fig. 5.14 Ash deposits at the south edge of Kilauea Crater, Hawaii. (Photograph by Macdonald, *U. S. Geol. Survey, Prof. Paper,* 214-D.)

well. For purposes of illustration the laws of settling velocities are being applied to those particles of volcanic ash carried away from the volcanic center and eventually dropped on the surface of some lake. Once on the water surface the deposition of particles on the lake bottom is controlled by the laws of settling velocities one of which is known as *Stokes' law* and the other known as the *impact law.* The resulting deposit is a variety of sedimentary rock called *tuff.*

STOKES' LAW.[5] If a large number of very small spherical particles all of common density are permitted to sink through a motionless

[5] G. G. Stokes, "On the Effect of the Internal Friction of Fluids on the Motion of Pendulums," *Cambridge Philos. Soc.,* Vol. 9, pt. 2, 1851, pp. 8–106.

See also W. C. Krumbein and F. J. Pettijohn, *Manual of Sedimentary Petrography,* D. Appleton-Century Co., 1938, p. 95.

body of water, the velocity with which an individual particle sinks through the water is directly proportional to its diameter. Stokes demonstrated this relationship by the following approach:

The resistance which a fluid exerts against the movement of a suspended sphere can be expressed as

$$R = 6\pi r\eta v \qquad (2)$$

where R = resistance in gram-centimeters per second per second.
r = radius of the sphere in centimeters.
η = viscosity of the fluid in dynes-centimeters per second per second.
v = velocity of the sphere in centimeters per second.

A small sphere sinking through the medium is acted on by the force of gravity acting downward ($\frac{4}{3}\pi r^3 d_1 g$) and buoyant force of the liquid acting upward ($\frac{4}{3}\pi r^3 d_2 g$) which leaves a resultant force acting downward of

$$\tfrac{4}{3}\pi r^3 (d_1 - d_2)g$$

where d_1 = density of solid.
d_2 = density of liquid.
g = acceleration due to gravity = 980 cm per second per second.

At the instant that the fluid resistance is equal to the net downward force the velocity of the particle is constant, and

$$R = 6\pi r\eta v = \tfrac{4}{3}\pi r^3 (d_1 - d_2)g \qquad (3)$$

Solving the equation for v yields the following expression:

$$v = \frac{2\pi g(d_1 - d_2)r^2}{9\eta} \qquad (4)$$

which is the equation of Stokes' law.

If the temperature of the water is held constant and the spheres have a uniform density Stokes' law is simplified to

$$v = Cr^2 \qquad (5)$$

in which C is a constant equal to $2\pi g(d_1 - d_2)/9\eta$ and has a value of 3.57×10^4 for water at 20°C and a specific gravity of 2.65 for the solid spheres.

IMPACT LAW. Stokes' law holds well for small particles but as the diameter is increased beyond 0.1 mm settling velocities are observed to be substantially slower than predicted by calculation. The deviation is sufficient to indicate that Stokes' law is no longer completely controlling the settling velocities. This apparent anomaly was demonstrated to be the result of the inertial impact of the fluid upon the particle.[6] The total impact expressed as $\pi r^2 v^2 d_2$ must be added to the viscous forces of the liquid opposing the effective weight of the particle, in order to obtain the actual settling velocity.

Hence,

$$\tfrac{4}{3}\pi r^3 (d_1 - d_2) g = 6\pi r \eta v + \pi r^2 v^2 d_2 \tag{6}$$

solving for v:

$$v = \frac{\sqrt{\tfrac{4}{3} g d_2 (d_1 - d_2) r^3 + 9\eta^2} - 3\eta}{d_2 r} \tag{7}$$

If the temperature of the water is held constant and the density of the spheres is uniform the equation is simplified to

$$v = \frac{\sqrt{c_1 r^3}}{c_2 r} = c_3 \sqrt{r} \tag{8}$$

which is the impact law in which c_1, c_2, and c_3 are constants of differing values.

The physical significance of these two laws will be discussed in later chapters. It should be understood, however, that they provide the basis for an understanding of the current velocities required to maintain a particle in suspension. Thus the rate at which silt and clay sink is controlled by the viscous resistance of the fluid, whereas the settling velocities of larger particles (primarily sand and pebbles) are responses to the inertial impact of the fluid. Very small particles are, therefore, held in suspension by currents of low magnitude, but large particles require very strong currents to maintain them in motion.

Intrusions

Tabular Bodies

Intrusions display a great range in composition, geographic distribution, and shape. They occur as individual bodies of dimensions ranging between a few millimeters in width and enormous irregular masses of total volume measured in hundreds of cubic miles. Tabular-shaped bodies with one thin dimension are most common. These are termed *dikes* or *sills* depending upon their relative position with respect to

[6] W. W. Rubey, "Settling Velocities of Gravel, Sand, and Silt Particles," *Amer. Jour. Science,* Vol. 224, 1933, pp. 325–338.

the stratification planes of enclosing sedimentary rocks. Similar intrusions also occur cutting larger igneous masses which were invaded by renewed igneous activity following initial crystallization. In such cases distinction is not drawn between the types of tabular intrusion, and all are called dikes. Metamorphic rocks are often profusely intruded by dikes and sills. Dikes are recognized as transgressing the rock cleavage, or planes of oriented crystals, whereas sills are intruded parallel to the cleavage.

Dikes. A dike is a transgressive tabular intrusion invading its host rock primarily by forcing apart the enclosing walls. These commonly display two parallel surfaces of which irregularities on one side of the dike have their counterparts on the other side. Normally the angle of intersection between the dike walls and the horizontal plane ranges between 90 and 45 degrees, but lower angles are not uncommon. The thickness of the thin dimension may be as small as a few millimeters, but widths up to 12 km are known. The average thickness is probably less than 5 m. These bodies rarely occur singly and discovery of one dike generally suggests the presence of another nearby. They may be connected at depth or are small offshoots from larger bodies. Oftentimes, they display similar or parallel trends or occur in two or more parallel sets intersecting one another. Hundreds of dikes may occur in local areas often with strong linear trend or radiating from some center. Such occurrences known as *dike swarms* are represented by the classic examples to be seen on the western coast of Scotland and in the United States in the Crazy Mountains of Montana (see Fig. 5.15).

Dike rocks are noteworthy for their diversity in chemical composition and mineralogy. Some bodies are almost entirely composed of quartz, and at the other extreme are those of ultramafic compositions of less than 40 per cent silica. Hundreds of exotic rock types occur as dikes, but the most common varieties appear to be either silicic or mafic in composition. (See Table 5.5.) Dike rock textures also run the gamut between extremes. Some are glassy others are aphanitic, porphyritic, phaneritic, and pegmatitic. No single texture appears to predominate, but where the dike is in contact with the intruded rock the texture is usually more finely crystalline than the central portion.

Sills. Sills are similar to dikes in shape and dimensions, but the term is restricted to bodies in which the igneous rock parallels the planes of stratification or the mineral lineation of the inclosing rock. (See Fig. 5.16.) Generally, the plane including the two large dimensions intersects the horizontal at low angles, but this is not essential to its definition. Igneous rock of the sill is emplaced by forcing the rock walls apart; hence intrusion is common along planes of weakness

Table 5.5. Summary Table of Characteristics of Igneous Bodies

Name	Orders of Magnitude of Emplacement	Shape Characteristics	Textures	Special Features	Per Cent Composition SiO_2:
	Depths: Shallow, Surface to 3000′; Intermediate, 3000 to 10,000′; Deep, 10,000′ or more.				Silicic +65% Intermediate 50 to 65% Mafic 45 to 50% Ultramafic ±40%
Dike	Shallow to deep seated, depth of intrusion indicated by texture	Tabular-discordant, thin dimension $\frac{1}{2}''$ to 1000′; long dimension from few feet to miles.	All textures between glassy and phaneritic.	Generally in groups of several, often radiating from a center or in parallel trends.	Greatest variability of rock types, many rare varieties but most common are silicic or mafic types.
Sill	Shallow to intermediate intrusion	Tabular-concordant, thin dimension a few feet to 5000′; long dimension from a few hundreds of feet to miles.	Dominantly porphyritic or aphanitic, glassy or phaneritic textures rare.	Generally in groups of two or more but not in large numbers.	Generally intermediate to mafic, the latter generally the larger.
Flow	Surface extrusion	Tabular to lenticular, lobate, commonly occupies old valley, lies on former surface of erosion.	Glassy, aphanitic, vesicular, amygdaloidal.	Mafic types cover enormous area. Usually consist of a thick series totaling several thousand feet. "Pillows," oriented vesicles, flow lines are typical structures.	Mafic in large plateau area, many flows. Silicic associated with mountain-building activity.

Laccolith	Shallow to shallow-intermediate intrusion	Related to sill, arched strata over roof, floored body, lenticular shape, maximum thickness several hundred to 5000′. Diameter 1 to 10 miles.	Porphyritic-aphanitic; porphyritic-phaneritic.	Generally in groups of several as mountains.	Intermediate.
Stock	Deep-seated intrusion	Generally conical discordant, may arch strata, no floor, surface area up to 40 sq. miles.	Phaneritic to porphyritic phanerite.	Metamorphoses surrounding strata, often has ore deposits associated; generally in areas of former mountain building, associated with dikes and sills.	Intermediate to silicic.
Batholith	Deep-seated intrusion	Elongate, in centers of mountain ranges, usually hundreds of miles in length, widen downward.	Phaneritic.	Metamorphoses strata for miles, associated with pronounced folding, boundaries gradational due to metamorphism.	Silicic to intermediate.
Ejecta	Surface explosion	Fragmental, ranges from large blocks to very fine ash.	Frothy, glassy, small fragments angular, larger fragments rounded.	Commonly stratified.	Silicic to mafic.

and where the overburden can be lifted by the pressure of the magma. Sills are preferentially emplaced at shallow to intermediate depths and

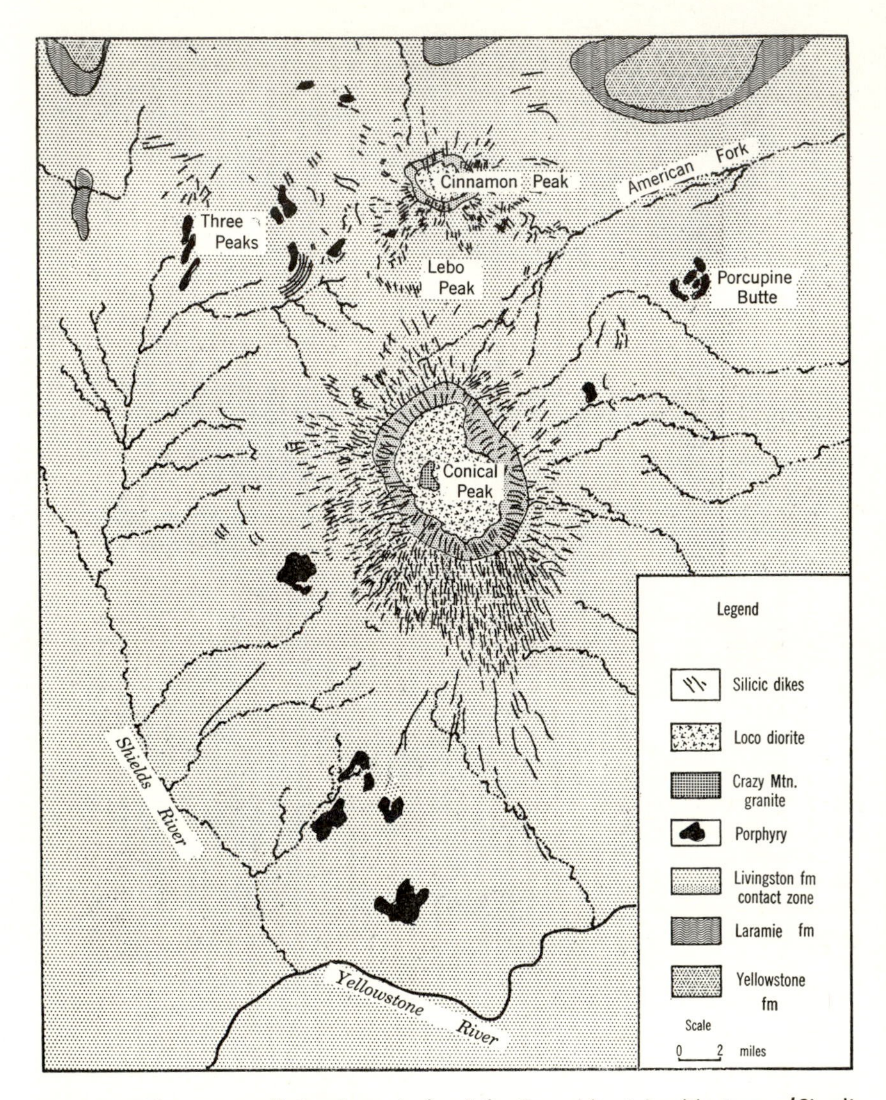

Fig. 5.15 Dike swarm radiating from stocks of the Crazy Mountains, Montana. (Simplified from *Livingston Folio #1* and *Little Belt Mountains Folio #56, U. S. Geol. Survey.*)

are not associated with intrusions typical of great depths. In this connection the general dimensions of the sill appear to be inversely proportional to the viscosity of the magma. Highly viscous magmas often develop pod-shaped bodies, whereas magmas of low viscosity develop bodies of uniform tabular shape. There is a preference for most sills to have a mafic or mafic-intermediate composition although the range in composition is large. Often sills are interconnected by dikes or feeding conduits which appear to have selected their position on the basis of zones of weakness or fissures in the host rocks.

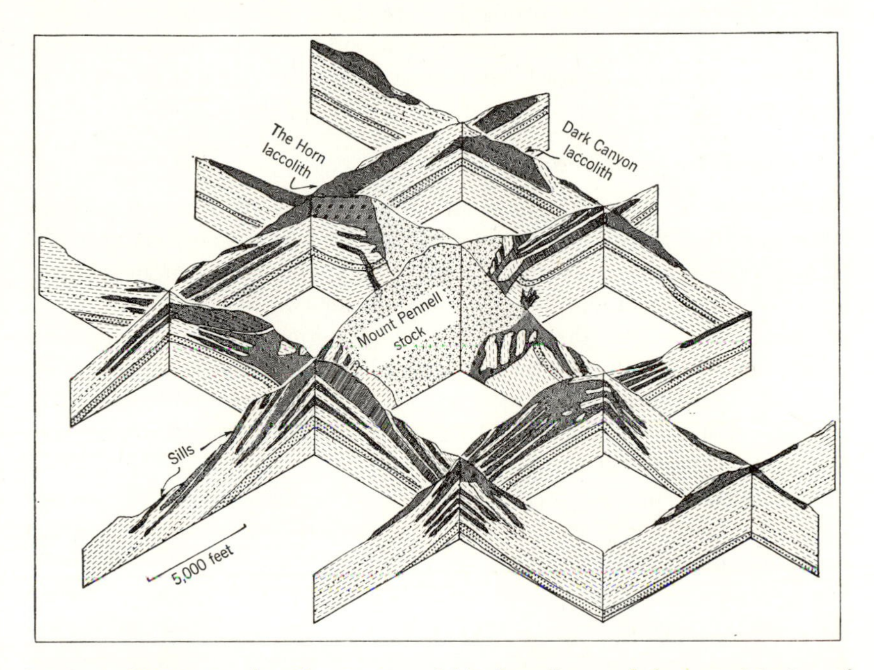

Fig. 5.16 Diagram to show the structure of Mt. Pennell, one of the large intrusions of the Henry Mts., Utah. Note that the Mt. Pennell stock is an intrusion which transgresses the surrounding strata. Laccoliths such as "The Horn" are lens shaped and arch the sedimentary cover forming the roof of the intrusion, whereas strata of the floor are virtually undisturbed. Laccoliths are sills of special form; note the tabular shapes of sills. (Simplified from Hunt, Averitt, and Miller, *U. S. Geol. Survey Prof. Paper*, **228**.)

Sills tend to be thicker than dikes and 30 or 40 m may be considered as the average. Maximum thicknesses of sill-like bodies are reported to exceed a kilometer, whereas the minimum thicknesses known to the writer are approximately 1 m. Sills occur singly and in groups, but large numbers interlayered one above the other are rare. Generally, a group will consist of five or less individuals, and ten should be considered greater than the average. Numbers comparable to dike swarms are unknown.

Sills do not show the textural extremes of dikes. Normally the former are aphanitic and often slightly vesicular near the top. Often the borders show chilling against the host rock, and the crystal size increases toward the center.

When the sill is 100 or more meters thick the mineral composition sometimes is different at the top than the bottom. This is attributed to crystal segregation during the crystallization process. Some crystals such as olivine or augite probably exceed the specific gravity of the melt and sink to the lower parts of the sill. Plagioclase appears to be less dense than the melt and often floats to the top where it is concentrated.

Bodies of Irregular Shape

Intrusions of irregular shape are much larger in dimensions than dikes or sills. They are also of such specialized outline that no completely satisfactory system of classifying them has been developed. In broad aspect they can be assembled under two major subdivisions: (1) large pod- or bun-shaped bodies whose major position of emplacement is parallel with the stratification planes of the sedimentary host rocks, (2) very large rudely conical or elongate bodies with markedly irregular boundaries transgressing the stratification planes of the sedimentary host rocks.

Laccoliths. The most typical of the pod- or bun-shaped bodies are *laccoliths* (see Fig. 5.16). These intrusions are fed from a central conduit from which the magma spreads laterally along a selected stratification plane and proceeds to arch the overlying sedimentary rocks. Sedimentary layers which form the floor of the intrusion are undisturbed, whereas those which form the roof are domed, sharply folded, or even broken by the continued injection of magma. Emplacement of such bodies must occur at shallow to intermediate depths where the directional release of pressure is in an upward direction, and the overlying rocks may be lifted to accommodate the influx of magma.

Deep erosion of these dome-shaped mountains (some have diameters approaching 1 mile) exposes the core and flanks of the intrusion to view, and the relationship to the overlying strata must be inferred from observations along the flanks. Here the textures and rock composition can be examined. Porphyries appear to be the prevailing textures, and there is a distinct preference for compositions to be intermediate. Some rough relationship between size of body and composition is noted, i.e., with decrease in silica there is a corresponding increase in size. Very large bodies covering many square miles and remotely related to laccoliths occur singly and are rare in geographic distribution. Typical laccoliths occur in groups of individual intrusions usually in clusters of four or five each. Such clusters are widely scattered geographically and are rare in occurrence.

Stocks. (See Figs. 5.15, 5.16, 5.17.) An irregularly shaped conical intrusion which abuts sharply against the stratification of the enclosing sediments is classified as a *stock*. Locally the force of the intrusion contorts or fractures the enclosing sedimentary rocks; large blocks of sediments, therefore, are dislocated in position, whereas others appear to have been completely engulfed by the magma. Where this type of intrusion has been exposed by deep erosion the contact of the igneous and enclosing rock is inclined from the horizontal at angles between 45 and 90 degrees and continues downward at such inclinations to unknown depths. The exposed erosion surface is a mile or more in diameter but seldom more than five. By definition the surface exposure of a stock must be less than 40 sq. miles.

Stocks are always of coarsely granular texture either phaneritic or porphyritic and silicic to silicic-intermediate composition. Commonly a zone as much as a mile in width of baked and metamorphosed strata borders the stock. Apparently much of the metamorphism was accomplished by high-temperature aqueous solutions escaping from the igneous rock following emplacement and during the progress of cooling. Such solutions often were rich in copper, gold, silver, lead, zinc, and other metals which were deposited primarily as sulfide minerals in the metamorphosed area. Dikes and, less commonly, sills are associates of the stock, and dikes radiating outward from the stock center are to be expected. The textures, metamorphic zone, and transgressive nature of the igneous rock indicate intrusion at great depths in the order of several miles below the surface, and with very few exceptions they occur in areas of present or former mountains.

Batholiths. Stocks are gradational with the largest of all intrusions, the *batholith* (see Fig. 5.17). This body typically occupies the core or central portions of mountains, is elongate parallel to the range, and extends the entire length of the range often for several hundred miles.[7] (See Fig. 5.18.) Like the stock its surfaces transgress the sediments and are accompanied by much folding in the sediments. The zone of surrounding metamorphism is extensive and intensive, gradation from igneous through metamorphic to sedimentary rock is complete, and in many localities no sharply defined boundaries are recognized. Rare occurrences of igneous-sedimentary rock contact of knifelike sharpness are to be observed also. Commonly large pendents of the roof rock are incorporated within the igneous rock. These show little evidence

[7] A batholith in the southern Andes is 700 miles long and 70 miles wide; one in British Columbia and Alaska is over 1200 miles long and 50 miles wide.

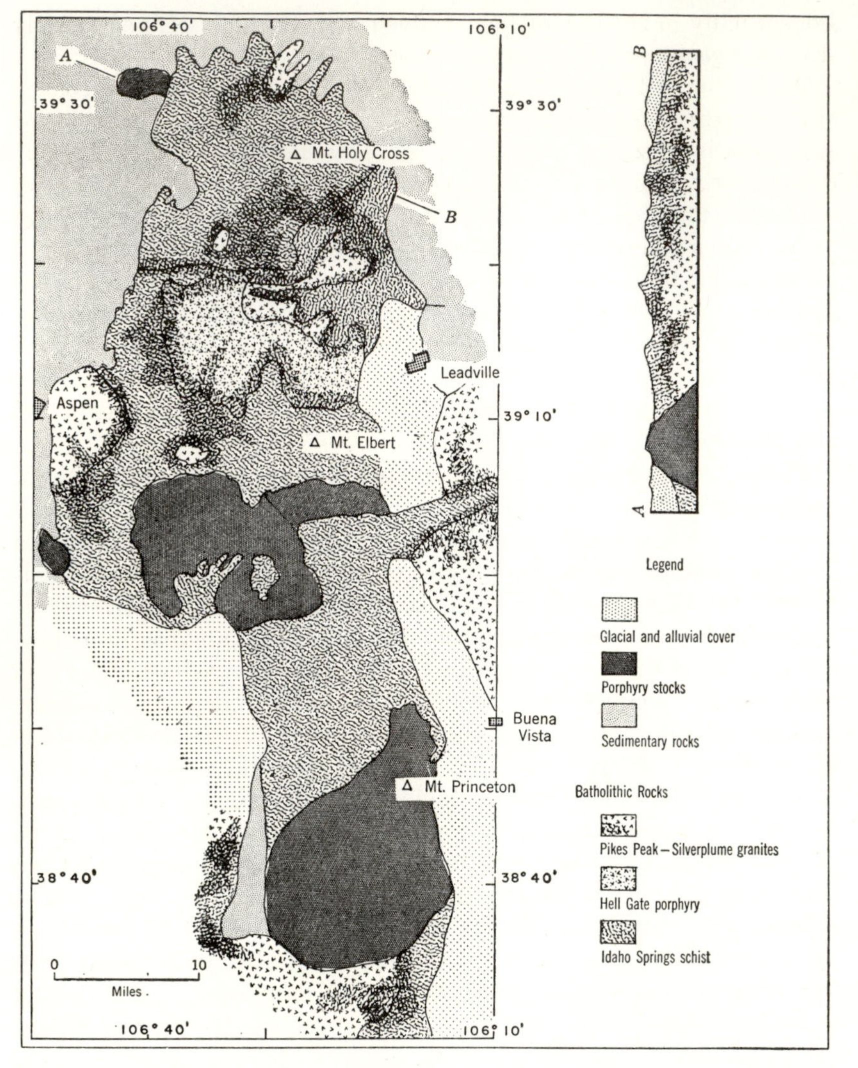

Fig. 5.17 Geologic map and cross section of a part of a batholith in the Sawatch Range, Colorado. Contrast the sharp boundaries of the stock with the irregular and gradational transition of schist into granite typical of the batholith. (Simplified from Stark and Barnes, *Colo. Sci. Soc. Proc.*, 1935.)

of being disturbed in position and maintain the lineation of the mountain structure but have undergone intense alteration and are metamorphosed thoroughly.

Batholiths are exclusively silicic or silicic-intermediate in composition and always coarsely granular in texture. Their composition and

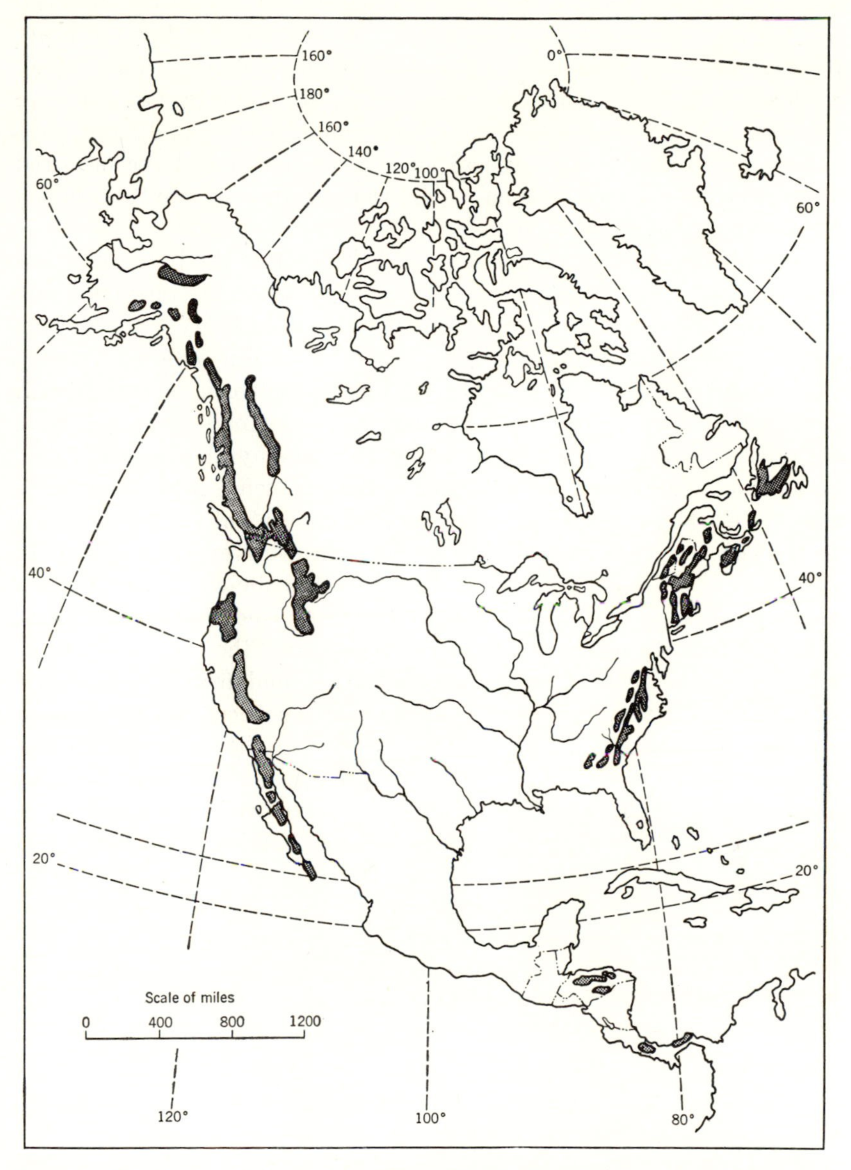

Fig. 5.18 Distribution of clearly defined batholiths in North America intruded during the Paleozoic or Mesozoic eras. Many other batholiths of Precambrian age are known in central and eastern Canada in particular, as well as in mountain areas of the United States, but their boundaries are much more nebulous. The diagram is intended only to supply magnitudes of dimension and localization in known mountain belts.

their enormous size suggests that the engulfed sedimentary material has been assimilated and incorporated into the magma, but the exact process involved is as yet still in the realm of speculation. Much evidence is being assembled which points to the transformation of sedimentary material into granite without a corresponding change from solid to fluid state. Where such is presumed to have occurred distinction between the magma which was intruded and that developed by metamorphism is not possible.

Dikes by the thousands are scattered throughout the batholiths and into the host rock. Stocks and other igneous bodies are offshoots, and the batholith is the source of intensive igneous activity. Restriction of batholithic emplacement to areas where mountain-building activity is in progress suggests a genetic relationship between the two and a deep-seated source for the magmas. Such magmas formerly were considered to invade the sedimentary rocks many thousands of feet below the surface, but there is some field evidence which indicates that the thickness of the roof rocks may have been only a few thousand feet. The position of bottom of the batholith, however, has not been determined, and it is likely that it extends into the earth for as much as 40 miles spreading outward with extremely indefinite boundaries.

Much remains to be understood concerning the origin of these primary igneous bodies, and their study is of fundamental importance to the field of igneous geology. Their primary concern to the engineer, however, centers about two items, namely: (*a*) they are tremendous areas of phaneritic, silicic, or silicic-intermediate rock, and (*b*) they are regions of widespread distribution of metamorphic rocks many of which are extremely variable in physical and chemical properties.

Selected Supplementary Readings

Daly, R. A., *Igneous Rocks and the Depths of the Earth,* McGraw-Hill Book Co., 1933, Chapters VI, VII, and VIII. Describes the shapes of igneous-rock bodies.

Grout, F. F., *Kemp's Handbook of Rocks,* 6th ed., D. Van Nostrand Co., 1940, Chapters I–VI. Detailed description of igneous-rock groups.

Iddings, J. P., *The Problem of Volcanism,* Yale University Press, 1914, Chapters I, VI, VII, and VIII. Observations and concepts concerning the intrusion and extrusion of igneous rocks.

Lahee, F. H., *Field Geology,* 5th ed., McGraw-Hill Book Co., 1952, Chapter 6. Describes the relationships of igneous bodies to the enclosing rocks in the field.

Pirsson, L. V., and Knopf, Adolph, *Rocks and Rock Minerals,* 3rd ed., John Wiley and Sons, 1952, Chapters VI, VII, and VIII. Detailed descriptions of rock-forming minerals and igneous-rock associations.

Spock, L. E., *Guide to the Study of Rocks,* Harper and Bros., 1953, Chapters 4, 5, and 6. Describes the development of igneous rocks, and outlines their characteristic features.

6

Soil-forming processes

WEATHERING

Definition

Rocks and minerals at or near the atmosphere-lithosphere interface are subject to pressures of 1 to 10 atmospheres and temperatures which may range between about $-50°$ and $+50°C$. Within a much more restricted span of temperatures rain water containing dissolved oxygen, carbon dioxide, and nitrogen moves through rock pores and

Soil profile in glacial sediment.

fissures and reacts chemically with some of the individual minerals. These reactions occur because many minerals are not stable when in contact with aqueous solutions at surface temperatures and pressures. Reacting minerals are largely silicates, those typical of igneous and metamorphic rocks.

All rocks at the earth's surface are also subjected to certain purely physical forces which tend to disrupt them into small fragments. As an individual particle is broken progressively into smaller pieces the area exposed for chemical reaction increases enormously, and the rate of decomposition is accelerated. Physical and chemical processes, therefore, are constantly operating to the same end, that of altering the aspect of rock material exposed at the earth's surface. A combination of all of the processes is implied by the term *weathering*.

Weathering is considered to represent the processes which tend to disrupt and decompose rocks and minerals in place. Purely physical weathering is restricted to breakdown without any corresponding chemical change. Where this process dominates the fragmentation is akin to crushing, and the particle size distribution is in accordance with the crushing law. Chemical weathering involves reactions which occur at the site where the rock is being decomposed.

Certain aspects of association lead to confusion of understanding processes grouped as weathering with processes classified as erosion which operate in the same direction. Erosion implies destruction of existing rock and removal of the products from the site of destruction. Transportation is the important aspect of erosion, but this does not involve alteration of the particle being carried away. Weathering tends to disrupt the rock, reduce the particle size, loosen the individual fragment, and lower the specific gravity and hardness of the original minerals by forming new products. All these attributes facilitate erosion.

Despite the interrelationship between weathering and erosion they are not interdependent. Each may proceed without the other and does so at different rates. In the humid tropics local, high, flat areas of very little runoff water may be weathered to depths exceeding 100 feet. Elsewhere in certain stream valleys or along cliffed shorelines erosion proceeds so fast that no soil mantles the surface. Fortunately, over much of the earth's present surface a steady-state condition exists between the rates of erosion and weathering, and a suitable soil cover prevails. This equilibrium results in removal of fresh, partially altered, and completely altered mineral grains from the weathering site. Somewhere else these are deposited in various degrees of admixture as a sediment. Upon burial and solidification this becomes a sedimentary rock, and later it may be uncovered and exposed to a second period of weathering.

Relatively few areas exist where crustal rocks are not subject to tensional and compressional forces developed as a part of the earth's internal behavior. Generally these forces are applied slowly over many thousands or even millions of years and eventually produce rupture of the rock. Oftentimes this rupture is associated with zones where rocks have yielded to the stresses by bending or folding, but more commonly the failure is in a system of parallel fractures called joints. Some joint systems are arranged with generally horizontal surfaces, whereas others are developed at steep angles to the horizontal and tend to break a layer or mass of rock into individual blocks of varying dimension. Oftentimes the spacing of individual joints may be several feet, but in other instances as in the case of slates they are separated by distances of less than an inch.

In some rocks the fracture system is incipient and is not observed until developed by other forces such as frost action. Hence, outcrops of rocks exposed in natural cuts or quarries generally show a well-developed joint system within a few feet of the earth's surface, whereas with continued depth joint spacing increases and becomes less conspicuous. The total of such rock fracturing is extremely important in increasing the surface area for further weathering and in providing openings along which water moves. Were joint systems absent many of the degradational processes would be very greatly reduced in rate of progress.

Physical Weathering

In a preceding paragraph physical weathering was defined as the disruption of rock by processes which are akin to crushing by a mechanical breaker. Thus the disintegration product has a size distribution in accordance with Rosin and Remmler's law of crushing. (See p. 121.) Samples of rock broken under the influence of physical weathering show a distribution in which the size range is large and skewed toward the finer sizes. If the cumulated per cent size frequency distribution of this rubble is graphed on specially ruled paper the points fall along a straight line. Such a relationship is shown by several examples in Fig. 6.1. Samples 1 through 5 represent the distribution of disintegration products of individual boulders of igneous and metamorphic rocks. Those boulders exposed on the soil have been ruptured by weathering in an environment of a generally cool and humid region (Illinois) where freezing is commonplace. Also shown are the distributions of two samples of weathered mica schist from North Carolina where the climate is humid but where freezing is relatively infrequent. The fragmented mica schist is considerably finer in grain size than the disinte-

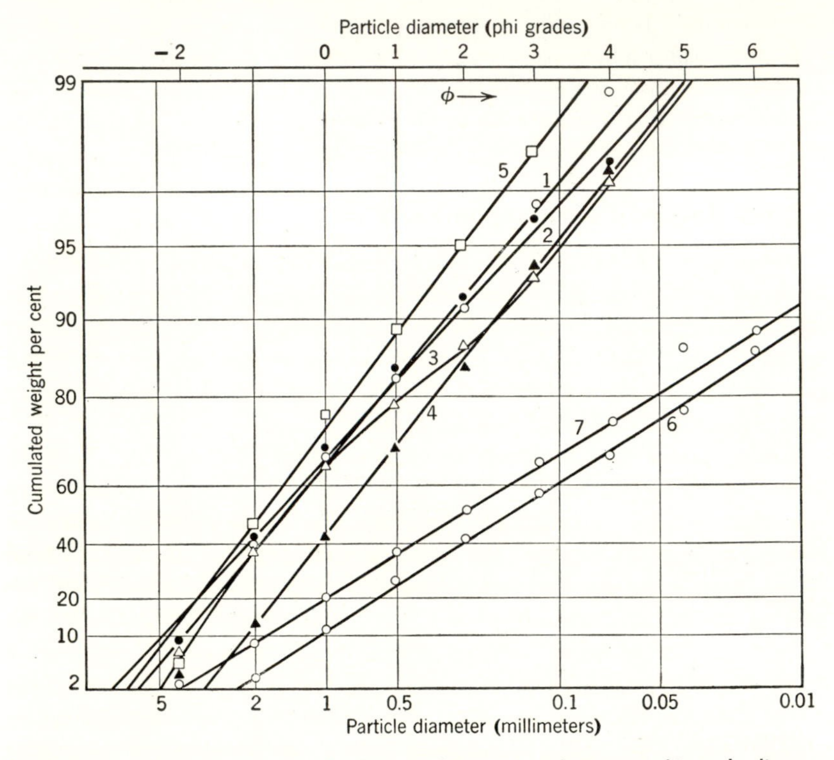

Fig. 6.1 Particle-size frequency distribution of igneous and metamorphic rocks disaggregated by the weathering process. Samples 1–5 are from boulders in glacial sediments near Cary, Illinois. Samples 1, 2, 3, and 5 are granites; sample 4 is a gneiss. Samples 6 and 7 are micaceous, clay soils forming in place near Spruce Pine and Burnstville, North Carolina, respectively. Points are plotted on the special base on which straight lines follow the particle distribution dictated by the crushing law. See also Fig. 5.13. (Reproduced from Krumbein and Tisdel, *Amer. Jour. Sci.*, 1940.)

grated boulders, but despite the presence of large amounts of clay the distribution obeys the crushing law. We can assume, therefore, that such a correspondence demonstrates beyond reasonable doubt that the processes of physical weathering involve application of mechanical forces and that these are not brought about by only the freezing of water but by other processes of disruption as well. Some of the forces involved are recognized as originating within the rock body, whereas others are applied externally, and, hence, a classification can be erected as follows:

1. Stress applied from internal conditions.
   a. Unloading effects (residual stress).
   b. Expansion of the crystal lattice.
2. Stress applied from external conditions.
   a. Growth of ice crystals.
   b. Growth of salt crystals.
   c. Wedgework action and soil stirring by organisms.

Stress Applied from Internal Conditions

Unloading. Fracturing of rock by unloading establishes the framework of rock breakage. An understanding of the underlying causes of unloading involves digression into some aspects of the earth's internal behavior. The point in question centers around the concept that the earth is mobile and that throughout time extensive areas rise or sink with respect to neighboring areas.

Sinking areas do not persist throughout time. Some sink for much longer periods than others, and some receive a greater sedimentary load than others. Eventually, however, the sinking ceases, and new forces are applied against the sedimentary column. Some of these forces have a strong vertical component which tends to elevate the sinking prism, drive it above sea level, and establish conditions for unloading. Other forces have a much stronger horizontal component which compresses the sedimentary block into folds. Altogether these forces have the effect of placing a unit of rock in a press and subjecting it to pressure which may amount to several thousands of atmospheres per square inch.

The loading effect is one of stress-strain relationships. All rocks are elastic to some extent and when lightly loaded are reduced in volume or changed in shape by an amount which is restored when the load is removed. Compression or shear of this magnitude is typical of the passage of earthquake waves through the earth shells to emerge at the surface. Thus the extent of distortion of the rock by wave energy is said to be within the elastic limit, and when the distorting force is removed a unit mass of rock returns to its former shape and volume. However, field evidence of folding and fracturing of rocks indicates that forces exist which are capable of causing rocks to yield beyond the elastic limit and produce permanent deformation. These conditions can be simulated in the laboratory by loading a cylinder of rock in a press. In such a procedure an interpretation of rock behavior is based upon a stress-strain curve. This is a plot of percentage of shortening or elongation (conventional strain) as abscissas and the differential stress as ordinates. An example (Fig. 6.2*a*) shows stress-strain curves for a dolomite under various confining pressures. Also shown are curves for the ductility (Fig. 6.2*b*) which is defined as

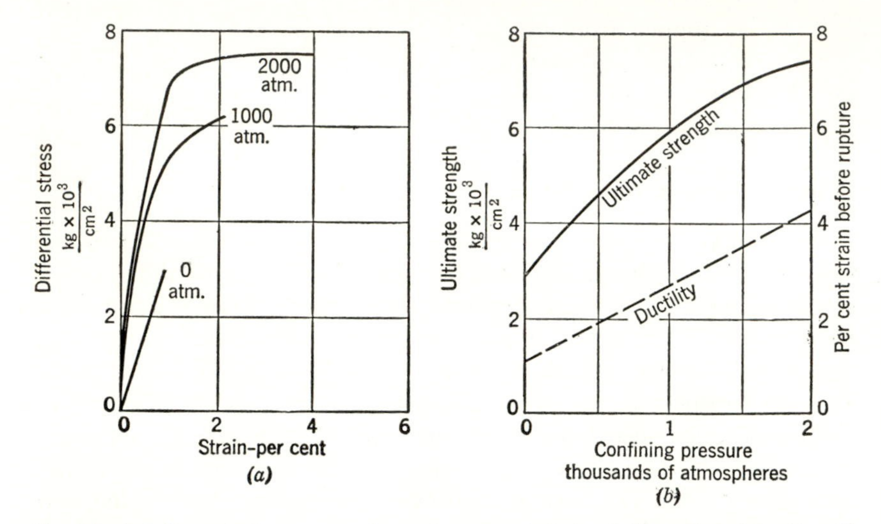

Fig. 6.2 (*a*) Stress-strain curve or change in per cent strain of Clear Fork dolomite with changing values of differential stress and under different loads of confining pressure. (*b*) Increase in ultimate strength and ductility of Clear Fork dolomite as confining pressure is increased. Note that rupture under atmospheric pressure occurs at less than half the differential stress than when the rock is confined by 2000 atmospheres. Note also that a rock becomes increasingly brittle (i.e., less ductile) as confining pressure is reduced. (From Handin and Hager, *Bull. Am. Assoc. Petrol. Geologists*, 1957.)

the total percentage of deformation before rupture, and for ultimate strength which is the maximum stress applied up to the point of rupture. These experiments show that as a rock is placed under increasingly greater amounts of confining pressures the differential stress required to produce the same amount of strain is greater, i.e., the ultimate strength of the rock is increased.

Applying the above information to rock buried in a sinking basin we can reason that as loading progresses the ultimate strength is increased and the percentage of strain measurable before rupture (ductility) must be increased as well. At considerable depths of burial ductility and ultimate strength are higher than at lesser depths. Thus as burial advances rocks tend to build up an internal stress in an attempt to reach equilibrium with the stress applied by loading. Let us assume that a unit volume of rock has been loaded gradually in a sinking basin but that the differential stress applied is insufficient to cause the rock to rupture. As the ultimate strength of the rock is increased under the greater confining pressure internal stress is increased also. Assume further that the sinking basin becomes a rising block and as erosion strips away the overburden unloading of the unit volume of rock proceeds. Unloading lowers the confining pressures and, hence, lowers the ultimate strength, and the internal stress is

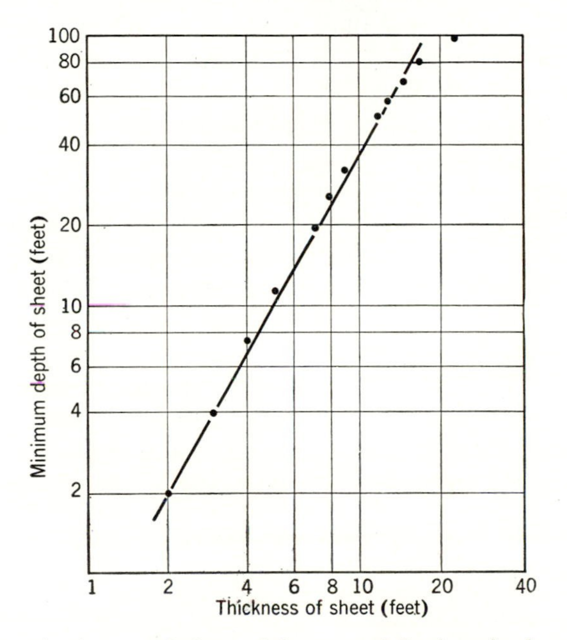

Fig. 6.3 Relationship between thickness of sheeting and depth to the sheet below surface in some granites in Massachusetts. The tendency to split in thin sheets near the surface is attributed to unloading. See also Fig. 2.15*a*. (Data from Jahns, *Jour. Geol.*, 1943.)

relieved by expansion of the unit volume of rock concerned. The net effect is to cause the rock to fail by tensional stress along a series of fractures many of which approximately parallel the earth's surface. Sedimentary rocks tend to part along the layering, and well-defined separation is typical in outcrop. Homogeneous rocks like granite break along large arcuate fractures essentially horizontal in position which are identified by the term *"sheeting."* (See Fig. 2.15*a*.)

At the earth's surface sheeting is accentuated by other processes of physical weathering, but the primary tendency is to develop progressively thinner sheets as the surface is approached and the load is released. This relationship is illustrated by Fig. 6.3 in which the thickness of an individual sheet of granite is plotted against its depth below the surface. Data for this illustration are from an area of granitic rock in Massachusetts where glacial erosion has stripped off the pre-glacial weathered zone and exposed unaltered rock. The thickness of sheet as against depth below ground surface is plotted as the logarithm of each to develop a linear relationship which holds for a depth of approximately 80 feet. Below that the thickness of the individual sheets becomes so great that the aspect of sheeting no longer is recognized. In the case illustrated by Fig. 6.3 the depth at which a sheet of given thickness is found varies approximately as the 1.5 power of the sheet thickness, but this value should not be considered as universally acceptable. Rather, the nature of the relationship is significant.

The effect of sheeting in homogeneous rocks is to break up large masses into units ranging between 2 and 20 feet in thickness. Blocks broken from individual sheets yield fragments of boulder or larger dimension but sufficiently small to be acted upon by other agents of weathering.

Expansion of crystal lattice. Formerly much emphasis was placed on expansion and contraction of rock masses due to alternate heating and cooling during the day and night. Recent experiments, however, appear to demonstrate that the forces generated are insufficient to cause the rock to fracture.

Stress Applied from External Conditions

Growth of ice crystals. In latitudes or altitudes where freezing of water occurs this process is very important in physical disruption of rocks. Joints developed through unloading or externally applied stresses are ideal channels for water entering the ground from the surface.

Water freezing in an open crack against solid rock walls appears to behave like a closed system in order to exert the pressure of expansion against the rock walls. This may be due to the super-cooling of water to temperatures of as low as $-5°C$ before rapid freezing occurs. The upper water surface may suddenly freeze over and close the entire system. When this occurs and freezing continues the increase of 9 per cent volume of this form of ice (Ice I) over water causes a pressure rise in the entire system. There is also the liberation of latent heat, approximately 80 calories per gram of water. As this heat is conducted away the system of water and ice is in equilibrium for the pressure which prevails. No additional freezing can occur until there is a further decline in temperature. This concept is illustrated in Fig. 6.4

in which is shown the relationship between freezing point and pressure necessary to maintain both the liquid and solid phases. For purposes of the discussion in question let us assume that the temperature has fallen below 0°C and that the surface water in the joints freezes. The system water and ice is now closed and in equilibrium at approximately one atmosphere pressure. Rock temperatures continue to fall, and additional ice begins to crystallize. The water is now under compres-

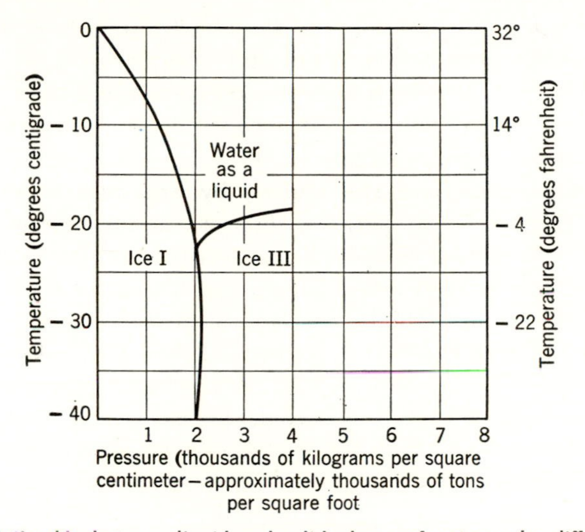

Fig. 6.4 Relationship between liquid and solid phases of water under different pressure and temperature conditions. Ice III, which is more dense than Ice I, is unknown in nature. (From Johnston, *Jour. Franklin Inst.*, 1917.)

sion, and the pressure in the system will build up to values expressed by the line between ordinary ice I and the liquid. If liquid continues to be present and freezing continues a pressure as high as 2115 kg per sq. cm (approximately 2,100 tons per sq. foot) can be exerted against the rock walls. This value represents the maximum theoretical amount because ice I occupies a greater volume than the liquid. In the laboratory under higher pressures, and temperatures of −22°C, ice I is no longer stable and alters to ice III. The latter has a higher density than liquid water and cannot exert a force against the enclosing walls. Ice III, however, does not develop in nature; hence we are concerned only with the relationships between ordinary ice and liquid water.

According to theory freezing of water under ordinary atmospheric conditions can exert pressures against the walls of a joint sufficient to rupture the rock or to wedge the fragment loose from its parent mass. This process is particularly important where the temperature crosses the freezing point often. Such is the case in the high mountains during the summer months and in the mid-latitudes during the winter season. These are the localities where frost action may dominate the weathering process.

Growth of salt crystals. In areas where evaporation exceeds precipitation, ground water in the soil moves to the surface by virtue of the capillarity gradient established. Surface or near-surface loss of water results in precipitation of the dissolved salts in the "soil." The restriction of this phenomenon is not only geographic, but also appears to be applied only in reduction in grain size in soils already reduced to small particle dimensions. The process is analogous to behavior of frost action in loose aggregate soils.

Wedgework action and soil stirring by organisms. Many references cite the importance of the wedgework action of root growth in crevices of rocks. Splitting of rocks is attributed to the continued expansion of the root system in joints. Although some prying action particularly in lifting slabs may occur the primary action appears to be one which permits water to permeate the joint. Generalized comparisons of rock breakage in heavily forested regions with that where no forests exist oftentimes show that the latter displays more rock rubble on the surface.

With the progress of other forms of weathering and reduction in the size of individual rock particles organisms in the weathering process become more important (Table 6.1). In humid areas an abundant fauna and flora inhabit the soil. Some of these organisms tend to keep it constantly stirred so that unaltered mineral particles are exposed to chemical attack. In other instances breakage and attrition occur as soil particles move through digestive tracts of worms. Organic acids released by bacterial as well as animal life increase the H ion concentration of the soil water. Aridity and low temperatures tend to reduce the number and activity of soil organisms; hence their importance is much less under such rigorous climatic conditions. The importance of humidity in the development of soil organisms is indicated by the relationship between the nitrogen content of soil and rainfall. (See Fig. 6.5.) Much of this nitrogen is the product of the activity of organisms and is combined in compounds many of which affect the acidity of the soil and hasten soil-forming processes.

Most organisms carry on their life activities near the surface of the soil and affect the small inorganic particles far more than the large particles. Their importance is, therefore, more in the nature of establishing certain biochemical conditions in the "A" and "B" horizons

Table 6.1. Important Groups of Organisms Commonly Present in Soils*

A. Animals
1. Subsisting largely on plant materials: earthworms, insects (ants, beetles, grubs, etc.), millipedes, mites, rodents, snails, woodlice.
2. Largely predatory: centipedes, insects (ants, beetles, etc.), moles, nematodes, protozoa, rotifers.

B. Plants
1. Algae (green, blue-green, diatoms), actinomyces, bacteria (aerobic and anaerobic), fungi (mushrooms, yeasts, molds).
2. Roots of higher plants.

* Generalized from T. L. Lyon and H. O. Buckman, *The Nature and Properties of Soils,* 4th ed., The Macmillan Co., 1943, p. 95.

of the soil than in providing new soil material. This is just one of the reasons why the "A" horizon is so well developed in podzol, prairie, and chernozem soils.

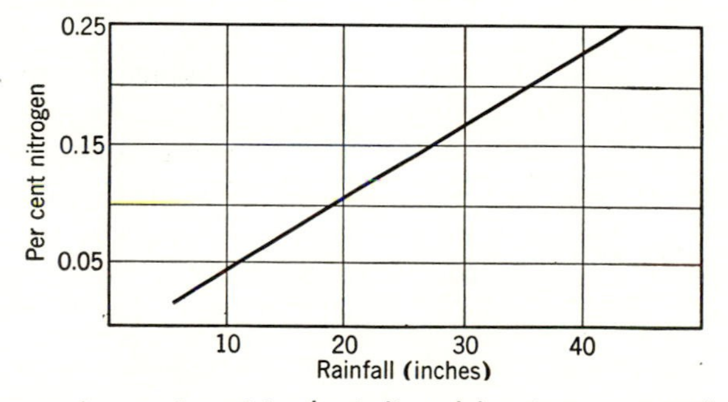

Fig. 6.5 Increase in organic activity (as indicated by nitrogen content) in soils with increase in rainfall. (Generalized from Jenny in Trask, *Applied Sedimentation,* John Wiley and Sons, Inc., 1950.)

Chemical Weathering

Type of Reactions

By far the most important group of processes which produce the soil result from chemical reaction between rock minerals and rain water moving downward from the atmosphere-lithosphere interface. Rain water is a solution in which nitrogen, oxygen, and carbon dioxide are important constituents. Carbon dioxide is particularly important because it tends to ionize and increase the H ion content above that present in pure water. Dissolved gases in rain water show a much different concentration than in the normal atmosphere (Table 6.2).

Table 6.2. Composition of Important Gases in the Atmosphere and Dissolved in Rain Water

	Nonvariable Components of Atmosphere* (Parts per Million)	Dissolved Gas in Rain Water† (Parts per Million, 20° C)
Nitrogen	±780,084	±636,900
Oxygen	±209,460	±341,700
Argon	±9,340	not determined
Carbon dioxide	±380	±2,140

* Data from E. Glueckauf, "The Composition of Atmospheric Air," *Compendium of Meteorology,* Amer. Meteorological Soc., 1951, p. 6.

† F. W. Clarke, "Data of Geochemistry," *U. S. Geological Survey Bull.* 770, 1924, p. 54.

Nitrogen is approximately 82 per cent as abundant in rain water, but oxygen is 163 per cent and carbon dioxide 700 per cent of the proportion in the normal atmosphere. Other important substances present in rain water are ammonia, nitric acid, sulfuric acid, sodium chloride, and chlorine. These are minor in volume but are so chemically reactive that they must be considered as effective soil-forming agents.

Reactions typical of the weathering process can be classified as follows:

a. Solution.
b. Colloid formation.
c. Base exchange.
d. Hydration and hydrolysis.
e. Oxidation and reduction.

Of these it is difficult to select the most important ones, but in some localities and on certain minerals one of these reactions may be favored. All are operating at the same time, but some proceed faster than others, and certain ones are more important in the alteration of some minerals. The products of these reactions are ions in groundwater solution and new minerals typical of sediments. Many of these ions are reprecipitated in the same, or new, environments when their solubility products are exceeded. Such minerals are also a part of those generally associated with sedimentary rocks.

Solution. All solutions concerned in weathering are aqueous and dilute. Also most of the solutes are highly ionized, but certain exceptions such as quartz may remain largely in the molecular state. The size of the particles which make up the solutes range between ions with diameters of 1 Ångstrom unit and loose molecular structures (colloids) of clay minerals about 20 Ångstroms across. Probably these larger particles are not strictly in true solution because the individual atoms may still maintain a geometric arrangement in the colloidal particle. The outer borders of the particle, however, are considered very irregular in arrangement, and some ions may be leaving the lattice. In the true solution there is scarcely any regular arrangement of ions, and they are constantly shifting position to maintain a general state of random distribution. When ionic concentrations reach the value of the solubility product for a given temperature and some precipitation is about to occur the un-ionized part may begin to stabilize into an orderly atomic arrangement of the crystal which is about to form.

Rain water moving downward from the atmosphere-lithosphere boundary is undersaturated with respect to most of the ions available in the common minerals. It is slightly acidic (10^{-6} mole per liter of H ion) a condition which tends to increase the solubility products of a number of ions. Certain ions, therefore, are more soluble under slightly acid conditions than under basic conditions. Other ions are more soluble under basic conditions and are removed much more slowly under acid conditions. (See Fig. 6.6.)

Solution is a process which centers around certain electrical properties of water. Pure water has a dielectric constant of about 80, whereas air has a value of about 1. This means that when water is placed between two plates of opposite charge the tendency to neutralize or reduce the applied field between the plates is 80 times that when air separates the plates. This is explained by the dipole structure of a water molecule which is such that the two H atoms are on the same side of the molecule. One side of the molecule, therefore, has a slight excess of negative charge from the O atom. An electrostatic field set up by two oppositely charged plates in water has a tendency to cause the water molecules to become partially oriented and to reduce the applied field between the plates.

The film of water surrounding the mineral establishes the physical condition where the space lattice of the crystal begins to undergo destruction. Sometimes this may be accomplished by water molecules actually entering the tunnel openings between atoms in the lattice. H or OH ions may enter and attract loosely held ions from the lattice. When water enters the openings of some crystals it appears to spread the lattice and separate it into units which become solute particles. In other cases outer portions of the lattice tend to split away. All of these processes are classified as solution and ultimately may result in causing a crystal to be completely dissolved.

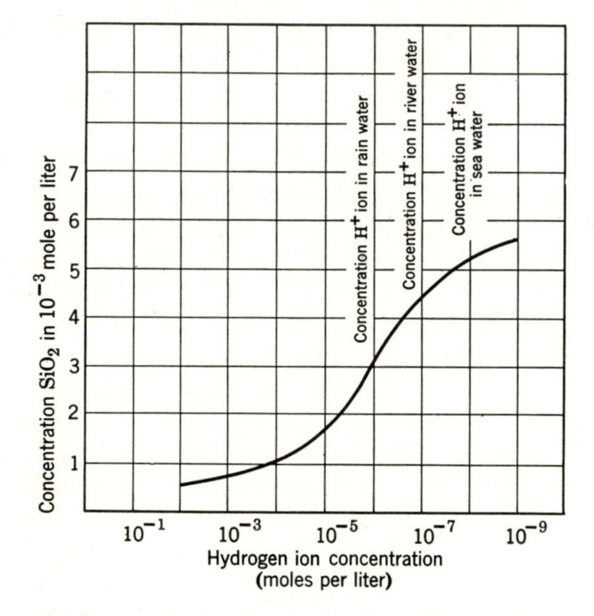

Fig. 6.6 Generalized curve of solubility of amorphous silica in waters of variable H ion concentration. (Modified from Mason, *Principles of Geochemistry*, 1st ed., John Wiley and Sons, Inc., 1952. Originally from Correns, *Einführung in die Mineralogie*, Springer-Verlag.)

Colloid formation. The mechanism which was described in the explanation of solution and destruction of the crystal oftentimes results in splitting off particles whose dimensions are large enough to possess at least a partial lattice structure. Such are colloidal particles.

The outer borders of large colloidal particles have many atoms with only partially satisfied electrical charges. These form active "handholds" for other ions which tend to neutralize these charges. Polarized water molecules, H and other metal ions, and OH ions fasten to the colloidal particles.

Attachment of ions to the surface of the colloidal particle appears to be of a generally loose form. Ions of opposite charges are held reasonably close to atoms in the lattice, but the former may be easily removed by some other ion with stronger electrical field. Other ions, however, penetrate deep within the crystal lattice and occupy holes within the lattice or may actually replace atoms in the lattice by virtue of their stronger electrical charge. Base exchange typical of the clay minerals appears to be an example of the type of loose attachment inasmuch as ions may replace one another with ease. The process of colloid development is dependent upon other reactions which disrupt the space lattice. Among these hydrolysis and oxidation and reduction are of primary importance.

Silica, iron oxide, and hydrous aluminum silicates (clay minerals) are the substances which tend to form colloidal particles during rock weathering. These materials also remain as soil residues because of their very low solubility. Nevertheless, dispersed particles of these substances move downward slowly as the soil forms and particularly during the late stages of soil development. Silica and iron oxide tend to form stable suspensions and are not readily precipitated by cations such as K, Na, or Ca which are present in the soil waters. The clay minerals, however, may have so many unsatisfied electrical charges on their surfaces that they tend to flocculate, particularly when many of the charges are satisfied by the loosely attached (adsorped) cations. During the late stages of soil development when the downward-percolating waters contain few cations some of the cations held on the surface of the colloidal particle tend to leave and become free ions. The clay particles then appear to deflocculate and move downward through the pores of the soil suspended in the water. At some lower depth where the concentration of cations is high the clay minerals are again precipitated as a zone. Generally this is near the base of the "B" zone, and as the process continues and the soil becomes old this is the zone of "hardpan." Hardpan is recognized as relatively impervious to water and upon drying becomes a hard "cemented" mass.

Base exchange. Considered relatively, colloidal particles are large in comparison with ions. Such large particles also are molecular arrangements in which many only partially satisfied electrical bonds hold the surface atoms to the molecular structure. These unsatisfied bonds offer "handholds" to ions of opposite charge, and they become attached to the surface of the colloidal particle. Attachments of this kind can be regarded as a form of adsorption. Still another and stronger form of adsorption occurs when a strongly charged ion may displace an ion of similar size but of lower electrical charge in the molecular, or crystal, structure. Both processes are known as base exchange. Exchange of bases is a reversible reaction, the direction of which is controlled primarily by the number of ions available for replacement and the number of replacing ions.

Hydration and hydrolysis. Hydration is an effect whereby positive or negative ions tend to hold the polarized side of water molecules to form a complex of the ion and water molecules. Positive ions such as Fe^{++} tend to hold the oxygen side of the water molecules in a complex ion $Fe(OH_2)_6^{++}$. Such forces holding the water molecules may be very strong and a fixed number may remain attached to a complex compound to form a crystal. In this structure the water molecules tend to occupy openings of the space lattice. Water so held is known as *water of crystallization* and in some minerals is released only at temperatures above the boiling point of water.

Silicates with ferrous iron as an important constituent in the lattice undergo hydration during the weathering process. Apparently the Fe^{++} ion leaves the lattice during solution and becomes surrounded with water molecules to make complexes such as $Fe(OH_2)_6^{++}$ or undergoes oxidation. The oxidized iron atom is Fe^{+++} which gathers OH ions from the water to form the stable hydroxide $Fe(OH)_3$.

One of the characteristics of iron hydroxide is its ability to change to a hydrated form on aging in the weathering environment. The final compound is $Fe_2O_3{\cdot}nH_2O$ (limonite) in which n commonly has a value of two. Silica and alumina appear to behave in a similar manner developing hydrates $SiO_2{\cdot}nH_2O$, and $Al_2O_3{\cdot}nH_2O$ respectively. On further aging some of the water of crystallization is lost, and in both iron and silica the anhydrous form (Fe_2O_3 and SiO_2) may develop eventually.

Hydrolysis as a weathering mechanism appears to be confined to the destruction of complex silicates typical of the igneous and metamorphic rocks. One concept of the mechanism is that H ions from the adsorped surface film of water move through the tunnel openings in the crystal lattice by virtue of their small diameter. The small size of the H ion gives it a stronger electrical field than larger ions. Therefore, when the H ion enters into the opening between O atoms occupied by large metal ions the H ion remains. The larger positive ion is forced out of the lattice because the electrical charge of the lattice must remain neutral and an excess positive charge cannot be tolerated. The effect is to change the complex silicates into a form which can be interpreted as resembling a variety of silicic acid. Hydrolysis of orthoclase for example, would undergo some sort of a mechanism as follows:

$$\underset{\text{(Orthoclase)}}{KAlSi_3O_8} + \underset{\text{(hydrogen ion from water)}}{H^+} \longrightarrow \underset{\text{(silicic acid-like structure)}}{HAlSi_3O_8} + \underset{\text{(ion in solution)}}{K^+} \quad (1)$$

This silicic acid-like molecule is unstable in the presence of water and undergoes further hydrolysis with rearrangement of the structure. The layered structure of clay minerals develops, hydroxyl enters the lattice, and the H ion is expelled.

$$\underset{\text{(silicic acid-like structure)}}{HAlSi_3O_8} + OH^- \xrightarrow{\text{(rearrangement of lattice)}} \underset{\text{(clay mineral—kaolinite in colloidal form)}}{(OH)_8Al_4Si_4O_{10}} + SiO_2 + \underset{\text{(ion or colloid form)}}{H^+} \quad (2)$$

Oxidation and reduction reactions. In the weathering process iron is the substance which primarily undergoes oxidation or reduction. Ferrous iron generally exceeds in weight per cent the ferric iron in most silicate minerals of igneous and metamorphic rocks. Upon weathering the Fe^{++} ion in the mineral loses electrons and is oxidized to the ferric state. During the process the Fe^{+++} ion becomes grouped with oxygens to form a stable anhydrous iron oxide (Fe_2O_3) or gathers OH ions to form the somewhat less stable hydroxide $Fe(OH)_3$. As has been stated above the latter undergoes hydration to form a stable hydrate $Fe_2O_3 \cdot nH_2O$.

Oxidation of iron in silicate minerals appears to proceed along with hydrolysis of the silicate structure. An example is the case of weathering of the mineral biotite.

$$\underset{\text{(biotite)}}{(OH)_2K(Mg, Fe)_3AlSi_3O_{10}} + \underset{\text{(ion)}}{H^+ - e^-} \longrightarrow \underset{\text{(silicic acid-like structure)}}{H_5AlSi_3O_{10}} + \underset{\text{(ions)}}{Fe^{+++} + Mg^{++} + K^+ + OH^-} \quad (3)$$

$$Fe^{+++} + 3OH^- \rightleftharpoons Fe(OH)_3 \rightleftharpoons \underset{\text{(stable hydrate limonite)}}{\tfrac{1}{2}Fe_2O_3 \cdot nH_2O} \quad (4)$$

$$\underset{\text{(silicic acid-like structure)}}{H_5AlSi_3O_{10}} + OH^- \xrightarrow{\text{(rearrangement of lattice)}} \underset{\text{(clay mineral—kaolinite)}}{(OH)_8Al_4Si_4O_{10}} + SiO_2 + \underset{\text{(ion or colloid)}}{H^+} \quad (5)$$

Reduction of ferric iron also occurs under certain special environmental conditions. These are under circumstances where an excess of partially decaying vegetation is present in the upper part of the soil ("A" horizon) or may be incorporated within a sediment during deposition. Carbonaceous material reduces the iron from the ferric to the ferrous state. This reaction results in removal of iron from soils inasmuch as the Fe^{++} ion is more soluble and is carried away in solution.

Oxidation and reduction of iron produces pronounced color differences in rocks and sediments. Oxidation develops the Fe^{+++} ion which in hydroxides, hydrates, or oxides has a brown to vermilion red color. All reddish-colored sediments or rocks attain this color by virtue of the presence of the Fe^{+++} ion. Ferrous ion, however, tends to develop compounds of characteristic gray, green, or even blue tints, and rocks high in this ion are characterized by such colors.

Iron-bearing sulfides are particularly sensitive to oxidation. The sulfide ion is oxidized to sulfate, and Fe^{++} ion is released according to the following reaction:

$$FeS_2 + O_2 \rightarrow SO_4^{=} + Fe^{+++} + e^- \tag{6}$$

$$Fe^{++} \rightleftharpoons Fe^{+++} + e^- \tag{7}$$

Role of Carbon Dioxide

Several of the reactions typical of weathering are dependent partially upon the amount of CO_2 dissolved in the downward percolating ground water. Reference to Fig. 6.6 shows that the H ion concentration of rain water is 10^{-6} mole per liter, an acidity developed largely by the reaction between CO_2 and water. A high H ion concentration (low pH) favors most reactions associated with the early stages of weathering when minerals of igneous and metamorphic rocks are undergoing decomposition.

The presence of CO_2 is important also in the solution of carbonate minerals such as calcite or dolomite. These minerals make up the great bulk of rocks classified as limestone or dolomite; hence weathering such rocks involves the reaction of calcite or dolomite with carbon dioxide. In the case of calcite this appears to be due to the following ionic equilibria (see also p. 340):

$$\underset{\text{(calcite—solid)}}{CaCO_3} \rightleftharpoons \underset{\text{(solution)}}{CaCO_3} \rightleftharpoons \underset{\text{(ion)}}{Ca^{++}} + \underset{\text{(ion)}}{CO_3^{=}} \tag{8}$$

$$H_2O + CO_2 \rightleftharpoons H_2CO_3 \rightleftharpoons H^+ + HCO_3^- \tag{9}$$

$$HCO_3^- \rightleftharpoons H^+ + CO_3^{=} \tag{10}$$

The effect of the equilibrium is to depress the carbonate ion. But to maintain the solubility product of $CaCO_3$ more calcium must go into solution to supply the deficiency. In the case of a closed system solution of calcium carbonate would end when enough Ca ions accumulated to effectively satisfy the solubility product of equilibrium 8. Weathering, however, is an open system, and, hence, all the ionic products tend to leave the locality as ground water moves downward. The equilibria are thus not satisfied, and continued solution of calcite progresses so long as there is movement of ground water.

Stability of Minerals to Weathering

The chemical reactions mentioned above affect different minerals at diverse rates of speed. Under some conditions oxidation and reduction will proceed more rapidly than solution, but in the case of the common carbonate minerals the opposite is true. Among silicate minerals there has been recognized a stability series in which some are more unstable than others in the weathering environment. As a general rule silicates in which the ratio of metal oxides to silica is high tend to decompose more rapidly than those in which the ratio is low. For minerals of the igneous rocks an order of stability has been described as shown in Table 6.3. The reader will note that the order

Table 6.3. Order of Weathering Stability of Minerals of the Igneous Rocks*

Increasing Stability to Weathering ↓		
	Olivine	Calcic plagioclase
	Pyroxene	↓
	Amphibole	Sodic plagioclase
	Biotite	↓
	Potash feldspar	
	Muscovite	
	Quartz	

* From S. S. Goldich, "A Study in Rock Weathering," *Jour. of Geology*, Vol. 46, 1938, p. 17.

of increasing resistance to weathering is the same as the order of crystallization from the melt. The primary control, however, is the ratio of metal oxides to silica. Olivine in which the ratio is high is very sensitive to weathering and decomposes rapidly, whereas at the other extreme is quartz which is one of the minerals tending to remain longest in the soil.

Minerals of the metamorphic rocks other than those which are shared in common with igneous rocks are generally rather resistant to weathering. Garnet, talc, and serpentine are perhaps even more stable than quartz, whereas chlorite is probably as stable as potash feldspar. In the case of metamorphic rocks their breakdown is controlled to a large degree by the percentage of the unstable igneous minerals which they contain. Of course, such metamorphic rocks as marble behave like limestone or dolomite although marble tends to be somewhat more resistant due to its lower porosity.

Table 6.4. Weathering Products of Common Minerals

Original Mineral	Products of Weathering Residual Mineral Products	Released Ions
Amphibole	Clay minerals, limonite, hematite.	K, Ca, Mg, Na
Biotite	Clay minerals, limonite, hematite.	K, Mg
Calcite	None for mineral; from limestones some quartz, clay minerals, and hematite as impurities.	Ca, HCO_3
Chlorite	Clay minerals.	Mg, Fe^{++}
Clay minerals	Under high moisture in tropics may develop bauxite ($Al_2O_3 \cdot nH_2O$).	SiO_2 (colloidal?)
Dolomite	None for mineral, from dolomite rock some quartz, clay minerals, and hematite as impurities.	Ca, Mg, HCO_3
Garnet	Garnet.	. . .
Gypsum	None for mineral; from gypsiferous shale or sandstone, clay minerals and quartz.	Ca, SO_4
Hematite	Hematite.	. . .
Limonite	Limonite.	. . .
Muscovite	Muscovite tends to remain, eventually alters to clay minerals and quartz.	K, SiO_2 (colloidal?)
Olivine	Clay minerals, limonite, hematite.	Mg, Fe^{++}
Orthoclase and microcline	Clay minerals, quartz.	K, SiO_2 (colloidal?)
Plagioclase	Clay minerals.	Na, Ca
Pyroxene	Clay minerals, limonite, hematite.	Mg, Fe^{++}, Ca
Pyrite and marcasite	Limonite, hematite.	Fe^{++}, SO_4
Quartz	Quartz.	Some SiO_2 (colloidal?)
Serpentine	Serpentine.	. . .
Talc	Talc.	. . .

With the exception of sulfides, carbonates, and sulfates the important and common sedimentary minerals are stable. These include quartz, clay minerals, and iron oxides which are extremely resistant to weathering. They are also the residual decompositional by-products of the weathering of most minerals as shown in Table 6.4. The small number of residual products tends to develop mineralogic and chemical homogeneity in highly weathered soils.

Environments of Clay Mineral Development[1]

In the United States clay minerals of the major groups appear to be reasonably stable in any of the climates which prevail inasmuch as recent information indicates that the predominant clay mineral present

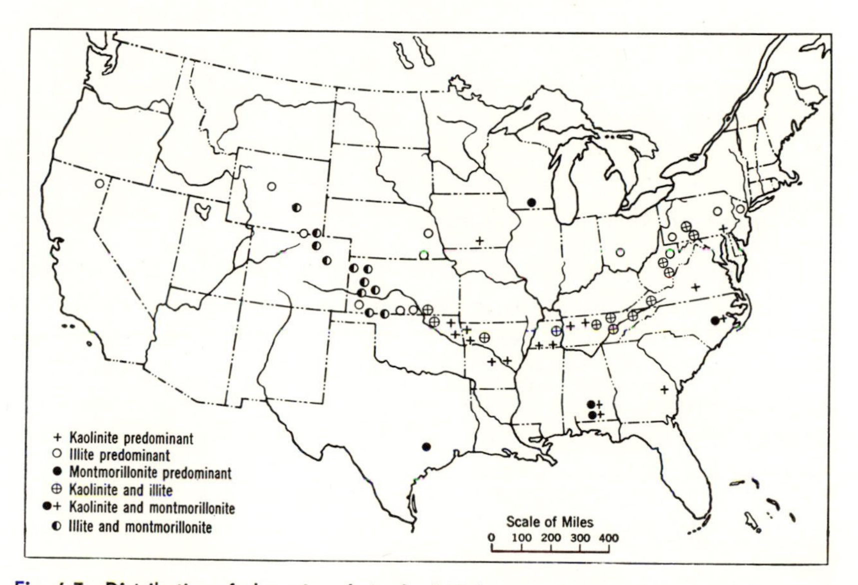

Fig. 6.7 Distribution of clay minerals in the "A" horizon of selected residual soils. Note the general restriction of kaolinite to the more moist regions and montmorillonite to the drier regions. (From Van Houten, *Amer. Jour. Sci.*, 1953.)

in the parent rock is also the predominant clay mineral in the "A" soil horizon. Illite appears to be dominant in marine shales. In fresh-water sediments kaolinite and illite are prominent, whereas montmorillonite is abundant only in relatively recent sediments (particularly volcanic ejecta).

Some general relationships between climate and clay mineral stability appear to prevail. Figure 6.7 shows the geographic distribution of the preponderant clay minerals in the "A" horizon of residual soils in a number of sampled localities. Illite is to be found widely distrib-

[1] The reader is referred to R. E. Grim, *Clay Mineralogy*, McGraw-Hill Book Co., 1953, Chapter 13.

uted. Kaolinite is restricted to zones of high rainfall, and to a lesser degree montmorillonite is associated with soils of dry climates.

The development of kaolinite seems to be favored by acid conditions (low *p*H) and soil permeabilities proper for good drainage and active leaching. Alkalis (K, Na) and alkaline earths (Ca, Mg) must be removed from the site as rapidly as they are liberated from the decomposing minerals inasmuch as calcium in particular appears to inhibit the formation of kaolinite. Crystallization of illite, however, is favored by the presence of potassium and alkalis in general. Montmorillonite develops in the presence of alkalis, but magnesium appears to be an essential element.

The Soil Profile

Progress of Rock Weathering

Destruction of rocks by weathering proceeds through a general order of events. When deeply buried rocks begin to approach the lithosphere-atmosphere interface by reason of removal of overlying rocks they gradually move into the realm where oxidizing conditions prevail over reducing, ground waters are unsaturated, and confining pressures are reduced. Initial manifestations of weathering are to a large degree dominated by internal stresses of unloading and development or enlargement of joints. Nearer the earth's surface chemical weathering begins to dominate, and as mineral particles are reduced in diameter more surface area is exposed for chemical attack.

With such a concept in mind the weathering of a given rock type may be considered as moving through a flow sheet (Fig. 6.8) in which at first physical processes dominate but later these are supplanted by chemical reactions. There are obvious weaknesses in this analysis in that both processes proceed together, but the flow sheet device permits visualization of the progress of weathering.

Intensity of Processes in Soil Horizons

Oxidation and hydrolysis of silicates appears to be most important in the early stages of weathering as the effects of these reactions are observed near the base of the "C" horizon. Near the top of the "C" horizon hydrolysis is more important than oxidation. This should be interpreted as indicating that oxidation reactions are completed largely before hydrolysis of the silicate minerals reaches its peak (see Figs. 2.13 and 6.9). If the rock is a limestone, dolomite, or marble these

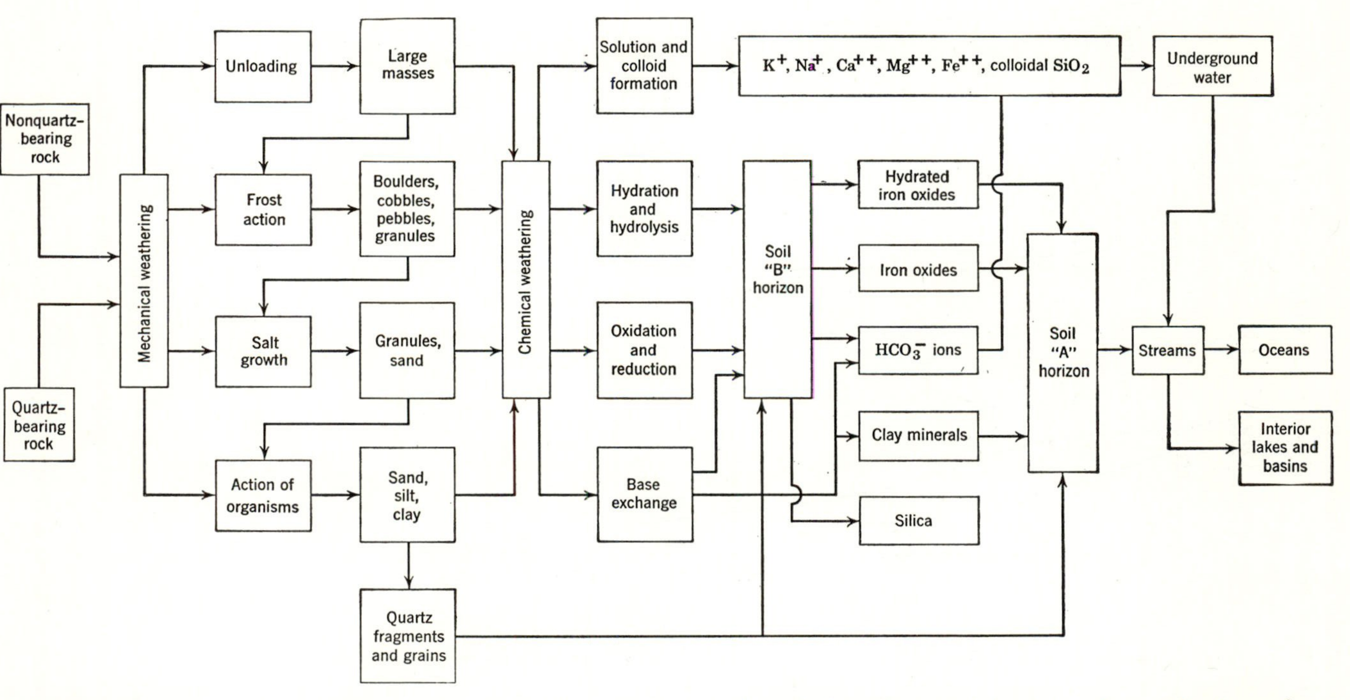

Fig. 6.8 Flow sheet of the weathering process showing the general routes of weathering of nonquartz-bearing and quartz-bearing rocks.

reactions are not prominent, and the "C" horizon is developed by solution. Glacial sediments which contain both silicate and carbonate rock

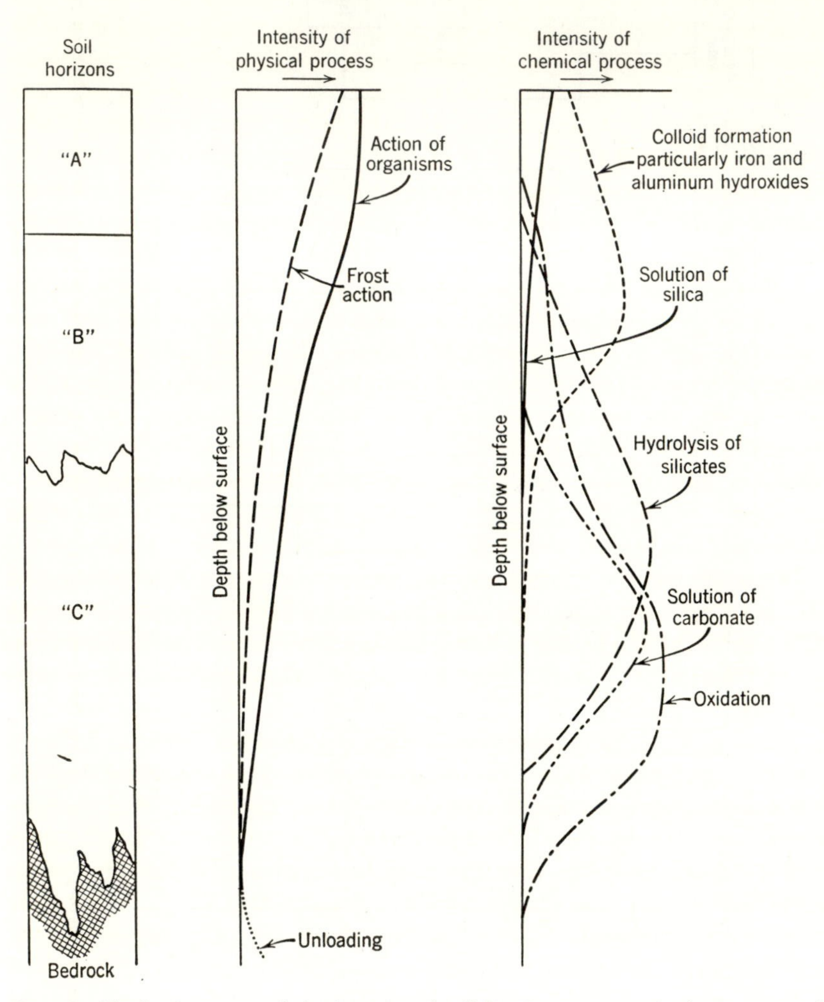

Fig. 6.9 Idealized concept of the intensity of soil-forming processes in the common soil profile. Downward migration of the positions of greatest intensity of chemical processes proceeds as each soil horizon increases in thickness.

fragments will show solution of the carbonates in the upper part of the "C" horizon as well as oxidation and hydrolysis of the silicate minerals.

Inasmuch as each of the soil zones is extended downward with the progress of time, conditions of the "C" zone are being superposed on fresh bedrock, and "B" zone conditions are being established in what was formerly "C" zone. If erosion strips off the surface soil at the same rate as the soil horizons are being extended downward, a condition of equilibrium is maintained, and the soil-forming processes continue indefinitely. Conditions of the "B" horizon are established in the upper "C" zone where rock fragments have been largely destroyed by oxidation, solution of carbonates, and hydrolysis of silicates. In the "B" horizon colloid formation is important, whereas reactions typical of the "C" zone are decreasing in intensity and tending toward completion.

Colloid behavior is the important aspect of the "B" horizon. In wet climates colloids of silica, clay minerals, and aluminum and iron hydroxides are moving downward through the pore space of the soil. Near the base of the "B" horizon these are often flocculated by metal ions which are being released from the decomposition of silicate minerals. As a result concretionary[2] growths of limonite are not uncommon, and concentration of illite and aluminum hydroxides is developed as a zone. All three of these substances tend to block the downward passage of water and reduce soil drainage.

In regions of dry climate ground water moves upward through the "B" zone between rains, and bicarbonate ion moves upward from the "C" zone. As the water is lost the reaction controlling the solution of carbonates is reversed and carbonates are precipitated as concretions.

Solution of silica, colloid formation, and intense organic activity characterize the "A" zone. Humus accumulates in the cold wet regions, and a distinctive soil color is developed. (See Table 2.1.) Soil residues which characterize the "A" horizon range in texture between small pebbles, sands and silts, and fine clays, the nature of the residue being controlled by the size of the resistant fragments. Thus, in the case of quartz-bearing rocks fragments of quartz residue may range between pebble and silt size. An "A" soil horizon will be characterized by a pebbly, sandy, or silty texture depending upon the abundance and size of the pieces of quartz. Granite, gneiss, sandy shale, and sandstone are typical rocks from which such soil textures are developed. Quartz-free parent rocks leave a residue which in theory is exclusively in the clay dimension. Typical among such rocks are gabbro, basalt, dunite, clay shales, some slates, and some schists.

[2] A concretion is a nodular or irregular concentration of mineral matter precipitated about a nucleus.

Major Soil Groups

In Chapter 2 some attention was paid to the classification of soils into groups typical of specific climatic conditions. The nature of these climates has much to do with the intensity with which physical, chemical, and biological processes alter the soil. For example in warm wet climates the intensity of organic activity is very high, whereas frost action is absent. In dry regions solution of carbonates is greatly reduced in magnitude, and carbonate precipitation is typical in the "B" horizon. On the basis of processes similar to those mentioned a soil attains certain attributes which characterize its aspect sufficiently to distinguish it from other soils. A summation of certain of these processes is recognized as *podzolization*, others as *calcification*, and still others as *laterization*.

Podzolization

Podzols are normal in the humid cold regions of heavy forest cover where humus can accumulate in the soil. Heavy rainfall removes the metal ions from the "C" and "B" zones about as fast as they are formed, and forest vegetation absorbs less of these ions than do grasses. For this reason as colloids are produced in the "A" and upper part of the "B" horizon there is less tendency for them to be flocculated until they reach the lower part of the "B" zone. Here silica, iron hydroxides, and aluminum silicates are flocculated and develop the "hardpan" previously mentioned. Humus does not move downward and is concentrated in the "A" zone so that a marked separation is often observed between the dark "A" zone and the light-colored "B" zone. Clay minerals, particularly kaolinite and illite, also form in the "B" horizon, whereas quartz tends to linger in the "A" horizon. For this reason the "A" zone is generally somewhat more sandy than the underlying "B" horizon.

Where kaolinite is the dominant clay constituent the permeability tends to be high (for clays) and the soil structure relatively open. If illite dominates, the permeability is strongly reduced, and a much more compact and plastic character is noted in the "B" horizon.

Calcification

This process is characterized by the precipitation of calcium carbonate and to a lesser degree magnesium carbonate in the "B" zone. Dry climatic conditions control this reaction, and soil water moves upward by capillary action and plant transpiration. All gradations of carbonate accumulation are recognized from a few scattered concretions in the "B" zone to the production of a solid layer of carbonate. Such a deposit is known as *caliche* and is observed commonly as a well-developed crust in desert soils. The depth below the surface at which carbonate precipitation occurs is a function of rainfall, being lowest in the higher rainfall areas and nearest the surface in the dry areas (see Fig. 10.19).

Soils representative of this process are alkaline, a condition which is suited to the growth of grass rather than trees. In consequence the more moist flat areas are grassland prairies. Continued growth of the grass leads to a mat of accumulated humus in the "A" horizon. The dark-brown color of this horizon in contrast to the lighter color of the "B" zone is typical of the prairie soils and chernozems.

Desert areas lack the moisture to maintain grasses, and decayed vegetation is essentially absent in the "A" horizon. The soil color is, therefore, controlled by the colors of minerals which are produced. With the exception of iron oxides the remainder are light colored. Addition of caliche tends to lighten them further. Locally, in volcanic areas where basic lava flows or ejecta constitute the parent rocks iron oxides are common residual products. These color the soils red because in the absence of vegetation the ferric form of iron is stable. The typical desert soil is gray or tan, but the occasional instance of a red or brown color must be recognized.

Laterizatior

Tropical regions of high moisture and high temperature establish special conditions of accelerated and intense weathering which lead to the development of a unique soil type identified as *laterite*. In such regions weathering activity is so intense that a tremendously thick soil (exceeding 100 feet) may be developed from the parent rock through processes collectively termed "laterization." These processes are peculiar in that they produce soils that lack the usual "A" horizon, but consist of great thicknesses of the "B" horizon. Here there is important fixation of ferric and aluminum hydroxides in the upper part to constitute one of the principal characterizing features. In the early stages of development concentration of iron and aluminum hydroxides is as small concretions of spheroidal shape, but in the late stages the concentration becomes so important that the soils become virtual ores of iron or aluminum. This tendency to concentrate insoluble hydroxides and oxides is the feature which has led to their name. From the indurated iron oxide "soil," bricks were cut to be used in construction of temples in India and Siam. Observing this process an English naturalist applied the term "laterite" from the Latin meaning "brick" to identify the peculiar soil type.

Laterites display certain special characteristics principal among which are the following:

1. Red, yellow, or light gray and white colors.
2. Absence of decayed vegetation in the surface soil.
3. Intense concentration of iron, aluminum, and titanium oxides and hydroxides. Oftentimes these form a well-cemented crust.
4. Important reduction in silica as compared to the original rock.
5. Development of kaolinite as the most important clay mineral.
6. Absence of plasticity and cohesion to the soil.
7. High porosity and permeability which creates a condition suitable for large infiltration of rain water. Soils absorb water so rapidly that they may be cultivated immediately following a rain.

The processes which lead to the appearance of the characterizing features enumerated above are not completely understood, but the general conditions which identify laterization are recognized. Under the intense temperature and moisture conditions which characterize the regions of laterite distribution (see Fig. 2.1—tropical areas of red and yellow soils) organic growth is rapid, and bacterial decay is accelerated. In the prevailing climates the decaying processes tend to release certain ions such as ammonium which depress the H ion concentration and increase the OH ion. For this reason downward percolating rain water is neutral or slightly alkaline ($p\text{H} = 7+$) in contrast to the more acid waters of the cold humid regions. Neutral or slightly alkaline water appears to be significant in the development of kaolinite as the most abundant clay mineral. High temperature and high moisture conditions tend to accelerate the rate of certain chemical reactions particularly solution, hydrolysis, hydration, and oxidation. Minerals such as biotite, plagioclase, orthoclase, pyroxene, amphibole, and olivine, all typical of igneous rocks, are unstable and undergo hydrolysis by some mechanism as illustrated by eqs. 1 and 2. There is no universal agreement that a silicic acid-like intermediate product appears such as indicated by the equations, but it is known that silica is removed during the process of lattice rearrangement into the layered structure of kaolinite. Such reactions tend to release Ca, Mg, K, and Na cations which tend to depress the H ion concentration somewhat more than already exists and to keep the solutions neutral or slightly basic. Waters of such low H ion concentration are capable of favoring the precipitation of aluminum and iron hydroxides and the removal of silica. The former become progressively concentrated in the upper part of the soil as the more soluble fractions are removed. Of these the behavior of silica is significant. In regions outside the tropics silica is extremely stable and constitutes one of the materials most resistant to weathering. Such environments are identified as having soil waters which are mildly or strongly acid. In neutral or slightly alkaline water silica is somewhat more soluble, and its removal from the site of weathering is to be expected in the case of the laterite soil. (See Fig. 6.6.) Removal of silica and magnesia and concentration of alumina and ferric oxide is demonstrated by analyses of samples taken at selected depths below the surface of a laterite soil in India as shown in Table 6.5. The parent rock is a somewhat metamorphosed ultramafic intrusion resembling dunite. (See Table 5.2, column 10.) At a depth of 29 feet the analysis reveals the

Table 6.5. Analyses of Laterite Soil from India at Selected Depths below Surface*

Approximate Depth in feet	Surface	2	4	9	19	29†
SiO_2	2.7	2.4	1.5	2.5	3.9	40.4
Al_2O_3	16.6	21.1	15.2	7.4	7.2	1.4
Fe_2O_3	70.0	65.5	73.0	77.2	75.2	10.2
MgO	—	—	—	—	tr.	34.2
H_2O	10.4	10.8	10.1	13.1	13.8	13.5

* Analyses recalculated to 100% from those given in J. S. Joffe, *Pedology*, Rutgers Univ. Press. 1936, p. 377.

† Generally unaltered parent rock of ultramafic composition similar to dunite. Compare with analysis 10, p. 96.

normal composition of the bedrock. Within a few feet above this depth profound alteration has occurred as indicated by the very spectacular reduction in the percentage of magnesia and silica. This change indicates that the lattice structure of such silicates as olivine and pyroxene has been destroyed and new lattices have been developed. Presumably some kaolinite is present at 19 feet of depth although this cannot be determined from the analyses alone.

Although hydroxides of iron most commonly are concentrated in laterization some parent rocks produce soils rich in aluminum and low in iron. Laterites of this type are white to light gray in color and are characterized by the ore of aluminum, bauxite. Several explanations have been proposed to account for the preferred concentration of either iron or aluminum. Two stages of alteration are recognized:[3]

[3] The stages in the development of laterites have been proposed by G. D. Sherman, "The Genesis and Morphology of Alumina-rich Laterite Clays," *Problems of Clay and Laterite Genesis*, Amer. Inst. Mining & Metall. Eng., 1952, p. 156.

1. Parent rocks are decomposed to certain primary clay minerals the principal one of which is kaolinite. Kaolinite maintains an open porous structure which permits deep penetration of water to the bedrock in which the initial stages of weathering are recognized. A few feet nearer the surface are spheroidal-shaped masses of outward appearance resembling bedrock but which have been altered completely to kaolinite.

2. Alteration of kaolinite to either hydrated iron oxides or hydrated aluminum oxides is dependent upon the local climatic condition. If there exists an alternate wet and dry season an iron-rich laterite crust is formed. Wherever the local climate is continuously wet a bauxite laterite crust is developed.

Outside the true tropics laterites grade into the red and yellow soils such as those found in the southeastern part of the United States. Here they display some of the characteristics of laterite development but the intensity is absent. In a broad way these red and yellow soils can be interpreted as developing in areas marginal to the climates where the ideal laterite is formed. There exists a strong resemblance, and the reader's attention is directed to the fact that in the United States the southeastern part also appears to be the region where kaolinite is preferentially formed in the soils. (See Fig. 6.7.)

Selected Supplementary Readings

Jacks, G. V., *Soil,* The Philosophical Library, 1954, Chapters V, VI, VII, IX, and XI. Concerns the properties of soils.

Jenny, Hans, *Factors of Soil Formation,* McGraw-Hill Book Co., 1941. The entire book is concerned with this subject.

Joffe, J. S., *Pedology,* Rutgers University Press, 1936. An excellent analysis and description of soil-forming processes and soils.

Keller, W. D., *Chemistry in Introductory Geology,* Lucas Brothers (Columbia, Missouri), 1957, pp. 35–59. An excellent presentation of the chemistry of weathering.

Lyon, T. L., and Buckman, H. O., *The Nature and Properties of Soils,* The Macmillan Co., 1943, Chapters XI and XII. A description of the weathering processes and development of soil materials.

Polynov, B. B., *The Cycle of Weathering,* Thomas Murby and Co., 1937. An advanced treatment of the chemical processes involved in weathering.

Reiche, Perry, "A Survey of Weathering Processes and Products," *Univ. of New Mexico Publication in Geology No. 3,* rev. ed., 1950. An excellent treatment of weathering processes.

Robinson, G. W., *Soils,* 3rd ed., Thomas Murby and Co., 1949, Chapters III, IV, V, VI, X, XI, XII, XIII, and XIV. An excellent treatment of soil-forming processes and soil types.

"Soils and Men," *Yearbook of Agriculture 1938,* U. S. Department of Agriculture, Government Printing Office, 1938, Parts IV and V. Selected articles on fundamentals of soil science and soils of the United States.

Streams

THE HYDROLOGIC CYCLE

Nature of the Water Cycle

Even the most casual observer of natural phenomena associates rainfall with a source of local water. Whenever a rainshower is light and steady he refers to this condition as a soaking rain. By this he means that the permeability of the soil is more than, or equal to, the rate at which a unit quantity of water falls upon a unit surface area. Therefore, nearly all of this water enters the rocky mantle of the earth. Such water, formerly a part of the atmosphere, now occupies a subterranean environment and is known as *ground water*.

In regions where the soil cover is sand the quantity of water furnished by exceptional downpours does not exceed the soil's permeability, and all such water enters the ground. As the percentage of clay in

The youthful stream.

a soil is increased its permeability declines rapidly, and clay soils pass only small quantities of water through a unit cross section. During a heavy rainfall over moderately clayey soil a condition is reached where the water which falls on one square foot is greater than the soil's permeability. Excess water must accumulate at the surface and generally flows from the spot. This water which flows to some locality of lower elevation is known as *runoff*. Under most circumstances

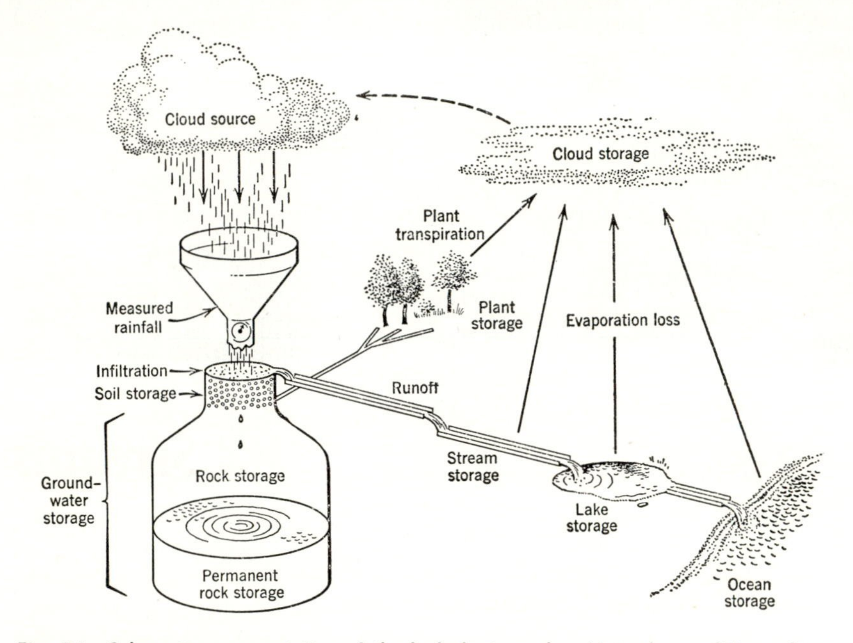

Fig. 7.1 Schematic representation of the hydrologic cycle. Note the conditions of temporary and permanent storage.

runoff becomes channeled and enters a stream system, but for short distances the runoff may flow from the site as a thin sheet.

Locally accumulated water enters the soil at the rate permitted by its permeability. Sometimes this is rapid, but in clay soils intake may be so slow that evaporation is the only method by which water is removed from the site. Where evaporation rates are high the pond of excess water leads an ephemeral existence and disappears between rains. Elsewhere, the quantity of rainfall may exceed the rate at which water is dissipated by evaporation or infiltration into the ground, and a permanent pond or swamp area develops. The level of such a pond will continue to rise until an outlet is reached, and all excess water is removed by means of the channel.

In every region there exists a temporary balance between the amount of water received by precipitation (either rain or snow) and the amount removed from the site as ground water, runoff, or by evaporation. This condition is part of the *hydrologic cycle* of the earth and is diagrammed in Fig. 7.1. Equilibrium is attained for local areas and for the earth as a whole although the nature of the equilibrium is not the same. Wherever and whenever an excess of precipitation occurs extra water is carried to some storage site. Part of the storage is ground water, some is held temporarily in streams and lakes, and most is held in permanent storage in the oceans.

Water levels in the storage sites vary as precipitation-evaporation ratios change. During epochs of continental glaciation much water was stored as ice on the land surface, and ocean levels fell approximately 300 feet below the present position. Continued rise in world average annual temperatures would cause additional melting of Greenland and Antarctic ice caps with a consequent increase in the height of sea level and invasion of some of the present land surface. Such world-wide (*eustatic*) sea level changes are believed to have occurred many times in the geologic past. Each time this took place a new equilibrium was achieved between precipitation, evaporation, and the level of the storage site.

Ground Water

As ground-water infiltration proceeds water moves primarily downward through the capillary openings between mineral grains of the soil. If the soil has uniform permeability water moves downward to the bedrock interface, or in glacial clays it may continue downward along irregular paths for many feet. Permeability differences within the bedrock result in certain layers or zones becoming the principal carriers of water, whereas others are essentially barren. Fissures are also important channels, and large volumes of water are carried in the network of openings. Nearly everywhere some water can be withdrawn from rocks beneath the surface, and the total amount stored in the ground must be a very large quantity.

Downward penetration of water comes to an end as rock permeability decreases with increased depths and as the geothermal gradient precludes the existence of fluid water. Ground water exists as an envelope surrounding the outer part of the earth. This envelope, having a somewhat irregular base, is known locally from mining and drilling operations which have passed beneath its lower boundary.

With the progress of time the level of ground water may rise locally until it coincides with the earth's surface where a lake or swamp will appear. Under other circumstances as local ground-water elevations exceed those of neighboring areas lateral discharge of ground water occurs to establish an equilibrium. In still other localities periods of infiltration are interrupted by long intervals of evaporation. Capillary water is gradually removed from the soil and uppermost rocks, and the upper boundary of the ground water may lie several hundred feet below the earth's surface. There exists an extremely irregular upper surface below which the rocks are saturated with water. This surface which can be extended from one local area to another is known as the *water table.*

In many places continued lateral movement of ground water is interrupted by a stream valley. Here ground-water flowage is directed into the stream channel and becomes a part of the runoff.

Runoff

Where clay soils prevail continued rainfall tends to deflocculate the particles. Water enters the spaces between the individual aggregates and also enters the lattice of the clay mineral. Illite and montmorillonite, in particular, swell and deflocculate into water-clay suspension. When this condition exists on uniform slopes *sheet wash* forms much of the runoff. The consistency of sheet wash ranges between water with very little suspended clay and a viscous, slow-moving *mudflow.* When the mixture is chiefly water nonchanneled runoff is called *sheet flow,* but when the soil particles dominate the suspension it is considered a variety of landslide and is known as a mudflow.

Sheet wash may proceed down slope for a distance of more than a mile before the water is effectively channelized. Elsewhere, sheet wash proceeds for a matter of feet before the water, and its suspended sediment is concentrated in some elongate depression and carried along a narrow and directed route. Channelized water is known as a *stream* and usually is a part of an integrated system which carries the runoff to some storage reservoir.

Factors Affecting Runoff

The quantity of water which flows from a unit area as runoff is a function of a number of variables of which soil permeability is only one. Another important variable is the angle of ground slope, i.e., runoff is increased with angle. Increase in slope is also related to stream density per unit area. For example, in New England as average land slope increases from 290 to 700 feet per mile stream density also increases from an average of 1.1 miles of stream per square mile of land surface to 2.1 miles of stream (see Fig. 7.2). Also, as additional water enters the channels they become progressively larger downstream and are capable of carrying much greater quantities of water.

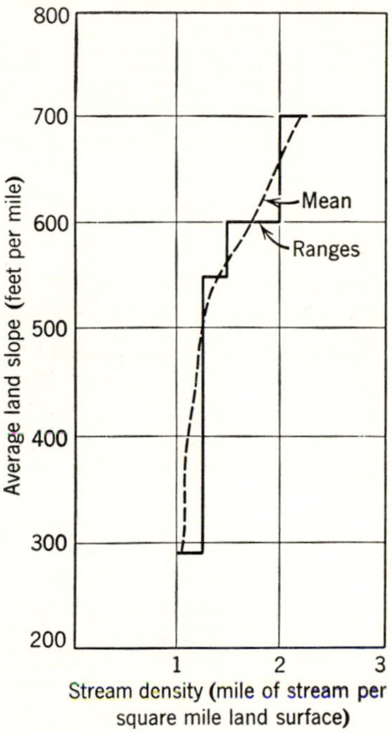

Fig. 7.2 Relationship between stream density and land slope in New England. This is a region dominated by metamorphic rocks of approximately uniform infiltration rates, and the increase in number of streams is a function principally of land slope. (From Langbein, *U. S. Geol. Survey Water Supply Paper*, **968-C**.)

Evaporation is important in reducing runoff, and inasmuch as evaporation rates are directly proportional to temperature the runoff varies inversely as temperature. Vegetation cover slows down the rate of runoff by impeding flow and by absorbing and transpiring large quantities of water. The vegetation cover also reduces the amount of clay which enters the water suspension. For this reason in arid regions where grasses are absent sheet wash is often loaded with clay, and mudflows are common occurrences. Conversely, in forested areas streams tend to be less loaded with clay-sized debris.

Sizes, slopes, and gradients of drainage basins are the most important controls of runoff.[1] This is illustrated by comparison of the Manasquan and Great Egg rivers in New Jersey. These two streams each drain about 50 sq. miles of coastal area. In June 1938 they were subjected to a flood caused by the same amount of rainfall in the two areas. Note in Fig. 7.3 how the Manasquan River rapidly discharged its water reaching a peak within one-half day after the heavy rain, whereas the Great Egg River did not attain its maximum discharge until $2\frac{1}{2}$ days had passed. The Manasquan River also discharged at five times the maximum rate of the Great Egg River and, hence, showed a rapid decrease in dis-

[1] Gradient is measured in feet per mile, or meters per kilometer, along the stream course.

charge once the peak was attained. This is a good example of a stream system which produces flash floods. Not so with the Great Egg River where discharge rates slowly increase and then decline. This stream stores water for days, and high water never reaches the extremes of flash-flood condition.

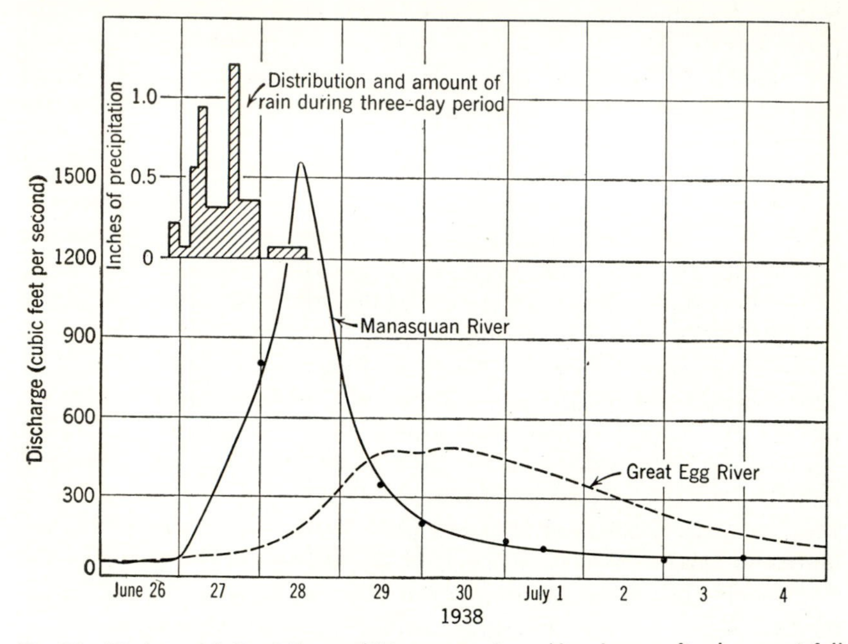

Fig. 7.3 Discharge of Great Egg and Manasquan rivers, New Jersey, after heavy rainfall of June 26–28, 1938. Note distribution of rainfall in inches during the three-day period and the differences in discharge between the two rivers. Both river systems have approximately the same drainage area, but the Manasquan is confined to a steep-walled valley system, whereas the Great Egg has a broad valley with the stream channel bordered by swamps. (From Langbein, U. S. Geol. Survey *Water Supply Paper*, 968-C.)

Why should these two streams be so unlike in behavior? The differences are attributed to stream gradient and steepness of valley-wall slopes, i.e., distribution of low land bordering the valleys. The Manasquan has twice the average gradient and about one-fifth the swamp land which surrounds the Great Egg River. The Manasquan must carry the water away rapidly because of its steep gradient and relatively narrow valley. The Great Egg River, however, has a broad valley with much swampy area bordering the channel. Here excess water is stored by raising the level of the water in the swamps, and it is only lowered again as the gentle stream gradient permits slow removal of the excess water.

Storage and Runoff

Small channels concentrating sheet wash and localizing moving water show a discharge pattern not unlike the Manasquan River (Fig. 7.3), i.e., a high discharge rate and rapid decline until the channel is free from flowing water. They are dry during all times except when rain is falling in the area. Such streams are known as *intermittent* and are distinguished from others known as *permanent* by virtue of the latter having some water flowing constantly in the channel.

Part of the water in a stream is being stored essentially in transport. Differences in storage capacities can be observed between the Manasquan and the Great Egg rivers. By reason of its slower rate of discharge the Great Egg River is temporarily storing water in its channel, and this storage is in slow transit to the ocean. Valleys of the type represented by the Manasquan River, however, are incapable of storing water and move it rapidly from the site.

In arid regions where the water table is some distance below the bottom of the stream channel, drainage basins having characteristics like that of the Great Egg River lose water by infiltration. Streams having permeable sands or gravels at their base tend to fill such gravels, and they act as temporary, or permanent, storage "tanks." In these areas the ground-water table is likely to be highest below the stream channel and slope away from this elevation as higher ground is reached.

A little thought on the matter will convince the reader that permanent streams are not maintained by storage within and along their own channels, but water must be supplied during periods when rain is not falling. This source is ground water which in higher and less sloping localities has built up to a level generally above that of the stream channel. Permanent streams are fed from the underground storage capacity, and their normal volume of discharge is maintained by the amount which the underground storage furnished the stream channel. In localities of high rainfall storage capacities are filled, and the constant supply of excess water from overflowing of storage keeps streams abundantly supplied with water. Moreover, drainage basins which cover extremely large areas have rain falling regularly somewhere on their watershed, and some of this runoff is added to the normal ground-water supply.

In the United States contrasting conditions in underground storage behavior exist between the eastern and western halves of the country. Figure 7.4 is a map of average annual precipitation and shows that the eastern half of the country receives 30 inches or more of rainfall, whereas the western half generally receives less. As a first approximation this rainfall value forms a boundary between two regions. In the eastern half the underground storage capacity is filled, and the excess is being fed to streams throughout the year. The western half with deficient storage supply is indicated where runoff gradually is reduced as water enters the ground-water system. In the wet area

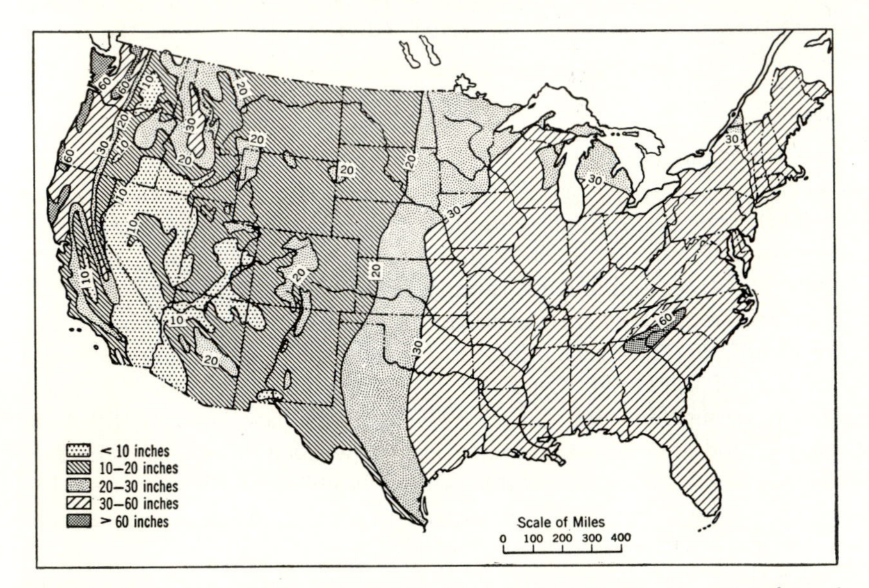

Fig. 7.4 Generalized average annual precipitation in the United States. Note the position of 30 inches of rainfall which serves to divide the country very approximately into an eastern (moist) and a western (dry) sector. West of the 20-inch rainfall line streams tend to feed the ground water rather than discharge from ground-water storage. (Simplified from *The Colorado River,* U. S. Bur. Reclamation, 1946.)

little change in the normal flow of streams is observed throughout the year. Except for floods occasioned by unusual and local rains the discharge remains relatively uniform. Most of the western half of the country experiences less than 20 inches of rainfall, and higher precipitation is confined to mountain belts. Here the supply to the streams is from melting snow, and that furnished by ground water is small. As the streams leave the mountains and flow across dry regions they tend to recharge ground-water storage along their bed, and in many instances stream discharge actually decreases downstream.

The conditions described above are illustrated by two streams, the Wabash River in Indiana and the Gila River in Arizona. Average annual discharge rates at stations along each water course are plotted in Fig. 7.5 for approximately the same distance along each stream. Discharge rates of the Wabash River progressively increase downstream particularly where important tributaries join the master stream. Near the headwaters of the Wabash at Bluffton, Indiana, the mean annual discharge is about 350 second-feet. At Mt. Carmel, Illinois, just above the junction with the Ohio River the discharge is approximately 29,000 second-feet. The Wabash drains an area in

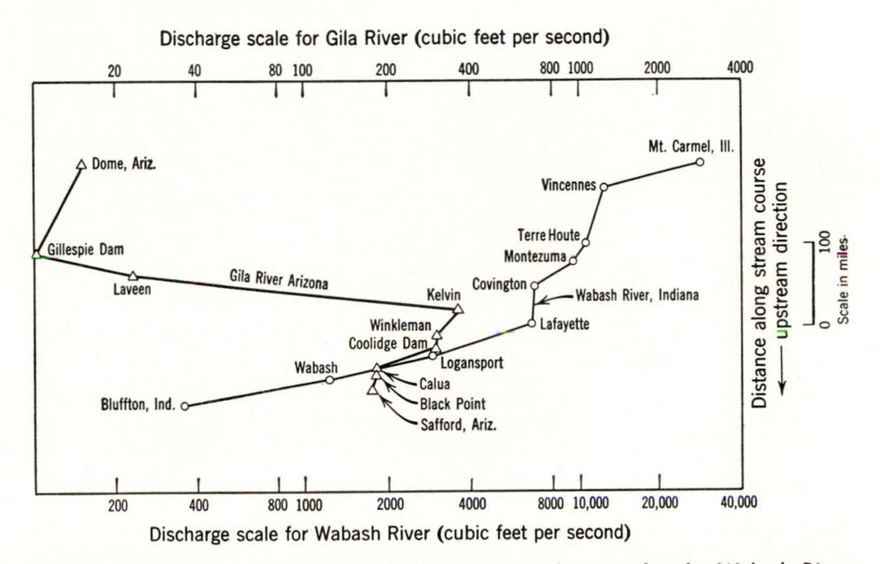

Fig. 7.5 Discharge rates plotted against downstream distances for the Wabash River, Indiana, and the Gila River, Arizona. The Wabash, typical of streams in wet areas, drains ground-water storage and increases greatly in discharge with progressive distance downstream. The Gila River, typical of streams in arid areas, adds to ground-water storage and tends to decrease in discharge downstream.

which ground-water storage is filled to capacity and runoff is accelerated by the entrance of tributaries into an already filled storage basin.

Characteristics illustrated by the Gila River are those in which the stream recharges the supply of ground water over part of its course. Between Safford and Kelvin, Arizona, tributaries contribute enough water to increase the discharge progressively downstream. However, from Kelvin to Gillespie Dam ground-water recharge occurs as indicated by the drop in discharge rate from an average annual value of 370 second-feet to about 10 second-feet below Gillespie Dam. Some of this loss is due to irrigation projects so that the drop in flow rate is not entirely representative of natural conditions, but the approximation illustrates the case. Farther downstream from Gillespie Dam to Dome some entering tributaries supply enough to increase the discharge rate once more. The volume, however, is insignificant as compared to the tributaries of the Wabash.

THE STREAM CHANNEL

Channel Growth

The most simplified form of a selected land surface can be imagined as a plane of uniform slope which intersects sea level, passes beneath the latter, and forms the floor of an ocean. Should rain fall with uniform distribution on this surface sheet flow would occur, and water moving with increasing velocity and thickness would merge with ocean water whereupon movement in one direction would cease. At each unit distance down the slope of the theoretical land surface the kinetic energy of a unit cross section of water would not be the same but would increase with distance downslope as the mass and velocity of water increased. This kinetic energy is employed in overcoming the friction at the water-land interface and within the water body, creating turbulent conditions in the flowing water, pushing solid particles downslope, and generating heat. Some of the demands on the available energy are small; others are large and are used in overcoming the internal and boundary friction forces. Transportation of solid particles appears to require only about 4 per cent of the total energy. In combination these forces are very effective in reducing the theoretical available energy.

In our imaginary model removal of solid particles from the land surface would increase as the water reached the lower slopes above sea level. The effect would be to alter the slope of the plane in some places more than others to produce a concave surface of asymmetric long profile. With the progress of time this profile would approximate that of the graph of a negative exponential equation in which land height (ordinate) is plotted against distance (abscissa). In the lower reaches the slope portion would approach sea level asymptotically. Comparison with actual long profiles of streams (Fig. 7.6) demonstrates that our theoretical analysis is substantially in accordance with observed conditions. In the development of such a profile the position of greatest stream energy has been migrating slowly upstream as the gradient has declined. The effect has been to move the zone of greatest energy toward the locality of maximum slope.

Our ideal model is unrealistic in one major respect, namely that an original land surface is not perfectly plane but has irregularities which tend to influence the direction of flow of water downslope. This will tend to concentrate water in some places at the expense of others. Such uneven distribution in flowing water produces an irregular distri-

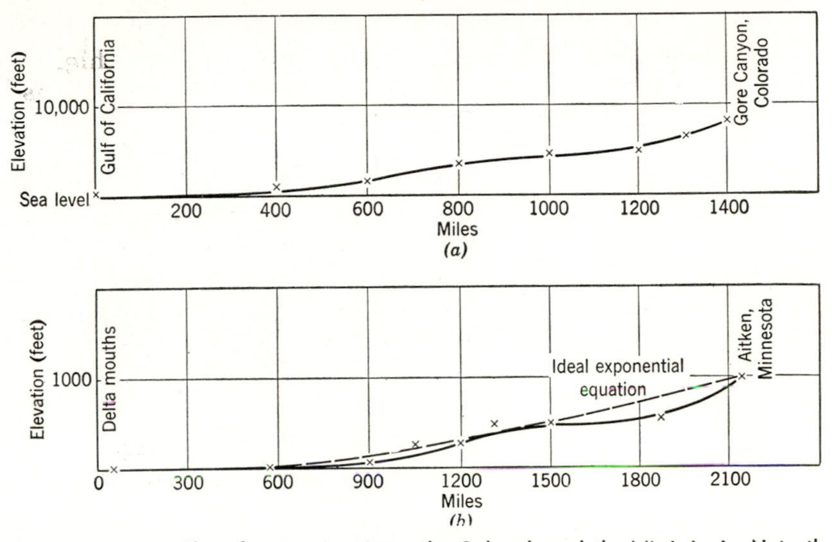

Fig. 7.6 Long profiles of two major rivers, the Colorado and the Mississippi. Note the approximation to the curve of an exponential equation.

bution of kinetic energy, and removal of solid particles from the land surface becomes unbalanced. Depressions and slopes are accentuated and joined together as channel systems. With the establishment of the system stream development proceeds rapidly, and changes in long profiles are confined primarily to stream channels.

Stream Flow

Stream erosion has been defined as the sum of the processes which are involved in removal of rock material from one site and its transportation and accumulation elsewhere. In order to understand the laws controlling transportation of solid particles by streams a thorough knowledge of the mechanics of stream flow is needed. For our purposes we can but acquaint ourselves with the bare essentials sufficient to provide the necessary basis of understanding the problems at hand.

As a liquid moves downslope frictional forces between particles must be overcome by the energy of motion. The internal resistance to such flow is ascribed to the viscosity of the fluid, and such forces are known as viscous forces. In very slow movement the viscosity of the fluid (i.e., the viscous force) controls the motion, and conversely highly viscous fluids control the velocity of flow. As the velocity of flow is increased forces due to viscosity assume less importance and eventually exert an extremely minor influence on the nature of flow. Under such conditions flow is controlled by inertial forces, and as these assume dominance the fundamental nature of flow changes.

Laminar Flow

With very slow horizontal movement of fluid one may visualize layers of sheets of fluid shearing over another. Here a shear stress controls the movement, but the internal friction within the liquid keeps individual layers of water as discrete masses. The force required to maintain motion is found to be proportional to the velocity change in the vertical direction, and the shear stress applied equals

$$\frac{\text{force}}{\text{area}} = \frac{V_2 - V_1}{Y} \mu \qquad (1)$$

where V_2 = the velocity of layer 2 (upper layer).
V_1 = the velocity of layer 1 (lower layer) and $V_2 \gtrless V_1$.
Y = a vertical distance between layer 2 and layer 1.
μ = a constant of viscosity for a given fluid, at a given temperature.[2]

Laminar flow exists when the internal friction in the fluid, i.e., the viscosity, is more important than the inertial forces in controlling the motion. This is a slow and steady motion in which the threads of the liquid maintain their identity during the flow. When such threads encounter obstacles they curve around or over them maintaining their relative positions one to the other. An obstacle such as a pebble merely warps the lines of flow over and around itself without disturbing the relative position of the water laminas. Increase in the velocity causes stream lines to approach one another over the top of the pebble and spread on the downstream side. So long as μ remains dominant in eq. 1 velocity differences between layers of liquid are low, the motion is laminar, and individual layers passing over an obstacle remain in fixed relative spacing.

[2] μ is called the dynamic viscosity because it has the dimensions mass/length × time. Another coefficient of viscosity represented by the Greek letter ν is the *kinematic viscosity*. This coefficient is obtained by dividing μ by the density of the fluid and has the dimension (length)^2/time. For water at ordinary temperatures the dynamic and kinematic viscosities are approximately the same.

As the velocity increase causes the stream lines to spread on the downstream side of the obstacle, eddies appear. These take paths upward and outward and destroy uniformity in the water layers. The viscosity term becomes negligible (in comparison to velocity differences between two vertically disposed threads of water) and inertial forces dominate the equations of flow. Motion of the fluid is now characterized by the development of eddies and is described as *turbulent.*

Turbulent Flow

Threads of liquid characterized by turbulent flow move in all directions upward, downward, forward, and backward, but forward motion is statistically dominant. Eddies have their own vortices of local motion, but the path of the eddy can be observed to move in any direction. Rough surfaces generate eddies which in streams originate principally along the wetted perimeter. An important feature of all turbulence is its high rate of diffusion for transfer of heat, momentum, and suspended sediment across the line of stream flow. Turbulence is important in particle transfer because particles can be lifted from the stream bed. Debris so lifted is carried upward and placed in the zone of maximum forward flow.

Transition from laminar to turbulent flow may be expressed in the form of a dimensionless number called the *Reynolds number.* Reynolds number, R, is defined as a velocity times a length divided by the kinematic viscosity, ν, i.e.,

$$R = \frac{VL}{\nu} \tag{2}$$

In a stream the "length" is the width of the stream, the velocity is the average velocity of the fluid, and the viscosity is the kinematic viscosity of the fluid. For flow in a pipe the transition from laminar to turbulent flow occurs when R approaches 2000. For water this is approximately 5 feet per minute. This rate is so slow that it is proper to assume that all streams exceed this value and, hence, are controlled by turbulent flow. The equation indicates that wide streams and fast streams are those which generate the greatest turbulence, and swirls of eddies can be seen to characterize such streams.

In the ideal stream of smooth bottom and sides eddies are generated at random and move in all directions. However, movement in a direction downstream is much more common than movement in any other direction. Where obstacles are present eddies may occupy a permanent position such as the lee side of boulders or submerged logs and near the stream margins in the case of certain rapids. These can be considered as standing eddies whose motion is not downstream at all.

Effect of Channel Shape on Position of Maximum Velocity[3]

Figure 7.7*a* is a diagrammatic sketch of the average downstream velocity distribution in a stream of V-shaped cross section. Values of this forward velocity are symmetrically disposed with the highest values near the middle of the stream. In detail, however, maximum average velocities are somewhat below the water-air interface (Fig. 7.7*b*) and decrease progressively from a maximum value as the bottom and channel walls are approached. Spacing of the velocity contours is a measure of the shear stress exerted by the water against the bottom, and those parts of the cross section where the velocity gradient is steep indicate highest shear stress. The slope of the vertical velocity distribution as shown in Fig. 7.7*b* depends upon the position in the stream where the values are determined. Vertical velocity distributions near the stream bank show lower values than near the stream center; hence, the slope of the curve is steeper, and the corresponding shear stress against the bottom is less.

Velocity declines in all directions from the maximum, even toward the upper surface of the water. This is indicated in Fig. 7.7*a* by a dashed line marking the boundary between the upward and downward directions of components of flow.

Along the wetted perimeter (Fig. 7.7*a*) can be visualized a series of uniformly spaced points $a,b,c, \cdots k$. From each point a perpendicular (orthogonal line) to the velocity contours can be constructed resulting in a series of quadrilaterals of differing areas. Each quadrilateral can be imagined as the cross section of a prism whose third dimension is the distance travelled by the water in unit time. The value of this dimension decreases from the zone of high velocity to zones of lower velocity. Based upon such a diagrammatic representation a general equation of the shear force exerted against the bed between equally spaced intervals such as *a-b, b-c, c-d,* etc., can be expressed by

$$F = MgJ \qquad (3)$$

[3] The treatment followed herein is that of J. B. Leighly, and the illustrations used are those which he presented in his article entitled "Toward a Theory of Morphologic Significance of Turbulence in the Flow of Water in Streams," *Univ. of Calif. Public. in Geography,* Vol. 6, 1932.

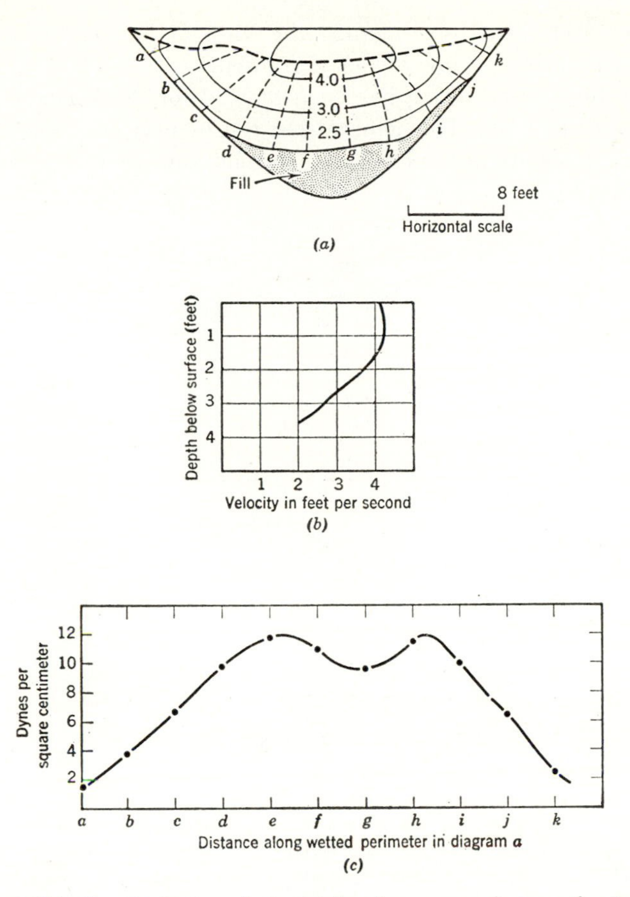

Fig. 7.7 (*a*) Velocity distribution of the intake of a power plant on the Adige River, Italy. Contours (in feet per second) are drawn through points of equal velocity. Points *a*, *b*, *c*, *d*, etc., are spaced equidistant along the channel perimeter. The dashed line indicates the trace of the surface of maximum velocity across the channel. (*b*) Velocity distribution of (*a*) in the center of the channel from the surface of water to the bottom. (*c*) Force exerted against the bottom at equally spaced positions along the channel perimeter. Note the tendency toward a bimodal distribution of force, with such maxima at the margins of the base of the stream channel. (Simplified from Leighly, *Univ. Calif. Publ. Geog.*, 1932.)

where F = shear force over the area of the prism.
M = mass of water in the prism.
g = acceleration of gravity.
J = gradient of velocity along the prism.

The total force is a product of the volume of the prism and the velocity of the water. A small prism of high velocity gradient may exert a greater shear force against the walls or bottom of the channel than a large prism of low velocity gradient.

Such a force distribution has been plotted for the V-shaped channel of Figs. 7.7*a* in 7.7*c*. This generalized representation shows the forces increasing from the air-rock-water contact toward the bottom of the

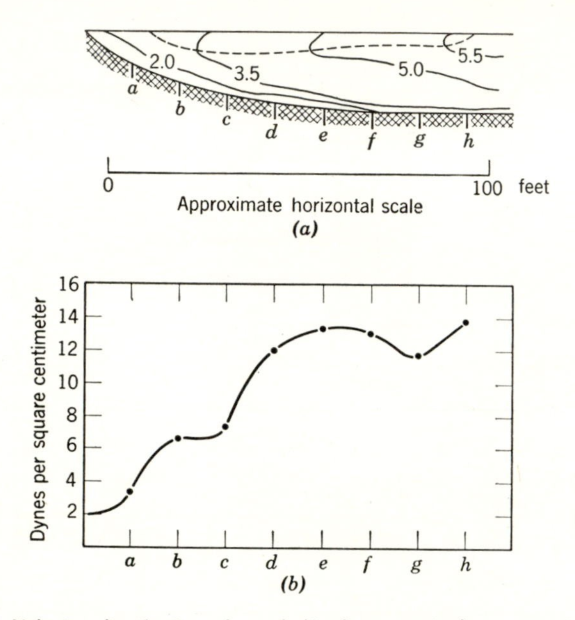

Fig. 7.8 (*a*) Velocity distribution of one-half of symmetrical cross section of Danube Canal at Vienna, Austria. Contours are in feet per second. Points *a*, *b*, *c*, *d*, etc., are spaced equidistant along the wetted perimeter. The dashed line indicates the trace of the surface of maximum velocity. Note that its position is lowest near the stream margins rather than along the central line as in Fig. 7.7*a*. (*b*) Distribution of force exerted against the stream bed in (*a*). Note that the force across the entire channel will show a bimodal distribution which is rather widely scattered across the broad channel rather than near the margins as in Fig. 7.7. (Modified from Leighly, *Univ. Calif. Publ. Geog.*, 1932.)

channel. The peak value, however, is not directly below the stream center but is disposed symmetrically on either side. These positions of maximum stress can be shown to be coincident with the zones of erosion or minimum deposition in the stream channel. Hence, the sites of maximum erosion in a V-shaped channel can be predicted.

A partial cross section of a velocity distribution in a canal of the Danube River at Vienna is illustrated in Fig. 7.8*a*. This is also a symmetrical cross section, but it is not a V-shaped notch. Vertical velocity gradients are essentially uniform, and the shear force directed against the bottom is proportional to water thickness as illustrated in Fig. 7.8*b*. Stream channels of such cross section are subject to erosion in about the same proportion as the distribution of the velocities, and the cross section maintains the same shape.

Streams whose cross section is clearly asymmetrical show velocity distributions similar to that illustrated in Fig. 7.9*a*. Here the fastest

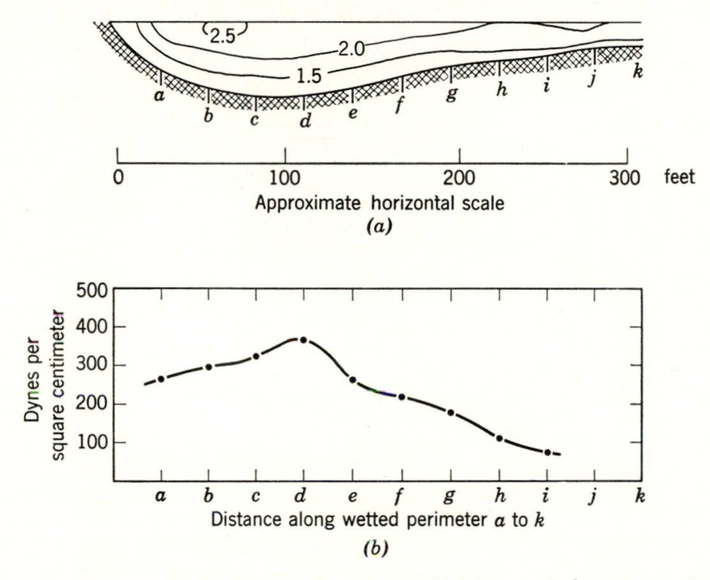

Fig. 7.9 (*a*) Velocity distribution (feet per second) of part of the cross section of the channel of the Klaralven River, Sweden. Note concentration of current near left-hand stream bank and at surface of water. Points *a*, *b*, *c*, *d*, etc., are equidistant along wetted perimeter of channel. (*b*) Force exerted on the stream bed over distance *a* to *i* illustrating the concentration of erosion in the zone of maximum force against the bed. (Simplified from Leighly, *Univ. Calif. Publ. Geog.*, 1932.)

water is near one of the stream banks. This is also the location of the most pronounced velocity gradient. Figure 7.9*b* shows the distribution of shear forces for this stream. These attain a peak value not near the stream center but near one bank. Note also that the zone of deepest water corresponds to the zone of greatest force, that the force exerted against the bank of deep water is near the maximum, and that the force gradient declines toward the shallow water. Erosion is not uniformly distributed but concentrated principally along one side of the stream channel, namely the outside bends of meanders.

Other observations which emerge from a comparison of streams of varying cross sections concern certain controls of velocity.

1. Maximum velocity values are moved upward with increase in roughness of the stream bed. Where slope and width are the same the streams having sandy or silty bottoms have higher velocity values nearer the bottom than streams of gravel or bouldery beds.

2. An obstruction to the free movement of water causes a marked depression of the line of maximum velocity downstream from the obstruction.

3. The zone of greatest velocity is depressed by steepening of the channel walls or an increase in the ratio of depth to width of the channel.

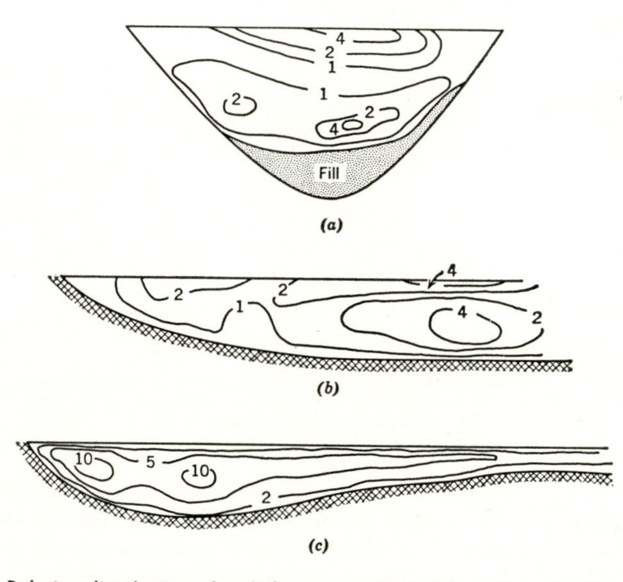

Fig. 7.10 Relative distribution of turbulence as indicated by values of eddy conductivity in (*a*) intake to power plant Adige River, (*b*) Danube Canal, (*c*) Klaralven River; see Figs. 7.7, 7.8, and 7.9. Note that the positions of greatest turbulence intensities along the stream bed are the positions of maximum erosion. (Simplified from Leighly, *Univ. Calif. Publ. Geog.*, 1932.)

An expression of the rate of transfer of water within any one of the theoretical prisms isolated in Fig. 7.7*a* can be calculated. This value known as *eddy conductivity* has been contoured in relative values for the stream channels illustrated in Figs. 7.7*a*, 7.8*a*, and 7.9*a* in Figs. 7.10*a*, *b*, and *c*. Observe the distribution of the turbulence values and in particular the positions of high turbulence concentration. In Fig. 7.10*a* high turbulence exists near the surface and near the lower part of the stream. Slight asymmetry in distribution of turbulence values can be determined. Generally, this asymmetry becomes more marked in streams of broader cross section, but in those of broad, but symmetrical, cross section (Fig. 7.10*b*) the highest values of turbulence are in a zone in the central part of the stream. Strong asymmetry of cross section Fig. 7.10*c* indicates marked asymmetry in eddy-conductivity distribution as illustrated by the positions of high turbulence concentration in the deep-water zone near the stream bank.

Examples of such distributions of eddy conductivity point out the importance of turbulence as a primary agent in stream erosion. Note that if sufficient energy exists to raise a particle from the stream bed the particle enters a zone of higher eddy conductivity irrespective of the nature of the stream cross section. Also, positions of high turbulence near the bed of the stream coincide with zones of deepest water. On the basis of the latter observation we must conclude that the pattern of stages in the progressive change of cross section of the stream channel must be strongly influenced by the distribution of turbulence.

Varieties of Channel Systems

Our theoretical model provides a useful mechanism for understanding the varieties of channel systems which will develop on the sloping surface. The model has been modified to show a few scattered departures from the plane surface and is so constructed that the shape or gradient of the surface can be changed at will. On each square foot of land surface an equal amount of water is allowed to fall as rain.

Assume an initial condition of a steep gradient as illustrated in Fig. 7.11*a*; the maximum slope is toward the sea, and this direction will be the principal component of flow. Primary irregularities will serve to channelize some of the water, but the main effect will be the development of a series of independent, parallel channels or drainage lines. This pattern observed frequently along road-cut slopes in clay is called *longitudinal* drainage. Local increase in steepness of irregularities initiates tributary drainage to the larger channels, and these will join with acute angles of small value as illustrated in Fig. 7.11*b*.

If the land surface is buckled into a series of folds as in Fig. 7.11*c* steep slopes are present on the sides of the folds. Tributary stream flow will be directly down this steep slope and will join the trunk stream with 90 degree junctions. This drainage pattern is known as *trellis*. Gently undulating land surfaces with a dominating slope direction develop patterns which are intermediate between longitudinal and trellis as illustrated in Fig. 7.11*d*.

If the model is deformed into a single conical hill a longitudinal pattern of a special type will appear radially distributed from the highest point (Fig. 7.11*e*). This drainage pattern known as *radial* is

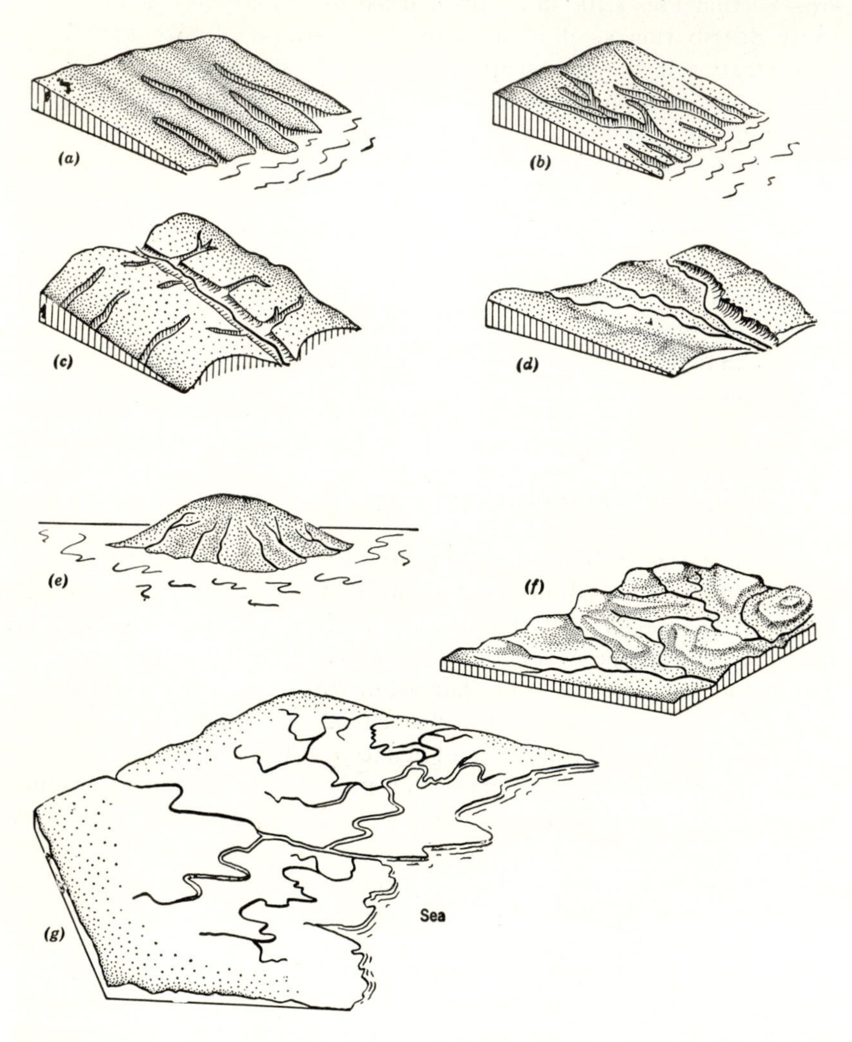

Fig. 7.11 Schematic illustrations of stream drainage. (*a*) Longitudinal, (*b*) advanced longitudinal, (*c*) trellis, (*d*) modified longitudinal, (*e*) radial, (*f*) dendritic steep slope, (*g*) dendritic low slope.

characteristic of volcanic cones and other domed hills or isolated mountains. Radial drainage also forms on gently domed elevations whose surfaces slope at gradients of as little as a few feet per mile.

As the primary slope of the land surface is lowered local slopes become more important in developing a stream pattern. Streams junction with larger angles as in Figs. 7.11*f* and *g* and individual channels vary in local direction. The general aspect is like that of branching vines, and because of this similarity the pattern is described as *dendritic.*

Dendritic stream patterns are by far the most common of all and differ somewhat in design depending upon the magnitude of primary slope. Where the slope is uniform all major streams and tributaries will drain in the same general direction (Fig. 7.11*f*). However, the most typical dendritic pattern is developed on surfaces whose local slope gradients exceed regional ones, and both are in terms of a few feet per mile. This pattern illustrated by Fig. 7.11*g* indicates that tributary stream flow in any direction depends upon the direction of the local slope. Note that streams in the same area may flow in opposite directions as the surface inclination dictates.

Stream patterns as described above are all developed on our model by first deforming its surface and later permitting water to flow over this surface. Any stream which develops under similar natural circumstances is designated a *consequent* stream because its pattern is controlled principally by irregularities of the initial topography. Once the stream begins to erode a channel the hardness of the material of the model affects the stream. This is particularly true when the model has been built of alternately hard and soft layers. As erosion proceeds the channel will respond to the position of these layers of differing hardness, and the stream course will be altered accordingly. Streams cutting channels in the surface material of the earth behave in like fashion. These influences will be discussed in later paragraphs.

THE STREAM'S LONG PROFILE

Laws Controlling The Profile

Short streams as well as long ones display a long profile whose lower reaches have a more gentle gradient than the upper headwater portions (Fig. 7.6). Also gradients in the headwater regions are likely to be more variable than in the lower reaches particularly in the case of large streams. Tributaries supply increasing amounts of water downstream so that the mass of water in the downstream parts is far greater than that upstream. However, steep gradients in the headwaters result in velocities which impart a high kinetic energy to the mass of water present. In the headwaters the stream channel is being lowered and lengthened so that the channel or valley is said to be increasing its length as illustrated in Fig. 7.12. Early recognition of this process led to the statement of one of the most fundamental laws governing stream behavior; namely, *each stream increases its valley length by headward erosion.*

In the same diagram the mouth of the stream is shown as emptying into a sea. The latter does not flow in a uniform direction downslope but remains as a body of standing water. Our theoretical stream loses velocity on entering this water body, and its kinetic energy approaches

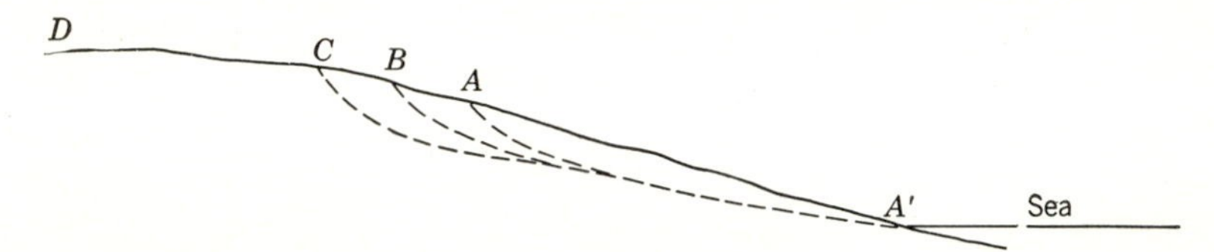

Fig. 7.12 Long profile of a stream at successive intervals of time. The mouth remains at *A′*, whereas the headwater channel shifts from *A* to *B* to *C*, and the profile shifts as indicated by dashed lines. The change in profile illustrates the fundamental law that a stream increases its length by headward erosion.

the limiting zero value. In this realm the stream channel cannot be deepened effectively as the condition of quiet water is approached. In the illustration employed sea level is described as the *base level* of the stream, and the concept just presented forms the basis of another fundamental qualitative law. This law can be stated as follows: *No stream can cut its valley below its base level.*

The base-level law can be applied to any body of water into which a stream empties. A lake far above sea level will act as a base level for any stream which enters so long as the lake exists. Any major stream is a temporary base level for its tributaries, for no tributary can deepen its channel below that of its master. But as the main stream valley is deepened the base level of a tributary is lowered, and further deepening of that valley is possible. Ultimately even this deepening ceases as the level of base level of the entire stream system is approached.

Waterfalls

In the upper reaches of the stream system tributaries frequently join the master stream by plunging over a waterfall or steeply inclined rapids. In the long history of a stream this condition is temporary as the site of a rapids or waterfall becomes the locus of attack by accelerated erosion. Waterfalls are most oftentimes the result of a hard layer of rock interrupting a series of softer rocks. Dikes (Fig. 7.13*c*), sills (Fig. 7.13*b*), lava flows, and irregular igneous intrusions into soft sediments are responsible for many waterfalls. Others result where capping layers of limestone or well-cemented sandstones overlie shales (Fig. 7.13*a*). Still other falls develop where glacial deepening of a

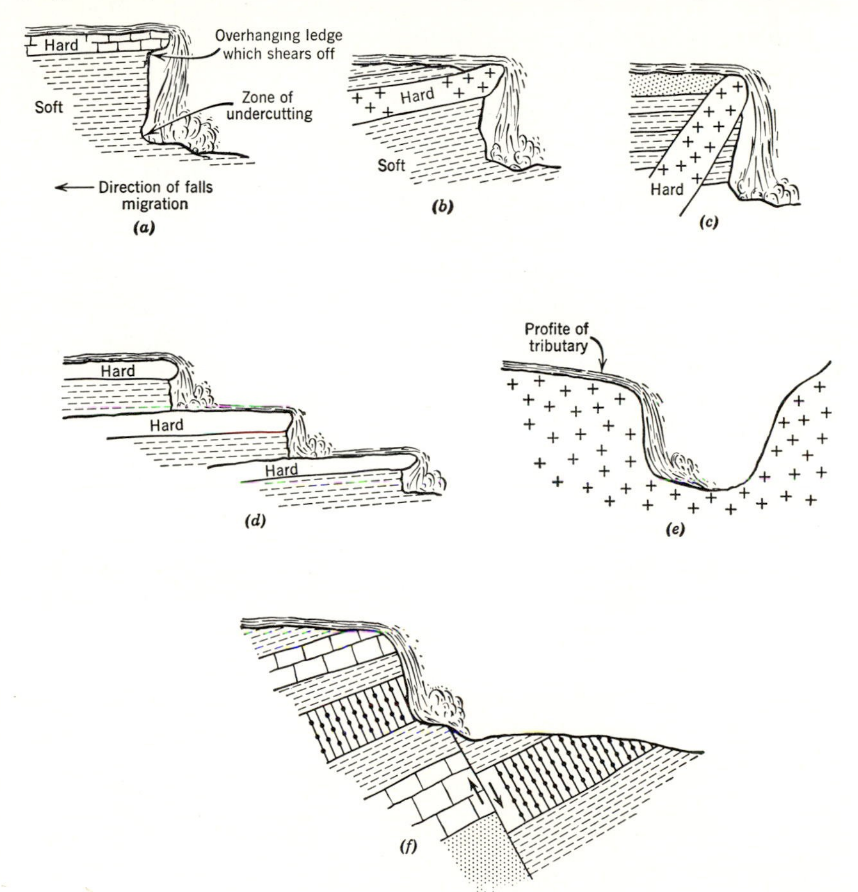

Fig. 7.13 Geologic conditions responsible for waterfalls. (*a*) Hard capping of a sedimentary rock or lava flow of horizontal attitude; (*b*) gently inclined hard layer interbedded with soft shales; (c) dike cutting sedimentary rocks; (*d*) alternating layers of hard and soft strata; (e) tributary left "hanging" as a result of deepening of a major valley by glacial scour; (*f*) displacement of strata by faulting.

master valley has left the tributaries "hanging" (Fig. 7.13*e*) or where fracturing of the earth's crust under stress has produced faults whose topographic expression is marked by a cliff (*scarp*, Fig. 7.13*f*).

As erosion undercuts the base of a falls the hard rock remains as an overhanging ledge. Eventually this can be supported no longer and shears from the face. By this process the waterfall retreats upstream and is gradually lowered to its eventual extinction. The famous Niagara Falls illustrates this condition, having migrated a distance of approximately $\frac{1}{4}$ mile in 260 years.

Rapids

Rapids occur where hard rock layers cross a stream such that local increases in gradient develop downstream or where any other obstruction such as boulders from a tributary stream produces the same effect. The condition approaches the waterfall except for degree of slope. Locally nonuniform water flow is produced and through the rapids the thickness of water is reduced as the water velocity increases. Above the rapids normal turbulent flow of water prevails, but as the water reaches the chute of the rapids the water becomes smooth, turbulence is reduced, and the condition of *shooting flow* is attained. At the lower end of the rapids where the gradient becomes lower an abrupt increase in the thickness of water occurs, and a standing zone of great turbulence develops, known as the *hydraulic jump*. In the zone of the jump as the water builds up to the height required by the downstream gradient much of the kinetic energy generated by the fall through the rapids is dissipated.

Swift-moving streams or areas of rapids and falls tend to produce eddies whose position may remain more or less fixed for much of the lifetime of the rapids or falls. Boulders, cobbles, and pebbles swirling around in these eddies erode a cylindrical hollow in the rocky bed of the stream or along the walls of the channel producing an erosion feature described as a *pothole*.

The Graded Profile

At all points where tributaries enter their trunk stream at a higher level accelerated erosion in the tributary eventually lowers the bed of the tributary to that of the master. Henceforth the two streams remain adjusted one to the other, and in essence their valleys are deepened and widened concordantly. This behavior expressed long ago as the *law of accordant tributaries* clearly demonstrates the principle that a stream system is a highly integrated veined network whose gradients are to an even-increasing degree controlled by that of the trunk system.

As erosion smooths the inequalities in gradient along a single stream the long profile achieves a slope which is adjusted to the sedimentary load it is carrying. Except for local conditions gradients are lower at the mouth of the trunk stream and increase gradually upstream but not at a uniform rate. The resulting profile as exhibited by the Mississippi and Colorado rivers (Fig. 7.6) is slightly concave upward and shows low rates of change in the lower course, but as the headwaters are approached the rate increases rapidly. Except for the headwater area such a profile is an equilibrium condition adjusted to available discharge and channel dimensions to provide the slope necessary to carry the load of sediment supplied from the upstream drainage area. That segment of any stream which has achieved such a condition is designated as being *at grade,* and that part of the stream is called a *graded or poised stream.* Headwater areas generally are not graded, and local segments in the downstream portion may be similarly ungraded.

As a stream system increases its length by headward erosion that segment which is poised gradually is lowered in elevation to coincide with that part of the stream profile which is approaching asymptotically the level of base level.

Controls of the Slope of the Graded Profile[4]

Despite the fact that several segments of a stream system may be poised or at grade, gradients in these sections are not the same necessarily. Actually the range in gradient may be considerable depending upon the part of the stream in which the poised sections occur. Even within the same segment the gradient may differ because of the variability in conditions in portions of the stream. The range in gradients of different graded profiles can be attributed to:

1. Increase in discharge downstream.
2. Changes in the sediment load.
3. Decrease in the particle size downstream.
4. Channel shape.

Collectively or individually each changing condition alters the graded slope. In the case of streams which drain into arid basins some of the graded slopes are as steep as several hundred feet per mile, and others can be observed as having gradients of a few feet per mile; indeed, the reader should not necessarily associate the graded profile with gentle gradients of very fine materials.

Increases in discharge downstream. Large poised streams usually have lower gradients than small ones, and a poised trunk stream has

[4] This section is based largely on ideas expressed by J. Hoover Mackin, "Concept of the Graded Stream," *Geological Soc. Amer. Bull.* Vol. 59, 1948, pp. 463–512.

a lower gradient than its graded tributaries. This behavior appears to be a response to changes in the cross section of the channel which in its most general form approaches a rectangle. The area of the cross section is approximately equal to the depth times the width, and the perimeter is twice the depth plus the width. Note that the cross-sectional area increases as the square, whereas the perimeter increases arithmetically. Because this relationship exists the frictional resist-

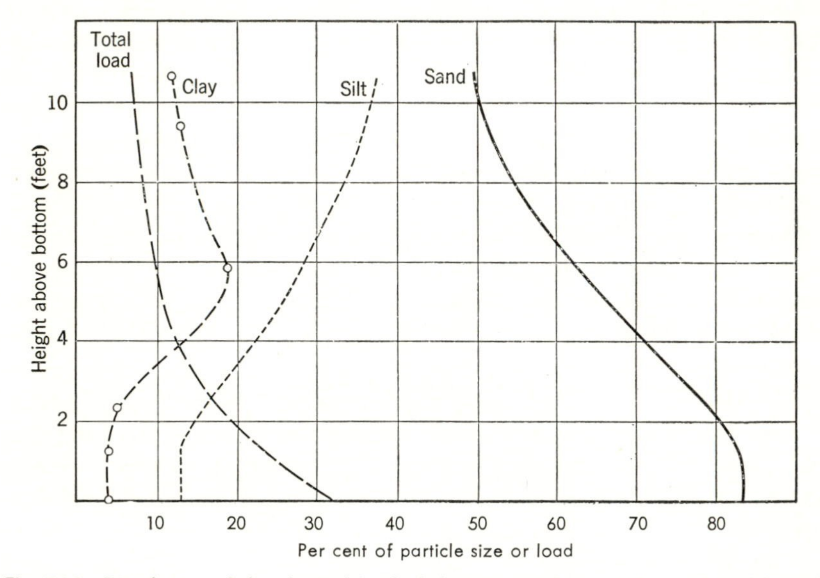

Fig. 7.14 Distribution of the physical load of the Missouri River at Kansas City, Missouri. The total load increases rapidly near the river bottom. Of the sand, silt, and clay fraction the sand is the primary constituent, but as the sand content diminishes in the upper water the silt content rises. (Data from Straub, *Physics of the Earth, Hydrology,* Dover Publications.)

ance exerted against the flow of water by the channel walls and stream bed is less in the trunk stream than in the total of the graded tributaries. A trunk stream is able to discharge the same amount of water with a lower gradient than its tributaries.

Changes in sediment load. Loads of transported sediment per unit volume of water vary along a stream course principally as a result of confluence with tributaries. Some tributaries are heavily laden and correspondingly increase loads in the master stream; others are largely clear water and serve to dilute the amount of sediment carried below the stream junction. Changes in gradient of the trunk stream develop accordingly, and local increases in sediment load indicate river segments of lower gradient than those adjoining.

Decrease in particle size downstream. Graded streams tend to sort out their sedimentary load in a downstream direction. Pebbles, cobbles, and even boulders are a common part of the load in the headwaters, whereas downstream sands, silts, and clays make up the load. During this transport the bulk of the coarse part of the load is concentrated along or near the bottom of the stream as shown in Fig. 7.14. With seasonal fluctuations in velocities a gradual winnowing process occurs, and periodically coarse material is dropped and fine material carried along. Gradually layer upon layer of deposits accumulate in irregular alternations of coarse and fine grain size. Where the sediment is coarse gradients are relatively more steep than where the deposits are of finer grain size. Thus the particle size distribution is diagnostic in indicating gradient distribution throughout the graded segments.

Channel characteristics. Variations in cross-sectional shape of the channel are typical of graded streams. Generally heavily loaded streams have a broad and shallow channel, and the cross section appears to be an adjustment to a form which will give the stream its maximum debris-carrying capacity. Also channel shape is controlled by the ease with which the bank can be eroded. Where banks consist of silts which are readily eroded the channel becomes wide and shallow and tends to shift its position at irregular intervals. Gravel banks appear to be much more stable, and the stream is more generally confined.

THE VALLEY CROSS PROFILE

General Statement

Cross profiles of valley and stream channels are important criteria to an understanding of the condition of the stream, its stage in development, and its past history. Headwater streams tend to occupy the notch at the base of a V-shaped profile. Stream banks are narrow or absent, and the water surface is in contact with the walls of the valley. Trunk streams of considerable length approaching a graded condition occupy broad valleys in which the stream channel fills only part of the bottom of the valley. Still further downstream where the long profile indicates an approach to base level valleys become extremely wide relative to the stream channel, and the stream wanders with great loops across the valley floor. Each profile is characteristic and independent of the actual dimensions of the stream and valley;

it is a response to conditions largely established by the position along the long profile.

Development of the Cross Profile

V-Shaped Notch

Initial runoff channels are likely to be reasonably symmetrical in cross section but variable in gradient. Channels of steep gradient permit the moving water to develop enough kinetic energy to remove soil particles and carry on erosion. Erosion is concentrated in the belt of maximum turbulence which is distributed more or less equally near the bottom and along the lower part of sides. The channel begins to be cut into a trench of rectangular cross section. With each succeeding rain the trench is deepened and widened by a process of undercutting. Depending upon the cohesive strength of the soil material and upon the vegetative cover a slope will be developed at which the soil will no longer maintain its steep walls and will slide into the trench. Progressive sliding results in a change in channel from a dominant rectangular trough to a small valley of V-shaped cross section. At the base of this notch flows the stream channel of rectangular cross section.

When the channel is deepened sufficiently to intersect the water table (Fig. 7.15*d*) water flows more or less continuously and erosion proceeds at an accelerated pace. Normally, during this stage of valley deepening bedrock is reached either above or below the water table. The case illustrated in Fig. 7.15 shows the bedrock lying below the water table. The permanent stream proceeds to erode its valley into the bedrock, but the tensile strength of the bedrock is greater than that of the overlying mantle, and the valley is correspondingly steepened, but the channel width is narrowed.

Irregularities in the V-shaped notch. In cases where the bedrock consists of alternating layers of hard and soft strata such as limestone and shale (Fig. 7.15) the valley tends to narrow and steepen in the hard layers and widen in the soft layers. The effect is to produce a series of steplike irregularities in the valley profile, recognized with particular ease in areas of scanty vegetation. Such *rock terraces* are typical of many canyons in arid regions of the world and are spectacularly displayed in the Grand Canyon of the Colorado River.

Still other modifications in valley width and steepness are responses to variations in rock hardness. Narrow gorges develop where hard rocks occupy both sides of the valley (Fig. 7.15*f*). Typically this condition is produced by the presence of igneous intrusions such as

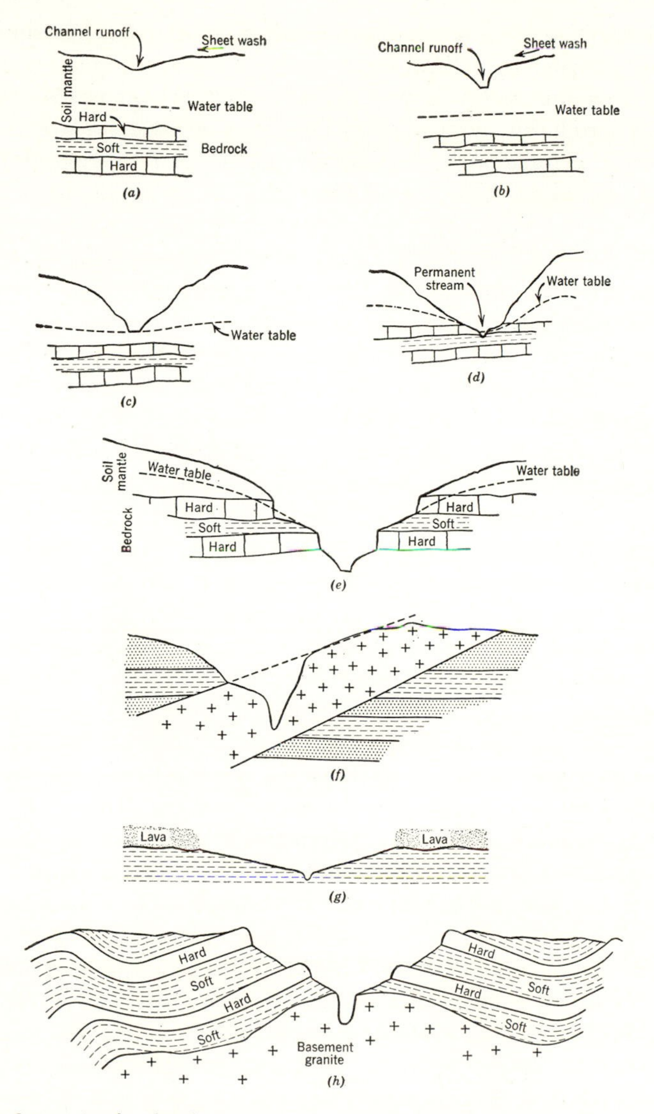

Fig. 7.15 Stages in the development of the V-shaped valley cross section. (*a*) Initial stage in which the channel develops in soil or sediment consequent to local surface irregularity. (*b*) Late initial stage (intermittent stream) of rectangular cross section of a small channel in the sediment. (*c*) The V-shaped valley is modified by the differences of resistance to erosion of hard and soft zones. (*d*) The permanent stream is initiated and the channel is cut into bedrock. (*e*) The shape of the V-shaped valley is modified by alternating hard and soft strata. (*f*) Valley cut in position between a dike and sediments illustrating the effect of hard and soft rocks on cross-sectional profile. (*g*) Broad valley developed in shale below lava caps. The shale is readily eroded, undermining the lava and causing rapid widening of the valley. (*h*) Common development of a gorge in igneous rocks where a stream has been superposed from overlying sedimentary rocks.

large dikes or stocks surrounded by softer sediments. As the stream cuts into the igneous rock a narrow gorge results, whereas in the sediments the stream valley is wider. The majority of cases of narrow gorges are produced by structural deformation. Strata are contorted into a series of folds or displaced by faults (Fig. 7.15*h*) so as to bring basement igneous and metamorphic rocks into position to be eroded.

Widening the Valley Floor

In a qualitative sense valley deepening ceases to be an important process as the valley becomes graded. Valleys carved in regions of low elevation arrive at this condition without ever having been deep or impressive. In mountainous or plateau regions gorges and deep V-shaped valleys are commonplace.

When the stream becomes graded a new stage in valley development is initiated, and valley widening becomes the important process. Valley widening is primarily a function of the meandering or winding stream. Meandering appears early in valley development and is initiated by original irregularities in the valley. Changes in the position of velocity and turbulence maxima which arise as the result of such irregularities cause differences in the rate of erosion on either side of the stream bank, and the meander begins (Fig. 7.16). Generally as the stream approaches the graded condition meandering is a very obvious and characteristic feature.

Meander development widens the valley by undercutting the banks and shifting the position of the zone of maximum erosion (Fig. 7.16). Figure 7.16*b* diagrammatically indicates the thread of maximum velocity and turbulence in the meandering stream channel. Note that to an observer facing downcurrent the right-hand side (outside bend) of a meander is the locus of greatest concentration of energy. In this region undercutting of the bank and deepening of the channel occurs. On the left-hand bank (inside bend of meander) current energy is least, and erosion is in effect nonexistent.

Inequality in current energy is attributed to cross currents. These appear as the centrifugal force, developed by the water rounding a meander, piles water along the outside of the bend and produces a lateral inclination to the water surface. Water flows outward from such loci of abnormally high elevation to establish equilibrium, and cross currents appear. Note in Fig. 7.16*b* that at the locality marked x' any developed cross current opposes the main current and at x the current velocities are additive.

During high-water stages the thread of maximum velocity appears to hug the inside bends of the stream. As the water recedes and velocities diminish the position of maximum flow is shifted to the outer bends.

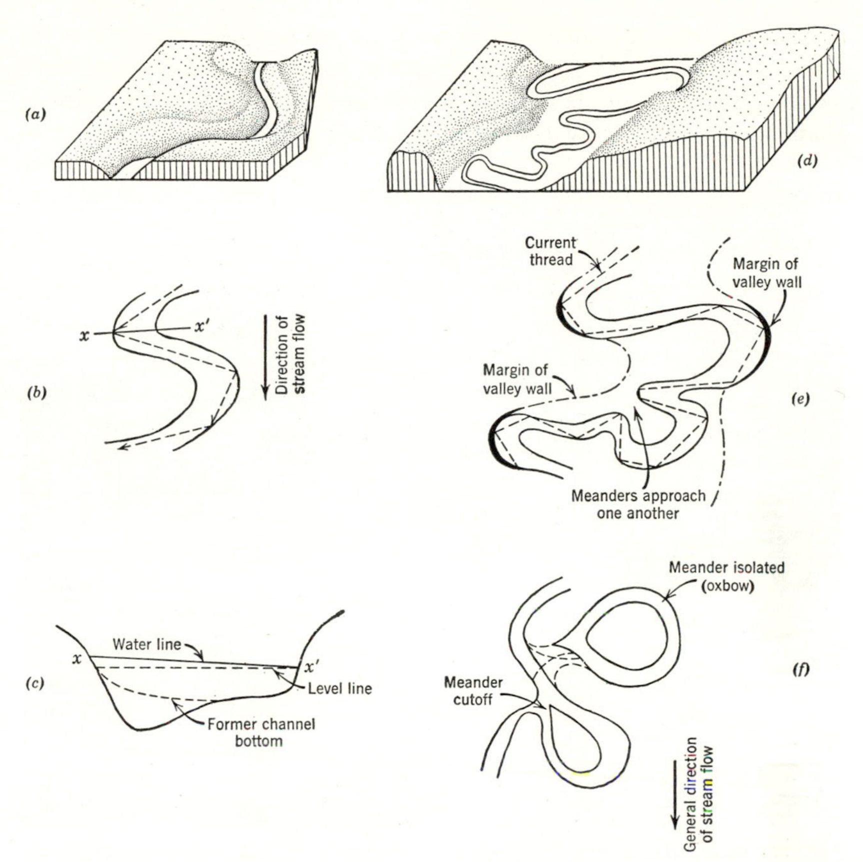

Fig. 7.16 Stages in the development of a meandering stream and an oxbow lake. (*a*) Meandering stream with current impinging against the outside bends. (*b*) The thread of the major current. (*c*) Inclination of the water surface due to centrifugal force as the stream swings around the meander bend and the location of the position of maximum erosion of the bed. See also Fig. 7.10c. (*d*) Meandering stream eroding valley walls at irregular intervals along the stream course. (*e*) Meanders approach one another as erosion of the valley alluvium occurs. (*f*) Cutoff occurs, the maximum current threads through the shortest distance, current in the bend is reduced, and the oxbow lake is isolated by deposition at the "neck" of the old meander.

Deposition of sand occurs in the form of bars projecting channelward from the inside bends of the meander. Bars of this type often transect one another in complex fashion as meander migration causes shift in the loci of bar deposition.

Undercutting of the outside bends of meanders increases the amount of meander curvature and causes meander loops to approach one another as illustrated in Fig. 7.16*e* and *f*.

As the valley walls retreat and the valley floor becomes extensive the meandering stream touches the wall at fewer and fewer points. The sinuous stream is confined largely to meandering within its own flood plain. Meander cutoffs (*oxbow lakes*) are common and the scars of abandoned channels are readily seen on aerial photographs and topo-

Fig. 7.17 Meander scars as seen from the air. Note how the progressive positions of the meander in the left foreground are indicated. The north shore of Hamilton Inlet, Labrador. (Photograph by L. H. Nobles.)

graphic maps (Fig. 7.17). Correspondingly the valley walls or bluffs are increasingly dissected by tributary streams and sheet wash whose total effect is to decrease the angle of slope of the valley. These processes lead to the development of the third major cross-sectional aspect of the stream valley.

Lateral widening of the valley walls and stream meandering across the valley floor develop a generally smooth surface sloping downstream with the gradient of the graded profile. Inasmuch as this valley floor is essentially at stream level, it is subject to floods during high-water conditions. During such times of flood much of the valley floor is covered with water which is in intense turbulent flow downstream with lower average velocity than that in the main channel. As the flood stage recedes water velocities outside the channel decline and eventually become stagnant. In such areas the solid load carried by the flooded stream is deposited on the valley floor, and a seasonal increment of sediment is added.

Filling the Valley Floor

Intermittent flooding of the wide valley is a common method of aggrading or filling the flood plain. Other conditions may accomplish the same result such as rise in level of base level, loss of channel water, or local damming due to landslides. In each case the process is one where there is local reduction in the stream's velocity and the supply of debris is more than the stream has capacity to carry with the available energy. The excess debris is dropped, particularly outside the stream's normal channel. Local deposition upsets the adjusted stream gradient, and debris begins to accumulate progressively upstream in an attempt to establish a new graded profile at a somewhat higher level than the one preceding.

Aggradation is particularly important in the lower reaches of the stream where the gradient is approaching base level. Valley filling in this area causes a substantial rise in the graded profile upstream, and in large river valleys several hundred feet of fill may occur between the present surface of the flood plain and the floor of the old valley cut into bed rock.

During floods overflow from the channel undergoes a rapid velocity drop when no longer confined between banks, and the heaviest deposition occurs adjacent to the channel. Gradually a narrow sinuous ridge is built on either side of the channel and rises above the general floor of the valley. Such ridges known as *natural levees* serve to confine the stream within its channel until breached or topped by the next important flood. Water between natural levees and the valley walls tends to remain generally stagnant as drainage from the levees is primarily away from the main channel. Such areas become great border swamps, and decayed vegetation is an important part of the deposits which gradually accumulate.

Periodically, local sections of the border swamps are buried beneath deposits of sand, silt, and clay which are supplied through a breach in the natural levee or by a new meander cutoff. The normal deposits of the valley bottom are an alternation of organic mud, sand, silt, and clay in an unsystematic order but in layers whose total thickness may exceed 200 feet. In the main channel throughout its sinuous and migratory route the deposits are much coarser and consist of gravels and sands distinct from the finer alluvium of the valley bottoms.

Tributaries to the main valley enter the border swamp areas, but because of the higher ground of the natural levees such streams are forced to flow in a route which parallels the main stream for miles before final junction. Thus, the broad valley becomes a complex network of winding streams all flowing in channels generally parallel, and not uncommonly each is contained within its own bounding natural levees. This parallelism in flow direction of trunk streams is characteristic of the lower reaches of the flood plains of major streams irrespective of their geographic distribution, and it is in this part of the valley where the most variable alluvial fill is to be found.

THE STREAM'S LOAD

General Statement

All naturally flowing water carries some sort of a load. Part of this is the rock debris which contributes to the turbidity of the stream and is designated as the *physical load.* Another form of load consists of the dissolved mineral and organic matter carried in ionic or molecular state and is identified as the *chemical load.* Ordinarily, when reference is made to the stream's load only the physical load is being considered because this is the visible part which actively affects the shape of the stream channel and valley. Statements concerning the chemical load are made primarily in connection with water supply for public or industrial use. For such demands the quantity and character of the dissolved material become very important.

The Chemical Load

The chemical load of stream waters is the primary source of salinity in oceans and evaporating lakes in arid regions. In some streams the dissolved solids are very small in amount, and in others the content is high. The normal association is for streams draining wet regions to have a low percentage of dissolved solids and generally for streams in dry regions to have high salinity. However, the total volume of salts carried to the sea is much greater from the wet areas of the world than the amount delivered from the dry regions. This apparent paradox can be attributed to the enormous volume of water transported by the streams of the drainage basins of humid regions. In terms of this great volume the quantity of dissolved salts measured in parts per million by weight is small, but the actual tonnage involved is enormous. For streams of arid regions where the water volume is small the relative percentage of salts carried is high, but the total annual tonnage is low. As a rough approximation of relative quantities of water being discharged by two streams compare the Wabash and Gila rivers (Fig. 7.5), draining humid and arid regions, respectively.

Among the most abundant ions are Ca, Mg, Na, K, CO_3, SO_4, and Cl. Silica is important, also, particularly in rivers draining wet tropical and subtropical belts. In these areas this substance may constitute half the quantity of dissolved salts.

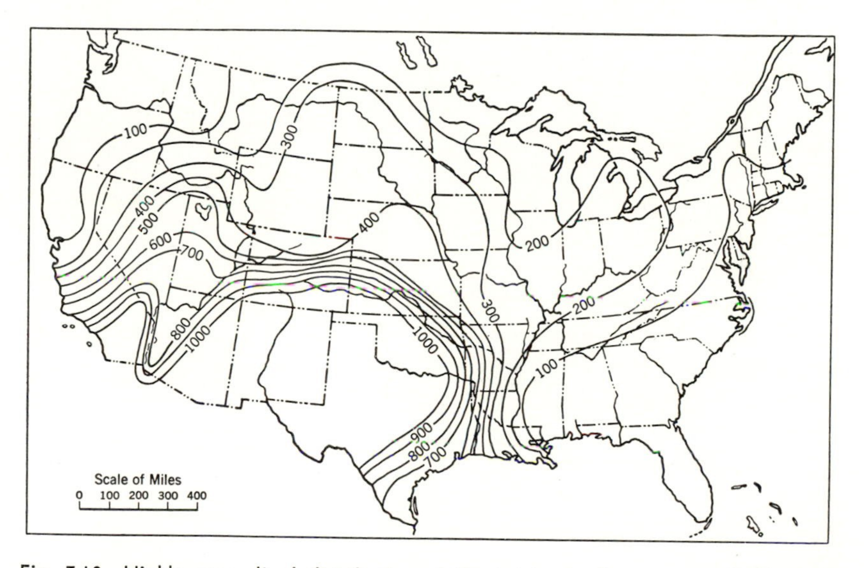

Fig. 7.18 Highly generalized distribution of the total salts (parts per million) in the river waters of the United States. (Data from Clarke, *U. S. Geol. Survey Bull.* **770.**)

Distribution of salinity of rivers for various sections of the United States is shown by the map Fig. 7.18. Here by generalized contours of total dissolved salts in parts per million is illustrated the concentration of areas in the southwestern states where rivers are exceptionally saline. By way of contrast the extreme northwest and southeast parts of the country are areas where the river waters have the lowest salinities. The belt of high river salinity confined to the southwestern quarter of the United States extends eastward into the more moist portion of Texas. This condition is attributed to evaporation from the eastward-flowing streams which do not decrease in salinity until the high rainfall belt is reached near the Mississippi Valley. Northward as evaporation rates progressively decline so also does stream salinity.

One of the most clearly defined responses to rainfall distribution is to be observed in the inverse relationship between the abundance of CO_3 ion on the one hand and chloride and sulfate on the other. In regions of high rainfall carbonate dominates. With increase in aridity the SO_4 and Cl ions increase at the expense of the carbonate. (See Figs. 7.18 and 7.19.) Rivers high in carbonate are always low in SO_4

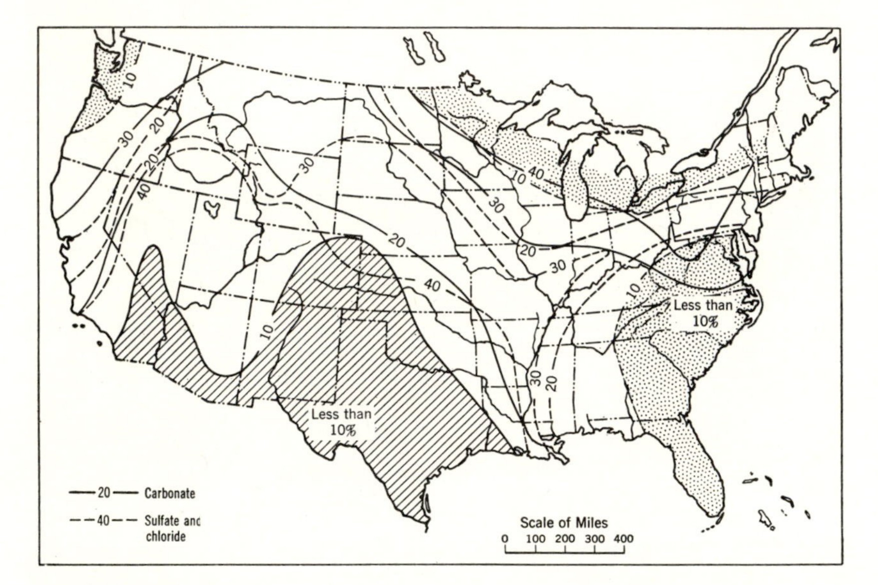

Fig. 7.19 Highly generalized distribution (parts per million) of the carbonates, and the sulfates and chlorides in the river waters of the United States. Note the inverse relationship between the sulfate and chloride content and the carbonate content. Carbonates increase in the direction of the moist areas, whereas sulfates and chlorides increase in the direction of aridity. (Data from Clarke, *U. S. Geol. Survey Bull.* **770.**)

and Cl ions. Such is the characteristic of rivers draining areas with abundant rainfall and heavy vegetation. Vegetation funnels abundant carbon dioxide to the ground water and in cool regions definitely increases the amount of CO_3 ion being removed from the subsurface and eventually by streams. In dry regions where vegetation is scanty carbonate tends to be precipitated in the soils, and the river waters are low in that ion but dominated by sulfate and chloride.

Chemical Erosion

Each stream carrying a chemical load from its drainage basin slowly but surely contributes toward erosion of the surface rocks and lowering of the topographic surface. The contribution by individual streams is relatively small, but for drainage basins the figure becomes significant. For example, the total drainage area of the Mississippi approximates 1,265,000 sq. miles from which 108 tons of dissolved salts are removed per square mile each year. Values for other drainage basins in the United States are given in Fig. 7.20. These range between as

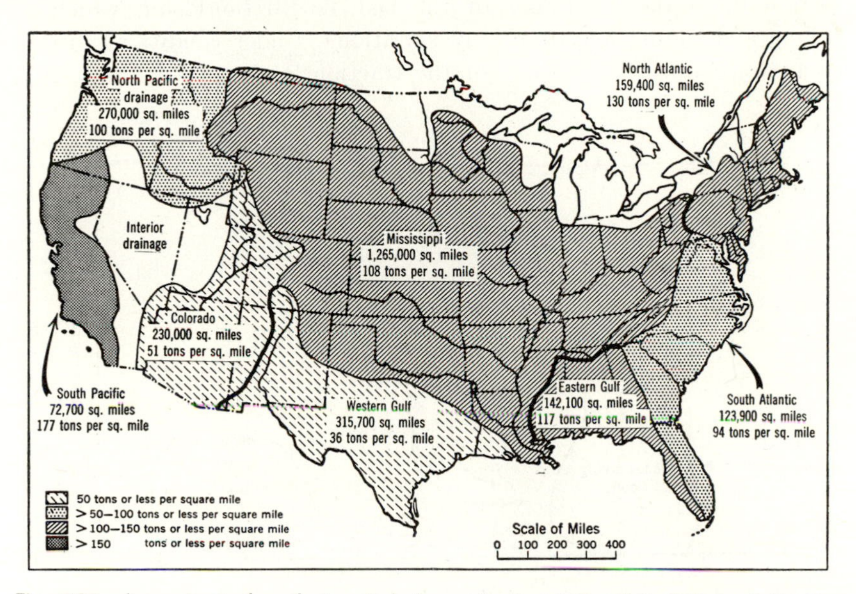

Fig. 7.20 Approximate boundaries of drainage basins of the United States indicating the area of the drainage basin (in square miles) and the amount of rock estimated to be dissolved yearly in each drainage basin (in tons). Note that despite the high concentration of total salts in the rivers of arid areas the principal solution occurs in the wet regions. (Data from Clarke, *U. S. Geol. Survey Bull.* **770.**)

low as 36 tons of dissolved salts in the western Gulf of Mexico drainage basin and a high of 177 tons annually per square mile from the South Pacific drainage basin. In a general way the correspondence between high rainfall and high solution is to be noted, and despite the higher concentration of salts in the streams draining dry areas the total chemical denudation is low per square mile.

The Physical Load

Primary consideration is usually given to the solid particle or physical load because of its importance in geologic processes. Fragments such as cobbles and pebbles generally slide and roll along the bottom, scarcely ever losing contact with the bed of the stream. Still smaller particles rise above the floor of the stream and by long or short leaps proceed downstream. Such a form of downstream progress is distinguished by its temporary suspension in the stream, and particles behaving in this way are said to be moving by means of *saltation.* At certain times the remainder of the bed, or *traction load,* which is rolling and sliding may move by saltation. Under conditions of decreased stream velocity some of the traction load stops moving alto-

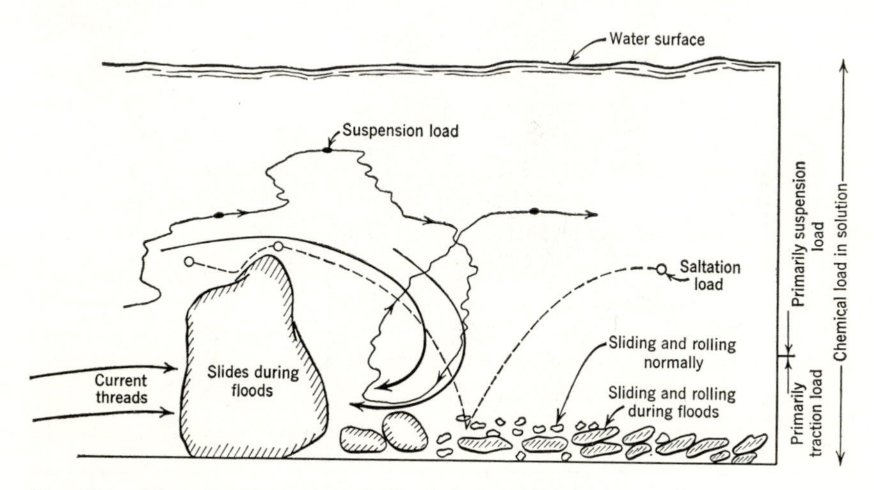

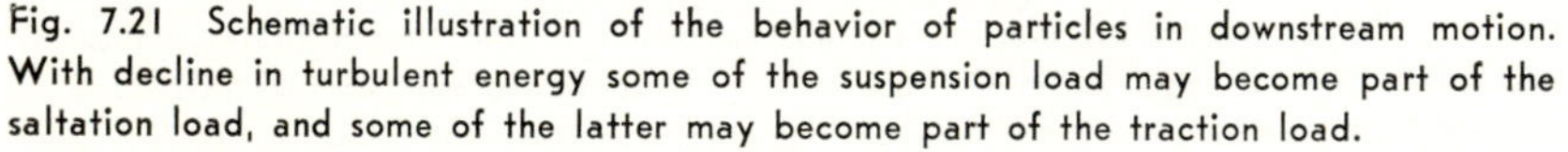
Fig. 7.21 Schematic illustration of the behavior of particles in downstream motion. With decline in turbulent energy some of the suspension load may become part of the saltation load, and some of the latter may become part of the traction load.

gether, and particles moving by saltation now proceed by rolling and sliding.

Particles of silt and clay size can be carried along by normal stream flow without coming to rest. Even low values of turbulence are sufficient to keep such small particles more or less indefinitely in suspension, moving up and down, from side to side, and backward as well as forward downstream. This is the *suspended load* as distinct from that moving by traction (see Fig. 7.21). Not all of the suspension load remains as such in its travel downstream. During floods coarse fragments are carried along with the fine, but the former settle out under lower velocity conditions. Sands and coarse silts belong in this category. Flood plains receive much debris of this size carried as suspended load and dropped as the river recedes to its normal channel.

Clay-sized particles probably are carried entirely by suspension and are distributed most uniformly throughout the stream. This is illustrated in Fig. 7.14 in which is shown the distribution of the load of the Missouri River at one sampling site (Kansas City, Missouri). The percentage of clay is less near the stream bottom than near the top because the coarse fraction is concentrated in the lower part of the stream. However, the actual amount of clay per unit volume of water is greater near the bottom than at the top and indicates the extent to which some clay settles from the suspension.

In the example shown silt is carried primarily in suspension and is more abundant than clay. Sand despite its abundance in suspension is moved primarily by traction and increases sharply in amount as the bed of the stream is approached. Still larger fragments spend most of their time in a fixed position and move as a part of the traction load by sliding short distances.

Every consideration of load eventually involves the particle size distribution of the material carried and the size limits of the largest particle moved by the stream at selected spots. The measure of the stream's ability to transport particles of a certain size is its *competency*. If a graph is prepared of stream competency with increasing distance downstream the resulting points will plot along a curved line whose slope generally simulates that of the long profile of the stream.

The quantity of load which a stream can transport past any position along the stream profile is termed its *capacity*. The capacity of a long stream increases rather rapidly at first and then levels off very slowly toward the mouth. Capacity is difficult to determine because of the uncertainty in the total quantity of load moved along the stream bed. Also, the amount of very fine clay which can be held in suspension appears to be much larger than most streams carry. The chief use of the term is to provide a concept that as the total energy of the stream increases its ability to carry a load is increased also. During times of flood the capacity is considerably greater than during low-water stages. As the flood recedes and the capacity falls some of the load is dropped on the flood plain. If the stream is graded this debris will gradually be moved downstream.

Moving the Physical Load

Erosion of valley walls and stream bottom requires expenditure of more energy to dislodge rock particles than is necessary to transport the particle when once in motion. In part this arises from the abrupt decrease in velocity which must occur in a very narrow zone between the water and channel wall along the perimeter of the channel. Here, the stream energy is low, and unless the threads of turbulence exert a strong upward and outward movement picking a particle from the

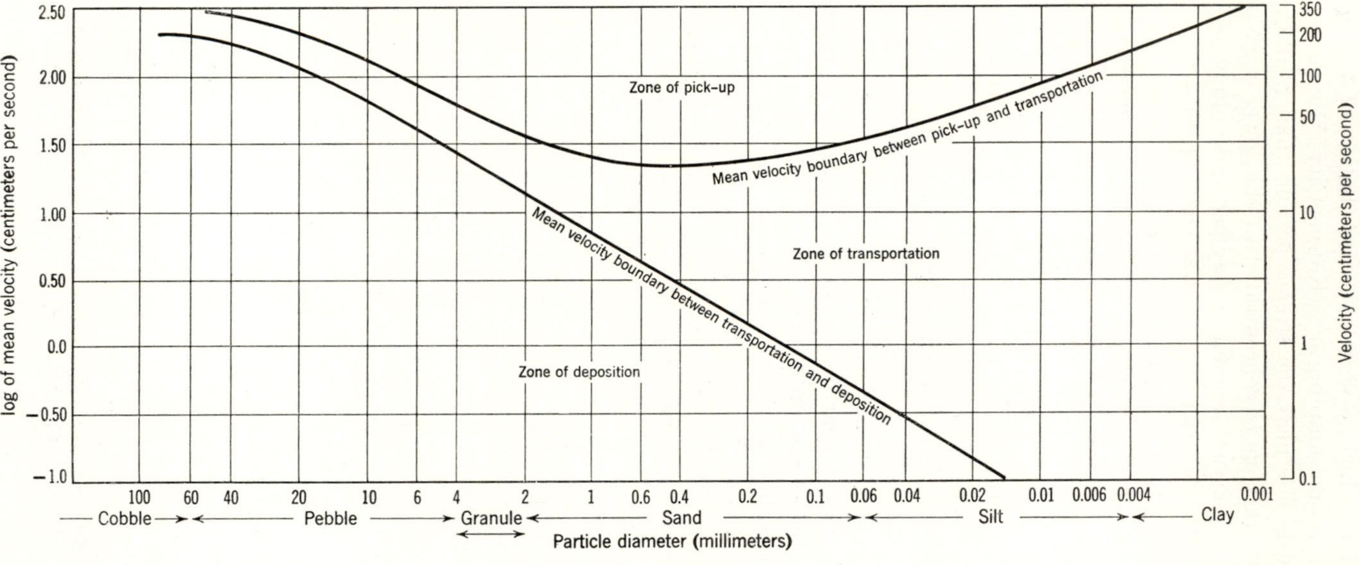

Fig. 7.22 Hjulstrom's diagram, somewhat simplified, showing the relationship between the average stream velocity and the movement of particles of specified size. The field of deposition indicates that below the boundary line current velocities are insufficient to propel particles downstream. In the field of transportation velocities are great enough to keep the specified particle in motion. Above the line marking the upper limit of the field of transportation, particles may be picked from the stream bed in accordance with the velocities indicated. Note that particles of medium sand size are caused to move at lowest velocities and that velocities needed to pick up clay (from a clay layer) are as great as those for pebble size. (Hjulstrom, *Recent Marine Sediments,* Am. Assoc. Petrol. Geologists.)

stream floor may require more energy than is available. Particles in saltation are important in this connection as they impart some of their kinetic energy to other particles struck during the leaping process. Also, sliding and rolling particles tend to break off smaller pieces as well as start other particles in motion. Information currently available indicates that considerably more velocity is needed to initiate particle movement among clays than to carry such a size in suspension. This concept is illustrated in Fig. 7.22 in which are shown velocities required to start in motion particles of various sizes resting on the stream bottom and velocities necessary to keep such particles moving.[5] Note the outline of the base of the field marked pick-up. This is the limiting boundary of velocities necessary to start particles in motion from a rest position. Among the coarse sizes extending into the fine sands the behavior is as anticipated, namely that large particles require greater water velocity to start them in motion than do smaller ones. Still smaller particles are not started in motion at such low velocities, and it is to be noted that velocities required to start a clay particle moving are approximately the same as those required to start pebbles. Sand is moved when clay and pebbles will remain on the channel floor. This behavior tends to concentrate sand in the stream's load and to accelerate erosion of sandy bottoms or banks. In contrast firmly packed clay- or pebble-armored banks are far more resistant to erosion and tend to remain fixed unless undercut.

The field of transportation is marked by a lower limiting boundary whose slope indicates that higher velocities are necessary to keep large particles in motion than are required for small particles. In short, the smaller the particle the less current velocity is required to keep it in transport once initial movement has begun. Thus, clay particles remain suspended in quiet water, whereas silt, sand, and granules require increasingly higher velocities to keep them in motion. The end result is concentration of sizes in the sediment which is deposited, and each individual layer tends to have a small standard deviation in size distribution. Great fluctuation in stream velocities causes layers of fine debris to alternate with those of coarse sizes in an unsystematic order.

Systematic Filling of the Valley

An interesting feature of the conditions illustrated in Fig. 7.22 is the departure in slope of the boundaries of erosion and deposition as

[5] Filip Hjulstrom, "Transportation of Detritus by Moving Water," *Recent Marine Sediments,* Amer. Assoc. Petroleum Geologists, 1939, p. 10.

current velocities fall. Note that wherever velocities are high and fluctuating coarse particles can be dropped to settle on the stream bottom but can resume movement with only slight increase in velocity. During spring freshets headwater regions tend to attain the high-water velocities of the previous spring and pick up particles dropped during the summer decline. In the lower courses of large streams where velocities are low, even during floods, silts and clays dropped from the bed and suspended loads are unlikely to be moved again except under high velocity conditions as indicated by Hjulstrom's diagram. Where streams are slow and sinuous the load accumulates in the channel particularly in the meander bends. From these loci of deposition sand may be removed during quickening of the current, but deposits of clay will tend to remain fixed.

Filling of the channel makes flooding easier when high water appears, and although some of the channel is deepened at that time the net result is to spread more water over the flood plain and add to the deposits. Natural levees are built higher, and the backswamp areas receive more silt. In the long run this process raises the level of the profile over the entire length of the graded part of the stream, and deposition occurs all along until the new profile is attained.

During the aggradational process the meandering stream changes its course across the flood plain. As a meander is cut off the channel shifts without system except to migrate by great loops in a generally downstream direction. Channels with their bed filled with coarse debris are cut off suddenly and left to be buried by finer deposits as the former channel becomes the site of marginal backswamps. Through this process the valley alluvium becomes a heterogeneous sediment which consists of a network of old filled channels, natural levees, and border swamps whose subsurface pattern is extremely complex. Except through information obtained by drilling the details of the valley fill cannot be determined.

Changes in Base Level

Large river systems whose history spans a significant part of late geologic time have experienced changes in base level. This has been particularly true during and since the last glacial stages when sea level fell and rose during the alternate periods of glaciation and melting. Lowering of base level establishes a new graded profile at a lower level, velocities increase, and downward cutting is accelerated or renewed. The bedrock channel is deepened, and sections of the old valley floor are left as a *terrace* along the walls of the new valley, or the terrace may be removed completely by erosion. Subsequent rise in base level necessitates filling of the new channel, and if the rise is great enough even the terraced remains of the old flood plain may become the site of additional deposition.

Stream terraces which represent partial dissection of alluvial fill are not uncommon features in the landscapes of today. They are particularly well developed in the semi-arid regions of western United States in basin areas adjoining mountains. In these localities they represent the dissected remains of extensive coarse alluvial fill which was carried from the recently elevated mountains by streams of high competency. Later, as the climate became more arid and the flow of water was reduced the streams were no longer carrying much debris from the mountain sites, and dissection of the fill to make a terrace took place.

Although the change in climate from humid to arid is responsible for some terrace development changes in base level or elevation of the land is a far more common cause of origin. Periodic elevation of the land area frequently is documented by a series of terraces one below the other in steplike arrangement, either spread over a large semi-arid basin or more characteristically within the valley walls. Individual terrace levels slope down the valley in accordance with relative levels of the old stream gradient, and partly isolated terrace remnants are identified on the basis of their elevation. By a systematic procedure of comparison of terrace levels the surface of the old valley floor can be reconstructed and compared with those of the present stream or pre-existing ones. Relative changes in the base level due to actual lowering of outlets or rise in land elevations also are reflected in the size distribution of deposits within the valley and the terraced remains. Coarse debris fills the old channels of the low stand of base level as the local competency of the stream is increased by the steepened gradient. Conversely, as the base level rises increasingly finer grained materials fill the valley.

Oftentimes a rise in base level lowers the capacity of the loaded tributary streams to such an extent that heavy aggradation ensues. The single river channel can carry its load no longer. As the latter is deposited the individual channel is forced to subdivide into a series of anastomosing smaller channels separated by many low islands which are the products of heavy deposition. This condition is known as a *braided* stream. Segments of streams which are braided are abnormal in many respects. For example, large-scale meandering effectively stops. Each individual channel is relatively straight, and the meander system as ideally demonstrated in the single channel is destroyed in essence. In and around the many channels most of the load is dropped, and at the downstream terminus of the braided section the suspended load is reduced to a minimum. Here, where the channels coalesce once more into a single channel normal behavior is resumed.

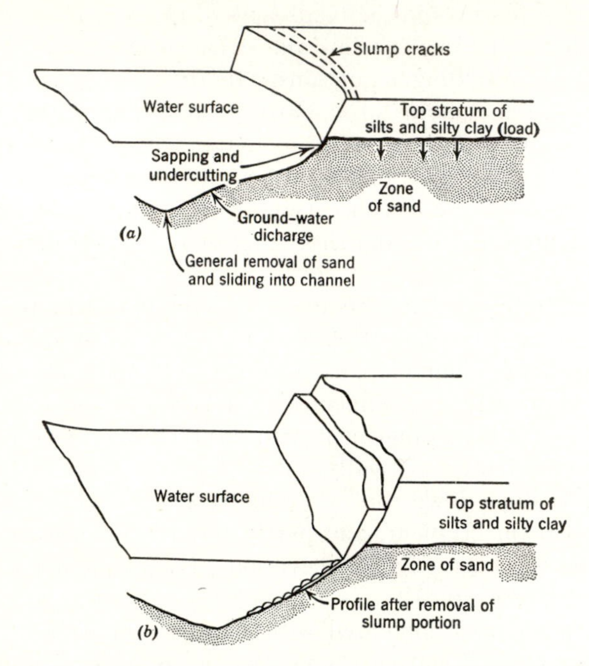

Fig. 7.23 Condition leading to the caving of channel banks in alluvium. The top layer of silts and silty clay rests upon water-saturated sand being carried off by prevailing current velocities. Undercutting results in slumping of large masses into the channel. (From Fisk, *Mississippi River Commission*, 1944.)

Meander Migration and Local Bank Recession

Pronounced meandering appears to be related to the generally or slightly aggrading stream rather than one which is overloaded locally way beyond capacity. When aggradation becomes very important the river course begins to straighten and to develop braided channels. Ideal meandering streams are those characterized by intense meander migration across the flood plain and cutting off of meanders. In such streams the customary pattern of meander migration is downstream, and the progressive position of old scars is a characteristic element of the flood plain. (See Fig. 7.17.)

In broad valleys where the meandering stream is cutting into its own deposits caving of channel banks is very well developed where the proper conditions exist. Such ideal conditions are to be found where fine silts rest upon a layer of sand in which ground water is moving toward the channel. In this condition sand tends to be mobile, and under the load of silts the sand beneath the water surface slumps and virtually flows into the channel producing a deeply undercut bank. The undermined top stratum of silt has no support, and collapse occurs into the channel along a shear surface of arcuate outline. These large "bites" into the side of a meander indicate a site of rapid erosion and accelerate meander cutoff and migration. (See Fig. 7.23.) Wherever the valley fill consists of alternate layers of sand and silt intense undercutting and meander migration are to be expected, and in these areas the river system can be considered extremely unstable.

THE DELTA

Definition

An alluvial tract spreading outward into the body of water into which a stream empties is a *delta*. Its precise shape and limits are not clearly defined because its deposits are gradational with others generally not considered to be part of the delta. In an upstream direction the deposits merge into the normal floodplain sediment, and the point at which the delta is said to begin becomes a more or less subjective decision. Downstream a similar indefinite boundary exists. The subaerial part of the delta merges into the subaqueous part which extends well beyond the visible shoreline. Gradually the delta sediments thin, and the nearshore sea or lake sediments begin.

The position and extent of a delta vary from time to time. When the stream load dumped at the mouth exceeds the ability of the waves and currents to distribute this load the delta is extended. Wherever and whenever the wave and current attack is sufficient to move all the load the stream carries delta growth ceases. If current and wave attack can transport more than the stream load the delta recedes or never develops more than a short snout. Small deltas or their absence, therefore, do not necessarily indicate that the stream's load is small although frequently this is the correct assumption.

Under all conditions, and whether large or small, certain external features characterize what is generally conceded to be a delta. These are:

1. Subdivision of the major water channel into several *distributaries*. These generally point upstream and join the main channel at large angles (i.e., as much as 90 degrees).

2. Projection into the body of standing water beyond the general shoreline.

3. Outward spreading of the alluvial fill into the body of standing water.

4. A series of low islands bordering the emergent part of the delta.

Delta Growth

Growth of deltas may be controlled by base level rise and fall. Lowering of base level accelerates delta growth enormously, whereas rise in base level decreases the load carried to the mouth by causing aggradation upstream. Also much of the emergent part of the delta is then submerged and subject to destruction by current and wave attack.

Most deltas are compound growths primarily the result of the development of new distributary channels. Ideally the delta begins with the growth of a lobe (Fig. 7.24) in which a single channel is restricted between its natural levees. These are gradually extended outward into the body of standing water. As the channel length is increased the gradient is progressively lowered; flooding of the natural levee results in a breakthrough and the appearance of a new and shorter distributary. The new route commands the largest discharge of water and carries most of the load, and a new delta lobe is built. In time this distributary becomes excessively long and another breakthrough produces a new distributary and a new delta lobe. Debris from each new lobe in part buries some of the old delta, and gradually the areas intervening between distributaries become filled and emerge from the shallow water.

Jet Theory of Delta Growth

Of late new concepts of delta growth have arisen based upon certain principles of current flow. One of these employing a theory of jet flow conceives a stream to be in equilibrium with its alluvial channel (i.e., its capacity and competence respond to any change in stream velocity).[6] Under the existing conditions a certain flow pattern is developed at its mouth. If this flow pattern is known its depositional pattern is subject to prediction as prescribed by theory. According to the concept the turbulent stream is considered to be a free jet flowing from a point source and can be related to two basic types of

[6] The concept of growth described is that proposed by C. C. Bates, "Rational Theory of Delta Formation," *Amer. Assoc. Petroleum Geologists Bull.* 37, 1953, pp. 2119–2162.

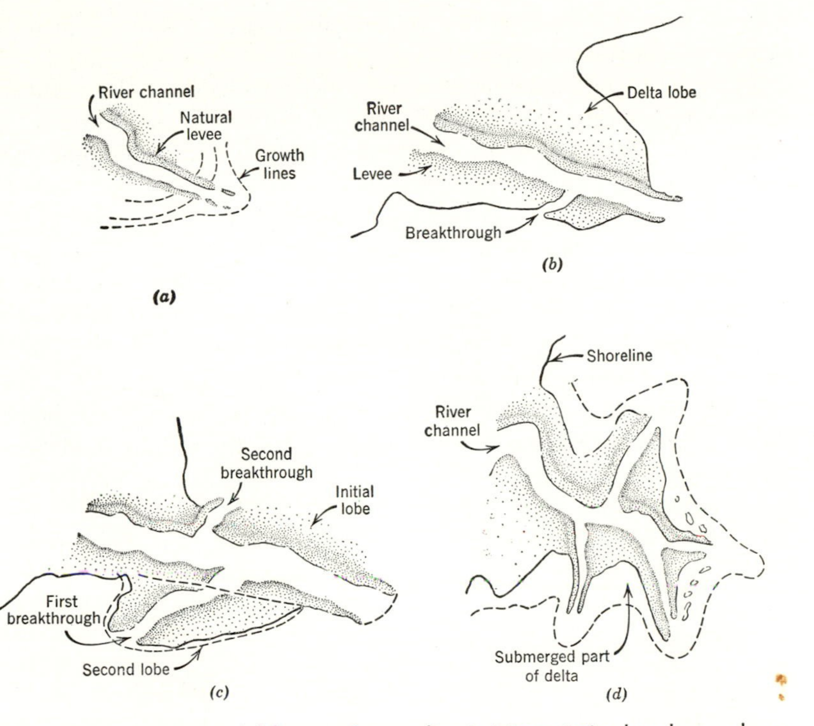

Fig. 7.24 Schematic stages of delta growth according to interpretation based upon change in gradients of distributary channels. (*a*) Natural levees extend the channel beyond the shoreline. (*b*) Breakthrough in the levee permits a new distributary of steeper gradient than that of the old channel to become the preferred water course. (*c*) Following the period of extension of the levees and the growth of the second lobe a second breakthrough establishes a new preferred channel. (*d*) Development of the delta continues as additional lobes are added through successive breakthroughs.

flow, identified as plane and axial. In *plane jet flow* mixing takes place in only two dimensions (i.e., along a horizontal plane), whereas *axial jet flow* involves mixing in three dimensions. An important difference between the two is that assuming purely inertial flow the lateral boundaries of the axial jet continue to spread downstream at a constant angle approximating 20 degrees. With plane jet flow spreading of lateral boundaries does not increase at a constant angle. Rather, the outline is a parabola in which the width of the jet of turbid water is of the order of three times the square root of the distance downstream from the mouth. Mixing in all three dimensions causes the axial jet to expend its energy and, hence, to deposit its load more rapidly than the plane jet. In plane jet flow the velocity drops uniformly with equal increments of distance from the point of issue, whereas with the axial jet the velocity drops as the square root of the distance from the mouth.

Either axial or plane jet flow may be developed at the mouths of streams depending upon the relative densities of the water of the stream and that of the standing-water body. Three different conditions can exist, namely:

a. The stream water is more dense than that of the body into which it flows; hence, entering water moves along the bottom. Vertical mixing is inhibited, and the flow is restricted to a plane. Only exceptionally turbid waters are sufficiently dense to sink in such a manner, and the condition is not typical of normal environments.

b. The stream water has the same density as that of the body of standing water. Mixing takes place in three dimensions, and typical conditions of axial jet flow prevail. This is the case of most streams entering into fresh-water lakes.

c. The stream inflow is less dense than that of the standing-water body, and the sediment-laden water floats outward upon the denser water below. Vertical mixing is minimized, and plane jet flow exists. Most streams entering into a saline body such as a sea are examples of this case.

Growth of the delta where plane-jet-flow conditions prevail is predicted by theory. Initial deposition of natural levees should occur beyond the stream mouth for a distance approaching four times the orifice width. Within this distance rapid decline of turbulent energy occurs along the flanks of the jet at distances only slightly greater than the width of the stream. Where this rapid dissipation of energy exists deposition will construct natural levees. In the core zone between the levees velocity deceleration is minor, and no deposition will occur. Seaward a distance beyond four times the diameter of the orifice turbulent energy declines as the velocity difference becomes smaller between the core zone and outer margins. Farther seaward four to eight orifice diameters deposition is established transverse to the jet stream. A bar is built which tends to connect the natural levees and block the outlet mouth. (See Fig. 7.25.) Still farther seaward deposition is spread apronwise within the limiting lateral boundaries of the parabolic outline of the jet stream.

Growth of the natural levees and transverse bar tends to restrict water flow and force the development of new outlets crossing the bar or natural levees. This is the condition which produces new distributaries and initiates the characteristic delta pattern. Each new distributary now acts as an independent orifice, and new jets come into existence to repeat the processes described above.

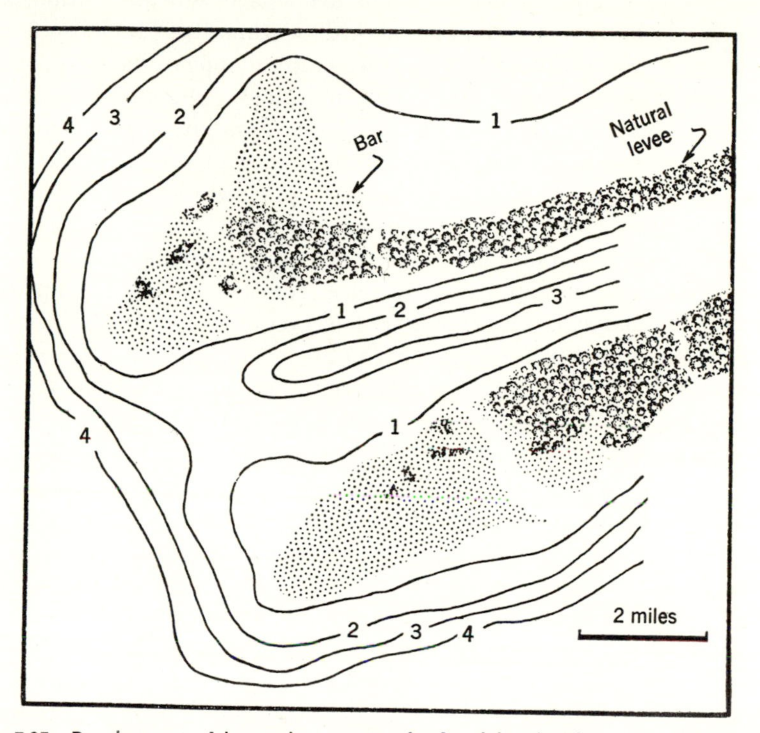

Fig. 7.25 Development of lunate bars at mouth of a delta distributary according to the theory stipulated by plane jet flow. Positions of levees and bars bear a relationship to the width of the orifice of water discharge. Levees extend four times the width of the orifice, and bars appear at a distance of eight times the width of the orifice, choking off water discharge and resulting in deposition and development of a new breakthrough. (Simplified from Bates, *Am. Assoc. Petroleum Geologists, Bull.*, 1953.)

Submergence in Deltas

Historical records of man's occupancy of the growing delta have been the first lines of evidence of the instability of the delta surface. In some areas near the outer edge islands have been observed to emerge from beneath the water in the space of a few hours. Such features designated *mud lumps* are known to be produced through compaction and dewatering of muds and local overloading of water-saturated layers. Overloading of highly fluid mud forces it to intrude and uplift zones of lesser overburden, hence the sudden appearance of such islands.

Apart from the local development of mud lumps the general characteristic is one of mild and slow submergence of the entire delta mass. Generally this submergence does not keep pace with deposition, and the delta is built upward and outward continually. In most deltas evidence for the sinking of the delta complex appears in the form of depressions along the exposed delta margins. Old channel mouths become drowned, and lake or marsh areas of circular or irregular outline occur within the subaerial part. Some of this depression can be attributed to compaction of muds and silts as they decrease in volume under the sedimentary load, but the major part is due to downwarp of the earth's crust in the area of heavy sedimentation. In the region of the Mississippi delta samples from bore holes indicate that much of the material presently 2000 feet below sea level once accumulated on the delta surface (see Fig. 7.26). Here in the downwarped segment has been accommodated, in continuous deposition, layer upon layer of natural-levee deposits, bar sands, backswamp silts and clays, and abundant organic debris in a typical heterogeneous association. Still deeper below the surface but not a part of the present Mississippi delta are thousands of feet of nearshore marine sediments intertongued with deposits of an ancient delta or deltas. The growth of such deltas (?) into the former Gulf of Mexico was punctuated by submergence, invasions of the sea, re-elevation, and deposition of characteristic deltaic nonmarine sediments.[7]

With such a complex history of submergence and sedimentary filling the delta's internal structure is not a simple layering of sands, silts, and clays, one upon the other. Rather, zones of preferred sand deposition seem to be distributed at the outer or seaward margins where prevailing currents and waves winnow out the fine clays. Nearer the shore beds of silts and clays are found mixed with sand. Growth of the delta in various directions produces an interfingered pattern of lenses of sand, silt, and clay, some increasing in thickness seaward, others thickening landward.

Structure of the Small Delta

Small deltas are built into lakes or stream beds, and these are sometimes observed upon disappearance of the lake. Examples are common

[7] A comprehensive report concerning the sediments of a representative delta and the associated alluvial valley is that prepared by H. N. Fisk, *Geological Investigation of the Alluvial Valley of the Mississippi River,* Mississippi River Commission, War Department, Corps of Engineers, U. S. Army, 1944.

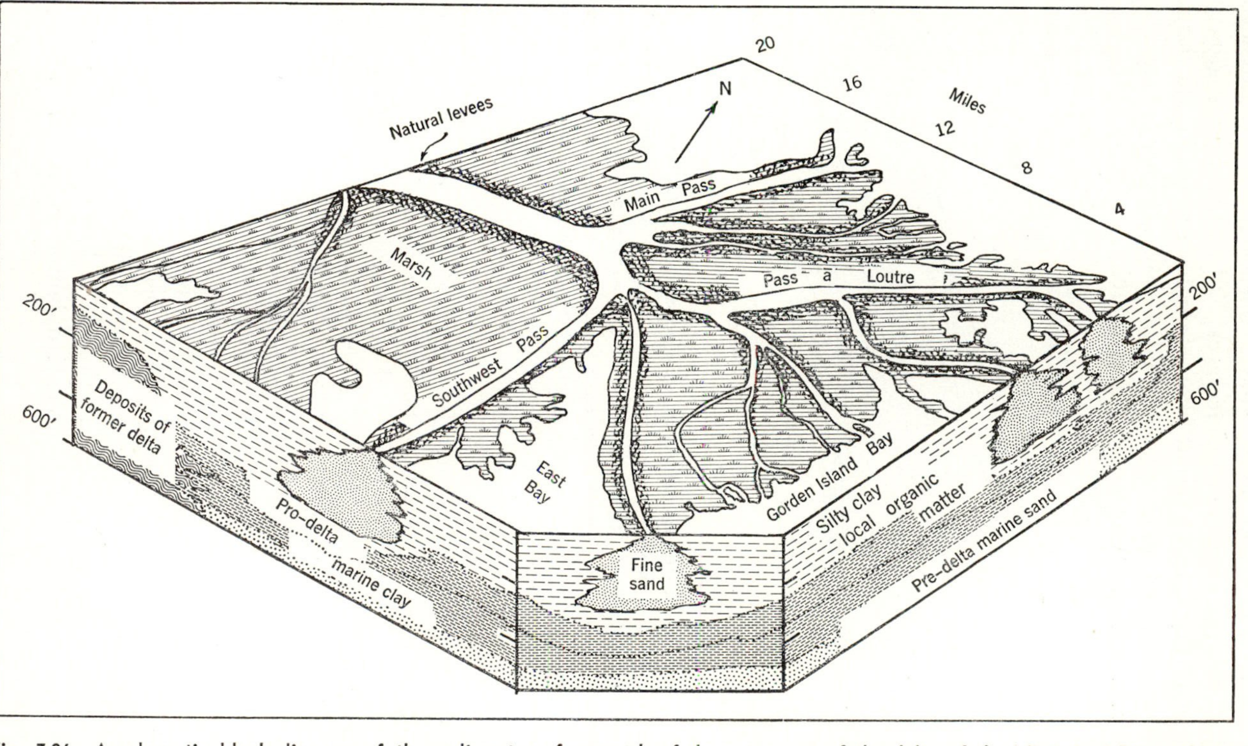

Fig. 7.26 A schematic block diagram of the sedimentary framework of the outer part of the delta of the Mississippi River. Note the distribution of predelta, prodelta, and delta sediments; also the position of sand concentration in the sites of distributary channels. (Simplified from Fisk, McFarlan, and others, *Jour. Sediment. Petrol.*, 1954.)

in former glacial areas where streams pouring from the melting ice carried sediment to a lake now drained. Such deltas rarely extended beyond a few square miles in surface area, and their period of growth was so short that their internal characteristics are unlike the large river deltas discussed above. Small deltas show a typical pattern of internal structure distinguished as *cross-bedding.*

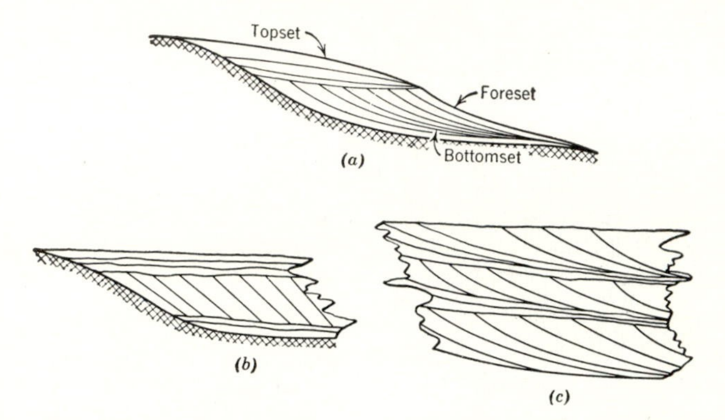

Fig. 7.27 Diagrammatic representation of cross-bedding. (*a*) Common structure of water-laid cross-bedding in which foreset beds are tangential with bottomset beds and are truncated by topset beds. (*b*) In "torrential" cross-bedding, considered to be developed by strong currents, the foreset strata are transected by both bottomset and topset beds. (*c*) Typical exposure of three sets of cross-beds showing that the topset of one group becomes the bottomset of overlying unit. Currents moved from left to right in all examples.

Cross-bedding consists of three fundamental units: a lower series of layers or laminas of horizontal position, an intermediate series of inclined layers, and an upper series of horizontal beds lying upon the inclined strata. The three units are indicated as *bottomset, foreset,* and *topset* beds in ascending order. (Fig. 7.27.) Ideally foreset beds are slightly concave upward and diminish in inclination until they merge into bottomset beds, but topset beds sharply truncate the foreset strata. A modification of this relationship appears in what is called *torrential cross-bedding* in which foreset beds meet the bottomset beds with sharp angular discordance in the same manner as between foreset and topset. Development of torrential cross-bedding is attributed to the action of strong currents moving in the direction of inclination of the foreset beds, whereas gradation of foreset into bottomset strata indicates lower velocities.

Existence of cross-bedding does not necessarily indicate the former presence of a delta inasmuch as this structure can be produced in other deposits such as bars, and in dune sand blown by the wind across the land surface. Nevertheless small deltas characteristically show cross-bedding.

The scale of cross-bedding is variable. In fine-grained silts individual units may be a fraction of an inch thick, whereas in sands and gravels they are measured in feet. Also, coarse sizes are more likely to display the torrential form. Within a geologic formation complete units of bottomset, foreset, and topset are piled one upon the other for many feet although individual "triplets" also occur.

ALLUVIAL FANS

Conditions of Accumulation

Certain tributary streams heavily loaded with sediment empty into another stream which does not have the capacity to carry all the supplied load. Deposition occurs near the point of junction, and ponding creates a lake upstream. Here, local small deltas are built as the dam increases in height. If the tributary has a much steeper gradient than the connecting valley much debris may be deposited at this junction, and an apron of stratified material is spread out at the mouth of the tributary. Where this is deposited beneath the ponded waters it is called a delta, but where tributaries emerge onto broad valleys a similar accumulation may be built subaerially on the land surface. Such a feature is designated an *alluvial fan* or *cone* depending upon the shape. Fans differ from small deltas principally by containing more mudflow (sheet-wash) deposits and less organic muds.

Although alluvial fans may form in any climate where a tributary of steep gradient empties into a valley or depression of low gradient, most alluvial fans are found in arid regions. The ideal conditions are where a mountain area abuts against a broad flat-floored valley or basin. In the United States this topographic condition is characteristic of the Basin and Range physiographic province. In this area precipitation is sporadic and is highest in the mountains; hence, streams carry their load to the basin and build alluvial fans as aprons along the mountain front. (See Fig. 7.28.)

Debris deposited in the fan generally is coarse near the headwaters and becomes finer in a basinward direction, and fine silts and clays are carried to undrained or poorly drained areas. Where the gradient suddenly slackens and the fan is accumulating a system of distributary channels develops as in the pattern of a delta. Development of the distributaries is aided by loss of water by the stream to the permeable

Fig. 7.28 Alluvial fans being deposited in basins between mountain ranges (Amargosa Range, California). Note individual small fans at mouths of major valleys in mountains near center of photograph and large fan being spread from drainage in foreground. Dunes have developed basinward from the small fans. The major part of the playa lies to the right off the photograph. A pediment surface is being developed by the stream which has cut back into the range in left foreground. (By permission from Spence Air Photos, Los Angeles, Calif.)

fan. As the stream water crosses the sandy and gravelly parts of the fan infiltration to the body of ground water occurs, and the stream capacity is reduced substantially. In many areas recharging of ground water by this method becomes so great that the distributary stream channels crossing the fan dwindle completely. In this manner fans become thoroughly charged with ground water, and the water table rises to a level where the excess water escapes from the surface in the form of springs. (See p. 327 and Fig. 9.16.)

Shape of Fan

Small fans a few hundred feet across show a cross profile convex upward sloping away from the stream channel toward the outer fan margin. Very large fans whose cross profile is measured in miles show the same relationships, but the convexity is not so readily observed. Long profiles of fans are concave upward, and the angles of slope are generally steeper in small fans than large ones. Slope angle is lowest at the stream mouth and increases toward the head or apex of the fan.

Fans coalesce to form a continuous apron along a mountain front, and the cross profile of such a composite (known as a *bajada*) is undulatory as the convex profiles of each fan merge into a single

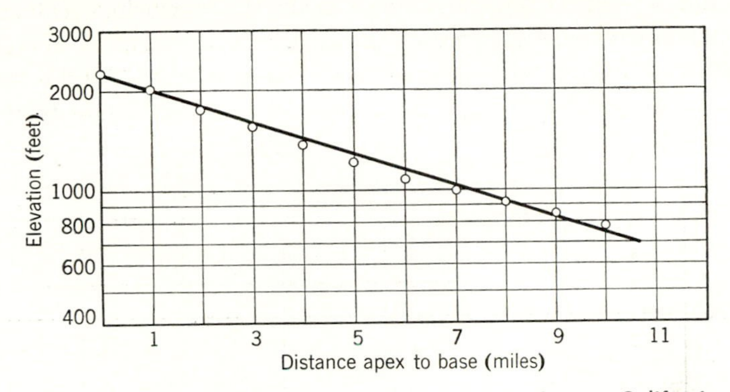

Fig. 7.29 The long profile of an alluvial fan at San Antonio Canyon, California. Elevation of fan surface plotted against distance from fan apex follows a slope described by a negative exponential equation. (From Krumbein, *Jour. Geology,* 1937.)

profile. Inclination of the long profile changes gradually from slopes of less than 2 degrees at the basinward limits to as much as 10 degrees near the apex.

Slopes have been observed to follow with good approximation a curvature prescribed by a negative exponential equation[8] expressed as

$$Y = Y_0 e^{-ax} \tag{4}$$

where Y = the elevation at distance x from the apex.
Y_0 = the elevation at the fan apex.
e = the base of natural logarithms.
a = constant which varies according to the general shape of the curve.
x = horizontal distance along fan surface from apex.

An excellent example of a fan whose long profile follows this law is one built outward from San Antonio Canyon along the Los Angeles–San Bernardino county line, California. (See Fig. 7.29.) The equation for this profile is

$$Y = Y_0 e^{-0.124x} \tag{5}$$

[8] Many geologic processes have been recognized to obey relationships which can be expressed by exponential equations. Among the first to propose this concept was W. C. Krumbein, "Sediments and Exponential Curves," *Jour. Geology,* Vol. 46, 1937, pp. 557–601.

Particle Size Distribution

Sizes of fragments constituting fan material range between large boulders and clay. In general the coarsest material is found near the apex and the finest silts and clays near the downslope extremities. Although this change is reflected primarily by the over-all increase in proportion of fine material downslope it is also indicated by the decrease in the largest particle size. Thus except for the mixing effects of mud-

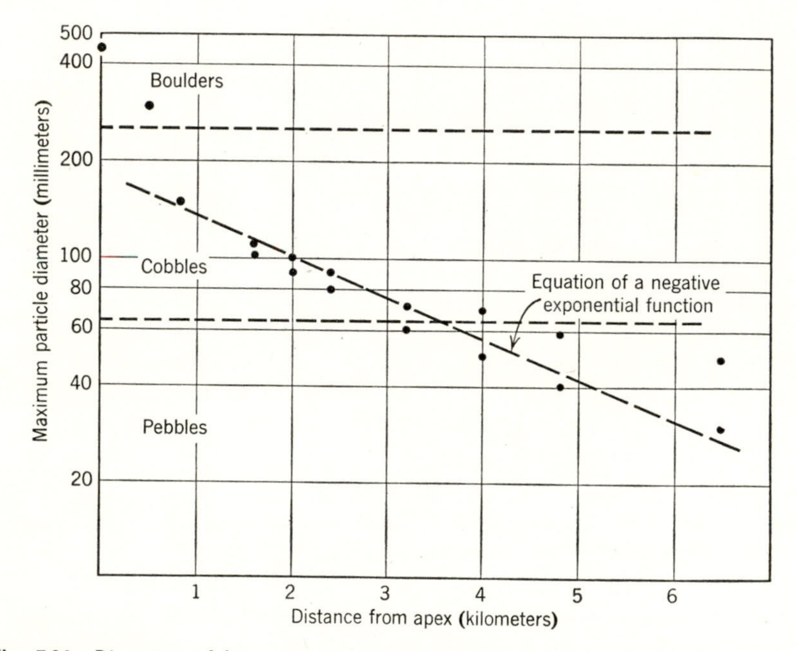

Fig. 7.30 Diameters of largest particles on surface of alluvial fan plotted against distance along fan (Santa Catalina Mts., Arizona). The general tendency to obey a relationship expressed by an exponential equation is present. (Data from Blissenbach, *Geological Soc. Amer. Bull.*, 1954.)

flows which carry coarse material in suspension the distribution of particle size follows the general laws of streams. Particles will tend to be dropped as the available velocity no longer is able to drive a coarse fragment downslope. This decrease in particle size with increased distance from the apex appears to be somewhat systematic, and particle size appears to be a function of downslope distance (in the lower part of the fan) as expressed by an approximation to an exponential equation. An example is shown by Fig. 7.30 in which the range of diameters of the largest particle found at a selected distance from the fan's apex is plotted against distance. The tendency for the largest particle diameter to show a distribution obeying some exponential function of distance is indicated by the diameters of the pebble and cobble size range. Decrease in dimensions of boulders with downslope distance does not follow the same equation although the general relationship may be similar.[9]

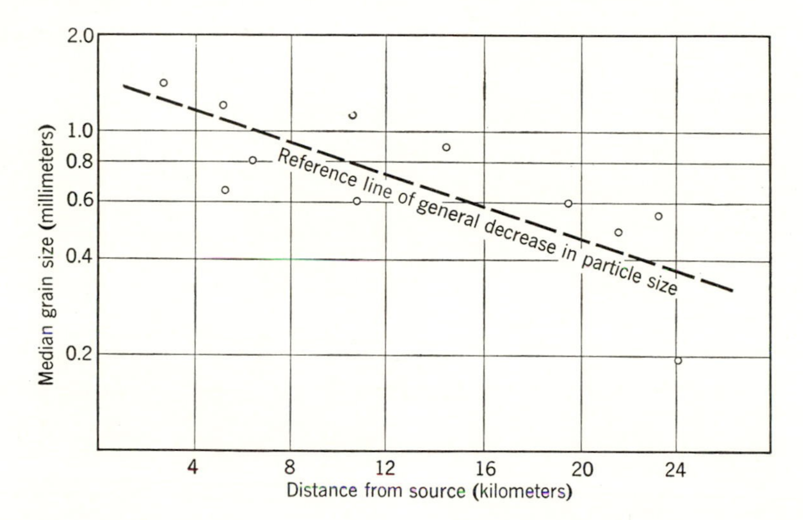

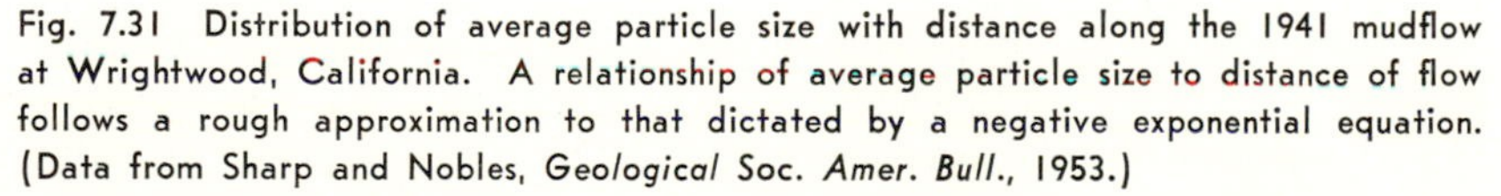
Fig. 7.31 Distribution of average particle size with distance along the 1941 mudflow at Wrightwood, California. A relationship of average particle size to distance of flow follows a rough approximation to that dictated by a negative exponential equation. (Data from Sharp and Nobles, *Geological* Soc. *Amer. Bull.*, 1953.)

A cross section of the fan shows it to consist of a series of layers, some of which are well-sorted beds of sand or gravel, but others are very poorly sorted and contain much fine material surrounding large particles. Also, individual layers show evidence of having been channeled, and such channels are filled with later stream deposits being carried down the fan. Variability both laterally and vertically is the rule rather than the exception.

Some of the layers consist of very poorly sorted debris characteristic of mudflows. The latter consist of solid material ranging in dimension between clay and large boulders made fluid by the addition of water. The viscosity of such a mixture, which contains about 20 per cent by

[9] Some general properties of alluvial fans are reported by Erich Blissenbach, "Geology of Alluvial Fan Deposits," *Geological Soc. Amer. Bull.* Vol. 65, 1954, pp. 175–190.

weight of water, is about that of freshly mixed concrete and is capable of carrying large boulders in suspension. Such flows move down the fan surface at rates ranging between about $\frac{3}{5}$ and 5 miles per hour.[10] Average particle sizes of mudflow material decrease downslope, also obeying a very rough approximation toward a negative exponential function. An example of this size distribution is shown in Fig. 7.31.

Most of the alluvial fan material consists of interlayered gravels, sands, and silts with some tendency for particles to statistically increase in coarseness from the surface toward the basal layers. Cross-bedding and ripple marks are common, the latter being commonly associated with the cut-and-fill of shifting stream channels.

STAGES IN THE HISTORY OF THE STREAM SYSTEM

General Statement

In much the same way that individual streams are genetically related to their tributaries so are stream systems oftentimes interrelated. These interrelationships affect the aspect of the landscape, and as each individual valley changes its cross-sectional profile so also is the topography altered. We have reasoned how an idealized landscape of uniform slope is gradually notched by the V-shaped profile of the initially developed stream. In the beginning only a few notches would be observed, but as time progressed more streams would carve additional valleys to develop a thoroughly dissected landscape. Geologists refer to a landscape in its early stages of stream dissection as *youthful*.

Still later the same landscape is altered by the process of valley widening as individual streams reach base level. Valleys are no longer narrow channels but tend to broaden and occupy more of the total land area. Uplands are drained completely by runoff which immediately finds a stream channel. Flood plains are to be noted along more prominent streams, and in some of them strongly looped meanders are apparent. Regions possessed of such aspects are called *mature*.

As valleys widen, interstream divides gradually are lowered, and individual valleys no longer are bordered by high or prominent walls. Flood plains are extensive, and a few trunk streams bordered by their natural levees meander at great length across the filled valleys. Oxbow lakes and extensive swamps are commonplace along the old stream courses. The subdued landscape is dominated by the river flood plain. Such an area is said to be in *old age*. (See Fig. 7.32.)

[10] R. P. Sharp and L. H. Nobles, "Mudflow of 1941 at Wrightwood, Southern California," *Geological Soc. Amer. Bull.* Vol. 64, 1953, pp. 547–560.

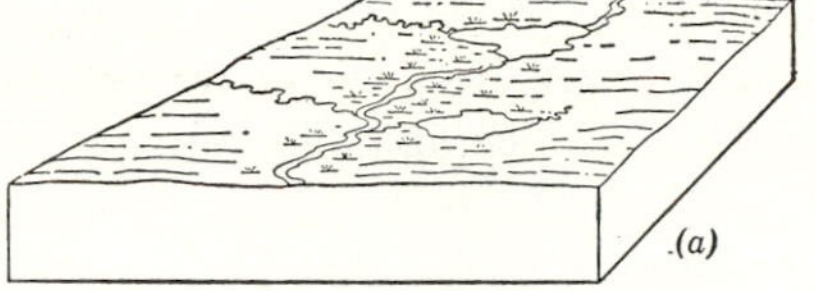

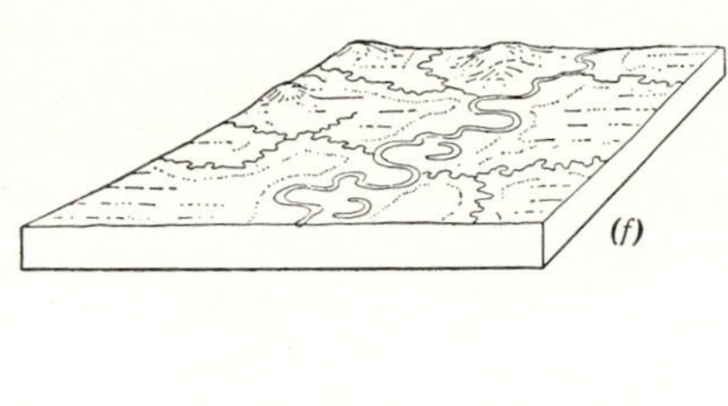

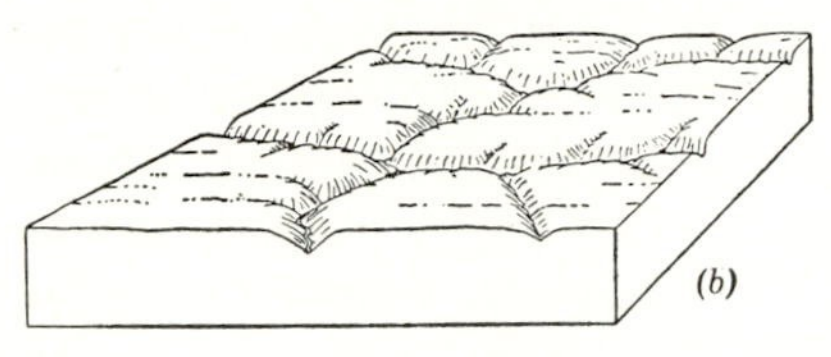

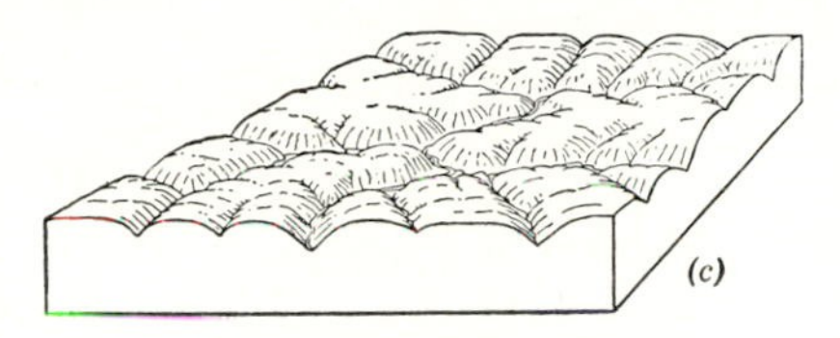

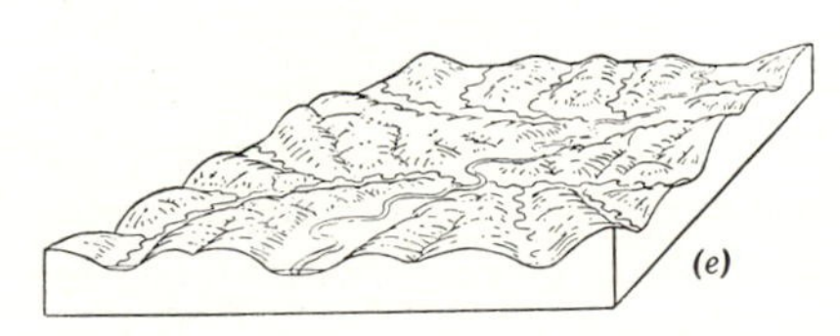

a. In the initial stage relief is slight, drainage poor.

b. In early youth stream valleys are narrow, uplands broad and flat.

c. In late youth valley slopes predominate but some interstream uplands remain.

d. In maturity the region consists of valley slopes and narrow divides.

e. In late maturity relief is subdued, valley floors broad.

f. In old age a peneplain with monadnocks is formed.

g. Uplift of the region brings on a rejuvenation, or second cycle of denudation, shown here to have reached early maturity.

Fig. 7.32 Idealized stages in the cycle of stream erosion in exterior drainage, i.e., humid region. (Reprinted with permission from Strahler, *Physical Geography*, copyright 1951, John Wiley and Sons, Inc.; drawings originally by E. Raisz.)

Localities characterized by streams in any of the stages described above are to be found where rainfall is ample enough to establish many permanent streams. Most of the earth's land surface falls within this category, and we refer to the stream stages as being developed in a *normal, or humid, region*. There are, however, extensive areas of semi-aridity where permanent streams also show the same stages in stream development. Actually, it is more appropriate to define an area showing "normal" stream stages as being controlled by drainage which by means of a trunk stream is directly connected with the sea. Such drainage is also called *exterior* because water which flows to stream channels eventually is carried completely outside of the drainage basin.

Distinct from regions of normal, or humid, conditions are isolated parts of the land surface which are deficient in moisture. These exceptional areas fail to accumulate sufficient water to fill large depressions which are the gathering sites of local runoff. Excessive evaporation removes much of the water from these basins, and no stream outlets occur. Landscapes developed by streams in this environment differ considerably from those of humid areas and are recognized as typical of stream erosion in arid basins. Regions in which such stream development occurs are more properly described as having *interior* drainage in which the runoff does not leave the basin except through eventual evaporation.

Stream Erosion and Crustal Deformation

Another concept that has dominated the interpretation of stream history concerns the relationship between the rates of stream erosion and deformation of the earth's crust. In the light of the length of geologic time local crustal movements are periodic and rapid with long intervals of stability separating changes in elevation. During the time of upwarp stream erosion frequently is incompetent to keep pace with the change in surface elevation, and conditions of gradient are altered. Streams meandering across broad flood plains are *rejuvenated,* and steep gradients of youth are re-established. Conditions for active valley deepening now exist, and the valley profile is altered accordingly. Relatively long intervals of crustal rest provide opportunity for streams to alter the landscape radically, and many areas are characterized eventually by mature or old age topographies. In some of these regions certain aspects of a former stream stage remain, and from these the past physiographic history of the area can be deciphered.

Widespread crustal instability is the rule rather than the exception, and some fluctuation of the land surface is to be recognized at most all localities. In some areas changes have been violent, whereas in others minor and gradual depression or elevation has occurred. Certain regions remained sufficiently stabilized for a stream system to proceed through its stages and reach old age before crustal uplift took place. An aspect of youth was then superimposed upon the old-age topography, and a *cycle of stream erosion* was completed.

This concept of the cyclical nature of the life history of a stream system or drainage basin as being controlled by fluctuations in the elevation of the land surface is fundamental to an understanding of the development of special features of landscape. Among the most important of these are gorges through mountains and widespread terraces along streams.

Characterizing the Cycle of Stream Erosion

Of primary importance in an understanding of stream history is the recognition of the several stages in the stream cycle. To this end the characteristics of each stage are described below. These have been grouped according to the nature of the cycle, namely whether the control has been developed by exterior or interior drainage. In each case, however, the characterizing elements are based upon relationships of the following features each of which is subject to some quantitative measurement of a low order of precision. Considered among the most important are:

1. Width of divides between valleys.
2. Steepness of stream gradients.
3. Cross-sectional profile of valleys.
4. Number of streams per unit area.
5. Shape of stream channel.
6. Number and disposition of lakes and swamps.
7. Special features of erosion.
8. Character of stream deposits.

Topographic maps and aerial photographs are suitable media for analyzing the features listed above. Generally, precise measurements need not be made on such maps, and a qualitative comparison is sufficient to catalog an area as to the stage of the stream cycle represented. A guide for this kind of analysis appears in Tables 7.1 and 7.2 which summarize the characterizing elements of each stage in the erosion cycle.

The ability to recognize the erosion stage which an individual stream has attained and that which is representative of a large area including several hundred square miles is important in characterization of a landscape. The observer must be mindful of the fact that a local stream or its segments may be in one stage of the erosion cycle, whereas the surrounding landscape may be in a different physiographic stage. The composite of all erosion characteristics of an area presents

Table 7.1. Cycles of Erosion in Exterior Drainage (Humid Region)

Stages	Divides	Stream Gradients	Valley Cross Profile	Number of Streams	Stream Courses	Lakes and Swamps	Erosion Features	Stream Deposits
Youth	Broad and extensive. Ratio of areas of upland to valley slope is >1.	Steep; falls and rapids; hundreds of feet per mile, irregular.	V-shape; stream occupies valley.	Few; tributaries few; uplands undrained.	Generally straight or slightly meandering.	If present they are only on uplands.	Streams may be in canyons; valleys are not gently sloping; valley deepening.	Thin veneer of gravels; much bare rock.
Mature	Become sharp ridges. Ratio of areas of upland to valley slope is <1 and of bottomland to valley slope <1.	Moderate; usually tens of feet per mile, stream grade regular.	U-shape; stream occupies only part of valley.	Many tributary streams. Uplands narrow divides. All land in drainage.	Sinuous. Stream swings from valley wall to valley wall.	Few swamps appearing in valley floor; few oxbow lakes.	Lateral erosion; valley walls subject to widening.	Sand bars in channel; deposits along inside bends of meanders; valley fill small; natural levees minor.
Old age	Low, between large stream courses. Ratio of areas of bottomland to slope >1 and of upland to slope <1.	Low, inches per mile.	Broad, low depressed area defined by meander scars and border swamps. Stream occupies part of external flood plain.	Large trunk streams, few tributaries.	Extremely sinuous; channel brimful; flooding regular.	Large swamps in valley; oxbow lakes large; numerous meander scars prominent.	None, except occasional rocky hill (monadnock); peneplain surface.	Extensive valley fill; natural levees prominent, streams may be braided.

Table 7.2. Cycle of Erosion in Interior Drainage (Arid Region)

Stages	Divides	Stream Gradients	Valley Shapes	Number of Streams	Erosion Features	Stream Deposits
Youth	Major interstream areas wide, streams generally small arroyos except in mountains.	Steep, hundreds of feet per mile.	Vertical-walled channel.	Streams numerous in mountains; many tributaries in regions of clay soils.	Canyons, badlands.	None in mountains; small fans.
Mature	Interstream areas narrow, many small arroyos and canyons.	Steep in upper courses, tens of feet on fans.	Many channels with vertical walls.	Many stream channels.	Highly dissected canyon land, small pediments.	Gravel deposits where tributaries enter main stream, fans coalesce into broad apron. Silts and sands in playa.
Old age	Interstream areas wide; no canyons, small arroyos.	Tens of feet per mile over entire course.	Valleys absent except for small channels.	Many small streams usually with small channels.	Broad pediments "inselberge," interconnecting basins.	Large fans, low-walled braided channels in playa.

an aspect which may be regarded as physiographic youth, maturity, or old age.

CHARACTERISTICS OF LANDSCAPES DEVELOPED IN THE CYCLE OF STREAM EROSION IN EXTERIOR DRAINAGE

Divides and Uplands

Landscapes in the youthful stage typically are those in which the imprint of stream erosion is slight. Monotonously flat surfaces are interrupted by few streams whose valleys may range between very deep chasms and inconsequential gullies depending upon the elevation of the land surface above base level. A rough approximation of the area in upland divided by the area in valleys yields a ratio whose value exceeds one. Locally, upland surfaces contain undrained depressions such as swamps, ponds, or lakes as the stream system is not ramified sufficiently to remove the runoff.

Typical mature regions are those of many valleys separated from one another by narrow divides. The ratio of the area in upland or divide to that in valley slope is now less than unity and ideally approaches zero. Ratios of the area in bottomland to that in valley slope are less than unity but tend to approach that value.

Old age areas have been very low divides separating broad valleys. Practically all formerly elevated prominences have been lowered to minor rises separating wide valleys some of which may contain several parallel flowing major streams. Bottomland dominates, and the ratio of the area considered flood plain to that considered valley slope exceeds one. Those few scattered upland areas which remain suggest the former height of the land. They constitute only an insignificant percentage of the landform distribution, and the ratio of such area to that of valley slope is always less than one.

Stream Gradients

The stream gradient is a sensitive indicator of stream stage, and with few exceptions low gradients indicate advanced stages in the cycle.

Valley Cross Profile

Cross profiles of streams in youthful areas usually are V-shaped notches. But in glaciated regions broad-bottomed valleys customarily are present, and in such circumstances the general rule is not applied properly. Except for such regions the stream effectively occupies the bottom of the valley, and valley walls slope from the stream edge.

Ideal streams in the mature stage have developed small but obvious flood plains and meander from one wall of the valley to the other. Cross profiles are U-shaped, and flat bottoms are typical. All indications point to valley widening rather than deepening.

Valleys of regions in old age are extremely difficult to delimit. In many cases no clearly defined valley can be observed. Flood plains are so extensive that only along the margin with an area in topographic maturity is the boundary to be established. Meander scars, border swamps, and natural levees dominate the landscape. In the zone outside the flood plain a few scattered rock hills, *monadnocks,* are erosion remnants that have escaped destruction.

Number of Streams

The number of streams draining a region is a good indicator of advance in the stage of the erosion cycle. It is strongly influenced, however, by the permeability of the rocks, and where shale is the bedrock runoff is extreme, and stream channels are initially very abundant.

Stream Courses

Stream patterns oftentimes are important indicators of regional age. The youthful stream tends to have a straight course with limited meandering. Exceptions are to be found among the streams draining level areas of glacial drift. Often such streams are highly meandering because of the very low gradients involved. In mature regions major streams are strongly meandering, and clearly defined loops are to be observed. The courses of old age streams are extremely sinuous with large meander loops whose radius of curvature tends toward uniformity. On aerial photographs the pattern of systematic migration of the meanders is striking, and by cross-cutting relationships indicates the direction in which the loops advance. (See Fig. 7.17.)

Lakes and Swamps

Shift in the position of undrained areas from upland to bottomland is an important attribute of the progress of the cycle of erosion. Mature regions are, therefore, localities of no natural lakes.

Erosion Features

Steep-walled chasms, canyons, narrow notches, and waterfalls are typical of youthful areas. However, many exceptions are found in regions of former continental glaciation or recently emerged coastal plains. Here stream erosion has been active for a short length of time, or the land elevation lies close to base level, and small sluggish streams characterize the youthful stage. Elsewhere, valley deepening is the primary function of the streams.

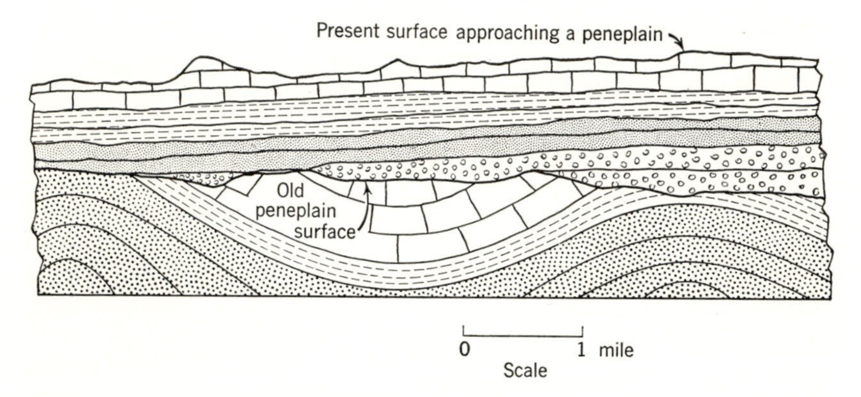

Fig. 7.33 A schematic representation of a geologic cross section showing a peneplain surface as preserved by burial under later sediments. Folds of beds in lower strata have been truncated by erosion to produce a generally flat peneplain surface. Later gentle subsidence and deposition have permitted burial of old surface of erosion. Gentle plateau-like uplift has restored the old peneplain to much the same position as it was before burial.

When maturity is reached major streams are actively widening the valley, and cliffs are the rule where the stream swings against the valley wall. Tributary streams are principally youthful, and the network is highly developed. All streams are in states of active erosion and land dissection is paramount.

Regions of old age are dominated by great flood plains, and erosion of the low divides is effectively completed. Activity of trunk streams is confined to redistribution of flood-plain deposits and enlarging the natural-levee pattern. The only evidence of the former higher elevation of the land surfaces is isolated monadnocks which by accident have escaped obliteration.

Monadnocks rise above the extensive low landscape the surface of which bevels all underlying rock structure which exists. Geologists refer to this surface as a *peneplain*. Peneplains formed in past geologic periods and now covered by younger strata have been recognized in many areas by means of tracing an old erosion surface of low relief. Such ancient erosion surfaces bevel strata which were tilted prior to peneplanation as well as those of horizontal structural attitude. By such evidence the erosion surface which may cover thousands of square miles is reconstructed, and the former existence of great stream systems interpreted. The peneplain surface is used by geologists to mark an important divisional boundary between units of strata. (See Fig. 7.33.)

Stream Deposits

Sediments laid down by streams become important only in mature and old age regions where flood plains are being aggraded. In youthful regions thin veneers of scattered boulders, gravel, and sand are present on stream banks and bottoms. This material is ephemeral and moves downstream during high-water stages. Bare rock is the typical aspect of the youthful stream channel. Mature streams tend to grade their valley floors primarily with silt and sand. Local pockets of gravel occur in the channel, and oftentimes these are preserved and buried beneath later-deposited silt and sand as the stream channel shifts. For this reason flood-plain deposits locally are heterogeneous although the bulk aspect is fine sand and silt. In old age clays and silts are the principal deposits although coarser material is laid down in the natural levees and wherever breakthroughs in the levees permit sand to be carried into the backwater swamp zone. Here much organic matter is buried with the silt. Also, during times of breakthroughs the flood load is dumped in the border swamp area, and lenses of sand or gravel may bury previously deposited organic muds.

CHARACTERISTICS OF LANDSCAPES DEVELOPED IN THE CYCLE OF STREAM EROSION IN INTERIOR DRAINAGE

Ideal young landscapes of arid regions consist of a broad basin defined by small mountain masses along one or more sides. The mountain areas are the sites of principal rainfall, and many stream valleys of youthful character cut the mountain slopes. Photographs of such slopes show a pattern of radial drainage of narrow straight channels and few tributaries. The broad basin is the base level for streams draining the mountains and gradually fills with debris removed from higher elevations. As a result the topographic aspect differs from that developed by the normal stream cycle although the actual processes of erosion and deposition are much the same. Desert landscapes can be classified as youthful, mature, and old on the basis of the same properties used to analyze the region of exterior drainage. (See Table 7.2 and Fig. 7.34.)

Divides

In the mountain areas stream activity precludes the existence of broad divides except for the mountains themselves, and in the youthful and mature stages streams are closely spaced and drain directly to the basin. Mountain valleys rarely are deep and radially distributed from the summit area. Along the mountain front valleys effectively end, and only small channels, called *arroyos* in southwestern United States, continue across the deposits of the alluvial fans. If the basin is extensive arroyos become progressively shallower and broader channels as the distributary system is developed, and several small (*playa*) lakes may be present in the lowest section of the basin. Divides in the basin area tend to be very broad and low and are not items of critical characterization although they do serve to isolate individual playas.

Stream Gradients

In the mountain sections gradients of youthful streams are steep, in terms of hundreds of feet per mile, but as the valleys open into the basin these gradients decline to the general slope of the basin surface. The divergence in slope is most marked in the youthful stage of the basin where streams are short and alluvial fans are small. As maturity is reached certain streams extend their canyons headward to intersect similar ones draining in the opposite direction, and a well-defined pass is cut through the mountain mass. Hence, gradients in the large valleys are reduced to values of tens of feet per mile. Concurrently, alluvial fans are increased in size, and gradients are steepened at the sites of former flat parts of the basin areas. The tendency is to establish a graded profile from alluvial fan into the mountain area. (See Fig. 7.35.) This condition is attained with old age, and stream drainage is radially distributed around the low domes which represent eroded remnants of the former mountains surrounded by large fans.

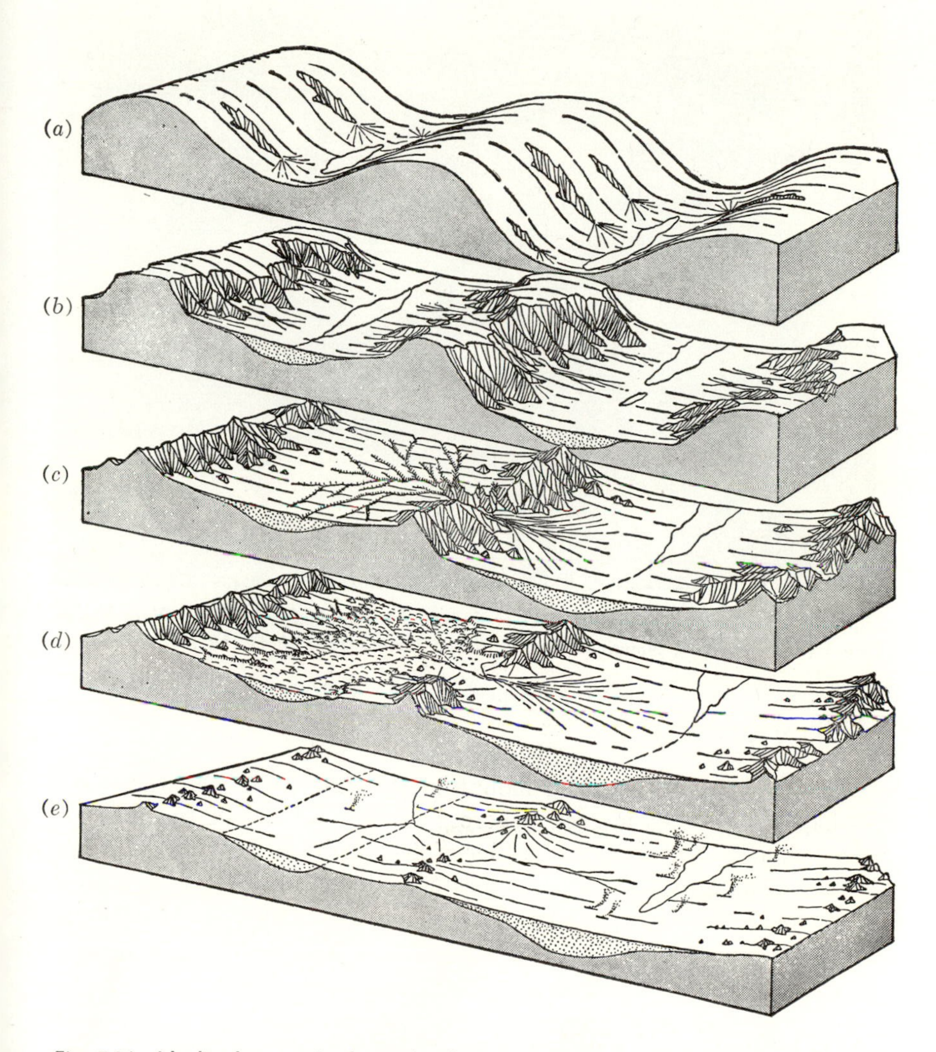

Fig. 7.34 Idealized stages in the cycle of stream erosion in interior drainage, i.e., arid region. (*a*) Initial stage following rapid uplift of mountain slopes. This stage is rarely observed in nature as erosion proceeds with great rapidity. (*b*) Youthful stage in which broad playas are developed and the mountain front has receded but the mountain tops are broad. (c) and (*d*) Mature stage of extensive alluvial fans and interconnection of playas; see Fig. 7.28. (e) Old-age stage, inselberge and interconnected basins. (Reprinted with permission from Longwell, Knopf, and Flint, *Physical Geology*, 3rd ed. Copyright 1948, John Wiley and Sons, Inc.)

Fig. 7.35 A dissected alluvial fan in an indented mountain front, Colea River Valley, Peru. (From Johnson, *American Geographical Society Spec. Publ.* **12.**)

Valley Shapes

Cross profiles of youthful mountain streams are the characteristic V-shaped notch cut in rock, whereas arroyos which cut the fans are steep, almost vertical-walled gullies of flat floor. As maturity develops in the mountains the valley floor tends to widen but to a limited degree, and the broad valleys observed in regions of exterior drainage are not observed.

Number of Streams

Where shales are the surface rocks or where the mountain mass is a large intrusion runoff is intense and the number of streams is increased materially over those cases where the surface rocks are permeable. Fans, however, are fundamentally permeable, and much ground-water infiltration occurs. On these areas the number of individual streams decreases over those in the mountain sections. In old age where the fans are very large and mountain areas are reduced to low bare-rock domes valleys are small but numerous in both mountains and fan areas.

Erosion Features

Canyons and badlands characterize youthful mountain areas, and basinward small fans are cut by shallow arroyos. Extensive flats characterize the major parts of the basins from which mountain fronts rise steeply and abruptly. In the mature stage when large fans are draped against the mountain front, slopes become graded between mountain mass and fan. The upper reaches of the fan merge into the rock surface of the mountain areas which have been lowered to the fan grade. Here is developed an extensive bare-rock surface of gentle slope particularly where large streams emerge from the more youthful mountain section. Such bare-rock surfaces known as *pediments* may be several miles in width and extend along the fanhead. They are erosion surfaces truncating hard and soft rocks alike and graded to the low fan slopes.

In old age pediments are very broad inasmuch as the mountain relief has been reduced to low domes, and erosion rates are scarcely measurable. Locally, parts of the mountain have not been lowered to the level of the pediment surface of fan and pediment. The observer gains the false impression that these local areas of topographic relief are mountain peaks buried beneath a thick mantle of alluvial debris. Such peaks are designated by the German term *inselberge*. It must be remembered that actually the broad pediment surface of rock forms a very extensive apron around the low dome and is mantled by no more than a few feet of sediments. The fill of the basin varies in thickness and rarely exceeds several hundred feet.

In the old-age stage basins on opposite sides of the low dome mountains are interconnected by a stream system which has gradually cut back from the lower basin to the higher one in an attempt to bring them to a common grade. Normally this does not involve cutting of any large valley but represents movement of debris through sheet wash or along numerous, but small, parallel channels lacking formal valley outline.

Stream Deposits

Throughout all stages of development deposition is concentrated in the fans and playa deposits. In youth fans are small and appear principally where the largest valleys emerge from the mountains. Here gravels intertongue with sands and silts, but coarse materials predominate. Basinward finer material is most abundant, and only silts and clays reach the playa lakes. In the mountains streams deepen their valleys, and only a temporary film of coarse gravel covers the bedrock locally.

As maturity is reached individual fans become intergrown, and an extensive alluvial apron blankets the mountain slope. Generally, fine debris characterizes this period of deposition, but lenses of gravels are commonplace. With the growth of fans of high permeability much water is removed from the runoff, and there is a tendency for silt and clay to be deposited as matrix material binding sand and pebble particles rather than being swept from the site of gravel and sand deposition. Deposition of this fine-sized matrix tends to reduce the permeability of the fan, runoff increases, and after heavy rainfall sands may be carried far out on the playa. The graded profile developed during maturity reduces the size of debris carried to the fan, and the upper deposits tend to be more sand and silt than the coarse gravels which characterize the initial fan. For this reason an over-all decrease in average particle size is to be expected from the base of the fan upward although layers of coarse and fine particles overlie one another in random order.

Old age brings about the condition where coarse material does not leave the mountain domes but remains to be weathered into small fragments. Eventually these move across the pediments to come to temporary rest on the fans. Winds tend to winnow out the sand and pile it into dunes which move across parts of the basin. Sheet wash restores the silts and clays, carrying the bulk of the latter principally to the center of the playa and the lake. All of these processes tend to redistribute and concentrate some of the debris which was carried to the basin. Shifting of sediments leads to segregation of certain particle sizes such as sand in dunes, silt and clay in dust storms, and lag gravel which armors the fan surface.

Drainage of one basin by the stream system of another or local uplift of a basin causes intense stream dissection. When this is partially completed important terrace surfaces are developed over much of the basin. The terrace levels slope with gentle gradient into the mountain areas, and their connection with former alluvial fans can be determined.

CLASSIFICATION OF PHYSIOGRAPHIC STAGE BASED UPON SELECTED MEASUREMENTS

Although the characteristics enumerated earlier have long been recognized as suitable for classifying landscapes on the basis of stages of the cycle of stream erosion, not all stream-developed areas are easily catalogued. Some features are not readily determined from topographic maps or photographs, and experience in the field has been a necessary prerequisite to interpretation. There is justification for approaching the problem of recognition of the physiographic stage of a local area on the basis of measurements of selected features observed on suitable maps. With this in mind certain data have been gathered from a series of topographic maps representing various stages in stream development. The intent has been to assemble sufficient information to indicate the ranges in values suitable to properly place the area in its physiographic stage.

Items selected for comparison and considered significant to interpretation of stream history are determined readily and are as follows:

1. Average topographic relief in the area (approximately 225 sq. miles).
2. Number of lakes and swamps per unit area (i.e., township or 36 sq. miles).
3. Number of major valleys per unit area (i.e., principal drainage lines on map).
4. Number of secondary (major tributaries) per unit area.
5. Average distance in miles between major and secondary streams.
6. Average angle of junction between major and secondary streams.
7. Average angle of junction between secondary (tributary) streams.
8. Average gradient (feet per mile) of major stream.
9. Average gradient (feet per mile) of secondary streams.
10. Average width of secondary valleys.
11. Average depth of secondary valleys.
12. Average width of major valleys.
13. Average depth of major valleys.
14. Ratio of width to depth secondary valleys.
15. Ratio of width to depth major valleys.

The results of an attempt to delimit stages of the stream cycle using the measurements described above are presented in Table 7.3. The intent has been to demonstrate the possibilities which such a procedure offers rather than to establish limits of characterization applicable to all stream systems.

Use of Table 7.3 as a device for the determination of the physiographic stage of a landscape should reduce uncertainties which result from qualitative appraisals. As a general rule no single observation is sufficient to catalog an area. Rather, measurements of all selected items should be made in order to assure the maximum accuracy of correlation. The reader should recognize also that the values listed

Table 7.3. Characterizing Features of Stream Cycles

Means / Cycle of Erosion and Stage	Average Relief (Feet)	Number of Lakes etc. per Township	Number of Major Streams per Township	Number of Secondary Streams per Township	Average Distance in Miles between Major and Tributary Stream Junction	Average Angle of Junction between Major and Tributary Streams	Average Angle of Junction between Tributary Streams	Average Gradient Major Streams (Feet per Mile)	Average Gradient Tributary Streams (Feet per Mile)	Average Width Tributary Stream Valley (Feet)	Average Depth Tributary Valley (Feet)	Average Width of Major Valley (Feet)	Average Depth of Major Valley (Feet)	Ratio of Width to Depth Tributary Valleys	Ratio of Width to Depth Major Valleys
Normal stream—youthful (relief <500′)	85	7 −	1 −	5 −	3 +	55°	60°	3 +	10	2200	50	8500	70 −	28 to 52 30 (av)	22 to 2000 600 (av)
Normal stream—youthful on glacial sediment	85	14	1 +	3	3	60° +	70°	3 −	25 −	1700	60	6300	100	8 to 60 32 (av)	38 to 400 110 (av)
Normal stream—mature (relief <500′)	250	<1	1 +	8	2 +	65°	55°	10	40 −	2700	190	8500 +	230	7 to 57 22 (av)	8 to 120 50 (av)
Normal stream—mature (relief 500′ to 2000′)	1100 +	<1	2	10	2 −	80°	55°	20 +	160	4100	500	9300 +	900 −	4 to 19 10 (av)	4 to 22 12 (av)
Normal stream—old age	70 −	1 +	1	3	13 −	35°		1 +	5	3300	35 −	23000 — — — 7000 channel	35 — — — 10	40 to 250 120 (av)	65 to 2000 780 (av)
Arid interior drainage—mature	1900 mts. — — — <200 on fan	1 −	1 +	7	1 +	80° range large	40°	25	160 mts. — — — 100 on fan	3700 −	750	10 +	25 +	3 to 8 6 (av)	1000 to 5000 3000 (av)
Continental glaciation in hilly and semi-mountainous areas	1000 +	3 +	1	8 +	2 −	70° +	65°	10	100	4800	480 −	7600 +	700 +	7 to 17 12 (av)	7 to 25 14 (av)
Mountain glaciation	3400	7 +	3 +	10 +	2 +	30°	40° +	130	390	7300	1600	16000	2000	3 to 6 4 (av)	4 to 7 5 (av)

as typical of each feature are orders of magnitude rather than precise values, and the extent of deviation from the average values is incompletely known.

Certain features appear to be more diagnostic than others, for example, the number of streams per unit area and the ratios of width to depth of valleys. Of these the latter has a wide range in values, but the overlap is relatively small. Currently, considerable interest is being shown by students of stream behavior in techniques of analyzing landscapes by measuring certain attributes similar to those listed above. The reader will find that most of the approaches being followed are controlled by rather advanced statistical analysis which permits a measure of interpretation heretofore impossible.

Selected Supplementary Readings

Cotton, C. A., *Landscape,* 2nd ed., John Wiley and Sons, 1949, o.p. Virtually the entire book is devoted to the influence of streams upon the aspect of the earth's surface.

Hinds, N. E. A., *Geomorphology,* Prentice-Hall, Inc., 1943, Chapter 16. An excellent treatment of the effect of running water on landscape development.

Kuenen, P. H., *Realms of Water,* John Wiley and Sons, 1956, Chapter VII. Concerns the balance sheet of terrestrial water.

Lobeck, A. K., *Geomorphology,* McGraw-Hill Book Co., 1939, Chapter V. A well-illustrated discussion of the characteristics of stream-developed landscapes.

Trunbull, W. J., Krinitzsky, E. L., and Johnson, S. J., "Sedimentary Geology of the Alluvial Valley of the Lower Mississippi River and Its Influences on Foundation Problems," in Trask, P. D., *Applied Sedimentation,* John Wiley and Sons, 1950, Chapter 12. Illustrates the problems encountered in a large alluviating stream.

"Water," *The Yearbook of Agriculture 1955,* U.S. Department of Agriculture, Government Printing Office, section on water and our soil. Contains a number of articles on water and runoff.

8 Shoreline processes

CONTINENTAL SHELVES

General Characteristics

Continents are bounded by a platform of extremely variable width, but whose average is about 40 to 50 miles. This more or less continuous platform slopes seaward very gently (of the order of 10 feet per mile). It is separated from the abyssal portion of the ocean by the continental slope which is inclined at relatively much greater

Ancient shoreline of Lake Bonneville, Utah. (From Gilbert, U. S. Geological Survey.)

values (as much as several hundred feet per mile) but which, nevertheless, is also of gentle grade. Much of the shelf receives a constant addition of sediments, accumulated as gravels, sands, muds, and limestone rock. Other areas receive no sediments at all, and still other parts rise above sea level as marginal islands. Certain characteristics can be ascribed to continental shelves in general, such as:

1. Their surfaces are not smoothly sloping but are terraced, irregular, and extremely variable from place to place.
2. Particle size distributions of sediments on the floor do not necesarily show a progressive decrease in average size seaward. In some areas the variability is great, coarse material being found near the outer margin.
3. The average sediment of the shelf appears to be more coarse than the average size being supplied by present-day rivers.
4. Shelves off most glacial areas have topographic characteristics which are unique.

The shoreward margins of the shelf are variable in topographic aspect and appear to be affected by conditions which have influenced the landward portion. They have been classified, therefore, into several groups based upon features which characterize the bordering land section.[1]

Shelves off Glaciated Areas

These are rather broad surfaces averaging 100 miles in width and upon which local topographic irregularities exceed the regional seaward slope. In some areas the bottom configuration is that typical of moraines, whereas in the more seaward portions basins and corresponding banks are the rule. Long steep-walled estuaries and fjords of the shoreline continue as channels across the entire shelf and in some occurrences continue across the continental slope as submarine canyons. On the shelf the troughs differ from submarine canyons in their great width, relatively straight sides, and basin depressions. The basin depressions are so large that they are comparable to the basins of the Great Lakes of North America. Elsewhere over the glaciated shelf, particularly some distance from the coast line, banks and shoals are typical. As a rule the nature of the bottom is variable, but sand or gravel predominates on the banks, mud covers the inner

[1] The classification and characterization used herein are those of F. P. Shepard, *Submarine Geology,* Harper and Bros., 1948, p. 145.

deeps and troughs, and rocky or bouldery bottoms are typical near the shore.

Shelves off Large Rivers

Shelves bordering mouths of large river systems do not average as wide as shelves off glaciated areas, but locally they may be extremely extensive. Their greatest width is not to be found where protruding deltas occur but adjacent to low-lying embayed coast lines. Broad horizontal terraces are found usually between 5 and 18 fathoms of depth, but others may be scattered at considerably greater depths. General flatness is typical of these bottoms. Sediments are predominantly muds, clays, and sands, particularly where the clay fraction has been winnowed out by current or wave action.

Shelves in Areas of Active Coral Growth

Platforms of active coral growth in warm clear water of normal salinity produce areas of very shallow water. Irregularly shaped shoals and reefs may be distributed widely over such shelves, the most notable example being the northeast coast of Australia. Calcareous sands, limy oozes, and reef rock constitute the bottom deposits.

Shelves off Young Mountain Ranges

Extremely narrow continental shelves, frequently restricted and locally absent, border young coastal mountain ranges. The narrow shelves seldom average more than 10 miles in width and are abnormally deep averaging approximately 45 fathoms. Their slopes are steep, particularly in the shoreward part, but tend to become more gentle seaward. Rocky banks and islands are not infrequent, but extensive areas of sand, clay, and muds are typical bottom materials.

Shelves Associated with Strong Currents

In certain areas where continental shelves are narrow or missing, strong permanent currents characterize the nearshore areas. The Gulf Stream near Cape Hatteras and the southern coast of Florida is an example of such a current sweeping a locally inconspicuous shelf. In the same region away from the concentrated current the shelf broadens and becomes shallow. This is true also of the emergent portion of the coastal strip adjacent to the locality where the current impinges against the shoreline. Such land is low and gently sloping in contrast to the steep slope of the submerged part beneath the areas of strong current. Bare rock surfaces and to a lesser extent coarse gravels and sand characterize these bottoms.

WAVES

Wave Action along the Landward Margin

As waves sweep across the continental shelf and approach land their influence is important in the shoal and shore areas. Such waves are responsible for much of the local current system and for erosional and depositional processes. So important is the influence of wave action on shoreline physiography that a general understanding of wave behavior is essential. Generally considered, waves can be classified into two rather distinct types, waves of oscillation and waves of translation, each fundamentally different in behavior.

Deep-water waves, or waves of oscillation, are generated primarily by the wind and speed across the body of water increasing in height with length of travel. Such waves are symmetrical in form and attain their greatest symmetry in deep water after the generating wind has subsided. These are identified as *swell* waves, and as they move into shallow water their character is altered very strikingly. At some particular position near the shore depending upon the depth of water and the height of the wave the form of the oscillatory wave is disrupted, and a *breaking* wave is formed. This is a wave of translation, and the entire mass of water rather than the form alone moves forward against the shoreline.

Oscillatory Waves

The ideal oscillatory wave is symmetrical with respect to a vertical plane passing through the crest, and its outline approaches the shape of a trochoid (i.e., a broad trough and narrow crest). It is the wave form which can be simulated by the path of a fixed position on the rim of a cardboard disc as the disc is rolled along a horizontal plane.

Waves of very low height (those whose ratio of length to height is of the order of 100:1) have an outline which approaches the profile of a sine curve. For the average trochoid outline still water is closer to the level of the troughs rather than the crests, and as the wave reaches the zone of breaking as much as three-fourths of the total height may be above still-water level.

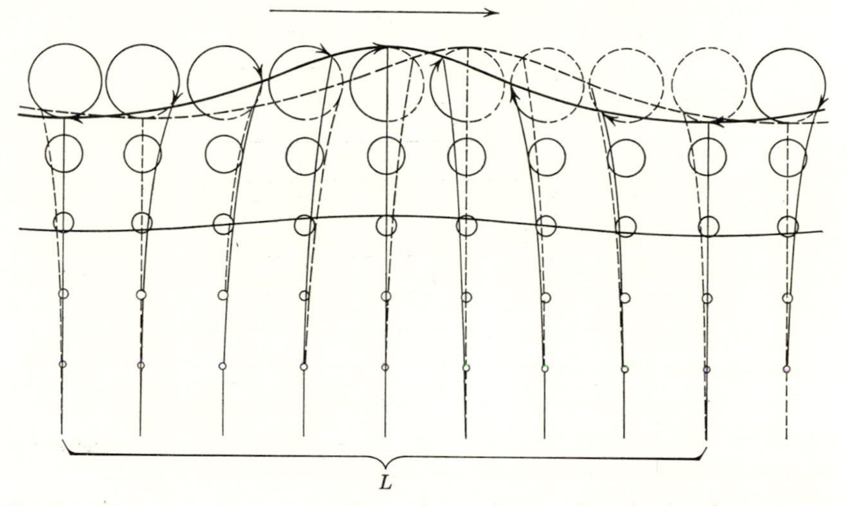

Fig. 8.1 Movement of water particles in the ideal wave of trochoid outline. Arrows on circles indicate paths and directions of particle movement. Decreasing size of the orbits is shown to scale. The wave line at approximately $\frac{1}{6}$ wave-length depth (L = wave length) shows the amount of vertical motion at that depth, and base of diagram shows movement to be expected at $\frac{1}{2}$ wave-length of depth. The dashed wave profile and corresponding dashed curving vertical lines are the position of the waves and particles $\frac{1}{8}$ wave period later than that illustrated by the solid lines. (Reprinted with permission from Kuenen, *Marine Geology*, copyright 1950, John Wiley and Sons, Inc.)

Within the wave form individual water particles are considered to revolve in circles (see Fig. 8.1). The effect is to produce a movement of water. In the troughs water motion is opposite to the forward movement of the wave, but at the crest water motion is in the direction of forward advance of the wave. With increased depth below the surface the diameter of the circles of particle movement decreases exponentially. Beginning with a diameter equal to the wave height, at a depth of one-ninth the wave length the diameter of circular orbit has decreased to one-half its original value. At a depth of two-ninths the wave length the diameter of the circular orbit is one-fourth, and at one-third the wave length in depth the diameter is one-eighth its initial length. With increased depth the relationship continues as illustrated in Fig. 8.2, the orbital path diminishing until imperceptible motion remains at a depth approximating one wave length of depth.

The relationship between wave length and wave height at the surface is such that for a trochoidal wave the height may theoretically

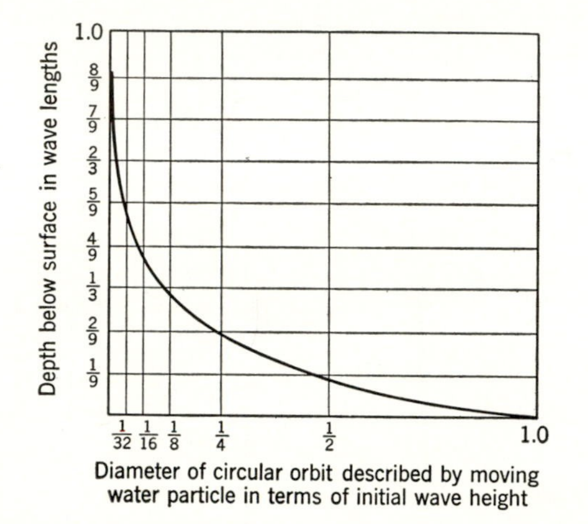

Fig. 8.2 Relationship between depth of water in terms of wave length and diameter of circular orbit described by particles during ideal wave motion (see Fig. 8.1).

attain one-seventh of the length. Generally there exists sufficient departure from the theoretical form to cause the wave to collapse before this height is reached. Ratios of one-twentieth to one-thirtieth height to length are common, and one-twelfth is extremely rare.

Deep-water waves (water depth exceeds one-half wave length) show relationships also between wave length, velocity, and period expressed by:

$$V = \frac{L}{T} \tag{1}$$

where V = velocity in feet per second.
L = wave length in feet.
T = period in seconds.
$L' = 1.56T^2$ (approximately) when L' is in meters. (2)
$L' = 5.1T^2$ (approximately) when L' is in feet. (3)

High waves result in some mass movement of water in the direction of forward advance of the wave. As the water particle describes its orbit above the mean water depth the particle moves forward, and when below that depth the particle moves in the opposite direction.

Deeper in the water, wave velocity decreases, and the total effect is to advance the particle forward somewhat more than backward. The net result is to move water gradually in the direction of the wave.

Height of open-water waves is controlled by the length of fetch during which time the wind is acting to raise the wave and the wave form is propagated forward. Wind of given velocity requires a minimum distance of fetch in order to generate a wave of maximum height. In small bodies of water the strongest wind can generate waves only of limited heights, whereas in the open ocean where the fetch is greater the same wind can develop much higher waves.

The Oscillatory Wave in Shallow Water

A wave striking against a cliffed shoreline where the water is deep is reflected without breaking. This rebounding wave interferes with incoming waves, producing steep-walled high crests which often collapse or which upon colliding with another wave are thrown upward. In this upward thrust considerable energy is imparted to the rising portion of the wave, and overhanging rock ledges are subject to destruction by the force of impact. Most shores, however, are gently sloping, and shallow water is found at the immediate coast line. As the wave advances into the shoals it reaches a water depth where the circular orbit of particle motion is restricted by the bottom. The effect is to retard the forward advance of the wave form and modify its shape. Suppose that certain groups of waves have traveled across deep water for days so that their form, period, and height are regular. As each wave reaches water depth of less than one-half wave length the velocity and length decrease, but the height increases. The slope of the swell wave is steepened, and ultimately it must break. Among short-period waves the increase in height is small, but long-period waves have been reported to rise as much as three times their deep-water height before breaking into surf. Despite the change in velocity and height the period of a given wave tends to remain constant as it approaches the breaking point.

Wave Refraction

A wave front approaching a shoreline at an angle less than 90 degrees does not enter shallow water uniformly but is refracted in such a way that the ray path swings to approach the shore at right angles. Bending of the wave front has its analogy in refraction of a light wave as it travels through a medium of one density into that of a different density.

Referring to Fig. 8.3, wave *a* and wave *b* approach depth contour

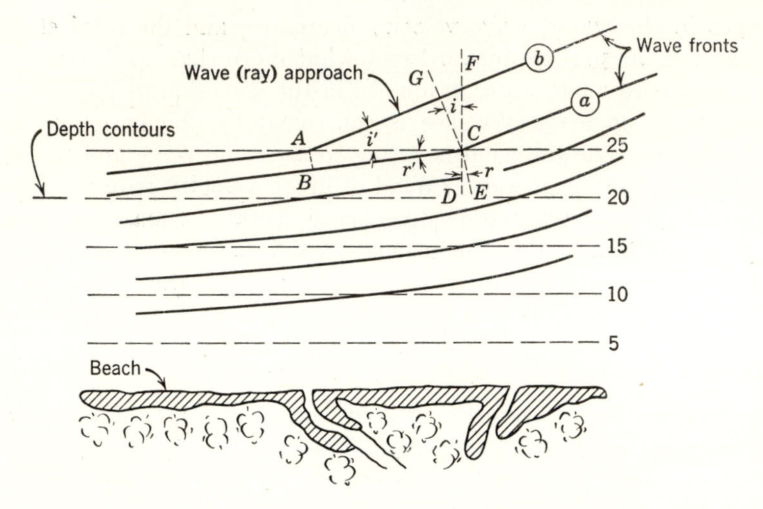

Fig. 8.3 Positions of successive wave fronts showing refraction as waves enter shallow water. Dashed lines are contours of water depth. Seaward from the contour of 25 value, waves approach with front of constant angle to the shore. The angle decreases progressively as refraction retards the velocity, and in shallow water the wave front becomes increasingly parallel to the shore. (Modified from Munk and Traylor, *Jour. Geology*, 1947.)

25 with uniform spacing and angle. Shoreward of this contour the wave front is bent so that the normals to the two fronts are GC and CE respectively.

$$\angle i = \angle i' = \text{angle of incidence}$$

$$\angle r = \angle r' = \text{angle of refraction}$$

i' and r' equal angles between depth contour and wave front

$$\sin i' = \frac{GC}{AC} \tag{4}$$

$$\sin r' = \frac{AB}{AC} \tag{5}$$

$$\frac{\sin i'}{\sin r'} = \frac{GC}{AB} = \frac{\text{velocity in deeper water}}{\text{velocity in shallower water}} \tag{6}$$

Under natural conditions the approach to the shoreline is very rarely uniform in progressive change in depth, and the refraction diagram is complicated, but the principle remains. A wave-refraction diagram is constructed by selecting uniformly spaced intervals along a deep-water wave and constructing normals to the wave front along these points.[2]

Such a construction schematically shown in Fig. 8.4 shows deep-water waves approaching a shoreline with wave front normal to the shore. Each front is represented as the position the same wave would occupy at successive intervals of time equal to one wave period. As the wave approaches the headland shallower water is encountered, wave refraction occurs, and the wave length is decreased over that of

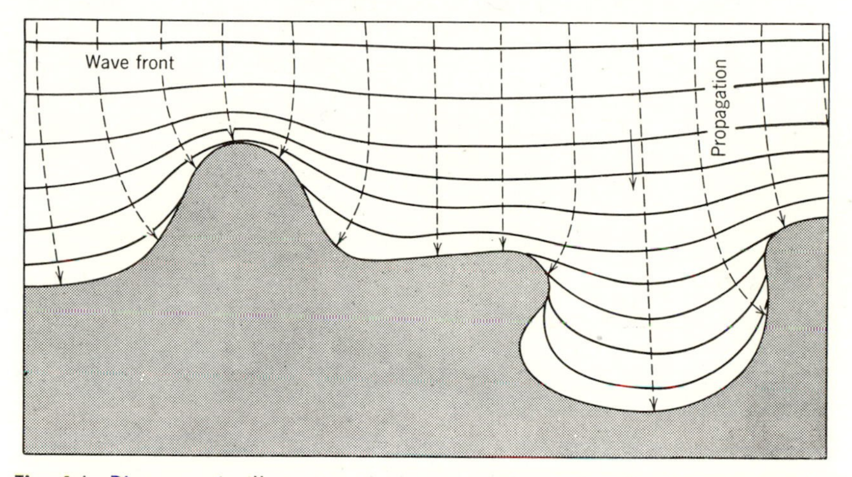

Fig. 8.4 Diagrammatic illustration of ideal refraction pattern for waves approaching an irregular coast. Note that the ray path is normal to the average trend of the shoreline (e.g., the center of the bay shore). Concentration of energy against the headlands is indicated. (Reprinted with permission from Kuenen, *Marine Geology,* copyright 1950, John Wiley and Sons, Inc.)

the deeper water zones in the bay. The dashed lines represent normals spaced uniformly along the deep-water wave front and are ray paths to the shore. Each ray crosses successive wave fronts perpendicularly and is, therefore, an orthogonal line. As the wave fronts are refracted spacing between orthogonal lines is closer along the headland than in the bay shores. If an equal amount of energy is capable of being released between any two orthogonal lines more energy is exerted against the headland zone than in the bay. Compression of orthogonal lines as the wave front approaches shallow water results in the wave energy being utilized in raising the wave height. In the bay areas where the orthogonals diverge, the energy is spread and wave height diminishes.

[2] For details of construction the reader is referred to W. H. Munk and M. A. Traylor, "Refraction of Ocean Waves," *Jour. of Geology,* Vol. LV, 1947, pp. 1–26.

Underwater shoals and ridges near shore have a similar effect upon the wave-front pattern. As the wave moves into shallow water it is retarded, whereas on either side the same wave form moves relatively faster. On the lee side of the ridge convergence occurs, and violent wave interference frequently results.

The effect of shoaling or deepening of waters is a relative one rather than an absolute relationship inasmuch as waves of small wave length are not influenced by the same bottom topography as those of long wave lengths. The ratio

$$\frac{D}{L} = \frac{D}{5.1T^2} \tag{7}$$

holds approximately where

D = water depth in feet.
L = wave length in feet.
T = period in seconds.

When $D/L = \frac{1}{2}$ the deep-water wave is said to have advanced into shallow water, whereupon wave refraction begins.

Then,

$$\frac{D}{5.1T^2} = \frac{1}{2}$$

$$D = \frac{5.1T^2*}{2} \tag{8}$$

When both short- and long-period waves approach the shore the above effect tends to separate waves of different wave lengths. This effect is particularly noticeable in cases where short- and long-period waves pass a headland at an acute angle. The long-period waves are refracted into the bay, whereas the short-period waves may continue on their course without bending of the ray paths.

Translatory Waves

As the deep-water wave grows higher in passing into increasingly shallow water it eventually becomes unstable and breaks. The position of breaking occurs when the orbital velocity of the water particles at the top of the crest exceeds the velocity of the wave form, as shown in Fig. 8.5. Theoretically, this condition of breaking occurs when the

* When T equals 10 seconds, D equals 250 feet approximately, and when T equals 5 seconds, D equals 60 feet approximately.

height to wave length ratio reaches one-seventh. Usually the theoretical maximum steepness is not attained, particularly when a strong wind tends to blow the crest forward. The "breaker," or "comber," is best developed on a broad shelving beach where as the deep-water wave advances into shallow water the velocity is checked, wave forms crowd together, and the height increases. Finally, the orbital velocity exceeds the wave speed and the wave develops a distinct "curl."

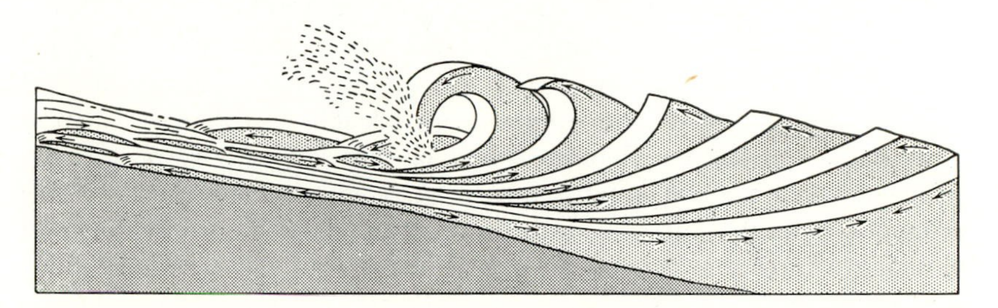

Fig. 8.5 Block diagram of a breaking wave showing paths of water movement. Note the general condition of eddy development which leads to turbulence. (Reprinted with permission from Kuenen, *Marine Geology*, copyright 1950, John Wiley and Sons, Inc. Originally from W. M. Davis.)

In Fig. 8.5 the successive stages of the advancing wave form are shown. As the wave height increases, the forward velocity of the orbit distorts the wave form which cannot be filled with water. The wave crest extends beyond support, dashes forward, and collapses. This is the wave of translation which is characterized by the forward movement of all particles of water at the same rate. All of the water, both top and bottom, is now in forward motion, and much energy is released as the wave climbs the beach. Such waves striking a cliffed shore exert energy at the water-level line and undercut the cliff. This is the wave capable of powerful erosion, and under this attack the shore line retreats.

In a general way the wave begins to break where still-water depth ranges between 1.3 and 2.5 times the height of the "roller," but swells with long period have been reported to break at even deeper water ratios.

A typical characteristic of the translatory, or solitary, wave is that despite its forward drive the water behind is at rest. Each wave is a separate entity and has no periodicity except as generated through some wave of oscillation. Perhaps the best form of the solitary wave is a tidal bore. The bore is generated by a mass of water in a bay mouth concentrated and driven during high tide into a stream channel of constricted walls. The wave moves upstream as an independent unit as though formed by driving a large board through shallow water. A necessary condition is that the speed of propagation be greater than

$$\sqrt{gd} \tag{9}$$

where g = acceleration of gravity.
d = average depth of water.

A somewhat specialized type of wave which locally, at least, behaves as a translatory variety is the *tsunami,* or "catastrophic," wave. The tsunami is generated almost exclusively by change in level of a localized part of the ocean floor through earthquake activity (faulting). As the sea floor is raised or lowered with respect to its adjacent portion along a fault line, a wave is produced as the water attempts to reach surface equilibrium. Landslides into water and plunging of great masses of ice from glaciers into the oceans produce similar but generally less impressive waves. The magnitude of the tsunami is directly proportional to the size of the fault scarp. Some of these waves have been reported to be of extreme height advancing to the shore as a wall of water 200 feet high. Fortunately this situation is rare, but waves of 15 to 20 feet of height are not uncommon along the shorelines of strong earthquake activity. Velocities attained by tsunamis are great. Their travel rate has been computed as exceeding 600 feet per second in crossing parts of the Pacific. Inasmuch as they are a variety of solitary wave they can be expected to generally obey eq. 9.

Sediment Transport by Waves

Advance of the oscillatory wave into shallow water produces a distortion of the circular orbital path of the water particle such that a to-and-fro motion is developed along the bottom. This movement is somewhat greater shoreward than seaward, and gradual movement of sediment toward land can be attributed to this action. The principal shoreward motion, however, is attained by the breaking wave, advancing as it does with strong movement along the bottom. (See Fig. 8.5.) Following initial collapse of the "breaker" the water is in a high state of turbulence, and much sediment is lifted from the bottom. Under the driving force of the wave some of this sediment is carried to the beach. Heavy particles settle to the bottom almost at once and are concentrated as finer debris is carried along. In regions where the sediment size distribution ranges between cobbles and clay the zone of average position of the breaking wave becomes a belt of gravel bottom.

Recent investigation has demonstrated the existence of two different types of wave behavior along the beach. Storm waves of small length, irregular height, and steep front tend to break near the still-water line. Such waves plunge vertically and produce swash of large volume but carry little sediment to the beach. Backwash from these waves is powerful inasmuch as relatively small amounts of water sink into the beach and the volume returning is great. The returning

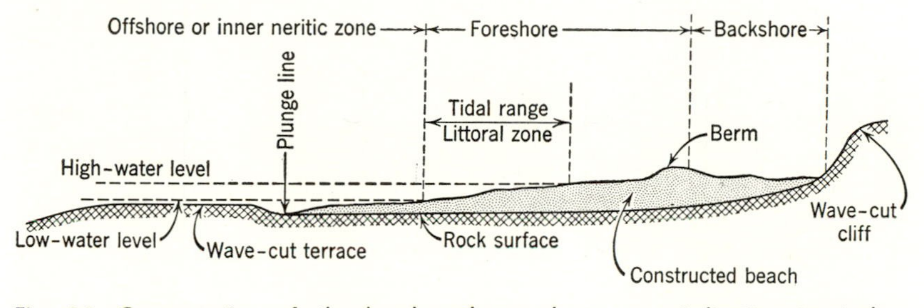

Fig. 8.6 Cross section of the beach and near-shore zones indicating terminology applied to various segments and principal features.

current gathers sand; erosion of the beach proceeds with surprising speed and often is indicated by distinct cutting of a *berm* (see Fig. 8.6). A somewhat different behavior is due to waves resulting from long fetch. Such waves approach with long, regular form and tend to break farther from shore. The shoreward sweep is with considerable power, and much sediment is carried to the beach. The swash, however, is widespread over the beach, and the volume per unit area is small. Moreover, the long period of these waves allows time for much of the water to infilltrate into the beach. Backwash is slight, and little debris is removed from the beach. Through this process, the *foreshore* is constructed, the berm is built higher, and beach aggradation is in progress.

The beach consists of several segments identified to some extent from the position of the berm. Terms such as *backshore* and *foreshore* refer to those parts of the beach which lie landward of the berm crest and seaward of the berm, respectively. Foreshore areas are restricted to the exposed part of the beach and extend seaward as far as the water line of lowest tides. Wherever the tidal range is great and the slope of the beach is gentle such as some sections of the coastline of the Netherlands the foreshore area is extensive, and this portion of the beach is called the *littoral.* Seaward from the foreshore area is that part of the offshore zone identified as *nearshore,* also called the *inner neritic zone,* which extends to water depths of approximately 20 fathoms (120 feet). Deeper waters extending to the margins of the continental shelf are called the *outer neritic zone.*

The beach is a constructed feature consisting of sediment lying upon an underlying layer, usually bedrock, whose origin is unrelated

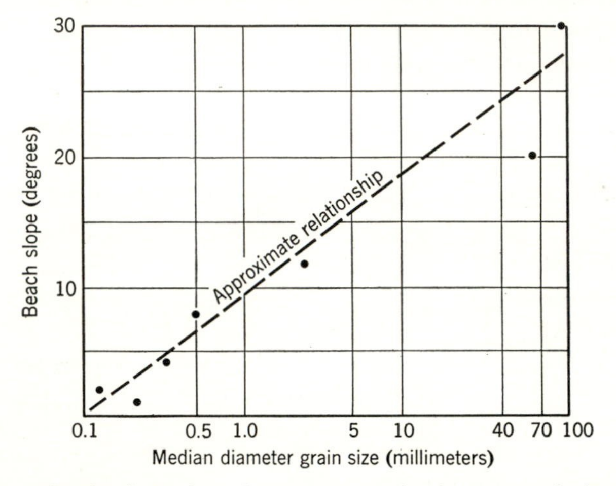

Fig. 8.7 Generalized relationship of average particle diameter and slope of beach. Note the very steep slopes which are typical of beaches whose average particle is in the cobble range.

to the existing sea or lake. Landward the beach gives way to a notch or cliff of height ranging between less than one foot and hundreds of feet. The notch identified as the *wave-cut cliff* marks the limit of the backshore area by virtue of its origin through the attack of storm waves.

Seaward from the foreshore the sediments of the beach can be traced underwater to a shallow trench, defined as the *plunge line,* which lies close to shore. The plunge line is the site of the most frequent breaking wave, and a trench has been developed by wave scour. Outward beyond the plunge line the sediments may disappear and be replaced by a shallow-water area. Here, the bedrock emerges to be beveled by wave attack into a bench called the wave-cut terrace. This erosional platform has an outer limit which is ill defined as its slope gradually joins the bottom of normal depth.

Oftentimes the slope of the foreshore is the identifying feature indicating whether erosion or deposition is dominating the condition of the beach. Steep foreshore slopes and prominently cut faces of the berm are typical of erosion. Flat beaches and low berms are usually indicative of building of the beach. In applying such generalizations attention must be directed to the average grain size of the beach sediment. Coarse material tends to rest with much steeper slopes than fine sand, and an aggrading gravel beach may assume a slope exceeding 20 degrees. (See Fig. 8.7.)

Longshore Drifting

Current action along the shoreline is recognized as being of several different types. Localities of high tidal range are usually those where sediment is moved from the shore to open water. Where tides are low currents moving parallel to the shore appear to be of greater importance. Except for certain areas of special geologic conditions most shore currents are direct results of wave action on the shore. Waves with ray paths oriented perpendicularly to the beach result in a seaward return of water whose velocity is controlled by the gravity drive down the slope of the beach. Normally, the resulting current (undertow) is weak. Under some conditions the wave pattern produces a large eddy system (rip current). Although this current is concentrated near the surface it carries fine sediment seaward.

Wave approach (i.e., ray path) is perpendicular to the shore only exceptionally, and usually the wave front strikes with an angle which under certain circumstances of wind direction is large. The effect is for the swash to advance up the beach and return describing a general parabolic path. This circuit causes the backwash return to be somewhat laterally distinct from the inrushing path. A sand particle carried inland from some offshore position is returned to the water somewhat farther along the beach than its position of initial entry. As the process continues the sand particle describes a zigzag path with a directional component parallel to the shoreline. Under the constant drive of waves sediment is shifted along a coast line, and temporarily may accumulate in some favored position. Over long periods of time this process, identified as *beach drift,* shifts sediment from beach to beach along the coast line. Where strong prevailing-wind directions dominate conditions along an extensive coast line particles migrate many miles to some site where current activity is incapable of moving them farther. An equally common situation is a continued shifting of sand and pebbles up and down a coast line as wind direction dictates the wave attack.

Selective transport is a prominent part of the process of beach drift, and particles of like dimension tend to be concentrated in certain areas. For example, gravels accumulate as residual debris at localities which are losing finer material. Other beaches accumulate sand, whereas silt- and clay-sized fragments tend to be carried seaward as well as long distances along the coast line. Coarse particles carried shoreward are deposited during the uprush of waves and are added to the foreshore portion of the beach. If the beach slope is steep and backwash is strong most of the pebbles are concentrated near the still-water line. Where the beach slope is gentle pebbles are spread widely and may eventually form a layer which is added to the beach. Also, a low berm ridge, often pebbly, is built near the limits of the foreshore. The over-all effect, however, is to create a gradual decrease in the average particle size in the direction of drifting. Certain beaches are in geographic localities where one general wind direction prevails over others. In this case a decrease in particle size distribution is to be expected in the direction of the prevailing beach drift and is particularly well demonstrated in those areas where the source of some of the sediment can be identified as having been derived from headlands which are being eroded.

SHORELINE PHYSIOGRAPHY

Features of the Shoreline

An important concept concerning the association of shoreline features was made many years ago in an effort to group them into some genetic relationship. The essence of the interpretation is that the principal features observed can be classified either as those associated with coast lines which have sunk relative to sea level or as those which have been relatively elevated with a portion of the former sea floor becoming land.[3] Coast lines of the former group are identified as *submergent,* whereas the latter are known as *emergent.* Studies carried on since the time of the original proposal have indicated that the concept is somewhat rigid and that in certain localities features of both occur together. Its use herein, however, is to permit assemblage of features into a series of progressive stages, best understood when learned in sequential order. The intent is to present the characteristics of submergent and emergent shores in order to describe

[3] This concept proposed by D. W. Johnson proved to be one of the great forward steps in understanding shoreline processes. See *Shore Processes and Shoreline Development,* John Wiley and Sons, 1919.

the origin of the features involved, but the classification suggested is one which follows in later paragraphs.

Shoreline of Submergence

In its ideal form, this shore is developed where a rough topography is submerged rapidly so that the sea invades the valleys and divides become headlands. Wave attack is concentrated on the headlands (see Fig. 8.4). They are undercut at the water line, and a wave-cut cliff appears. With continued retreat of this cliff there remains at the base a terrace, slightly below water level, and identified as the wave-cut terrace or bench. The wave-cut bench grows as the cliff retreats, and, locally, where rock of special resistance exists (particularly in bedded rocks) small island pinnacles project above the water level. Such features identified as *stacks* add to the picturesque coast.

Debris broken from the headlands remains only temporarily at the base of the cliff. Here, in due time, it is shattered into small pieces which can be transported by current action. Longshore drifts move this sediment bayward where it is deposited as it enters the deeper water of the bay where wave and current energy are reduced. The effect is to produce first an underwater bar and later an emergent beach projecting from the headland. Such depositional features are identified as *spits,* and the headlands are called *winged* when spits project from either side.

Spits develop from bars as the water becomes sufficiently shallow to affect the oscillatory wave. Large waves are caused to break, gather sand from the bar, and pile it higher and slightly landward as the wave form collapses in the deeper water of the bay. Once the spit emerges above the water it behaves as a beach. Translatory waves breaking along its shore add more sand to its surface and sweep more coastwise to lengthen the spit. Spits grow toward one another from headlands on either side of the bay, and the respective length of each spit is a rough approximation of the direction of prevailing wind. The spit represents only the subaerial part of the deposit which extends farther across the bay and may cross it entirely. The latter is termed a *bay-mouth bar* and serves to reduce wave advance into the bay. Under initial conditions of shallow-water bays and long headlands the spits will grow, move shoreward as the headlands retreat, and eventually partially close the opening to the bay. At localities where the tidal range is great sufficient water will move in and out of this restricted section during the ebb and flow to maintain an open channel and prevent additional elongation of the spits.

Gradual reduction in the length of the headland forces continued landward shift of spits as the sediment does not remain fixed against wave attack. Short spits or those exposed to strong wave attack in open water migrate landward differentially, effecting a curved outline identified as a *hook*. The fact that a hook can be maintained indicates an abundance of sand being moved by the shoreward drift. Thus as sand is lost through beach drift a new supply is received, and the hook is stabilized or may actually increase in size.

Shifting wind directions sometimes prevent extensive spit development, and sand migrates along the beach line from headlands into the bays. By this process small round bays tend to accumulate broad beaches at their heads, but estuaries or bays connected with a river system do not behave in like manner. In these cases river silts are deposited along the sinuous shoreline, and extensive mud flats are characteristic.

In the ideal case of the submergent coast line, bays are not extremely deeply indented, beaches accumulate at the head, and spits disappear as the headland and bay head approach one another. Eventually a straight shoreline is developed, and much of the bay-head beach is shifted. In fact, whether a beach remains static is dependent upon the supply of sand migrating along the coast. In most cases beach width is reduced, and the steady-state condition which continues to exist has typically narrow beaches.

Under constant storm-wave attack the straight shoreline continues its retreat leaving a wave-cut bench of ever-increasing width. The eventual width of this wave-cut bench has been a matter of conjecture among geologists. Considerable evidence exists that entire islands have been planed off to a common level. There remains a matter of interpretation whether given sufficient time large areas of a continent could not undergo such marine planation.

Shoreline of Emergence

Contrasted with the rough initial topography of the submergent coast line is the low-lying strip of the ideal shoreline of emergence. Uplift of the coast has exposed an extensive belt of former sea bottom, and the old shore position is now elevated and considerably farther inland than the new. The immediate shore is rather straight and is sinuous only where the few stream channels exist. The beach is broad and covered with fine sand except at stream mouths where swampy parts of deltas may be exposed. Water depths are abnormally shallow, and storm waves reach their breaking position several miles from shore. Wave refraction is not locally controlled; storm waves are gradually refracted to a position where their ray paths advance at approximate right angles to the shore. Great long breakers result, and strong waves of translation move across the shallow water.

Turbulence developed at the position of initial breaking distributes sediment from the bottom into the overlying waters which gradually lose much of their turbulent energy as they advance shoreward. Much of the sediment settles quickly and in time constructs a bar in the already shallow water. This bar shifts shoreward as wave attack drives it in that direction. The migration is accompanied by removal of sediment on the seaward side and addition of this material to the general area of the bar. Through this process the bar increases in size, and as it moves shoreward erosion on the seaward margin produces deeper water in that area. The principal position of wave breaking is maintained, therefore, near the site of the bar. Under this constant wave attack and shoreward advance the bar eventually emerges from the water, whence it first appears as a series of low-lying elongate islands described as *barrier beaches.*

Waves sweep across the barrier islands, drive them inland, and continually add sand to their bulk. Through this process they must continue to increase in dimension not only in length but also in width. Islands coalesce into narrow barrier beaches miles in length. At this stage a body of water identified as a *lagoon* is isolated from the sea and separates the land section from the barrier beaches. Most lagoons are connected to the open ocean by tidal channels through the barrier beach and are subject to a change in water level as tides rise and fall. The lagoons also receive a supply of fresh water from streams and, hence, are much less saline than normal sea water. For this reason they are sites of dense marsh vegetation, principally grasses.

Winds sweep sand across the surface of the barrier beaches and generate small dunes. These begin as low hillocks along the backshore and migrate landward increasing in stature as they do so. When the sand supply is ample they may build to heights of from 10 to 20 feet before they reach the lagoon and are destroyed. In the lagoon the sand is spread, adding to the width of the barrier beaches and making the water more shallow.

The lagoon varies in size reaching its greatest width when the shore of the mainland turns inland to follow a river channel. Throughout its length, however, it is extremely shallow, and what may appear on maps to be a potentially fine harbor is in fact not one at all. Vegetation thrives in these shallow waters, and large swamps develop. Sand, silt, and clay are gradually added as streams contribute their sediment; this debris is shifted about and locally sorted. The total area suited for swamp growth continually increases. Gradually the lagoon fills through deposition by streams and invasion by dune sand. Except for channels maintained by tidal scour the barrier beach is joined to the mainland in whole or in part.

Along the oceanward side the shoreline is straight and continuous, and wave attack is directed against the beach, whereas dune sand migrates across the former site of the lagoon. Some sediment still reaches the beach from the open ocean, but this supply is reduced as the water is deepened by erosion of the bottom. In essence the principal supply of sand has been greatly reduced. Equilibrium conditions are reached as sand is shifted along the long expanse of beach. Certain local areas become zones from which sand is removed and shifted along shore to new sites of temporary deposition. These portions are not fixed, but from time to time they also migrate back and forth along the many miles of coast line. A condition of equilibrium is, therefore, a state of continual change, and the characteristics of one year are not necessarily those of the next.

Filling of the lagoon with sand, silt, and organic matter produces a condition of unstable fill. Where silt and sand have been swept to rest upon swamp deposits a layer of water-saturated peat is produced. In its saturated state this material yields under very small stress in contrast to the overlying sand which when properly confined is capable of supporting very heavy loading. Sometimes the covering of sand is twenty or more feet thick, and no surface evidence of the existence of the plastic peat layer is to be found.

Classification of Coast Lines

Any attempt to group shorelines into common and distinct varieties meets with certain difficulties principally in connection with the genetic interpretations involved. Generally considered, the shore is defined as that portion between lowest low tide and the landward edge of the wave-transported sand, whereas the coast includes much of the landward portion including sea cliffs, elevated terraces, and lowlands. Some indefinite distance landward the aspects of the coast line are no longer visible, and this position can be regarded as the inner limits concerned. Sea cliffs and terraces formerly known to have been once at the shoreline are recognized in some localities to be as much as several hundreds of feet above sea level and some miles inland. Elsewhere, the coast line shows an invasion of the sea, and large river valleys obviously excavated by streams are partly submerged below present sea level. The student of geology cannot examine such conditions without recognition that the boundary between sea and land has shifted both upward and downward over some period of time.

The rate at which such shifting occurs is not precisely known, but along coast lines of active faulting almost instantaneous changes have been recognized following an earthquake. In other areas such as where deltas are building seaward the change is slow and of a different nature. Drowning of river valleys is also a slow process involving rates estimated to be of the order or magnitude of a foot per thousand years. Even slower rates are believed to have prevailed throughout the geologic past, but with sufficient time enormous parts of the continents are known to have been submerged eventually.

Coasts in areas of Pleistocene glaciation show a remarkable correspondence of geology between the submerged portion and the land area such that there exists no doubt that some parts of the continental shelf were once out of water and subject to glaciation. Calculations of the actual lowering of sea level, which would be involved in the process of storing upon the land all the water necessary to provide the ice of the continental glaciers, have produced figures of as much as 300 feet. These values are to be interpreted as representing the order of magnitude of fall and rise in a cyclical change in sea level from its pre-glacial position, through a lowered stage, and a return to some "normal" position represented by the present. Worldwide raising or lowering of sea level is described as *eustatic,* whereas local change in land level which involves part of the continent is diastrophic change. In the long run diastrophic movements are capable of producing the most profound alterations as thousands of feet of change in level are known to have occurred. Nevertheless, the recent eustatic shift in sea level has impressed itself upon coast lines the world over, and its importance must be considered in erecting a classification of such physiographic areas. The aspect as now observed, therefore, represents a summation of all effects, diastrophic, eustatic, and depositional, but the features characteristic of one may dominate the others.

Following is a classification which has been designed to embody the genetic aspects mentioned above but is descriptive in its detailed form. It is representative of one of the most recent attempts in this connection.[4]

Primary, or Youthful, Coast Lines

Shorelines so classified are those where the influence of wave attack against the shore is minor, deposition is insignificant, and the outline

[4] This is part of the classification proposed by F. P. Shepard and is presented in detail in *Submarine Geology,* Harper and Bros., 1948.

is the result of physiographic development on the land. The type of landform, therefore, is fundamentally the basis for subdivision.

Drowned river valleys. These are principally the result of downwarp along the coast line or eustatic sea-level rise following deglaciation. Whichever has been the cause the result is the same inasmuch as there is a marine invasion and valleys become deeply indented bays and estuaries. Drowned river valleys are recognized by shallow waters in the estuary, the dendritic form of the tributary valleys, accordant bottom levels, valley cross sections, and the extension into an existing river system landward.

Drowned glacial valleys. Coast lines associated with drowned glacial valleys are typically complex. They consist of a network of principal fjords which often extend inland for as much as 100 miles and a maze of islands and channels which constitute an extremely irregular, if not an indefinite, shoreline. The fjords are characterized by extremely steep walls, with hanging valleys well up their sides, and in the polar areas glaciers at the valley heads. The typical fjord has a U-shaped cross section and has unusually deep waters sometimes in excess of several hundred fathoms. Perhaps the most characteristic feature is an underwater sill which rises above the main floor of the fjord at its mouth. Here occur some of the most shallow waters. Frequently, a narrow strip of flat land occurs along the shore only slightly above the water level. This strip known as the *strandflat* constitutes the only section suitable for harbor installations and communities. The origin of the strandflat appears to be related to recent uplift and is attributed to rebound of the rock surface to its former level after removal of the load of ice.

Sills at the entrance to fjords create a condition of water stagnation in the deep waters. In the fjord proper, fresh water from streams dilutes the salt water and reduces its density. Stratification of this brackish water with lightest at the surface and heaviest at depth prevents turnover (see page 274), and the sill proves to be a barrier to inflow current action despite the outward flow of surface water. The net result is stagnation of the deep saline waters and movement only of the upper waters. Oxygen in the bottom waters is gradually depleted completely, and generation of H_2S by sulfate-reducing bacteria is the characteristic condition. Muds from these bottoms are black and emit a strong sulfurous odor.

Coasts of deltas. Large and small deltas invading the continental shelf alter the coastline by causing outward bends and important but low-lying seaward projections. The deltaic section is recognized by the system of river distributaries, natural levees, swamps, and scattered small islands inland of the area where the delta is being extended.

Drowned alluvial plains. These are not commonly occurring features and are recognized by a relatively straight shoreline and an inland section of an alluvial plain sloping gently seaward. The southern portion of the Texas coast line is principally a representative, but it has been modified in part by a eustatic drop in sea level which has left marine clays at the immediate shoreline.

Coasts of glacial deposits. Coasts developed by incomplete drowning of moraines tend to be irregular and as a whole embayed. Perhaps the best means of recognition outside of actual field observation is in tracing, by means of topographic maps or air photographs, the moraine section from the land into the water.

Coasts of wind deposition. Obviously these are localities where a low-lying coastal strip borders a desert section from which dunes are migrating shoreward. This is a very unusual occurrence inasmuch as the condition requires the prevailing winds to be seaward.

Coasts extended by vegetation. Certain tropical areas support shoreline vegetation characterized by the mangrove. These are plants whose roots, which join the trunk above the water line, must extend into waters of high salinity. The nature of their interlocking root system forms an impressive barrier marginal to the shore and acts as a breakwater. Sediment accumulates in the projecting root system. The mangrove barrier grows seaward, and a straight shoreline is extended. Such shorelines are recognizable only by means of aerial photographs or by direct observation.

Coasts shaped by volcanism. Coast lines in waters of the Pacific have been affected commonly by volcanism. Lava flows spreading down the slopes of volcanoes have entered the sea and built important promontories. When associated with currently existing volcanoes the gently sloping lava flows continuing underwater are readily recognized. Local volcanic cones rising above the water are characterized by their symmetry. Explosion, or collapse, craters (*calderas*) produce nearly circular, deeply indented bays often with narrow inlets. These are the scenic, nearly enclosed craters typically displayed in travel folders of the region. Any such bays in a volcanic area or concavities in the side of a recognizable volcano should be interpreted as having originated in the manner outlined.

Coasts shaped by major earth movement. Recent faulting tends to produce straight shorelines with marginal deep water often identified by narrowing of the continental shelf. Submarine slopes are known to continue steep as extensions of the land slopes. Coasts developed along strongly folded belts are extremely variable in aspect. Generally they are sinuous with embayed sections representing downwarps and promontories representing upfolds. Depending upon the relative trend of the shoreline to that of the folding the aspects of the bays and prominences may be symmetrical or not. If folding parallels the shoreline the result may be a steep straight shore not unlike that produced by faulting. Coastal islands are likely to be present, and their long axis will parallel those of the land ridges. If the trend of the folding is inclined to that of the shoreline a deeply embayed coast typical of the ideal condition of the shorelines of submergence is to be observed.

Secondary, or Mature, Coast Lines

So long as the conditions of the primary shoreline continue to exist the shore is characterized by such features, for they have been able to dominate over the influence of wave attack. But nearly all of the processes listed as responsible for primary shores are episodic in development; the initial condition is established, and the process is halted. After this stage wave attack exerts its full influence, and eventually the shoreline is altered accordingly. Features typical of marine erosion and deposition make their appearance, and in time characterize the physiography. At this stage features typical of the shoreline of submergence or emergence are recognized. The progressive sequences described above make their appearance and eventually establish the steady-state condition.

An exceptional condition occurs in clear tropical marine waters where colonial corals and algae construct reefs in the shallow water fringing the mainland and islands. The reef is extended seaward by the growth of these organisms, and a rock platform at sea level is constructed. Broad shoals appear, sometimes associated with calcareous-sand beaches in the bays, where sediment broken from the reef is carried inland or moved along the coast.

LAKES

General Characteristics

A lake is defined as an inland body of standing water isolated from the sea. This very general definition encompasses waters of great range in dimensions and salinity, from small ephemeral ponds to very large basins. Lake Superior (31,000 sq. miles) and the Caspian Sea (170,000 sq. miles) are recognized as the largest of the fresh-water and salt-water bodies, respectively. Lakes occupy basin depressions which have become filled with water because the inflow temporarily exceeds the amount lost through evaporation, ground-water infiltration, and overflow. Most lakes have surface elevations above sea level, but near the ocean coast lines they approximate sea level. In some desert areas evaporation keeps the basin only partly filled, and the surface elevation is well below sea level. Inflowing streams deliver sediments to the lake, whereas the outlet is being lowered steadily by erosion. Eventually the combined effect of these two processes tends to destroy the lake either by filling or draining. When this has occurred, there remains only the sedimentary fill and shoreline topography to identify its former existence.

Generally accepted classifications of lake basins have been based principally upon their origin, and although several such classifications have been proposed the one listed in Table 8.1 is preferred.

Wave activity along the shoreline of the lake follows the same broad pattern typical of oceans, and its intensity is a function of the size of the body of water. Storm waves generated on Lake Superior and Lake Michigan are comparable in height to those occurring in the Mediterranean. Consequently in these bodies longshore drift and wave cutting are important processes. As lakes decrease in length wave activity is strongly reduced and in some occurrences is of extremely minor proportion. In such cases wave-generated currents and erosion due to wave action are essentially nonexistent. For these small bodies of water the principal shoreline changes are brought about by ice push, biological activity, and sediments introduced by the inlet streams.

Ice push produces a rather conspicuous rampart along the shores of small lakes. The locality must be in an area where winter temperatures permit growth of a thick layer of ice across the entire lake. Expansion of this ice mass forces a gravel ridge to appear bordering the shoreline. Over a period of time this ridge becomes stabilized at a height of several feet above the local ground level and remains as a conspicuous topographic feature.

Biologic activity is extremely important in most all small lakes, and the residuum of this activity is one of the principal substances accumulating on the lake bottom as annual or seasonal layers. Bottom materials principally are: (*a*) bodies of floating organisms, (*b*) plant and animal remains of the littoral zone, (*c*) wind-blown materials both organic and inorganic, (*d*) silt, clay, or coarser sediment introduced by streams, glaciers, from slope wash, or eroded from shore, and (*e*) marl (largely $CaCO_3$) produced by plants and animals and mixed with clays and silts.

Table 8.1. A Classification of Lake Basins*

1. Basins formed by crustal movements.
   - A. Rift valley or fault block. These are great linear depressions formed by downdropping of local parts of the earth's crust.
   - B. Broad downwarp of river basin. Gentle subsidence of a segment of a river system results in flooding of the area and development of a lake.
   - C. Sinking of valley alluvial fill due to minor faulting (earthquake activity). Extensive alluvial fills are subject to compaction under the vibrating influence of earthquake action. Water is forced from the pore space and important subsidence of the valley fill occurs.
   - D. Upwarp of an area isolating an arm of the sea. Long inland extensions of the sea become isolated by gradual rise of the intervening area.
2. Basins formed by streams.
   - A. Meander cutoff, oxbow type.
   - B. Valley border lakes located between valley walls and natural levees.
   - C. Stream dammed.
   - D. Stream quarried.
3. Basins of marine origin.
   - A. Lagoons developed by barrier beaches.
   - B. Irregularities in newly uplifted sea floor.
4. Basins developed by solution (limestone cave areas).
5. Basins of glacial origin.
   - A. Scoured basins.
   - B. Valleys blocked by glacial debris.
   - C. Irregular deposition of glacial debris.
6. Basins formed by the wind.
   - A. Shallow scoured basins occupied by ephemeral lakes.
   - B. Shallow depressions isolated by dunes.
7. Basins of volcanic origin.
   - A. Valleys dammed by lava flows.
   - B. Lakes occupying extinct craters or calderas.

* This classification is essentially that of C. R. Longwell, Adolph Knopf, and R. F. Flint, *Physical Geology,* 3rd ed., John Wiley and Sons, 1948, pp. 142–146. It was proposed originally by I. C. Russell in 1895.

In some cases the annual increment is readily distinguished by color contrast, whereas in other cases the principal change to be noted is in the remains of the organisms. An example of the former is the annual silt and clay layer (*varve*) deposited in a glacial lake which represents seasonal and annual change by color contrast and grain size of the very fine sediment carried to the quiet waters. Pollen and spores carried by the wind from land plants marginal to the lake are representative of variation in organic remains as seasons change. As the flora around the lake changes so likewise does the fall of pollen into the lake. Indeed, this was the pattern in the geologic past, and studies of ancient lake sediments have been particularly useful in reconstructing the changing environment of nearby land areas on the basis of the sequential order of pollen and seed changes.

Lake currents are much more restricted in variety than those of the oceans. Some large lakes display prevailing directional currents, but these are not in operation continuously such as the major oceanic

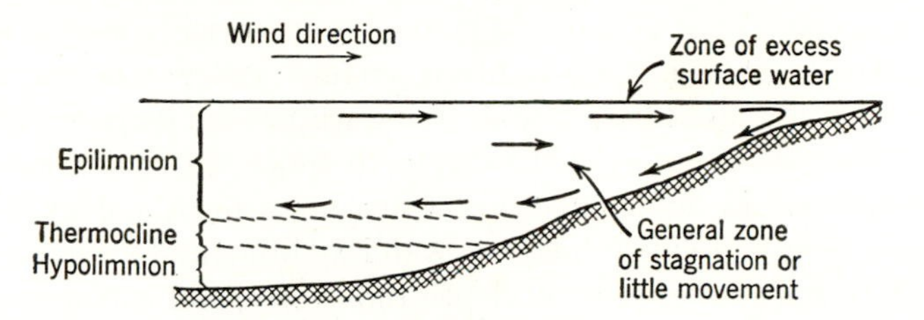

Fig. 8.8 Thermal zones near shore of fresh-water lakes where excess water has been driven by winds. The return currents move along the upper boundary of the thermocline causing the hypolimnion and parts of the epilimnion to remain stagnant. (By permission from Welch, *Limnology,* copyright 1935, McGraw-Hill Book Co., Inc.)

currents. Tidal effects are insignificant, and except in the large bodies beach drift, longshore currents, and undertow are only locally existent. Lakes in tropical regions have very little circulation, but those in cold climates tend to exchange bottom and surface waters as temperature differences cause water of greatest density to sink.

Important but sporadic movements called *return currents* are formed when water is driven to a far shore by a strong wind. This local rise in water level causes a return current of bottom water to move into areas of low level. During the summer when a lake is thermally stratified the upper portion defined as the *epilimnion* is approximately at the same temperature. Excess water driven to one shore returns to restore surface equilibrium via the boundary of the colder and denser water in the region of intermediate depths of the *thermocline.* (See Fig. 8.8.) Return currents move along the boundary of this dense layer leaving a mass of water in the epilimnion zone lying stagnant between the two currents. Below the thermocline lies the cold water of the deeps, the *hypolimnion,* which remains quiet, stratified by density throughout much of the year.

Thermal Stratification

If a vertical series of temperature records is taken at regular intervals of depth from surface to bottom when ice covers a lake, it will be found that the temperature of water immediately below the ice layer is very near to freezing. Lower, the temperature rises slightly but steadily to the temperature of maximum density near 4°C. Colder water has lower density and, therefore, remains stratified near the surface. Slowly, under the warming influence of spring, surface water rises in temperature until maximum density is attained. Suppose now that the series of temperature readings is repeated again. The surface water will be approximately 4°C, and the underlying water will be somewhat colder. Still deeper the temperature will rise to approximately 4°C. During this condition surface waters are warmer but heavier than the immediately underlying water, and density instability exists. Threads of heavy water begin to sink, and colder water rises to the surface. In due course this water is warmed, and by this constant exchange the entire body of water is brought to a temperature of 4°C. In some lakes of great depth the overturn of surface and deep water does not extend entirely to the bottom. Shallow lakes exchange water completely, and surface waters carrying a full complement of dissolved oxygen are carried downward to replenish the supply required by bottom-dwelling organisms.

During the condition of the spring overturn, mixing of water even in deep lakes appears to be carried out by winds driving water toward one shore and unbalancing the water level. In this case the return water cannot follow along a boundary between light surface water and the more dense water of the thermocline but must move laterally along the bottom for at least part of the journey. General circulation thus has been accomplished, and the oxygen supply of deep waters is once again renewed.

The period of overturn comes to an end when surface-water temperatures gradually rise above 4°C. Such water is of lower density than the deeper water, hence it remains at the surface. As summer approaches, surface waters continue to increase in temperature and also become increasingly isolated from the bottom water. If the series of temperature readings is repeated the records indicate a distinct upper layer of high temperature, the epilimnion. Here the water temperature is approximately uniform. Next deeper is a layer, the thermocline, characterized by relative rapid temperature drop (i.e., steep thermal gradient). The lowermost layer, hypolimnion, is nearly uniform, and the temperature at its upper surface is only one or two degrees warmer than the very bottom waters. (See Fig. 8.9.)

Sometime during the fall season as air temperatures fall the water temperature becomes uniform from surface to bottom once again through overturn. During these brief periods rarely exceeding a few weeks in the spring and fall the lake consists of one water zone only, and thorough mixing of water occurs in shallow lakes from top to bottom.

For purposes of demarcation the thermocline is distinguished from the epilimnion when the temperature gradient exceeds 1°C per meter of depth. Similarly, the hypolimnion is recognized when the tempera-

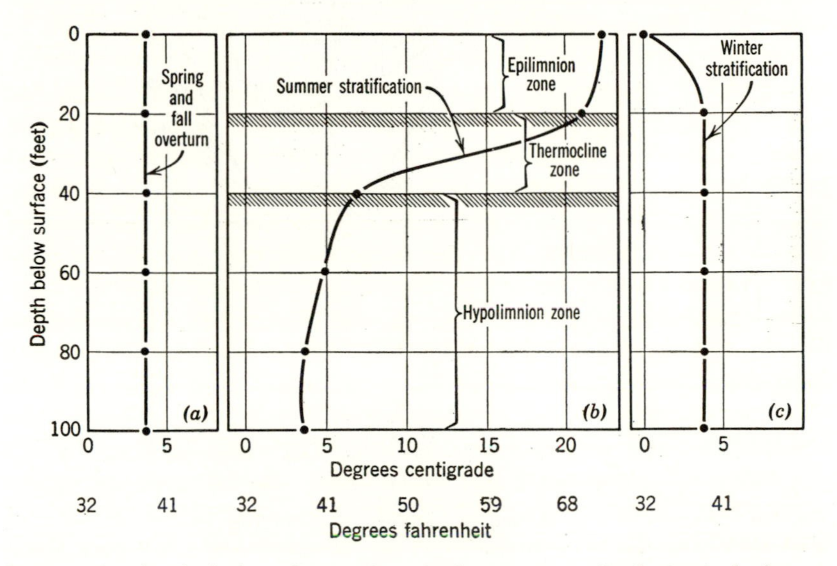

Fig. 8.9 Graphs of ideal conditions of seasonal temperature distribution in fresh-water lakes. (*a*) Spring and fall overturn; uniform temperature of 4°C is found from surface to lake bottom. (*b*) Summer conditions showing pronounced gradient in the thermocline, warm water of epilimnion zone, and cold, dense water of hypolimnion. (*c*) Winter conditions with coldest water stratified at the surface and uniform temperatures of 4°C below the surface waters.

ture gradient returns to less than 1°C for each meter of depth. In the winter season water at freezing temperature accumulates and remains stratified at the surface. The epilimnion is then limited to the frozen layer, and the thermocline level has moved surfaceward. (See Figs. 8.9 and 8.10.)

One of the principal influences of the thermocline is the isolation of the hypolimnion. During the summer months this barrier limits circulation to the upper waters of the epilimnion, and bottom waters remain in a condition of effective stagnation. More or less the same condition exists during the winter season when the thermocline rises to the surface and stagnation of the lower layer is brought about. In larger lakes that do not completely freeze the winter period is a condition of rather continuous turnover in the upper layers, and in the deeper parts of such lakes the thermocline must be nonexistent to all intents and purposes.

Tropical lakes show well-developed temperature zones throughout the year, and in some stagnation of the hypolimnion exists permanently.

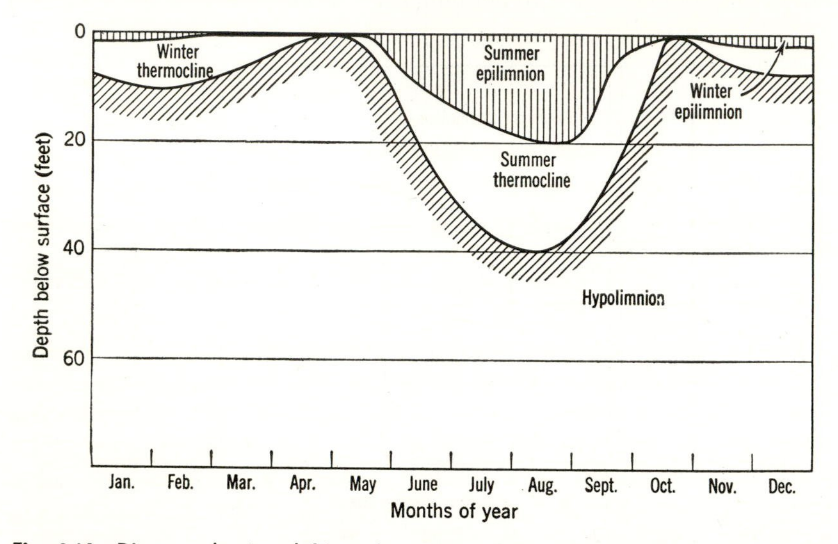

Fig. 8.10 Diagram showing shifting of positions of zones from month to month in the ideal fresh water lake of approximately 45° N. latitude. Note that the hypolimnion rises to the surface during the spring and fall overturn, that the zone of the summer thermocline is much thicker than the winter thermocline, and that the winter epilimnion is virtually nonexistent.

For this reason the oxygen supply in many deep tropical lakes becomes depleted, and decomposition of dead organisms leads to release of much hydrogen sulfide. Black highly organic muds accumulate in these oxygen-depleted waters, and only selected organisms, particularly anaerobic bacteria, can exist.

Organisms of Fresh-Water Lakes

Plankton

Aquatic organisms can be classified as *plankton, nekton,* and *benthos.* Plankton are organisms chiefly of microscopic size which more or less drift in the waters. Such floaters usually are concentrated near the surface where their life cycle is influenced principally by light and temperature. The nekton consists of swimmers, principally relatively large organisms which can control their position in the waters by swimming to the selected locality. Benthos are organisms dwelling on the bottom. Some are crawlers or burrowers, and others such as some shellfish remain fixed in position of their attachment.

In lakes planktonic organisms consist of algae, fungi, certain protozoa, some worms, and the larvae of many animals which when adults become swimmers or bottom dwellers. So far as can be ascertained generally, all natural waters in all latitudes and altitudes support plankton although the quantity ranges between wide limits. One of the striking features is the similarity of planktonic life over much of the world. The essence of this general uniformity is in the occurrence of closely allied species in approximately the same environmental associations. This is particularly true of diatoms, blue-green algae, green algae, and certain protozoans.

Higher Aquatic Plants

Plants of higher orders, principally bryophytes (mosses), pteridophytes (ferns), and spermatophytes (flowering and seed bearing), are common in most lakes. These higher plants do not occur in deep water but are restricted to the shallow waters of the vicinity of the shoreline. In the shallow water is to be found well-defined horizontal gradation in zones between plants living in air with only their roots in water-saturated sediment and those which have developed adaptations enabling them to live completely submerged. Here the line of demarcation between such simple plants as algae and the highly developed seed-bearing varieties is not clear to the layman.

In a broad way the higher aquatic plants are classified as:

1. Emergent, i.e., those rooted at the bottom and extending above water for most of their length. Bulrushes, cattails, wild rice are examples. Water depths up to 6 feet.

2. Floating, i.e., those floating on the surface of the water but which have some parts projecting out of the water. Most of these organisms are rooted to the bottom. Water depths up to 8 feet. Water lilies are examples.

3. Submergent, i.e., continuously submerged and rooted. Water depths to about 20 feet. Eel grass is an example.

Bog Lakes

General Characteristics

In regions which were exposed to recent (Wisconsin) glaciation and whose surface topography has not been modified by stream action there is to be found a special type of fresh-water lake called a *bog, moor,* or *muskeg.* They are best developed in localities of abundant precipitation where freezing temperatures prevail throughout much of the year. The term "bog" is not uniformly defined, but its connotation herein is to apply to lakes or ponds having the special characteristics outlined below:

1. These lakes are small in dimension, generally only a few hundred yards in diameter, but have been reported to be as long as 2 miles.
2. The shoreward margins are characterized by a semifloating mass of vegetation composed of characteristic plants, the remains of which produce a mat extending toward open water and developing a false bottom.
3. Plant debris constituting the false bottom is characterized by being largely peat.
4. Waters are of extremely low salinity but range from alkaline to rather strongly acid.

The false bottom. Of all the characterizing features of the bog the false bottom is the most typical. It extends from some distance landward of the present shore into open water where it terminates with a zone of water lilies or similar floating hydrophytes. At its landward margin its boundary is not clearly defined, and its existence is recognized only by the presence of a peat soil upon which grows a thick mass of shrubs such as willow, larch, juniper, or coniferous trees. Locally, tamarack and black spruce and some pines grow upon this vegetation mat which during spring months is thoroughly saturated with water but in the late summer is dry and hard. Following the mat lakeward one is aware of an increase in saturation and a general spongy quality to the saturated mass. At still greater distances lakeward the observer is aware of water rising through the spongy vegetation under the weight of his body. A rod driven through this mass reveals the presence of a layer of water beneath the vegetation and extending downward several feet to the true bottom of the lake. The observer is also aware of clearly marked zonation of vegetation from the shoreward section to the extreme outer limits. (See Fig. 8.11.) Study of this plant distribution has revealed that the false bottom is progressively extended lakeward as plant debris accumulates until the eventual outcome is to completely overwhelm the pond. The bog is then covered by the false bottom, and no indication of the underlying water is evident. With the progress of time additional plant growth is anchored to the mat of the false bottom, and the annual accumulation is added to increase the thickness of the mat.

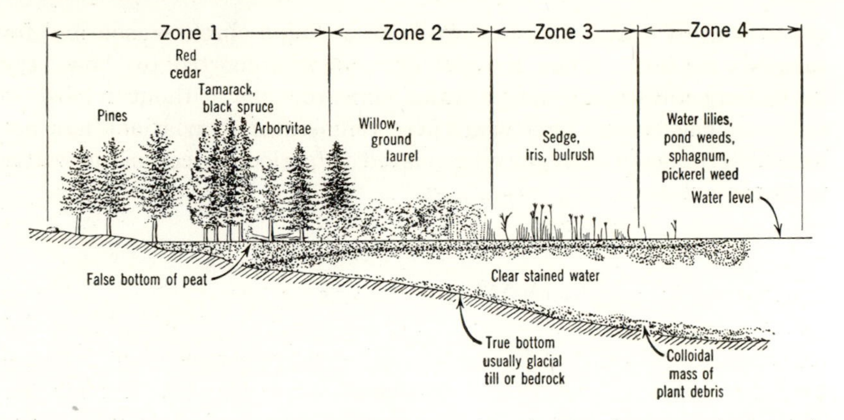

Fig. 8.11 Cross section of a typical bog or muskeg showing plant zones between the shore and open water. Zone 1 is characterized by swamp trees; zone 2 by tall shrubs; zone 3 by sedge; and zone 4 by floating hydrophytes. The false bottom of floating peaty vegetation is extended lakeward and interwoven with sphagnum moss.

The false bottom approaches the actual bottom, and there remains a thick mass of water-soaked peat to indicate the former presence of the lake.

Biologic conditions. Certain specific biologic characteristics serve to distinguish the bog from most other lakes. These are:

1. Very poorly developed benthonic fauna but abundant micro-organisms and snails inhabiting the outer limits of the vegetation.
2. Well-developed marginal flora producing the vegetation mat.
3. Decomposition of plant debris to form peat.

The water. Beneath the floating mat of vegetation is rather clear water, usually with the color of strong tea. This is an organic stain derived from the vegetation during the peat-forming process. Streams draining these lakes are typically dark waters and often are said to be high in iron content. This is seldom ever correct. Analyses of these waters show them to be very low in mineral matter except for calcium carbonate in those cases where the surrounding glacial sediments contain much finely comminuted carbonate. The acidity of the clear water is rather variable. A series of measurements of the pH of water in thirteen Michigan lakes showed a range between 4.2 and 9.0, the mean being 6.6. Should this figure prove representative it would be reasonable to assume that an acid environment is more typical than one which is basic or neutral.

Temperature stratification is observed in the deeper and larger bog lakes. But in shallow bodies beneath the false bottom very marked temperature variations are known both during a single season and for succeeding years. Some European moors are reported to show very large temperature differences from time to time without regard to season. The cause underlying these temperature variations has not been determined, but it may be related to periodic ejections of water from the extensive vegetation mat.

Sediments of Fresh-Water Lakes

Inorganic sediments of fresh-water lakes range between boulders and very fine clays in dimension. In small lakes coarse sediments are limited generally to the immediate area of supply, but in very large lakes sand, pebbles, and cobbles are moved long distances by the longshore drift, and particle size distribution is similar to that along ocean shores. Deep-water sediments are fine clays ordinarily mixed with some organic debris, principally derived from plankton and nekton inhabiting the overlying waters.

Sediments of organic origin are marls and muds containing large amounts of organic depositional products chiefly plant debris. The muds are colloidal suspensions frequently in a unit volume of over 90 per cent water. This is a sludge reaching a thickness of as much as 25 feet and forming an ill-defined boundary between water and solid sediment. *Marls* are calcareous clays, silts, or sandy deposits usually of light color but sometimes nearly black which upon drying frequently become light gray. Lakes in which carbonate is an important ion in the water produce a characteristic variant in the marls. Some are filled with fragments of snails or entire shells, whereas others contain few shells but abundant carbonate in the form of small crystals of calcite. A gradational relationship exists between the muds and marls. As the sludge undergoes compaction the water content is reduced somewhat, but the important change is the reduction in amount of organic debris. Examples are known of black, highly liquid muds (whose solid material is 50 per cent organic) which when buried to a depth of 10 feet still contain 60 per cent water. Their color, however, has been altered to nearly white, the organic content is reduced to less than 10 per cent, and the sediment is a typical marl.[5]

Alteration of the sludge in lakes whose waters are high in CO_3 ion

[5] These are values determined from bottom sediments of Lake Mendota, Wisconsin, by W. H. Twenhofel and V. E. McKelvey. (See "Sediments of Fresh Water Lakes," *Amer. Assoc. Petroleum Geologists Bull.*, Vol. 25, 1941, p. 842.)

is strikingly different from those in which the carbonate is low. One typical high carbonate lake has a sludge whose initial $CaCO_3$ content was determined as 26 per cent, whereas in the underlying compacted, light-colered marl it has risen to 73 per cent of the solid matter.[6] Other lakes are low in carbonate ion, and the accumulating sediments are correspondingly low in carbonate. Black sludge from one such lake averages 50 per cent organic matter, but 8 feet deeper into the compacted mass only 8 per cent organic matter remains.[6] Both lakes represent clearly defined examples of destruction of organic debris and consequent concentration of the inorganic fraction with increase in length of time of burial. Those lakes rich in CO_3 ion display noteworthy concentration of carbonates. Lakes low in CO_3 ion show relative increase in clays and quartz.

Elimination of organic matter is attributed to the activity of bacteria. Samples of mud from Lake Mendota, Wisconsin, indicate the aerobic varieties to be about 15 times more abundant than the anaerobes which destroy the organic compounds to extract the oxygen. One of the products of aerobic bacterial action is carbon dioxide. As this is released from the sludge undergoing alteration it dissociates in water and increases the H ion concentration which results in solution of carbonate. Lakes high in CO_3 ion are those in which the calcium carbonate content of the buried sludge is increased markedly by concentration. In this case the escaping CO_2 appears to be buffered by the alkaline solution of the water, and carbonate is not dissolved. The thickness of the zone of concentration is impressive and has been reported to be over 200 feet below large lakes.

The relationship of zonation in water, plant life, and bottom deposits is illustrated in Fig. 8.12 which represents diagrammatically the conditions as observed in a very small lake in a cold climate. Along the shores but above water level are deciduous trees whose roots penetrate a water-saturated zone. Shoreward the trees disappear and are replaced by the zone of hydrophytes, and conditions exist not unlike the bog already described. Sediments in this zone are accumulating in the waters of the summer epilimnion and are rich in shell material although of somewhat dark-gray color. In deeper water the shell life is absent, and the sediments are green-gray oozes. Oozes of the hypolimnion zone are alternately dark and light in well-developed laminas, suggesting that the colors are due to seasonal differences in the accumulating material. The summer layer probably is dark due

[6] These values are for Crystal Lake, Wisconsin, and determined by W. H. Twenhofel and V. E. McKelvey, *op. cit.*, p. 843.

to its high organic content, and the winter layer is to be expected lighter in color. Under compaction and with aging as the organic content is reduced the ooze has undergone alteration and has become green-gray in color throughout. This is the layer which when underlying waters are high in CO_3 ion displays concentration of calcium

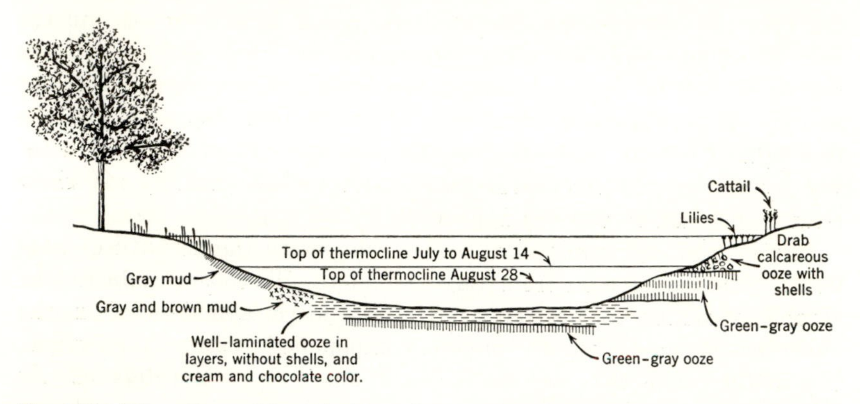

Fig. 8.12 Schematic position of thermocline and distribution of bottom sediments in McKay Lake, Ontario. Note appearance of laminated clays as typical deposits of the hypolimnion. (From Kindle, *Royal Soc. Canada, Trans.*, 1927.)

carbonate with depth as the organic debris is consumed by bacterial activity. Gray and brown muds are those darkened by accumulated organic residue which when of very high concentration produces the black sludge described above. Drab calcareous oozes are typical marls. Sometimes these are somewhat sandy or silty and are characteristically variable in particle size distribution.

Saline Lakes

Precipitation of Salts

In the course of geologic time the isolation which creates a lake establishes some very profound changes in the body of water. Where the climate remains generally humid levels of the lake rise and fall with cyclical fluctuation in rainfall or as new outlet channels appear. Other lake basins exist in belts of dramatic climatic change where precipitation varies between that typical of grass plains and that of pronounced aridity. In these areas the basin is alternately filled with fresh water or lowered by intensive evaporation. High-level stages are recorded by beaches and wave-cut cliffs, whereas low-level stages may be indicated by deposits of clay and salts precipitated from the waters. Some lakes such as the Caspian began as isolated arms of the sea; hence, they contained saline water originally. Others of which Great Salt Lake, Utah, is an example have become salty through the continuous evaporation of fresh water carried into the lake by the inlet streams. In the Caspian Sea precipitation of salts occurs in certain localized areas, and fresh water introduced by the Volga has reduced the salinity below that of normal sea water. Great Salt Lake is a remnant of a very much larger body of water, Lake Bonneville, which existed in Pleistocene times. As evaporation reduced the volume of this lake and inlet streams introduced salts washed from the surrounding rocks concentration of salts within the lake rose. Finally, the solubility products of certain salts were exceeded, whereupon they were precipitated. Ideally, the order of precipitation of salts in a lake undergoing concentration parallels that of evaporating sea water: beginning with $CaCO_3$ and followed by $CaSO_4$, either as anhydrite or gypsum ($CaSO_4{\cdot}2H_2O$); NaCl; Na_2SO_4 and $Na_2SO_4{\cdot}10H_2O$; double salts, principally sulfates of sodium, potassium, and magnesium; and finally KCl and $MgCl_2$. The most common precipitates are $CaCO_3$, $CaSO_4$, $CaSO_4{\cdot}2H_2O$, and NaCl, whereas the more soluble salts accumulate under conditions of very exceptional evaporation and concentration.

Rhythmic precipitation resulting in thin layers of salts is clearly related to seasonal and climatic changes which affect the volume of water in the lake and hence its concentration. For example, in 1897 the level of Great Salt Lake stood at 13.5 feet (Saltair gauge), and the salinity of the water was 13.8 per cent. At this high-level stage $CaCO_3$ and $CaSO_4$ were the only salts precipitated.[7] By 1935 the lake level had fallen to −2 feet, and correspondingly, the salinity had risen to 27.6 per cent. During the years of decline in water level the zero level was reached at which time NaCl and $Na_2SO_4{\cdot}10H_2O$ began to precipitate. As the level continued to fall to −2 feet there was little change in salinity as the removal of salts from solution kept the concentration approximately constant. Should the lake level return to its 1897 level as a result of localized rainfall NaCl and

[7] In Great Salt Lake very little $CaSO_4$ is precipitated either as anhydrite or gypsum because of the very low concentration of Ca ions in the water. Precipitation of $CaCO_3$ removes most of the Ca which is present in solution.

The most comprehensive study of the sediments of Great Salt Lake has been prepared by A. J. Eardley, "Sediments of Great Salt Lake, Utah," *Amer. Assoc. Petroleum Geologists Bull.*, Vol. 22, 1938, pp. 1305–1411. Most of the information reported herein is from his article.

$Na_2SO_4 \cdot 10H_2O$ would no longer precipitate, and a new layer of $CaCO_3$ and $CaSO_4$ would accumulate to complete one cycle of a banded "*evaporite*" deposit.

Despite the great areal extent of some salt lakes they tend to be extremely shallow, and wave action is restricted. Agitation is thorough, and denser water locally formed at the surface by evaporation sinks and is replaced by lighter water. Only in local bays are significant differences observed in concentration from the main body of the lake; usually restricted bays have a higher salinity than open water. Fall and spring overturn, typical of fresh-water lakes, is absent, and although some thermal stratification is observed during the summer it is restricted to the extreme upper surface. Important current activity is limited to the shoreline where beach drift and longshore currents are influential in mixing sediments. The major aspect, however, is to find little lateral mixing of sediment types, and extensive areas are characterized by surprising uniformity.

Sediments

Sediments of Great Salt Lake are good examples of restricted lateral mixing. (See Figs. 8.13 and 8.14.) Alluvial fans with slopes of 12 degrees at their apex extend to the water's edge where the slope approaches 2 degrees. Beyond this limit the coarse material scarcely extends at present, and the sediments are replaced lakeward by calcareous and argillaceous marls, clays, and oozes. These interfinger into and cover fan material, yet their slope is no more than 5 feet per mile. Locally, reefs constructed by calcareous algae separate coarse gravels and sands from limy clays. Such rapid changes in sediment type are the counterparts of those where alluvial fans merge into playa-lake deposits; similar processes of gradual burial of alluvial fans by lake clays and marls characterize large salt lakes.

The process of burial by impermeable sediments serves to enclose the fans with water-tight barriers except in their upper reaches where they blanket rock slopes. These are the infiltration sites where water enters the fan which gradually fills to capacity. Saline waters from the lake are "perched" upon the impermeable clays; hence, most of the ground water is uncontaminated by brines. Instead, localized supplies of fresh water are available, sometimes in large quantities, wherever such fans are tapped by wells.

Clays are the most widespread sediments of saline lakes. Their extensive distribution is the result of having most of the coarse particles deposited on the alluvial fans or on lowlands which border the lake basins. Great Salt Lake is no exception; sandy alluvium covers the low areas between the water and the mountain uplifts (see Fig. 8.13). Within the lake particles of clay and silt dimension are

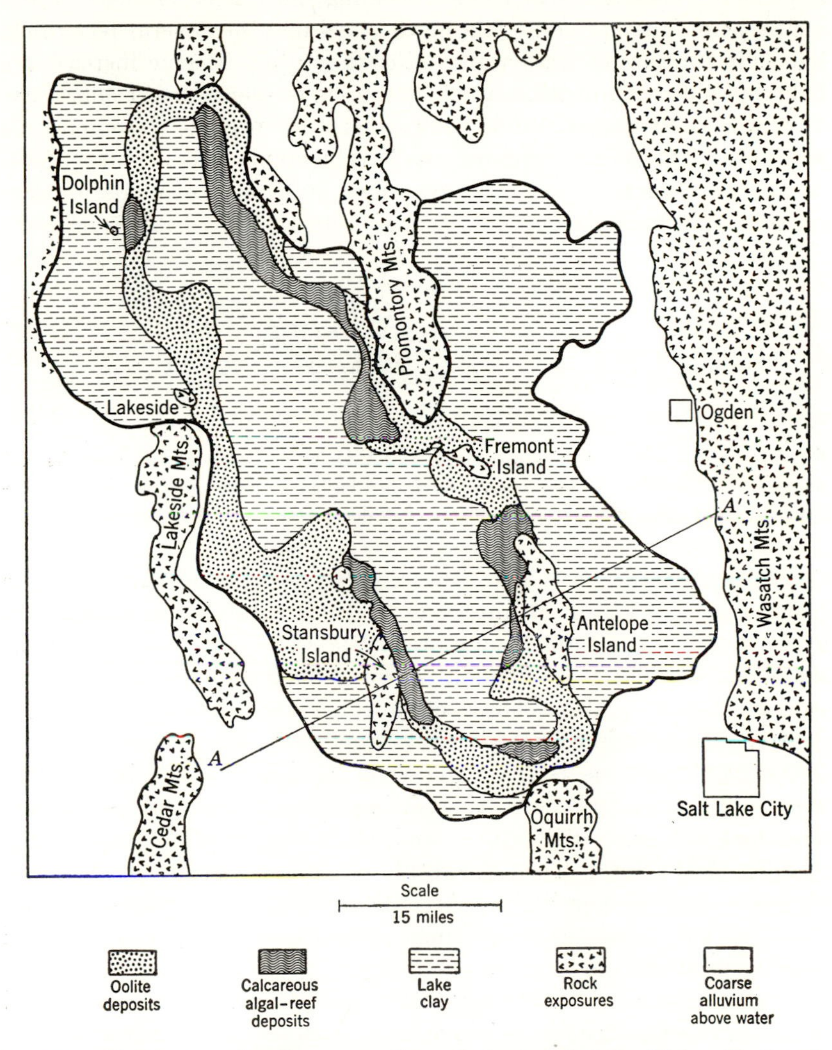

Fig. 8.13 Distribution of sediments on floor of Great Salt Lake, Utah. Algal reefs and oolite sands are related to occurrences of rock prominences and near shore areas. Lake clay is the typical unconsolidated sediment; this becomes black a short distance beneath the surface. Oolites are formed in some of the more aerated localities. (Simplified from Eardley, *Amer. Assoc. Petroleum Geologists Bull.*, 1938.)

carried long distances before final deposition. Local areas may show gradation of sand to silt to clay outward from the shore. However, the usual condition is to find clay widely distributed, occurring near the shore as well as near the lake center. At the water interface the clay is generally light gray in color, becoming dark gray or black a few inches below the surface. The black clays may be several feet thick, but as in the case of fresh-water lakes they often become increasingly light in color with depth. Most of the clays release fetid odors, one of which is hydrogen sulfide, and others are organic decomposition products. The former is known to be a common product of sulfate-reducing bacteria. These are anaerobes living in the muds and are capable of extracting oxygen from gypsum or anhydrite in order to

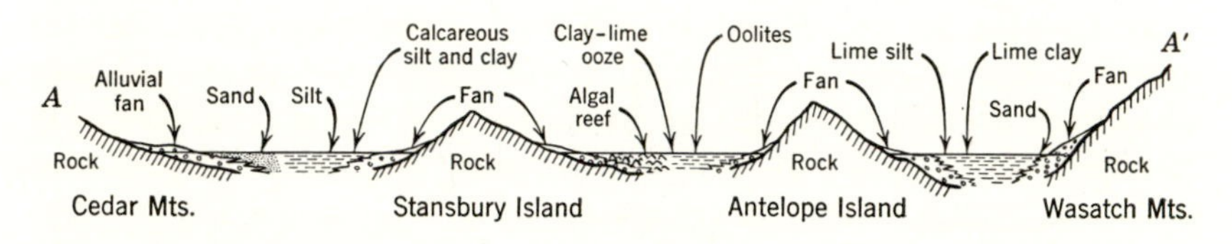

Fig. 8.14 Cross section *A–A′* across Great Salt Lake, Utah. Alluvial fans from the rock slopes merge rapidly into lake clays. Reefs are buried by lake clays as these fill the basin. (Based upon Eardley, *Amer. Assoc. Petroleum Geologists Bull.*, 1938.)

carry on their life processes. Principal by-product of this reduction reaction is H_2S which saturates the mud. If soluble iron is available in the water pyrite forms as minute crystals or concretions as the soft sediments become consolidated. Aerobic bacteria inhabit the upper part of the muds also. These organisms direct their efforts toward decomposition of the incorporated vegetable and animal material. Much of this debris is the remains of plankton which inhabit the waters in large numbers. Brine shrimp, for example, are extremely abundant in Great Salt Lake, and their fæcal pellets are readily visible. As bacteria digest the residuum carbon is concentrated, and the sediments become black. At still greater depths continued bacterial action consumes increasing amounts of the partly digested organic products, and the colors become increasingly light.

Montmorillonite is a characteristic part of the clay fraction and because of its strong base-exchange tendencies becomes either a calcium or sodium clay as the law of mass action dictates. Some lakes are extremely high in Ca ion, and in these cases calcium montmorillonite clays form. Currently Great Salt Lake is high in sodium, and the montmorillonites have exchanged much of their calcium for sodium. The effect is a marked decrease in the permeability of the clay to downward-moving water. Consequently there is no recharge to the ground water from this source.

After initial compaction additional loading by sediments does not appear to destroy the water-clay suspension. The low permeability of sodium montmorillonites does not permit water to be "filter pressed" and move under an hydraulic gradient. The clay deposit, therefore, remains as a semimobile mass unable to eject its interstitial water. This is a sediment which has very little ability to withstand differential stress. For example, rock fill sinks through the clay mixture forcing it aside in much the same manner as viscous fluids yield under differential stress. Piling sinks with very minor resistance, and support of construction on such foundations is most difficult and expensive.

Reefs. Calcareous sediments are typical shore marginal deposits of saline lakes despite the presence of finely divided calcite in the clay fraction. Most conspicuous are the reef deposits precipitated by algae which inhabit shallow water, particularly near rocky prominences or islands. Algae fasten to some solid object, usually a rock surface, upon which they expand their colonies as gelatinous masses and filaments. During photosynthesis aragonite ($CaCO_3$) is precipitated as a film within the cell structure and eventually becomes deposited as a single lamina upon a previously deposited layer. The resulting rock mass has a rounded or curved surface and in cross section displays concentric growth structure. Some algal deposits are ball-shaped or nearly spherical masses between one and several inches in diameter. (See Fig. 12.7.) As the colony expands, individual balls, discs, or curvate forms merge together into increasingly large masses constituting a reef of a few to hundreds of feet in length. Some of the reef rock stands as impressive rounded masses 20 to 50 feet high draped upon the underlying rock foundation. Algal colonies prefer certain water conditions as well as solid bottoms upon which to build their reef; hence, zones of reef may extend for many miles. (See Fig. 8.13.) Long stretches of reef growth in more or less discontinuous bands constitute solid foundation positions in an otherwise soft clay bottom.

Deposits from individual colonies become cemented together into an extremely porous and permeable rock resembling travertine or tufa. Also aragonite alters to its more stable form, calcite, and this process of recrystallization results in increased interlocking of crystals. Reefs appear to be draped over other rock masses as rounded porous deposits allied in appearance to the tufa accumulated near hot springs. For this reason the algal reefs have been mistakenly called tufa and described as inorganic precipitates from waters saturated with calcium carbonate.

As the water level falls in evaporating lakes algal reefs form at progressively lower levels; hence, exposed reefs can be seen standing well above the present water surface.

Oolites. Another limy sediment frequently associated with zones of reef growth is a sand consisting of small spherical pellets called *oolites.* (See Fig. 12.4*d*.) The individual oolith is a concretion of sand-size dimension and develops in such large numbers as to produce important deposits. Each oolith consists of a series of concentric shells of microscopic size usually deposited around some nucleus such as a quartz grain or fossil fragment. Once the oolith attains the size of a sand grain it behaves as any other clastic particle and is swept along under wave and current action to form an extensive subaqueous bank. Locally, under the driving force of the wind oolites are swept inland to make beaches and dunes. Under burial and compaction oolitic sand becomes cemented into a bed of limestone which preserves such original sedimentary structures as ripple mark and cross-bedding. (See Figs. 7.27 and 12.1.) In Great Salt Lake (Fig. 8.13) oolith deposition is more prevalent in the western part of the lake than in the eastern half. The association, however, is principally with reefs and rocky promontories or islands.

Oolites are not restricted to saline lakes but are found as well in shallow warm ocean waters of somewhat higher than normal salinity. Their occurrence in banks and shallows of present-day coral reef areas is typical. At these sites the environment of development is the shallow turbulent water of the reef bank. From this position of origin the individual pellets are transported into the lagoons or deposited as extensive aprons flanking the actively growing reef. In the saline lakes oolith growth appears to be associated with minor but effective wave activity against the algal reefs and rocky shores. From these sites of origin the oolites are moved along the shallows to the beaches. Lakeward, gradation into clays is extremely rapid wherever wave action is unable to move oolites.

Precipitation of calcium carbonate as concentric shells of aragonite and calcite around the central nucleus does not appear to be directly the product of organic activity. Rather, the mechanism is considered to be brought about by removal of carbon dioxide from water of abnormal salinity during a condition of wave agitation. Sudden reduction of the H ion concentration in a localized position is believed to force precipitation of calcium carbonate on some available nucleation center. As each shell is added in sequence to the oolith, the particle becomes increasingly spherical, and the preferred shape is attained.

Selected Supplementary Readings

Clarke, F. W., "The Data of Geochemistry," *U. S. Geological Survey Bull.* 770, 1924, Chapter V. A summary of the chemical character of waters of saline lakes.

Fenneman, N. M., "Lakes of Southeastern Wisconsin," *Wisconsin Geological and Natural History Survey Bulletin VIII,* 1902, Chapter II. Develops the theory of wave form and shoreline development of inland lakes.

Johnson, D. W., *Shore Processes and Shoreline Development,* John Wiley and Sons, 1919, o.p. One of the first attempts to classify and describe the processes of development of shorelines.

Kuenen, P. H., *Marine Geology,* John Wiley and Sons, 1950, Chapter I. Concerns physical oceanography. Chapter V describes the environments of deposition of sediments of the oceans.

Mason, M. A., "Geology in Shore Control Problems," in Trask, P. D., *Applied Sedimentation,* John Wiley and Sons, 1950. Concerns the information to be extracted from geologic examination of shore conditions.

Shepard, F. P., *Submarine Geology,* Harper and Bros., 1946, Chapter IV. A chapter on the classification of shorelines.

Welch, P. S., *Limnology,* McGraw-Hill Book Co., 1935, Chapter IV. An excellent treatment of the physical conditions in fresh-water lakes.

9 Ground water

INFILTRATION OF WATER

The Ground-Water Reservoir

Within the soil is a transient zone known as the *belt of soil water,* generally increasing in thickness after a rain but gradually decreasing and migrating in the interval between rains (Fig. 9.1). Following rainfall soil water tends to move downward, eventually stabilizes, and moves upward during periods of surface evaporation. Beneath the

Onandago Cave, Missouri. (Photograph by Massie, courtesy Missouri Geological Survey.)

zone of soil water is another zone, normally partially dry but which may become locally saturated following a period of rainfall. Later, this portion of the *vadose zone*[1] loses water to the belt of soil water and also to the underlying permanently saturated zone. The lower surface of the vadose zone is not sharply defined but is transitional

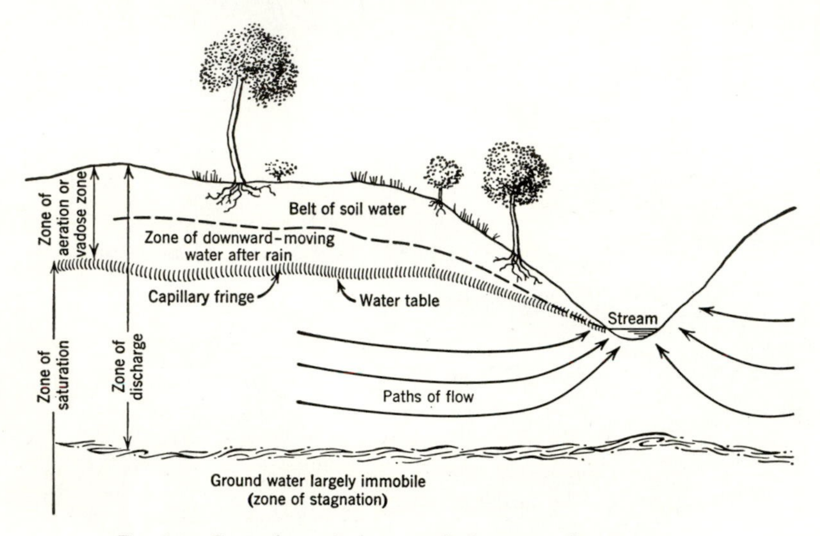

Fig. 9.1 General terminology applied to ground-water zones.

with the water table through the *capillary fringe.* Here water rises through capillary openings in the rock or unconsolidated sediment and marks the bounding limits between the *zone of saturation* below and the *zone of aeration* above. (See Fig. 9.1.)

In the saturated zone all pore space in the rocks is considered filled with water. Above this position openings are filled with air and some water. Within the zone of aeration water movement tends to be downward generally, whereas in the zone of saturation lateral movement may be the dominant direction of motion. A strong lateral component of movement is observed particularly in the upper part of the zone where water is moving toward river channels. Still deeper beneath the surface waters tend to remain fixed in position. This is identified as the *zone of stagnation* and is particularly recognized in some of the thick sedimentary basins. Here, several thousands of feet below the surface, water fills all openings in rocks but is immobile.

[1] The vadose zone is defined as that belt extending from the surface downward to the water table.

Some of this water may be highly saline, a feature which in deep basins is the rule rather than the exception.

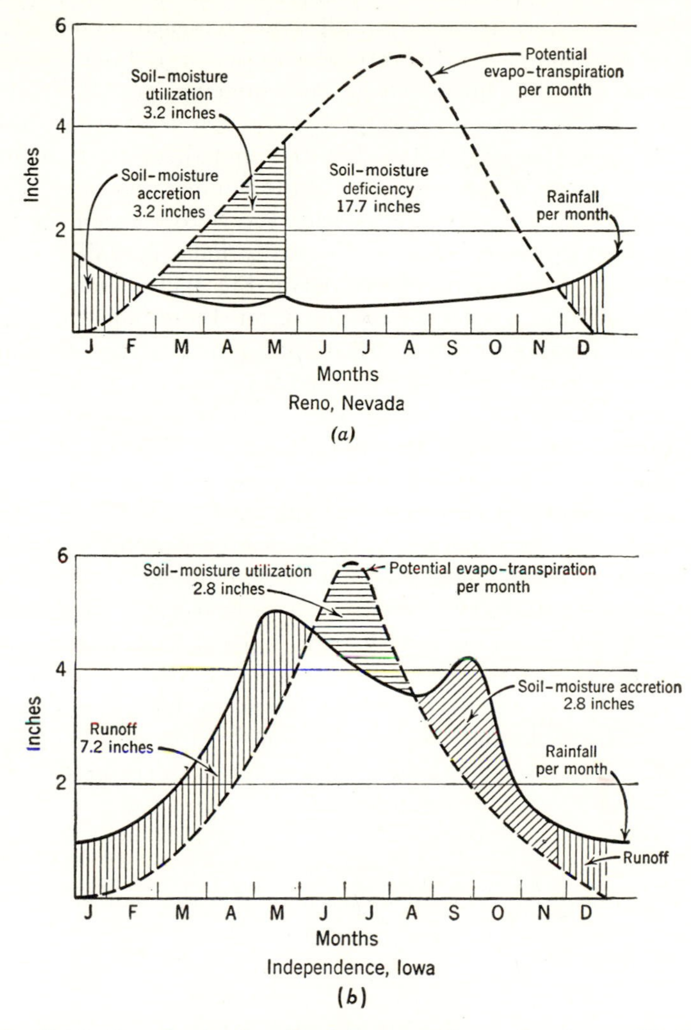

Fig. 9.2 Distribution of rainfall, soil-moisture accretion, soil-moisture utilization, soil-moisture deficiency, and runoff in (*a*) a desert region and (*b*) a humid region. Contrast the lack of moisture deficiency at Independence, Iowa, with the large moisture deficiency at Reno, Nevada. (From Thornthwaite, *Water and Man,* 1950.)

Recharge of water to the soil occurs in both wet and dry climates, but the amounts supplied are not the same, and the periods of upward movement of water to the soil surface differ in length of duration.

Examples illustrating the conditions of each environment are shown in Fig. 9.2 which indicates rainfall distributions, soil-moisture demands, and soil-moisture recharges for Reno, Nevada, and Independence, Iowa. Note the average low rainfall value for each month at Reno (Fig. 9.2*a*), and compare with water loss as shown by the evaporation-transpiration curve. During December, January, and February the demand is less than the rain supply, and water is added to the soil cover. By June this supply is exhausted, and there exists a marked deficiency of water for the remainder of the summer and fall. Without recharge to the soil from some exterior source (irrigation or underground supply) desert conditions eventually must prevail.

A contrasting situation is illustrated in Fig. 9.2*b* which shows the rainfall distribution and water demand for Independence. In this locality ground-water storage is filled until June, and runoff is important. Summer months require more moisture than is furnished from rainfall, and utilization of stored soil moisture is indicated. With temperature decline during the fall season and reduction in plant transpiration and evaporation rainfall once more exceeds moisture demand. Soil moisture accretion prevails until saturation is attained, and runoff begins once more during the winter season.

Contrast in soil-water conditions in the two climatic zones is striking and doubtless the prime control on ground-water accumulation. Nevertheless, other factors exert their influence on water entering underground. A list of the most important of these includes land slope, distribution of soil cover, variety of soil, grain size, soil temperature, type of bedrock, and the magnitude of fissures or other openings in this rock. To some extent these factors are interdependent, whereas in other respects they operate independently. The degree of this relationship concerns the manner by which water is held in the soil or rocks.

Two fundamentally distinct occurrences of water can be described. The most important of these is found in interconnected openings. These range from tiny pores between individual mineral grains to large systems of tunnels and caves. Water occurs also in disconnected openings such as that dissolved in the melts of magmas or held within the lattice of minerals such as clays. This water does not migrate from one pore opening to an adjacent one but remains fixed in position and is removed only by rigorous activity such as heating above the boiling point of water. Water so "fixed" is not considered as part of the ground water and for this reason is omitted from further discussion in this chapter.

Movement of Water through Pores

Capillary Effects

Water seeping into dry soils after a rain enters an undersaturated zone where it behaves in one of the following ways:

a. Some is heated and becomes vapor, whereupon it is controlled by vapor pressure equilibrium and moves from higher to lower pressure.

b. Some wets the surfaces of particles and is called *pellicular or film water.* This water moves slowly under the influence of molecular forces of cohesion and in a direction from thicker to thinner films. Movement of film water is important in dry or partly saturated soils in connection with capillary effects.

c. Some moves as a particle or thread of water through pore space under a gravitational drive, a pressure head, or the influence of capillary forces.

Water moving through interconnected openings is controlled by certain laws of flow which are not necessarily the same. For example, water rushing through large fissures is unaffected by capillary forces, whereas water in the interstices between mineral grains is influenced to a considerable extent by the laws of surface tension. In sand and gravel where interstitial openings are large excess water moves freely, influenced by gravitational forces and differences in water heads. In clays surface-tensional forces tend to offset such effects, and pore water can move several feet upward in opposition to the gravitational force.

Wetting surfaces of coarsely granular sediments such as gravels and sands does not require large quantities of water. However, thorough wetting of all particles in clays requires large amounts because of the enormous total surface area present. Most of the pellicular water is unavailable except to plants and soil organisms and insofar as is known tends to move relatively short distances.

There are two kinds of water surfaces in the soil: one is in contact with the mineral grains, and the other is in contact with air. Forces at these two surfaces behave very differently. Water in contact with a soil particle spreads over the particle as a film under the molecular attractive forces of the solid. A drop of water in contact with air tends to pull into a sphere as the molecular forces within the water exert a greater attraction than do those in air upon the surface-water film. In a pore between mineral grains the air-water boundary tends to become convex toward the water as those molecules exert a pull on the relatively small number in the surface film, and the walls of the grains attract a water film upward. Thus, the pore and its walls are to be compared to a capillary tube, and the water rises along the surfaces of the grains approximately as the standard equation for capillary rise.

Upon entering dry soil from a saturated portion water films spread over the grains and join in the capillary openings. Here, the water is placed under tension and rises into higher capillary spaces. By this process capillary rise continues until the tension developed within larger capillary pores is no longer able to lift the water against the gravitational force, or until the thinness of the film, the weight of water to be lifted, and the resistance to flow become so great that the capillary rise effectively stops.

Capillary potential.[2] Fifty years ago there was introduced a concept that the movement of water through soil could be compared to the flow of heat through a metal bar or electricity through wire. The driving force is considered as the difference in attraction for water between two parts of the soil of different moisture content, and the term *capillary potential* was applied to express a measure of this attraction at a selected point. The value of the capillary potential is known to vary inversely as the particle size, but if one compares soils of the same particle size distribution the value will also vary inversely as the moisture content. Thus the energy required to pull a given amount of water from a soil that is nearly saturated is less than that required to extract water from a soil of low moisture content. Also, a certain amount of work is required to shift a unit mass of water from one position in the soil to another which is less saturated inasmuch as the measure of work is the product of the force per unit area and the distance the water is moved. The capillary potential is a tensional force and can be considered as a negative pressure required to move a unit mass of water from a saturated, or free-water, surface to a higher level through capillary openings in the soil.

In a column of soil which is saturated the capillary potential is zero as virtually no work is done to remove a unit weight of water. Higher in the soil where the moisture content is lower the capillary potential is larger, and the work done in shifting water to a higher level is greater. Still higher in the column where the moisture content is much lower a much higher capillary potential is required to extract a unit weight

[2] An excellent book on general soil properties presenting the concepts relative to soil water is L. D. Baver, *Soil Physics,* 3rd ed., John Wiley and Sons, 1956, Chapter VI.

of water. Thus between points of different heights above a free-water surface the capillary potential is greatest at the highest, or least saturated, point.

The most widely known method of measuring the capillary potential is based upon the so-called tensional force of the unsaturated soil for water. The essential features are an apparatus called a *tensiometer* consisting of a porous clay cup partly filled with water. A manometer is sealed to the cup. The cup is placed in the soil and allowed to reach equilibrium. As the soil dries water leaves the cell under tension, and the mercury rises in the manometer indicating a drop in positive pressure (i.e., there has been an increase in capillary potential in the soil). As the soil is moistened from some outside source the tension of water in the soil is less than that in the cup. Water moves through the pore space to the cup, and the manometer level falls, indicating an increase in positive pressure. There has been a decline in capillary potential in the soil.

Pressure Gradients

Two separate conditions exist between the saturated and unsaturated portions of the soil, and these can be expressed in terms of the pressure gradients which exist. In the undersaturated portion the pressure gradient may be considered as controlled by the capillary potential. This value increases with increased height above the free-water surface but represents increasingly greater values of tension. The value of positive pressure must decrease in the same direction, and a gradient must exist of higher positive pressure at the static water surface to lower positive pressure in the unsaturated soil higher in elevation. All points along the static water surface have zero capillary potential, and no pressure gradient can exist between such points. Below the static water level the soil is saturated, the capillary potential is zero, and movement of water is controlled by differences in hydraulic head. *Hydraulic head* is defined as the difference in elevation between the water level in a manometer connected to a point in the saturated soil and some reference level. Water moves from a position of higher hydraulic head to one of lower head, i.e., along an *hydraulic gradient* which is defined as the difference in hydraulic heads between two points divided by the distance of flow between them.

Under certain conditions water moves from the free-water surface in opposite directions, moving upward into the unsaturated zone along the capillary potential head and in the saturated portion downward along the hydraulic gradient. Other conditions may exist where the hydraulic gradient declines in an upward direction and water moves from an underground supply toward the surface, or rain may enter the surface and move downward to the water table. If a free-water surface is established separating saturated from unsaturated soil continued upward movement will occur along the capillary potential gradient.

Movement of water in conditions of both saturated and unsaturated soils can be illustrated by the three hypothetical examples given below.[3] In the situations illustrated porous cup tensiometers are placed in vertical position in the unsaturated portion of the soil, but they are considered as connected to water manometers for comparison with those placed in vertical position in the saturated section.

Figure 9.3*a* illustrates a static-water condition. The example selected has as its analogue a paved highway covering a porous subgrade fill which rests upon bedrock or a clay bed of low permeability. Evaporation at the surface is prevented by the pavement, and water infiltration has been by lateral movement until equilibrium conditions were attained. The six manometers connected to the six positions in the soil, three above and three below the level of saturation (free-water surface), stand at the same level relative to this surface. Static equilibrium exists in the saturated zone, and water in the manometer rises to a level equal to the distance between the point of measurement and the free-water surface. No hydraulic gradient exists, and the water is stationary. Likewise, in the unsaturated soil water levels in the manometer stand at heights controlled by the capillary potentials. Levels of water in the manometers stand at positions below each respective porous cup by a length equal to its distance above the free-water surface, and only partial saturation of the soil is indicated. In all of the six manometers the hydraulic head is the same. The hydraulic gradient must be zero, no movement of water can take place, and condition is one of static equilibrium.

The example illustrated by Fig. 9.3*b* is a condition of water infiltration from rainfall upon a soil which has been developed from an underlying sandy sediment. Water percolates downward from the surface into the permeable subsoil and gradually moves outside the area. In the upper part of the soil the water moves through a zone which is partially saturated with water, but about half way through the soil column standing water is reached. The surface of the saturated zone represents the level of the water table. Below the water table downward movement of water continues through the remainder of the soil, and water is in motion throughout the entire soil column in contrast to the static condition of the preceding example.

[3] The examples are taken from L. A. Richards, "Retention and Transmission of Water in the Soil," *Water, Yearbook of Agriculture,* 1955, U. S. Department of Agriculture, pp. 144–151.

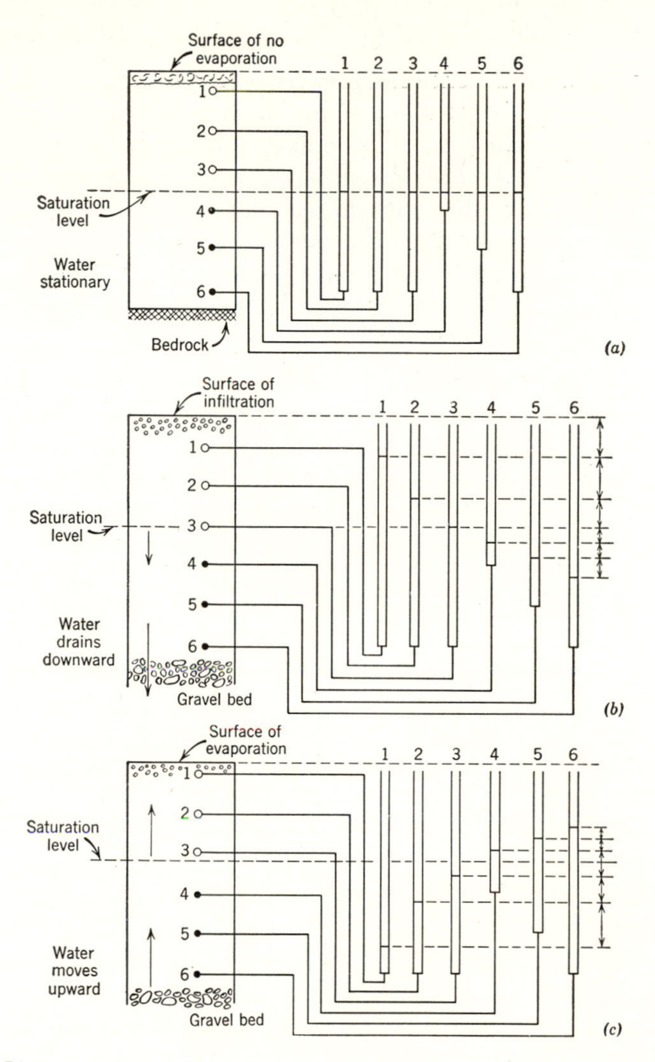

Fig. 9.3 Diagrammatic representation of three different conditions of hydraulic gradient. (*a*) Static condition. Water moves neither downward nor upward but remains in a fixed position. The surface of no evaporation at the top stabilizes the negative pressure gradient upward; downward percolation is inhibited by the impermeable bedrock. The water level is the same distance below the reference line (top of soil) in each manometer. (*b*) Condition of downward water infiltration. Note that water levels in the manometers indicate the downward gradient. The relative drop in level between manometers 1 and 2 and 2 and 3 is greater than between 3 and 4, 4 and 5, and 5 and 6. This defines the level of saturation as between the positions of manometers 3 and 4. Note that in the saturated zone the position of water in the manometer is higher than the point of recording from the reference surface. (*c*) An upward gradient from the gravel bed to the surface. Spacing of relative heights of manometer readings indicates that the level of water saturation exists between the position of manometers 3 and 4. (Modified from Richards, *U. S. Dept. Agriculture Yearbook*, 1955.)

The difference in condition between example *a* and example *b* is demonstrated by the positions of the water levels in the manometers. The level of standing water is indicated by manometer 3 in which the water level stands at the position of the porous cup. At this position saturation prevails, and the capillary potential is zero. Note also that the water level in the manometer recording pressure of the uppermost porous cup (1), is slightly lower than the distance between the tensiometer and the free-water surface. This indicates that a capillary potential of small value exists at the position of the tensiometer. A similar situation is to be observed at the next lower tensiometer, (2), where the capillary potentials are about the same, indicating a nearly saturated condition of the soils. During rainfall (condition of Fig. 9.3*b*) water draining downward from the saturated portion at the surface moves under an hydraulic gradient as indicated by the difference between the hydraulic heads between the upper and lower tensiometers. Along the gradient water moves into the partly saturated section which is gathering water under the capillary potential as well, but as this value approaches zero the hydraulic head becomes important. Each manometer recording pressure at successively lower positions in the soil shows a decline in hydraulic head, and the gradient established near the surface is continued below the level of soil saturation.

There is a difference, however, in the amount of head as indicated by the heights between manometer readings spaced equidistant in the soil. Note that the drop in water level between the soil surface (reference line) and 1, 1 and 2, and 2 and 3 is more than the drop between 3 and 4, 4 and 5, and 5 and 6. The hydraulic gradient between 3 and 6 is less than between the surface and 3. This condition is interpreted as indicative of water saturation existent between the positions of manometers 3 and 4. Above 3 the soil is undersaturated, a capillary potential greater than zero exists, and the drop in head is greater per unit distance than at depths below tensiometer 3 where water is occupying all the pore space.

The condition illustrated by Fig. 9.3*c* is that of a soil being fed by leakage from some *aquifer*,[4] and water rises through the soil to be removed by evaporation. The hydraulic head as illustrated by the manometers is greatest at the base of the soil and is progressively reduced upward. Pressure drop between 6 and 5 and 5 and 4 is approximately the same, whereas between 4 and 3 the drop is greater. This in itself is not necessarily an indication that the zone of water saturation lies beneath the position being recorded by manometer 3.

[4] An aquifer is a layer of high permeability carrying water.

Rather, the location of the water table below porous cup 3 is indicated by the presence of a value greater than zero for the capillary potential, and water stands in the manometer below the position of the tensiometer. The drop between 2 and 1 is large and indicates a high degree of undersaturation and a large capillary potential. Water moves along this hydraulic gradient in an attempt to reach the surface for evaporation.

Hydraulic gradient and suction head. In the cases illustrated by *b* and *c*, water is moving with a velocity controlled by a driving force which can be expressed by the equation

$$Q = Ki \tag{1}$$

where Q = volume of water crossing a unit area in unit time.
K = constant varying for each soil type.
i = hydraulic gradient.

The factor of proportionality, K, is called the *hydraulic conductivity* or in unsaturated soils the *capillary conductivity*.[5] K has the dimensions of velocity, is the ratio Q/i, and can be considered as the calculated value of a flow velocity for a net driving force equal to gravity.

Figure 9.4 illustrates infiltration of water following irrigation of a dry-climate soil and reversal of hydraulic gradient after cessation of irrigation.[6] Graph *b* is a plot of hydraulic-head values measured with reference to the soil surface and plotted against depth in the soil. Diagram *a* represents the bore hole through the soil showing vertical spacing of porous cups at 10 to 50 cm depth. Spacing of these positions is along the x-axis of diagram *b*, whereas the corresponding hydraulic-head values are plotted along the y-axis. The dashed line is the slope of an hydraulic gradient of unity, i.e., the level of water in the manometer is equal to the depth of a porous measuring cup below the surface.

At the close of irrigation, water was infiltrating at a rate of 2.6 cm of water per hour per unit area. In the zone 10 to 30 cm of depth the difference between manometer readings was 46 i.e. $[+3 - (-43)]$. This value divided by the difference in depth (20 cm) indicates an hydraulic gradient of 2.3. The ratio $Q/i = 2.6/2.3 = 1.1$ cm per hour represented the hydraulic conductivity in this interval. Between 30 and 50 cm of depth the hydraulic gradient equaled the slope of unit hydraulic gradient (i.e., 1 and had a driving force of 1 g). Over this

[5] The capillary potential gradient can be substituted for i in eq. 1 to express movement of water in capillary pores.

[6] The example described is from L. A. Richards, *op. cit.*, p. 148.

depth range the hydraulic conductivity was 2.6 cm per hour per unit area (2.6/1 = 2.6). The day following irrigation the hydraulic conductivity remained only slightly less than 2.6. Three days after irrigation the average hydraulic gradient from 10 to 50 cm of depth approached zero as did the corresponding hydraulic conductivity. The

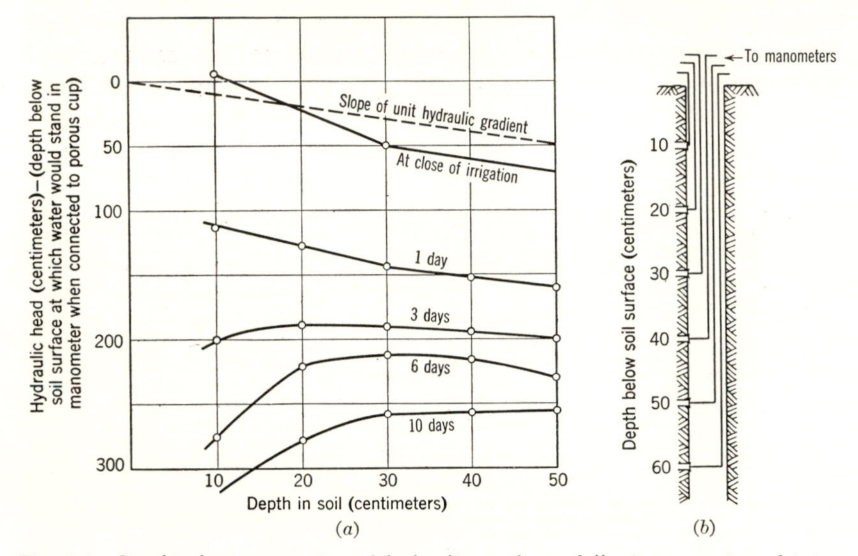

Fig. 9.4 Graphical representation of hydraulic gradients following cessation of irrigation of a sandy loam (Pachappa). (*a*) Hydraulic head plotted against depth below soil surface showing change in direction of gradient developed as downward infiltration is replaced by an unsaturated-soil condition. Note that 3 days after irrigation the head is nearly the same at all depths. This is a static condition such as *a* of Fig. 9.3 Later, the capillary potential is much greater at the soil near the surface and a gradient is established toward the soil surface. (*b*) Bore hole in sandy loam showing location and depth below surface, in centimeters, of porous cups connected to separate manometers whose readings are indicated in graph *a*. (From Richards, *U. S. Dept. Agriculture Yearbook*, 1955.)

situation indicated a static condition and cessation of any downward water movement. Actually, the precise static balance must have been attained during the interval of time between the first and third days when the hydraulic head was the same at all points of depth between 10 and 50 cm. This is the condition analogous to Fig. 9.3*a*. Six days following irrigation an hydraulic gradient was established flowing toward the surface (negative pressure head in partly saturated soil) and also downward from the general position of 30 cm of depth. By the tenth day most of the water was drained from the soil, and the negative pressure gradient was upward, particularly from 30 cm depth to the surface (condition of Fig. 9.3*c*). Meanwhile, the static-zone position continued to migrate below the depth of measurement.

Negative pressure head is frequently expressed as *suction head,* a term which is generally equivalent to capillary potential. The value is calculated by subtracting the depth of the measuring cup below the soil surface from the relative hydraulic head at that point.[7] Table 9.1 indicates the calculated suction heads at various depths on successive days following termination of irrigation (see Fig. 9.4). Values which are approximately the same for different depths indicate a static condition. A gradient toward successively greater negative values indicates direction of water movement in unsaturated soils.

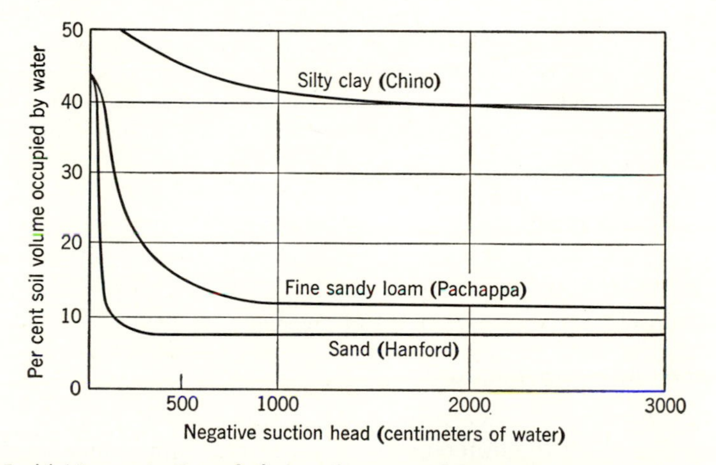

Fig. 9.5 Moisture retention of three soil types with increase in suction head. Sand loses almost all of its water with a very low suction head, whereas clay has retained approximately 40 per cent of its pore water when the suction head is 3,000 cm of water. (From Richards, *U. S. Dept. Agriculture Yearbook,* 1955.)

Suction head and suction gradient are important in controlling moisture retention in the soil. This is illustrated by Fig. 9.5 which is a graphical representation of change in the percentage of water occupying pore space in three soil types with increase in suction head. Sand loses nearly all of its pore water with a very low suction head. For example, with suction heads of approximately 100, as developed one day after irrigation of Pachappa sandy loam (see Table 9.1), sand (Hanford type) would yield all but 10 per cent of the water theoretically held in the pore space. Additional suction head does

[7] Suction-head valves can be determined from a graph such as Fig. 9.4*a* by subtracting from the observed value of hydraulic head, at a selected position, the hydraulic head represented by unit hydraulic gradient.

Table 9.1. Suction Heads in Centimeters Water Following Irrigation of Sandy Loam (Pachappa)*

Depth below Surface (cm)	Days Elapsed Following Termination of Irrigation				
	0	1	3	6	10
10	10	−100	−190	−270	−300
20	0	−110	−160	−200	−260
30	−20	−120	−155	−190	−240
40	−20	−120	−155	−180	−220
50	−20	−120	−150	−190	−200

* These values are to be compared with their respective positions in Fig. 9.4*a*. A negative suction head indicates that a force exists which will cause water to move in a direction of increased negativity. Where values are uniform at successive depths on the same day a static condition prevails. Note, until the end of the first day following irrigation, the suction head is downward, whereas from the third day onward there is an increasing suction-head gradient toward the soil surface.

not alter the amount of moisture retained. The fine sandy loam (Pachappa) loses all but 20 per cent of its moisture with a suction head of about 250 cm, whereas silty clay retains 40 per cent of the potential pore water at 3,000 cm of suction pressure.

Factors Affecting Infiltration

Recharge of soil water throughout a selected watershed area can be determined by the difference between rainfall and runoff loss after a storm has thoroughly soaked the soil cover and some water is accumulating in puddles. The amount of infiltration, however, is dependent upon many variables, for example:

1. Nature of the vegetation cover, i.e., whether in grass, forest, or crop land.
2. Internal characteristics of the soil such as size of pores, thickness of soil, degree of swelling of clays, amount of organic matter present.
3. Moisture content and degree of saturation preceding rain.
4. Duration of rainfall.
5. Season of year and particularly temperature of soil and water.

Open cultivated fields have infiltration rates which are much lower than grasslands. This is due in part to greater runoff than in grass-covered areas but also because the latter contains more organic material which tends to keep the pore structure open. Initially, the intake of a soil which is grass covered and that which is cultivated and bearing corn may be the same, but after the end of the first hour of rainfall the meadow may continue to absorb water at twice the rate of open land. (See Fig. 9.6.)

Infiltration must be determined principally on the actual soil under consideration. For example clay soils with a high degree of swelling can be expected to have a much lower rate of intake when wet than

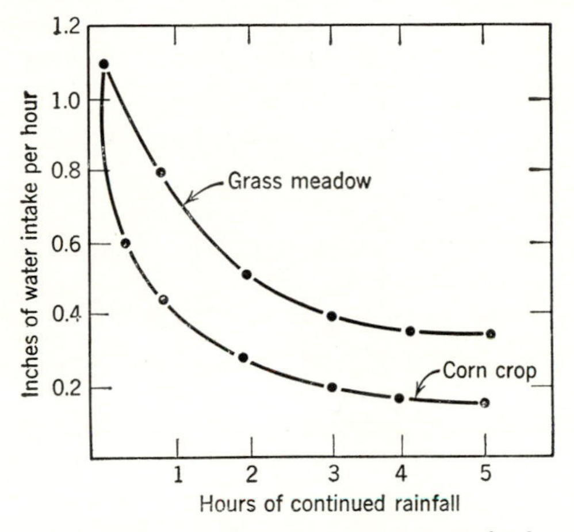

Fig. 9.6 Comparison of rates of infiltration between grassland and corn-crop land. Initially the intake of the soil is the same, but the permeability of the meadow remains higher than that of the open-crop land. (From Musgrave, *U. S. Dept. Agriculture Yearbook*, 1955.)

when dry. Clay soils which tend to crack on drying may have a very high rate of infiltration until swelling closes the open cracks. Thick sands, low in organic and clay content, may transmit nearly as much water when saturated as when dry. They lack material which causes swelling and clogging of pore space, and water moves rapidly through interstices between grains. Laterites or those soils from which colloids have been removed tend to maintain a high rate of permeability after being wet. Silty loams with water-stable aggregates, i.e., aggregates which do not deflocculate, also display these characteristics.

First approximations of infiltration behavior can be determined by comparison of the soil in question with each of four selected standards.[8] (See Fig. 9.7.)

Group A soils are those of high permeability and generally consist

[8] The grouping of soils is that of G. W. Musgrave, "How Much Water Enters the Soil," *Water, Yearbook of Agriculture*, 1955, U. S. Department of Agriculture, pp. 151–159.

of thick sands and certain silts. The latter is thick loess (wind-blown silt) that contains small amounts of interstitial clay and sufficient organic matter to maintain an open structure.

Group B soils are sandy soils or silty loams of moderate thickness. Soils developed on thin sheets of loess or sandy beds of old lakes fall into this group.

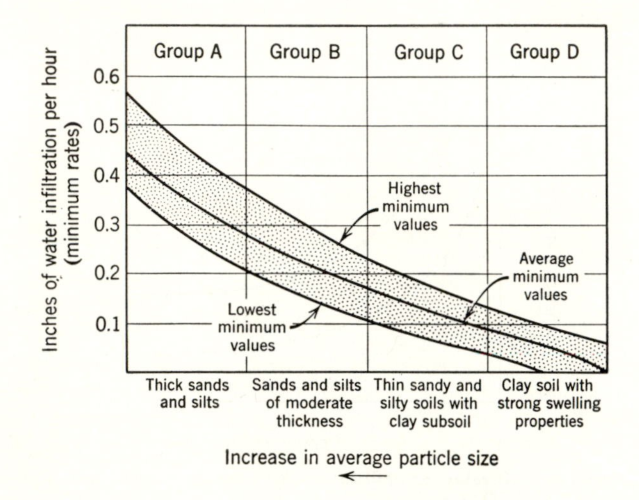

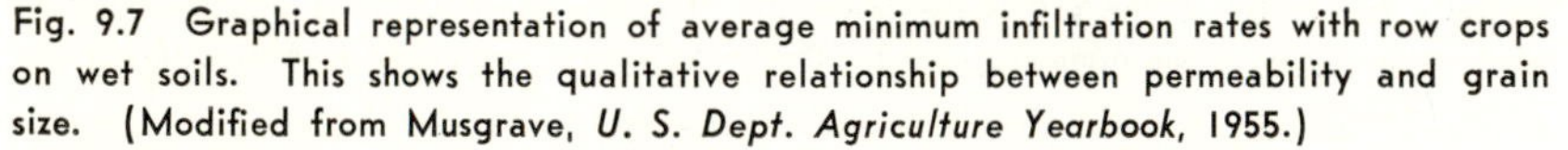

Fig. 9.7 Graphical representation of average minimum infiltration rates with row crops on wet soils. This shows the qualitative relationship between permeability and grain size. (Modified from Musgrave, *U. S. Dept. Agriculture Yearbook*, 1955.)

Group C includes several varieties of thin soils. They are clay loams, thin sandy loams, clay soils, and clay soils low in organic content. Their range includes glacial clay (till), some river bottom silts, and sandy clay soils of the Atlantic Coastal Plain.

Group D soils consist of those with high swelling rates, principally heavy clays. These include gumbotils (weathered glacial clay), marine clays such as occur along parts of the nearshore Gulf Coastal Plain, some of the polders of the Netherlands, and playa-lake deposits especially those containing salts. Soils in this category seldom are able to permit downward passage of more than 0.1-inch (average) rainfall per hour. Hence, in those areas which experience heavy rainfall runoff characteristically is very large. Thick sands, however, can absorb water at the rate of approximately $\frac{1}{2}$ inch per hour at least. This is to be interpreted as indicating that within the rainfall belts of the United States runoff from thick sand is virtually impossible.

The effect of soil temperature on infiltraton rate is rather pronounced.

Obviously, during the winter months soils in which pore water is frozen tend to become effectively impermeable. This is true of sand as well as clays. But even when the soil temperature exceeds the freezing point of water the infiltration rate is directly proportional to tempera-

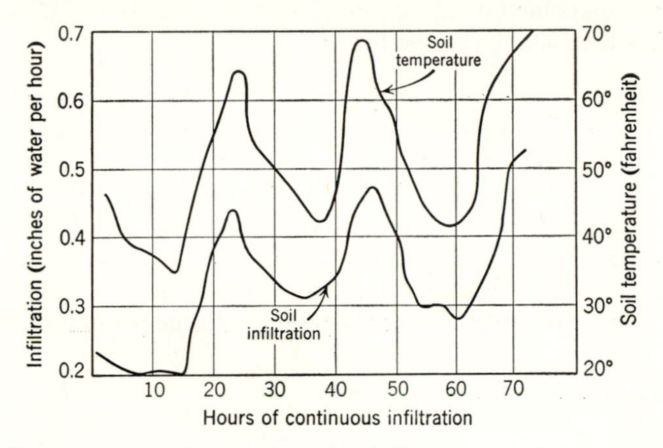

Fig. 9.8 Temperature of soil plotted against infiltration rate of water during a continuous test of 72 hours, near Colorado Springs, Colorado. Note that the infiltration rate rises and falls with corresponding rise and fall in soil temperature. (From Musgrave, *U. S. Dept. Agriculture Yearbook,* 1955.)

ture. This relationship is attributed to the increase in viscosity of water as the temperature approaches freezing. Figure 9.8 is a plot of a soil-infiltration test at Colorado Springs, Colorado, where during a continuous test of infiltration rate the soil temperature was varied. As the temperature was raised infiltration increased, and with decline in soil temperature there was a corresponding drop in intake rate.

FLOW BEHAVIOR[9]

Darcy's Law

In a preceding paragraph reference was made to a simple expression (1) controlling the flow of water through soil, namely

$$Q = Ki$$

which can be stated also as

$$Q = K \frac{H}{L} \tag{2}$$

[9] The reader who wishes an advanced treatment of this subject is referred to M. King Hubbert, "The Theory of Ground-Water Motion," *Jour. Geology,* Vol. 48, 1946, pp. 785–944.

where Q = water volume crossing an area per unit of time.
H = difference in head between two manometers in the path of fluid movement.
L = distance along the path of movement between positions of manometers.
K = a constant varying for each soil.

This equation is known as Darcy's law. The essence of Darcy's statement (1856) is that the flow of water through a column of soil is directly proportional to the difference in pressure at the ends of the column and inversely proportional to the length of the column.

Now if A equals the cross-sectional area across which discharge is being measured

$$\frac{Q}{A} = q = \text{quantity discharged per unit of time per unit of area} \tag{3}$$

Then for a selected hydraulic gradient

$$q = \frac{K}{A} \frac{(h_2 - h_1)}{l} \tag{4}$$

where $h_2 - h_1$ = hydraulic head (i.e., difference in manometer readings)
l = distance between manometers.

Or

$$q = -K' \frac{dh}{dl} \tag{5}$$

K' which is the capillary conductivity or hydraulic conductivity can be shown to be dependent upon the permeability of the rock or soil in question, the density and viscosity of the moving fluid all involved in some such relationship as

$$K' \propto \frac{k\rho}{\eta} \tag{6}$$

where k = coefficient of permeability of soil or rock.
ρ = density of fluid.
η = viscosity of fluid.

When water moves through rocks in the zone of saturation the factor H/L becomes the slope of the water table. The rate of water movement is slow, and hydraulic gradients of 10 to 20 feet per mile are exceeded rarely. Velocities of 30 to 60 feet per day are known (in the laboratory) through well-sorted gravel with hydraulic gradients of 5 to 10 feet per mile, but through rock pores velocities of 5 feet per day are exceptional. Rates of water movement under selected hydraulic heads are best known from studies of wells. Here, under controlled conditions discharge rates can be determined.

Flow from Wells

Discharge from a well is directly related to the permeability of the material which it penetrates. Pumping lowers the water table in the immediate vicinity of the well site, and the local water level falls. The rate of flow into the well, however, will increase as the slope of the water surface increases (i.e., with increase in hydraulic gradient).

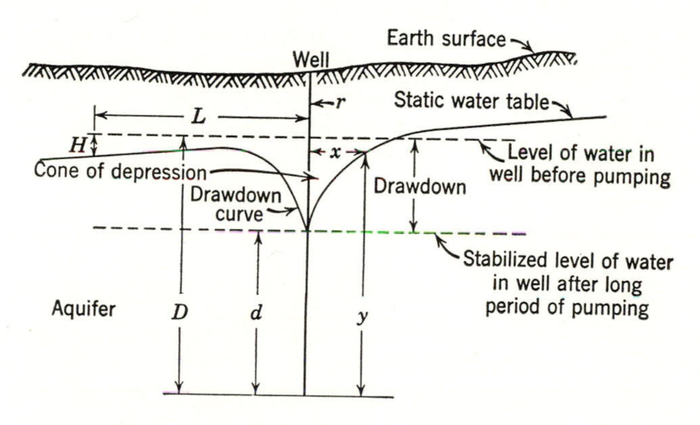

Fig. 9.9 Diagrammatic representation of a well penetrating an aquifer. Pumping has lowered the water table locally to develop a cone of depression around the well. Terminology is as indicated. (Modified from Tolman, *Ground Water*, copyright 1937, McGraw-Hill Book Co., Inc.)

Eventually a steady-state condition will prevail, and flow into the well will equal the withdrawal. At this time the boundary of the water surface will be described by an inverted cone known as the *cone of depression* (Fig. 9.9). Theoretically the base of this cone extends outward from the well to the limits of the water-bearing bed. Actually the cone of depression is insignificant in most cases beyond a radius of approximately $\frac{1}{4}$ mile.

Calculation of actual amounts of water which can be furnished by pumping a well requires specific information concerning the behavior of water in the formation from which it is being drawn. An approximation can be determined by means of equation (14).[10]

[10] The following solution is from C. F. Tolman, *Ground Water,* McGraw-Hill Book Co., 1937. p. 386.

With reference to Fig. 9.9 let

r = radius of well in feet.
d = thickness of water in well in feet (when pumping is in progress) measured from base of aquifer being pumped.
D = thickness of aquifer below static water level in feet (measured to base of well when entirely in aquifer or to base of aquifer if well is deeper).
$D - d$ = drawdown in feet.
x and y = coordinates of any point on the drawdown curve.
q = quantity of water pumped in gallons per day.
P = permeability coefficient (rate of flow in gallons per day through a cross section of one square foot under an hydraulic gradient of 100 per cent at a temperature of 60° F).

The total area through which water is passing into the well at any distance, x, from the center of the well is $2\pi xy$, and the slope at any point is

$$x, y = \frac{dy}{dx} \tag{7}$$

Then,

$$q = 2\pi Pxy \frac{dy}{dx} \tag{8}$$

$$q \frac{dx}{x} = 2\pi Py\, dy \tag{9}$$

integrating,

$$q \ln x = \pi Py^2 + C \tag{10}$$

when $x = r$ and $y = d$

$$q \ln r = \pi Pd^2 + C \tag{11}$$

$$C = q \ln r - \pi Pd^2 \tag{12}$$

substituting, and solving for y in eq. 10

$$y^2 = \frac{q}{\pi P} \ln \frac{x}{r} + d^2 \tag{13}$$

Substituting R for radius of base of cone of depression, x, the corresponding value of y becomes D (i.e., thickness of aquifer below water table), and solving for q in eq. 13

$$q = \frac{\pi P(D^2 - d^2)}{2.3 \log_{10} \frac{R}{r}} \tag{14}$$

The drawdown $(D - d)$ may be calculated from eq. 14 by rearrangement of terms whereupon:

$$d = \sqrt{\frac{D^2 - 2.3q \log_{10} \frac{R}{r}}{\pi P}} \qquad (15)$$

Values of log (R/r) vary only between small ranges; hence, the effect of large "circles of influence" is not important in altering the value of q. On the other hand P varies enormously even for the same variety of sediment, and this value has profound effect upon the value of q (see Tables 9.2 and 9.3).

Porosity of Sediments and Sedimentary Rocks

Loose gravel and sand contain pore space which is easily observed with the unaided eye. This is true also of certain rocks such as vesicular basalt, reef limestone and dolomite, sandstone, and conglomerate. Where particles are very small as in silts or clays pore space becomes increasingly difficult to see except under high magnification. For loose aggregates porosity varies with grain packing, shape, and perfection of sorting of the grains. When sorting is poor small pieces occupy pore space between larger fragments and tend to clog the openings. Certain sandstones and siltstones are extremely well sorted indicating that in each case grain dimensions are more or less uniform. Under these circumstances, porosity remains approximately the same for all grain sizes larger than clay.

Particles of natural soils or sediments are not the ideal spheres generally represented in diagram. Some sand grains approach spherical outline but are not uniform in size. Other sands are composed of sharply angular fragments closely interlocked. Practically all sandstones have their grains cemented together by calcite or silica. These minerals occupy part of the interstitial space, and, hence, porosity is strongly reduced (see Table 9.4).

Clays are extremely variable in porosity and frequently contain much greater amounts of water than are theoretically permitted in normal pore space. All clays should be regarded as sediments of high porosity, but when cemented into shale their pore space is reduced drastically. Clay minerals consist of aggregates of tabular crystals which predominantly settle from suspension with their large surfaces on the sedimentation plane. This sedimentary fabric tends to produce an initial structure which contains much water between individual

Table 9.2. Relationship between Well Diameter and Circle of Influence

(Values of $\frac{1}{\log R/r}$ for Use in Eq. 14)*

R in feet	Well Diameter ($2r$) in Inches 4	6	8	12	24	48
100	0.36	0.38	0.40	0.43	0.50	0.58
200	0.32	0.34	0.36	0.38	0.43	0.50
500	0.28	0.30	0.31	0.33	0.37	0.41
2,000	0.24	0.25	0.26	0.27	0.30	0.33

* Part of table from C. F. Tolman, *op. cit.*, p. 387.
R = radius of cone of depression
r = radius of well

Table 9.3. Approximate Values of Permeability Coefficient (*P*)*

Sediment	Permeability Coefficient (Estimated) Average	Range
Very fine sand	50	10–300
Fine sand	300	50–2,000
Medium sand	600	100–5,000
Coarse sand	1,500	300–10,000
Fine gravel	5,000	1,000–20,000
Medium gravel	15,000	3,000–50,000
Gravel lens in alluvial silt, Tongue River, Montana	68	
Sand above shale (Fort Union formation), Montana	132	
Sandstone, poorly cemented (Fort Union formation), Montana	77	
Sandstone, poorly cemented (Judith River formation), Montana	13	
Coarse sandstone (Raritan formation), New Jersey	3,788	
Fine sand, lower part "800 foot" bed, Ocean City, New Jersey	2,609	
Red clay, Dubois, Idaho	20	
Loess from Platte River valley, Lexington, Nebraska	2	1–4
Sand and gravel, Platte River valley, Nebraska	3,000	1,000–4,150

* Values are in terms of gallons per day through a cross-sectional area of 1 sq. foot under 100 per cent hydraulic gradient.

Table 9.4. Porosity of Selected Sediments and Rocks*

Material	Porosity (Per Cent of Total Volume)
Sand (stream)	48
Sand and silt (glacial outwash)	36
Sand and gravel with large pebbles (glacial outwash)	25
Glacial clay (till), sandy and gravelly	21
Clays, assorted	45 (av)
Silt (lake)	36
Silt and clay (Mississippi River delta)	80–90
Sandstone (gas bearing, Bartlesville, Okla.)	23 (av)
Sandstone	16 (av)
Limestone, marble, dolomite	5 (av)
Granite, schist, gneiss, quartzite	<1

* Data principally from O. E. Meinzer, *The Occurrence of Ground Water in the United States,* U. S. Geol. Survey, Water Supply Paper 489, 1923.

crystals. In fact, in many clays the solid particles are held in colloidal suspension within the water phase. Such clays display a very large initial porosity. When additional layers of sediment are loaded upon such a colloidal suspension compaction occurs, water is ejected, and the porosity is reduced markedly. Many fine-grained silts and clays have an initial porosity which exceeds 75 per cent, a condition which as mentioned above consists of a suspension of solid particles in a water medium. Clay colloidal suspensions are extremely unstable, and flocculation may occur by the simple process of introduction of salt water. Strongly ionized solutions are very effective in destruction of the colloidal state due to the effect of neutralization of charged particles by an oppositely charged ion. Marine clays deposited in the neritic zone, therefore, are initially much more firmly compacted than fresh-water ones, and extremely high porosities can be considered exceptional. Deltaic, lagoonal, and silty clays accumulating in harbors where brackish water exists have a much more open structure and are reported as having porosities of 50 per cent or more. Fresh-water delta, lake, and river clays display a depositional fabric in which the water-clay-silt suspension is partially supported by the irregular framework of solid particles resting upon one another as well as a continuous film of water separating particles. As additional sediment is piled upon the water suspension the cell structure developed by irregular piling of particles collapses in part, and water is expelled.

Effect of Compaction of Clays

Permeability in suspensions of water and clay is not high, and water is expelled slowly as loading proceeds. Excess pressure is developed higher than normal for the hydrostatic column at that specific depth. The process produces areas of high potential differential pressure, and water moves very slowly along this gradient. During the early stages of compaction the direction of this movement is upward, but as loading is increased migration of water is accelerated in a lateral direction. This is the response to the normal anisotropism which develops in sediments as they begin the lithification process. Some of this anisotropism results from the range in size distribution and the tabular shapes of clay minerals piled one upon another. Together these attributes reduce permeability in a vertical direction as compared to the horizontal. Also, bedding planes begin to appear, and water is directed to move along these natural surfaces of separation. Under a pressure gradient some of this water travels rather long distances, all laterally, whereas some may actually move in a zigzag manner up, down, and horizontally, following the direction dictated by the decrease in potential.

Layers of sand which interrupt a clay sequence undergo slight compaction and maintain a high permeability relative to the clay. Water which is squeezed from the clay moves into such sand beds which if they are very extensive may permit migration of fluids for long distances. Important occurrences of salt water in sedimentary basins are attributed to such an origin. Also there is good reason to believe that this process is the mechanism whereby oil is squeezed from dark shales to accumulate in reservoirs of sandstone. A problem of interest to civil engineers is land subsidence in some localities where oil and water are being removed. Here removal of the interstitial fluids permits collapse of openings between grains with a reduction in porosity and sediment volume. The resulting differential subsidence at the earth's surface has created conditions where various surface installations have been damaged.

Compaction of clays when saturated with water create certain special conditions in high-latitude regions where frost action is very important. Alternate freezing and thawing of water poses many special problems. Still deeper below the surface permanently frozen water fills interstices. Large areas are underlain by these conditions, and special attention to the characterizing features is warranted.

PERMAFROST[11]

Occurrence

Soil or rock in which the pore space is permanently occupied by ice (permanently frozen ground) is defined as *permafrost.* Ground of temperatures below 0°C, but containing no ice as a cementing substance, is called *dry permafrost.* This low temperature dry condition is found in sands or other easily drained, coarsely granular, soil materials. Permafrost exists only where the temperature continues to be at 0°C or below for most of the year, but local fluctuation across the freezing point of water is characteristic for the upper part of the soil.

In the northern hemisphere existence of permanently frozen ground is considered by some to be related to Pleistocene glaciation which established the initial condition for extensive development. Thousands of years of continued refrigeration produced a widespread and thick zone of permafrost over much of the northern hemisphere. Although amelioration of the climate has doubtless reduced the total permafrost area and confined the occurrence to regions of high latitudes the total area now in existence amounts to approximately 20 per cent of the land surface. There is no requirement, however, that permafrost conditions can exist only in localities of Pleistocene glaciation. Rather, the condition needed is currently intensive refrigeraton.

Considerable variation in thickness of permafrost is typical within the same climatic zone. Generally considered, however, it is thin (i.e., a few millimeters) in the warmer sections, and thickest (230 m) where temperatures are persistently low. The upper zone known as the *active layer* melts during the summer season and freezes again during the cold months. As a rule this zone is thinnest in the cold areas and is thickest near the permafrost margin, but more important controls appear to be the permeability and porosity of the ice-bearing zone. (See Fig. 9.10.) This is illustrated by the data listed in Table 9.5 which shows that poorly drained soils and those of high porosity and organic content are those displaying the thinnest active layer. Near the limits of the permafrost zone the base of the active layer (*permafrost table*) reaches levels deeper than those which are

[11] The reader is referred to an excellent summary of the subject by S. W. Muller, *Permafrost or Permanently Frozen Ground and Related Engineering Problems,* Spec. Report Strategic Engineering Study 62, Office of Chief of Engineers, U. S. Army, 1945.

attained by average present-day winter freezing, and the active frozen layer rests upon a zone of unfrozen ground above the main mass of permanent permafrost. In lower latitudes the zone of permanently frozen ground decreases in thickness, and where it no longer exists

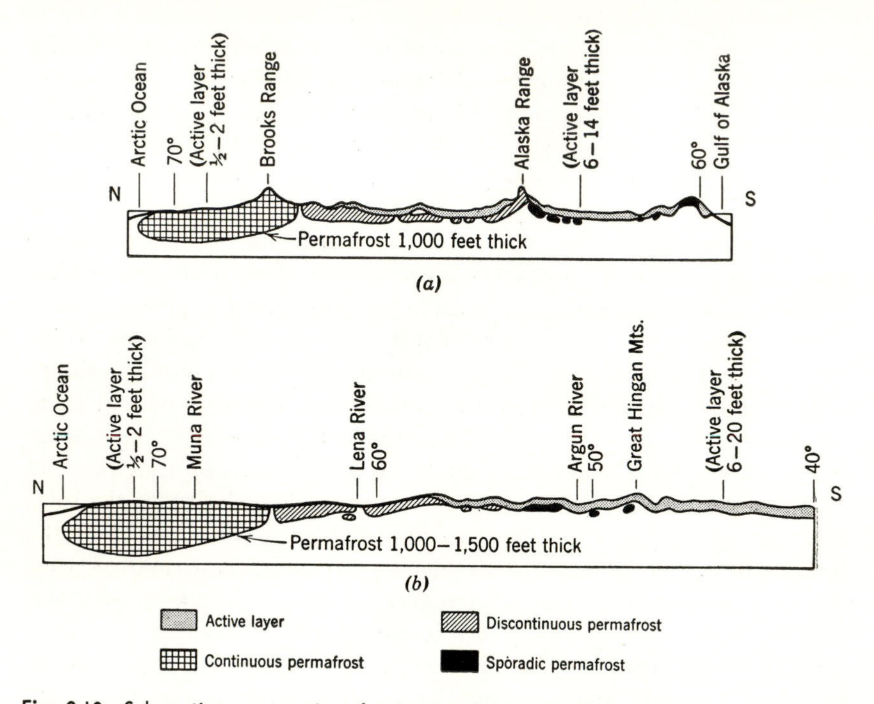

Fig. 9.10 Schematic cross section showing conditions of permafrost in (*a*) Alaska and (*b*) Asia along longitude 120° E. Note that the active layer thickens near the temperate margin of permafrost; note also the great thickness of permafrost in the polar regions. (Reprinted with permission from Black, in Trask, *Applied Sedimentation*, copyright 1950, John Wiley and Sons, Inc. Original drawing of *b* by I. V. Poiré.)

permafrost conditions are not recognized. Within the permafrost belt zones of unfrozen water confined between layers may be under considerable hydrostatic pressure, and these may also constitute zones of water supply. Wherever such conditions exist on slopes gradual downhill movement of the soil zone occurs frequently accompanied by numerous but small mudflows.

Growth of Ice Crystals

Permanently, or temporarily, frozen ground contains disseminated ice crystals or masses of crystalline ice between layers primarily com-

Table 9.5. Thickness of Active Layer of Permafrost*

Soil—Type and Plant Cover	Geographic Locality: Siberia South of Lat. 55° N.	Siberia Lat. 62° N.	Coast of Arctic Ocean Lat. 70° N.+
Sandy soil	3–4 m	2–2.5 m	0.7–1.0 m
Clayey soil	1.8–2.5 m	1.5–2.0 m	0.5^{+} m
Peat swamp soil	0.7–1.0 m	0.7–1.0 m	0.2–0.4 m
River terrace, sandy clay (peat moss)	0.5–0.8 m		
River terrace, sandy clay (birch and poplar)	1.5–2.5 m		
River terrace, well-drained sandy soil (pine forest)	2.5–3.5 m		

* Data from S. W. Muller, *op. cit.*

posed of soil particles. The amount of ice that can form in the soil depends principally upon the amount of water available. This water may exist either in local interstitial spaces or in the adjacent ground from which moisture may be drawn. Where the hydraulic gradient is toward highly permeable sediment thick layers of ice can form. Growth of ice crystals is accompanied by an increase in soil volume which results in heaving of soil. This phenomenon designated as *ground swelling* is particularly destructive to roads and other forms of surface construction. Swelling takes place principally under three conditions:

1. Where the upper soil consists of clay saturated with water but is relatively impermeable. This is the case of local interstitial space being filled with water which upon freezing causes local hydraulic pressure gradients. Disseminated ice crystals are common in these soils.
2. Where water-bearing fissures or layers are present within the active layer. Differential freezing of this soil condition results in movement of water to the zones of ice-crystal growth. Large individual crystals and thick layers of ice intercalated with soil are typical.
3. Where upper layers are permeable (i.e., sand or gravel) and are underlain by an impermeable zone such as clay, bedrock, or permafrost. Water is ponded on such an impermeable barrier and as differential freezing begins moves freely under hydraulic gradient. Frequently, the pressure head is sufficiently great to heave the overlying soil whose weight is unable to equalize the upward pressure exerted by the water.

Icing and Soil Heaving

As ice crystals increase in dimension or as water is initially converted into ice, pressure is exerted on the water system due to volume increases. This pressure head may cause liquid water to move laterally as well as upward under the pressure gradient. The result is the appearance of soil mounds, some of which are small, but others may reach as much as 100 m in height. Such mounds somewhat analogous in development to laccolithic intrusions of magma are broken by radiating fissures. The fissures frequently are the sources of springs

Table 9.6. Classification of Materials according to Swelling Tendency*

Character of Swelling	Type of Material
Ground swelling absent.	1. Bedrock. 2. Rubble of pebbles and boulders with interstices filled by smaller fragments of same material.
Ground swelling minor.	1. Gravel and coarse sand (small amounts of silt and clay) which may be saturated but which have no additional supply of water.
Ground swelling important.	1. Gravel and sand interlayered with clay and silt, and with access to additional supply of water. 2. Sandy clay, clayey sand, and clay. 3. Fine sand, sandy silt, silt, saturated "mud." 4. Organic decomposition products principally peat.

* Simplified from Muller, *op. cit.*

of fresh water, indicating more or less continuous flow of water toward the mounds. In cold weather water issuing from these springs freezes into large masses of ice which may block road cuts or fill foundations.

The relative intensity of swelling and heaving of ground during freezing, as indicated in Table 9.6, is controlled principally by the amount of liquid water which is available to move to the sites of crystallization. In fine-grained material such as clay water does not freeze in the capillary openings at the usual temperature but is known to remain liquid well below −20°C (−4°F). Where interstices are larger such as in sands water freezes at higher temperatures. As ice forms in the larger capillaries liquid water is drawn from finer capillaries to add to the growing ice crystals. As long as the water supply exists larger and larger crystals of ice develop until very extensive intergrowths are formed. This mechanism is particularly satisfactory in explaining the strong ground swell which exists in those localities where silts and clays are underlain by beds of sand. Here, important

Fig. 9.11 Cross section showing conditions of permafrost leading to icing in dwellings. Heat released from building prevents freezing of underlying active zone. The vicinity of the building becomes the locus for outlet of liquid water as freezing occurs in the active layer outside. Water underlying the freezing active zone is placed under hydraulic pressure and is forced to the surface beneath the building.

ice growth is not in the silts and clays but in the permeable bed. Water is transferred through the capillary openings without freezing, eventually reaches the locus of crystallization, and is added to the mass of ice. A very severe variety of swelling results from the hydrostatic pressure of ground water trapped between the frozen active layer and the impervious permafrost zone beneath. As freezing proceeds pressure on the water is increased until in some instances the overlying surface is bulged upward or cracks open.

During the winter months masses of ice are formed by successive freezing of water that may seep from the ground, rivers, or springs. This phenomenon called *icing* creates serious problems in connection with dwellings and other forms of construction. Icing consists of irregular sheets or incrustations piled one upon another. Such masses create nearly impassable conditions when spread across roads, as mounds of 20 m in height and 50 m wide are known to have formed.

The condition develops during periods of thin snow cover when winter freezing occurs in the upper part of the active layer and confines liquid water between this frozen zone and the underlying main body of permafrost. Ground water is gradually forced from the active layer and caused to spill out over the surface to freeze. A particularly hazardous situation develops at the site of heated buildings which tend to create a condition of thawed ground. Beneath the building water is not confined by overlying ice and becomes a locus where water is forced to the surface as a constant flowing spring. Such water gradually invades the dwelling and by progressive freezing may completely fill the structure with a mass of ice. (See Fig. 9.11.)

Upon thawing considerable settling takes place in the swollen ground. Part of the reduction in volume is from melting of the ice layers, but some is due also to actual flowage of water-saturated sediment under the residual hydraulic pressure generated by the freezing. Pronounced landsliding and creep are characteristic of the period of summer thaw.

FLOW THROUGH ROCKS

Types of Flow

Previous mention has been made that water which enters rocks either by infiltration through the soil or through stream channels in dry regions can be classified as moving through three distinctly different varieties of openings. These are pores, fractures, and caverns or large channelways. When water moves through rock pores and along fissures or bedding planes whose separations are of the order of a millimeter or less the quantity of water passed through a large unit cross section is in accordance with Darcy's law. Sands, gravels, and well-sorted but poorly cemented sandstones are the nearest approach to the ideal case; the flow can be identified as homogeneous-laminar, and the general direction of movement is indicated by the slope of the ground-water table.

Wherever large openings are wide fissures, joints, and caverns ground-water movement is much more rapid, and the laws controlling flow are similar to those which apply to small and large pipes. In these cases flow velocities are sufficient to cause movement to be turbulent and complex. Flow is not uniform in all directions but is preferentially directed and heterogeneous. Channelways are large in cavernous limestone or basalt, and water movement is essentially that of a large pipe network. Where such an interconnecting network system exists hydraulic pressure is important in directing the path and rate of movement, and configuration of the water table is inconsequential in controlling the actual hydraulic gradient.

Much use can be made of a qualitative appraisal of the flow conditions expected to exist in sediments or rocks which are being investigated. The condition may be inferred from the varieties of the rocks concerned, and Table 9.7 is an attempt to assemble some of the desired information. Selected sediments and rocks are listed to indicate the general water-bearing attributes of commonly found materials. As the reader inspects the table he should recognize that in loose sediments where fissures do not occur the existence of large pores and a size distribution of small standard deviation indicate a potentially favorable aquifer. Sediments in which openings are capillary in size yield only small quantities of water.

Wherever the sediments have been cemented into rock some of the pore space is filled with mineral matter, and the interstitial openings are reduced in size. For this reason where water is carried through interconnected pore openings the amount which can be furnished is less ordinarily than the corresponding sediment. However, if the rock is highly fractured the actual permeability of coarsely granular rocks may be considerably greater than their uncemented counterparts.

With few exceptions shales, igneous, or metamorphic rocks are notably poor water-bearing materials due to the extremely small size of the pores, and their water-carrying capacity is controlled by fissure systems. Exceptionally great aquifers are cavernous limestones and vesicular and scoriaceous basalts which through a complex system of channelways direct enormous volumes of water to certain localized points. These rocks should be regarded as the greatest class of water carriers.

Still another group of rocks exists of very low porosity and permeability and which because of their ductility tend to remain unfractured. Beds of coal, salt, gypsum, and soft shales fit this category. Also, there are certain highly metamorphosed rocks (gneiss and schist) which under the pressure conditions to which they were subjected tended to yield by plastic flow rather than fracture. Each rock type constitutes an impermeable barrier to the movement of water which tends to be ponded upon the surfaces of these barriers.

Table 9.7. Water-Bearing Properties of Sediments and Rocks*

Name	Desirable Specifications	Degree of Porosity	Degree of Permeability and Type of Opening	General Characteristics	Maximum Rate of Supply (Gallons per Minute for Drilled Well of 1 ft. Diameter)
Gravel	Clean and well sorted; pebbles should be principally granite, quartz, quartzite, or less desirable limestone or dolomite.	High	High (pore)	Typical occurrence in glacial outwash and alluvial fan.	1,000
Sand	Clean and well sorted; widespread bed.	High	High (pore)	Coarse sand yields more than fine sand. Poorly sorted sands or those with high clay matrix are undesirable. Very fine sand is a poor source of water.	Several hundred
Alluvium	Minimum amounts of clay matrix; alluvial fans should contain gravel lenses.	Extremely variable	Extremely variable (pore)	Poorly sorted clayey alluvium is undesirable, but where gravel or coarse sand layers are present makes a desirable producer.	Extremely variable 500 to 1,000
Volcanic sediments	Poorly cemented agglomerates and coarse deposits.	High	High (pore)	Ash is a poor producer but ponds water on its upper surface.	Up to 600

* Data principally from O. E. Meinzer, *op. cit.*, p. 117–148.

Loess	Not an aquifer.	High	Low to moderate (pore)	Openings collapse on pumping.	Estimated ± 1
Till	Must be interbedded with gravel or sand.	Extremely variable	Very low except in gravels (pore)	Till itself is a very poor source of water, but it is generally associated with outwash which carries water.	$\frac{1}{4}$ to 5 average, range to 100
Clay	None	High	Very low (pore)	Does not transmit water but acts as ponding bed.	Less than 1
Conglomerate	Clean and well sorted; poorly cemented.	Generally high	Generally high (pore and fissure)	Porosity and permeability dependent upon degree of cementation.	Lower than gravel
Sandstone	Clean, coarse grained, and well sorted; poorly cemented; extensive occurrence.	Generally high	Generally high (pore and fissure)	Porosity and permeability dependent upon degree of cementation.	Several hundred
Limestone or dolomite	Must be well jointed and cavernous or very porous as in reef or oolitic textures.	Extremely variable depending upon character of pore space	Generally high but extremely variable (pore, fissure and cavern)	Dense and poorly jointed limestones carry little water. Cavernous ones carry enormous quantities. Water may be ponded on upper surface of limestone bed.	Up to 350,000; 450 common

Table 9.7. Water-Bearing Properties of Sediments and Rocks* (*Continued*)

Name	Desirable Specifications	Degree of Porosity	Degree of Permeability and Type of Opening	General Characteristics	Maximum Rate of Supply (Gallons per Minute for Drilled Well of 1 ft. Diameter)
Chalk	Should be well cemented and very porous.	High	Low to high (pore)	Openings tend to collapse reducing permeability.	Variable, depending on cementation, usually less than 100
Shaly limestone	Should not consist of alternating thin beds of shale and limestone but should contain thick beds of porous limestone.	Moderate	Moderate to low (fissure)	Generally poor water sources but where bedding is irregular and well developed may produce water.	Extremely variable but generally less than 100
Shale and slate	Must be highly jointed, well bedded, or sandy.	Moderate for shale, low for slate	Low to moderate (fissure)	Slates are better aquifers than shales and also better than granites. Large diameter wells recommended.	Up to 50, 1 to 10 average
Gypsum and salt	None	Low	Very low (pore)	Tend to be interbedded with red shales and siltstones or thick limestone and dolomite.	None
Coal	Must be well jointed. Should not be considered an aquifer.	Low	Low (fissure)	Usually dry but ponds water on upper surface. Water generally sulfurous.	Very small

Basalt	Cavernous and vesicular lava flows, interbedded with agglomerate.	Variable	Often extremely high (pore, fissure, and cavern)	Generally a great water producer unless dense and nonvesicular.	1,000 to 450 common
Rhyolite	Must be highly jointed and vesicular, otherwise a poor aquifer.	Low but variable	Low but variable (pore and fissure)	Contains few openings unless vesicular.	Locally can exceed 1,000, but usually 1 to 25
Granitic rocks	Poor aquifers, must be jointed or weathered.	Fresh rock very low	Generally very low (fissure)	Unaltered massive granitic rocks are extremely undesirable.	1 to 10
Quartzite	Must be well jointed.	Low as a rock, but may be high, due to fractures	Depends upon degree of fracturing (fissure)	In folded regions quartzite is generally fractured.	10 to 60
Gneiss and schist	Poor aquifers, but cleavage makes them better than granitic rocks.	Low	Low (fissure)	Schists tend to have openings closed at depth.	Less than 12

The Surface of the Water Table

Flow of ground water through intergranular pore space is controlled so frequently by a gravity head that movement toward stream valleys in wet areas results in the depressed water-table surface as shown in

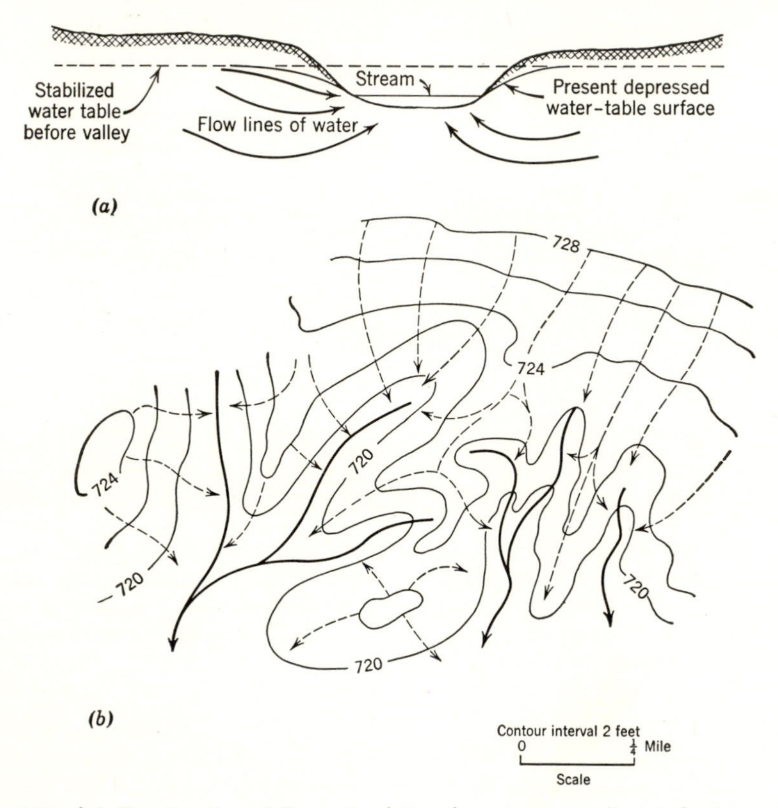

Fig. 9.12 (*a*) The direction of flow network in a homogeneous sediment showing paths taken by fluid particles. Note that an upward component of movement exists beneath the stream bed. Here the column of water in the saturated sediment is less above some underground reference level than the adjacent column outside the valley. (*b*) Contours on the surface of the ground-water table in a stream system within a humid area. Orthogonal lines (dashed) show paths of flow to the channels. Note the higher elevations underlying topographic divides. Contrast the irregularity of the water table in the area of stream valleys with the tendency toward uniform slope in the upland area.

Fig. 9.12*a*. This depression is identical to the draw-down curve of a well undergoing pumping, and the stream tends to remove water from the site as fast as it is supplied. Directions of flow are shown by orthogonal lines concentrating on the perimeter of the stream channel and indicating the direction of the hydraulic gradient. Immediately beneath the stream bed where the local hydraulic head is less than that at depth or in the surrounding area the flow will be toward the channel. If the stream channel is dry the condition is analogous to the negative pressure head or suction head developed in partly saturated soils and

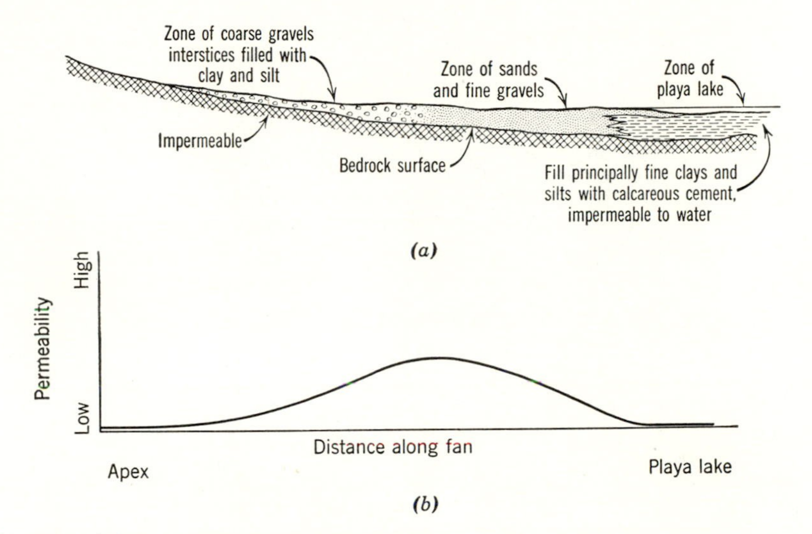

Fig. 9.13 (*a*) A cross section of an alluvial fan showing an idealized distribution of sediments which lead to filling of fan with water. (*b*) A general plot of the relative permeability of the sediments in cross section (*a*) with distance downstream from the fan. Note that the position of the zone of maximum permeability is intermediate between the playa lake and apex of fan.

results in upward movement of water. In areas of rough topography the water table has marked undulation, and the flow network is a pattern similar to that illustrated in the lower half of Fig. 9.12*b*. Note that in the upper part of the figure beyond the influence of stream valleys the water table is much more regular in elevation and has a regional rather than a local gradient.

Wherever a permeable bed rests upon an impermeable one there exists a tendency for water to be ponded on the surface of the impermeable bed. The ponded water increases until it can flow down an hydraulic gradient and emerge as a spring or along some stream valley. This condition is well illustrated by alluvial fans which absorb water carried to their surfaces by the mountain streams. In the apex where coarse gravels are mixed with fine material permeability is lower than part way down the slope where the bulk composition of the fan consists of gravels and sand generally free from much interstitial clay. This is the sector of greatest uniformity of grain

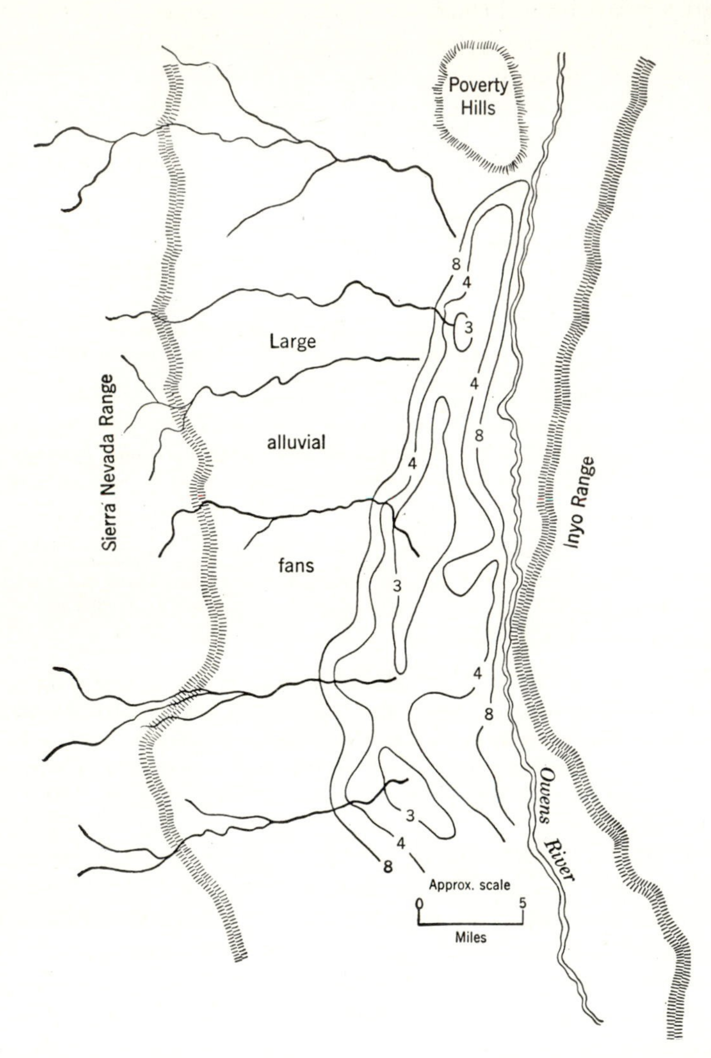

Fig. 9.14 A diagrammatic sketch of part of Owens Valley, Inyo County, California, showing by contours the depth (in feet) to the water table. Note the decline in level of water as the Owens River is reached, indicating withdrawal of water by the stream. Compare with Figs. 9.13 and 9.15. (Simplified from Lee, *U. S. Geol. Survey Water Supply Paper* **294.**)

size within individual beds, infiltration is high, and the water table rises to a position very close to the surface (see Figs. 9.13, 9.14, and 9.15).

Downslope in the low area of the playa fine clays and silts which settle from suspension in the lake waters produce a permeability barrier. In this zone the ground-water table declines rapidly toward the bed rock surface. Water standing in the playa lake is "perched"

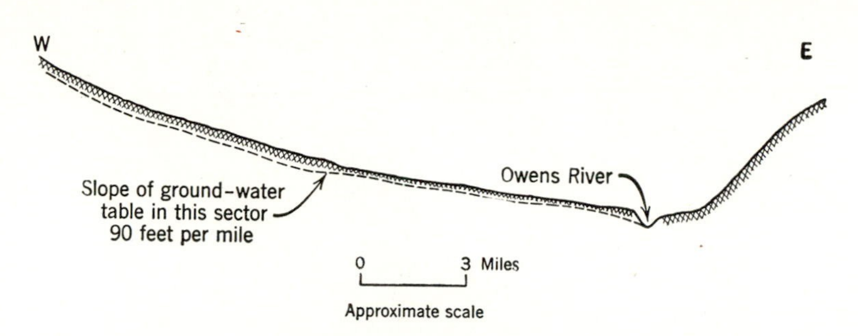

Fig. 9.15 West-East cross-sectional sketch across center of Owens Valley, California (Fig. 9.14), showing the slope of the water table, the rise of the water table near the base of the alluvial fan, and the rapid drop in level along the river. Proximity of the water table to the surface indicates that the fan is nearly filled to capacity with water.

and prevented from downward migration by the impermeable clays. Height of water above the bedrock surface is, therefore, not indicated by the elevation of the playa lake. It can be determined by drilling to the proper depth or indicated by a "line" of springs where water is leaking from the saturated fan. Frequently such springs are upslope some distance from the floor of the playa and are marked by zones of grassy or bushy vegetation (see Fig. 9.16).

Artesian Conditions

Infiltration and downslope flow through an alluvial fan is clearly defined evidence that some hydraulic gradient must exist. Water moves toward the playa and reaches an impermeable barrier in the fine silts and clays which block further movement. Similarly confined water occurs in gently inclined rock aquifers where an hydraulic gradient is developed independent of the water-table slope. Although such conditions are typical of sandstone or limestone strata interlayered with impermeable beds, cavernous lava flows, and even strongly fissured metamorphic rocks may exhibit the same features. The fundamental framework is a permeable zone inclined from its surface outcrop within which water is confined by impermeable zones above, below, and downslope. Eventually, the permeable zone becomes

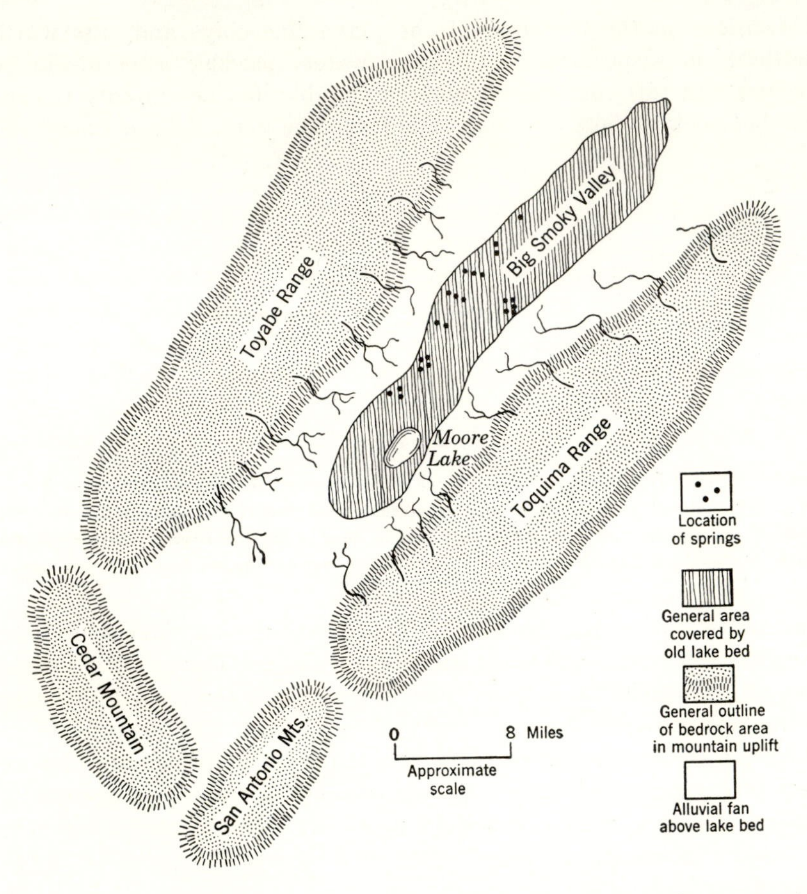

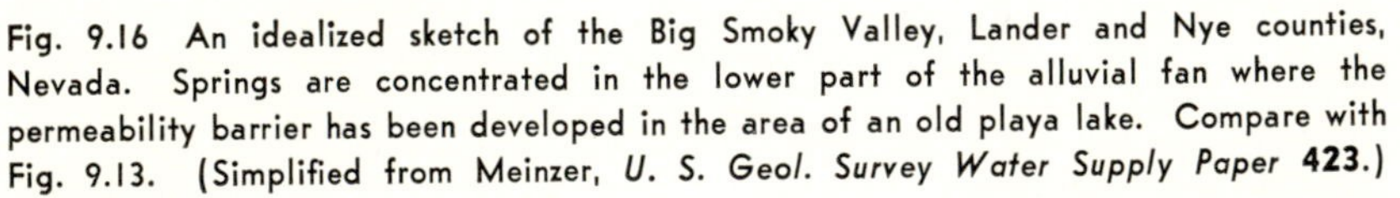

Fig. 9.16 An idealized sketch of the Big Smoky Valley, Lander and Nye counties, Nevada. Springs are concentrated in the lower part of the alluvial fan where the permeability barrier has been developed in the area of an old playa lake. Compare with Fig. 9.13. (Simplified from Meinzer, *U. S. Geol. Survey Water Supply Paper* **423**.)

filled, and water is under pressure developed by the hydraulic head available. Any well which taps this source becomes a virtual fountain. The term *artesian water* was first applied to such flow where there existed sufficient head to cause water to rise above the level of the local ground surface. Artesian springs often carry water from deep zones to fountain at the surface via some system of fissures which provide the channelways under the existing pressure head. Such springs usually are distinguished by a constancy of flow from season to season and by uniformity of water temperature throughout the year in regions of pronounced differences of seasonal temperature.

Certain geological conditions are typical of the ideal, extensive artesian supply. These conditions are described below and are to be regarded as virtual requirements:

1. The aquifer must crop out over a broad and widespread area where it may gather water by infiltration. This recharge belt must occur in a climatic zone where rainfall exceeds local demands. Total infiltration should keep the reservoir filled.
2. Surface elevation of the gathering ground must be substantially higher than the area where artesian wells are to be drilled. The water column representing the difference in height between the two areas provides the pressure head necessary to cause water to flow within the artesian sector.
3. The aquifer must dip away from the outcrop area and continue in the subsurface as a permeable zone. Some distance downdip the permeability must be strongly reduced to impede further water movement. The same effect is produced if the dip of the aquifer is reversed in the center of the basin and the strata rise toward the surface once more on the opposite side. In either case the aquifer fills with water, and a pressure head is developed. Under natural conditions this water escapes only at artesian springs or remains effectively stagnant within the water-bearing formation. Excess water is rejected in the surface recharge area.
4. Except for the outcrop area the aquifer must be overlain by an impervious zone to prevent upward escape of water. Immediately underlying impermeable zones are desirable although not absolutely necessary inasmuch as impermeable rock is encountered with depth sooner or later.

Artesian Basins

Artesian conditions are the rule rather than the exception in the regions of large sedimentary basins. In these areas sediments have accumulated in a slowly subsiding ovaloid-shaped depression in which rather extensive sheets of sandstone or limestone, covering thousands of square miles, are interlayered with shales. Wherever erosion has exposed the margins of such basins and the aquifer crops out over an area suitable for large-scale water catchment artesian conditions prevail. In some desert areas, bordered by mountains, where rainfall is ample part of the year an aquifer is exposed along the mountain front and is inclined toward the topographically lower desert basin.

Here artesian water supply may be large and can support an important irrigation system. Such conditions offer much promise in creating agricultural economies in areas now too dry for normal growth. It must be emphasized, however, that the water demands must be balanced by the recharge supply.

Important artesian conditions occur also in areas of high rainfall such as Florida. Here a cavernous limestone (Ocala limestone) is the principal water carrier of the entire state. The aquifer is not uniformly porous or cavernous but contains layers of impervious cherty limestone. The effect of the cherty zones is to provide a seal which prevents upward escape of ground water by confining it to the permeable layers. Gentle inclination of strata from the structural arch which is highest in the general northwestern part of the peninsula but trends southeast toward the center of the state provides the framework for the principal hydraulic head. The northwestern section is also the major area of water intake. The magnitude of the local hydraulic head developed and the hydraulic gradient over the state are shown in Fig. 9.17. This is a contour map of an imaginary surface to which water would rise in a well drilled into the Ocala limestone at any designated locality. The imaginary envelope called the *piezometric surface* attains its highest elevation above sea level near the center of the peninsula and slopes outward in all directions reaching low values along the Keys and western shoreline.

In the topographically highest sections water will not rise in wells to the ground surface, but in the marginal areas sufficient head exists to cause similar wells to fountain. The area within which flowing wells are to be expected is indicated by shading in Fig. 9.17. Elsewhere in the state water rises in wells only part way to the surface, but certain artesian springs discharge enormous volumes because they are connected by an intricate fissure and cavern system to regions of higher piezometric surface. Some of the springs have been located on the map (Fig. 9.17) where their distribution can be seen to be confined principally to the northern half of the state.

An important aspect of artesian conditions is to be noted in the area of the Keys. Here the piezometric surface indicates that water will rise as much as 10 feet above sea level in areas completely surrounded by the sea. On such otherwise waterless islands where artesian conditions exist a supply of fresh water is available from deep wells despite the absence of a local watershed.

As regions have become increasingly populated demands for artesian water have also expanded. In most areas the demand has exceeded the annual recharge, and the piezometric surface has fallen. Wells

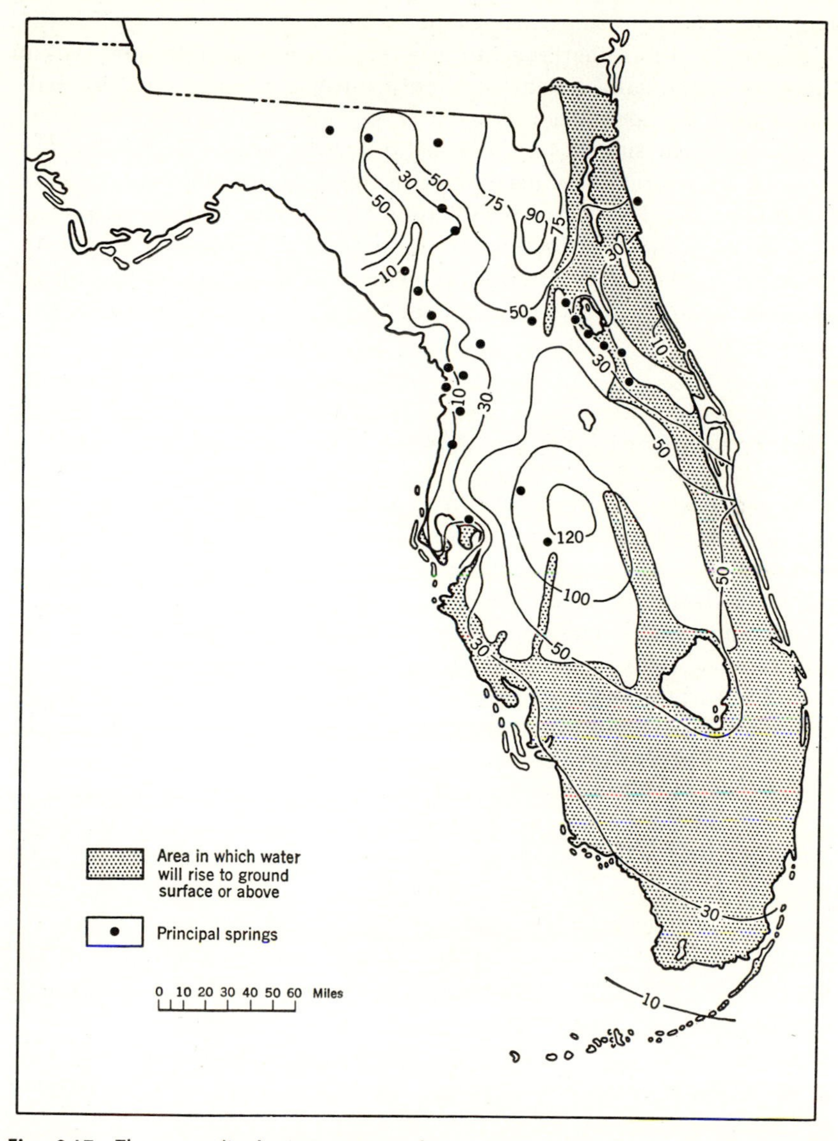

Fig. 9.17 The generalized piezometric surface of artesian water in the Ocala limestone in Florida during 1934. The contours represent the theoretical height in feet above sea level to which water would rise in a well. Within the dotted area water will rise to the ground surface or higher. Elsewhere, water fountains at the surface only in the localities of certain springs. (Generalized from Stringfield, *U. S. Geol. Survey Water Supply Paper* **773-C**.)

formerly flowing ample volumes stop altogether, and sufficient supply has been available only through pumping.

Many localities marginal to the sea have begun to observe an increase in the salt content of their fresh-water supply, as excessive withdrawal has permitted salt water to enter the aquifer. Invasion of the artesian supply in the coastal sectors is now in progress in Florida. The area in which ground water contains 100 or more parts per million of Cl ion corresponds approximately to the shaded portion of Fig. 9.17, and enlargement of the area is in progress. Salt-water pollution of the fresh-water supplies creates many serious problems, the solutions of which do not appear to be forthcoming in the immediate future.

Encroachment of Salt Water into Shallow Aquifers

Along coastal margins salt water moves in and out of inlets with the tides, invades bordering marshes and lagoons, and penetrates sedimentary formations, particularly those inclined inland and cropping out beneath the sea. The effect is for salt water to occupy pore space in sediments and sedimentary rocks inland from coastal areas. Under natural conditions in areas of abundant rainfall fresh water infiltrates from the surface and rests upon underlying saline waters. The actual relationship is one in which fresh water virtually floats upon the more dense salt water. Normal sea water has a specific gravity of 1.025 and hence is 41/40 heavier than fresh water. This difference in weight accounts for the tendency for fresh water to float upon heavier water. Similar stratification holds also in streams where a wedge of fresh water flows seaward displacing salt water even beyond the limits of the tidal channel. Correspondingly, wedges of salt water restricted to the bottom of the tidal stream channels invade some miles inland beyond the continental margins. Here, the effect of high salinity is to gradually pollute shallow underground water supplies particularly when the "blanket" of fresh water is being removed through wells.

In reclaimed areas such as the coastal polders of The Netherlands salt water is present initially at the surface. As rain falls on this reclaimed land fresh water begins to displace salt water by forcing the fresh-water–salt-water interface downward. Depression of this boundary surface is related to the densities of the two waters, i.e., a column of fresh water of 41 feet will balance a column of salt water 40 feet high. Thus, for every foot that fresh water extends above sea level it continues downward 40 feet below sea level. This is the condition which leads to the existence of a zone of fresh water beneath the surface of islands in the sea and enables plants to exist along their margins. In Fig. 9.18 a fresh-water zone is shown floating upon salt water in a small tropical island. Accumulation of fresh water has gradually caused a depression in the salt-water–fresh-water interface, and equi-

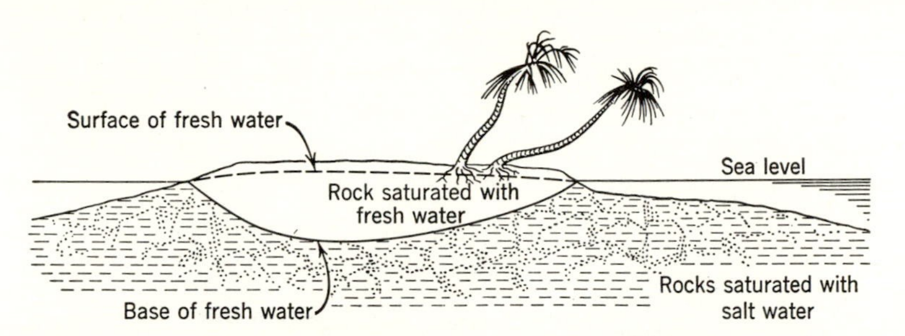

Fig. 9.18 A lens of fresh water superposed on salt water will cause depression of the surface of the salt water because 41 parts of fresh water will balance 40 parts of sea water. In the illustration the fresh-water table on the island stands at a maximum elevation of 2 feet above sea level; hence the thickness of the column of fresh water at maximum must be 82 feet (i.e., it exists to a depth of 80 feet below sea level).

librium has been attained where the water-table surface stands at a maximum height of two feet above sea level. Disregarding frictional effects this height of two feet above sea level indicates that the fresh-water column balances the salt-water column in the ratio of 41/40, and the fresh-water column must be 82 feet thick at its maximum.

In an open system where fresh and salt water are in direct connection the thickness of the fresh-water zone can be approximated by the equation

$$F = H\ (40 + 1) \tag{16}$$

where F = thickness in feet, of fresh-water zone.
H = height in feet of water table above sea level.

Assuming a flat topography and temporary cessation of rainfall recharge the fresh-water–salt-water interface will become a level surface as illustrated in Fig. 9.19*a*. If recharge to the fresh-water system should cease for a long time ionic diffusion from salt water into fresh water would eventually cause the fluids to reach uniform composition, and the boundary between the two fluids would disappear. Presumably this condition does exist in certain low-lying coastal islands which are unable to capture any rainfall. In most regions, however, there is a periodic addition of fresh water, and the salt-water–fresh-water interface is maintained. Local valleys may channel enough water to supply some to the subsurface. Later, after the water table has been built to a level higher than the valley floor flow of fresh water to the valley system takes place. With rise in water-table level the salt-water–fresh-water surface is depressed differentially to a pattern simulated by Fig. 9.19*b*. Peaks of salt water underlie stream valleys and are maintained at certain elevations by the height

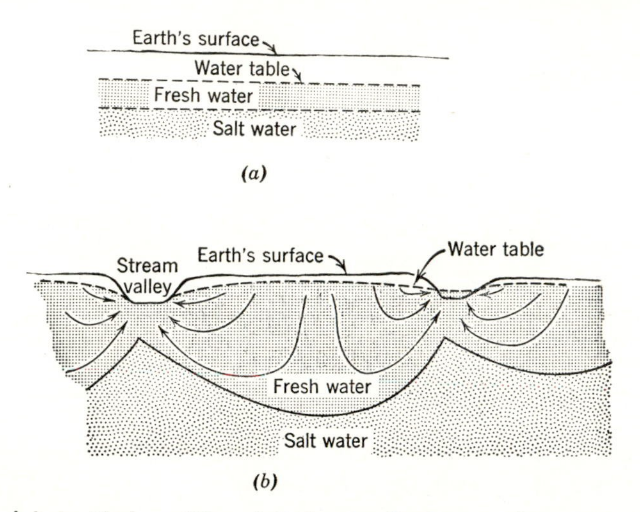

Fig. 9.19 (*a*) An ideal condition of fresh water floating on salt water in an area of temporary cessation of rain-water recharge. (*b*) The flow lines show the paths of movement of fresh water resting on salt water in a coastal area cut by streams. Lowering of the water-table level will result in emergence of cones of salt water beneath the stream channels. (From Parker, *U. S. Dept. Agriculture Yearbook*, 1955.)

of the water table in the upland areas. Flow from the zone of elevated water table toward the streams appears to follow paths indicated by the flow-line pattern in Fig. 9.19*b*, i.e., principally downward and upward along the salt-water–fresh-water interface to emergence along the valley floor. Any significant lowering of the water table in the upland area will result in general rise of the salt-water–fresh-water boundary and introduction of salt water into the streams.

In deeply embayed coastal areas where tides are high salt water flows long distances inland in the stream channels, and a brackish-water regime prevails in the stream. Here, a narrowly defined strip of salt water exists beneath the stream channel and separates the masses of fresh water on either side of the channel. The condition is shown in Fig. 9.20*a*. In the case illustrated a thick layer of fresh water has produced a depression in the salt-water–fresh-water interface, and large amounts of excess fresh water flow to the tidal channel diluting the normal marine water. Vegetation dependent upon fresh water may grow along the stream banks, and fresh water may be

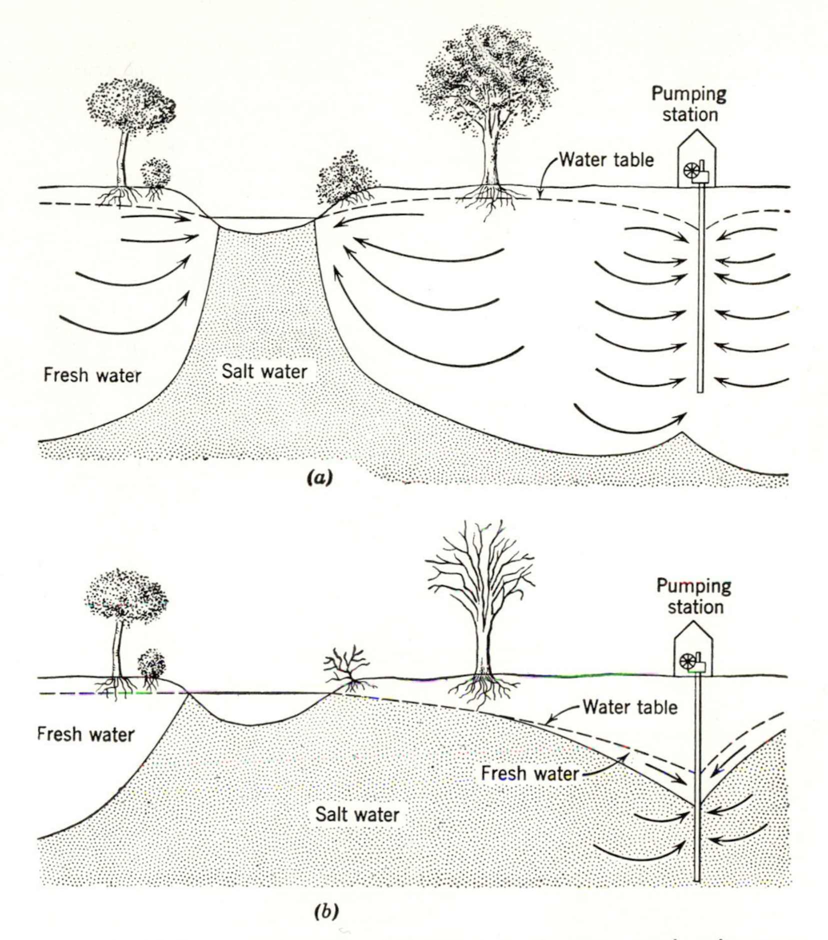

Fig. 9.20 An idealized represenation of conditions in the water supply adjacent to a tidal stream where a "cushion" of fresh water is thick enough to hold back salt-water invasion. (*a*) Disposition of salt and fresh water below tidal stream. The well is pumping small quantities of water, and the zone of fresh water is sufficiently thick to depress the level of salt water. (*b*) Rise of salt water as a result of excessive nearby pumping which has effectively removed all the fresh water. The well is pumping water of very high salinity. The vegetation is destroyed by salinity of water in contact with roots.

withdrawn from nearby shallow wells. Providing the volume of water withdrawn from the well does not exceed the normal fresh-water recharge the situation will remain static as illustrated. If, however,

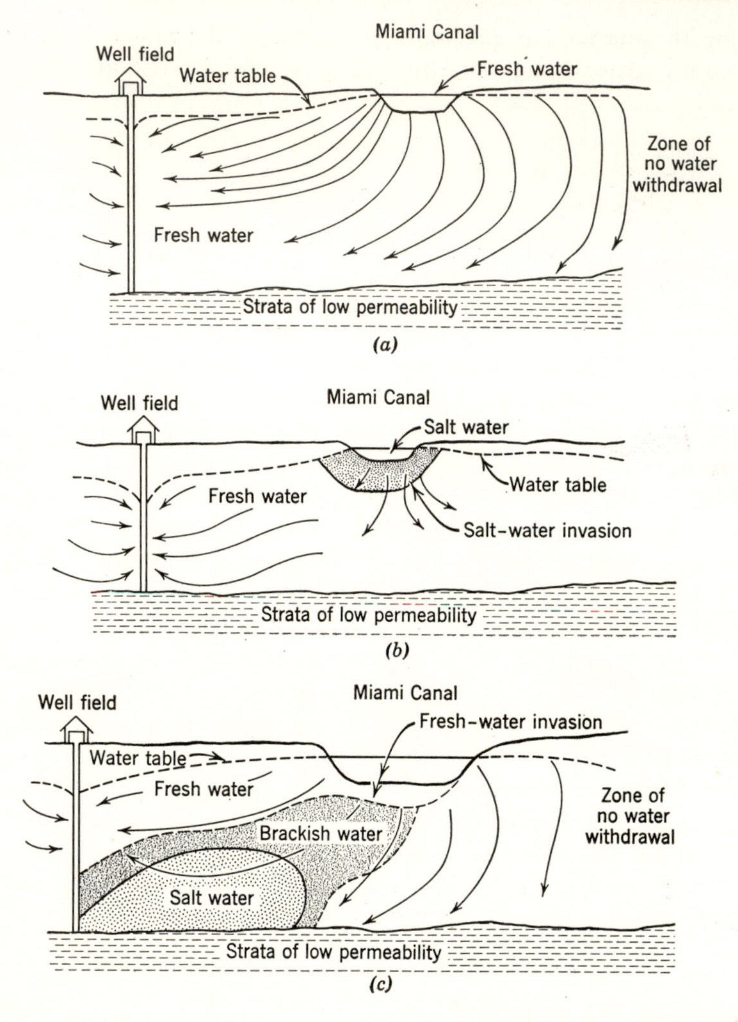

Fig. 9.21 Simplified illustration of condition in Miami, Florida, water supply (1939). (*a*) Fresh water from the Miami Canal (draining the Everglades) supplies water to adjacent wells. Flow of fresh water is from the canal into permeable strata adjacent to the canal. (*b*) A period of drought results in removal of fresh water from the canal, and salt water from the ocean moves inland. As the supply dwindles the fresh-water surface falls below the canal base and permits invasion of salt water into the underground system. (*c*) After the normal rainfall regime is renewed fresh water returns to the canal in quantities sufficient to force withdrawal of salt water. Fresh-water recharge to the ground is resumed. Pumping causes heavy salt water to flow downward toward wells and to contaminate the fresh-water supply. As pumping is continued the zone of mixed brackish water gradually replaces the water of high salinity; the system is flushed as increasing amounts of fresh water are introduced from the canals. (Modified from Parker, *U. S. Dept. Agriculture Yearbook*, 1955.)

excessive pumping is carried on the fresh-water "cushion" is removed, and salt-water invasion becomes pronounced (Fig. 9.20*b*). As the salt-water mass expands it eventually reaches the well, bringing fresh-water pumping to an end. Also, the roots of trees reach salt water, and such vegetation is destroyed. On the opposite side of the channel, however, the undisturbed lens of fresh water remains, and in this locality salt-water invasion does not occur so long as the equilibrium is not disturbed.

A very interesting record of salt-water encroachment has been provided by a circumstance affecting part of the Miami, Florida water supply.[12] Much of the fresh-water recharge to the underground supply takes place along the Miami Canal which drains fresh water from the Everglades region. Normal shallow ground-water flow to the well system is illustrated in Fig. 9.21*a*. Withdrawal of fresh water from these wells establishes the flow and maintains a general gradient in the water table from the canal toward the wells. In 1939 a pronounced drought restricted flow of fresh water from the Everglades region, and salt water from the ocean invaded the canal to a distance of about 10 miles inland. Salt water began to enter the underground supply as shown in Fig. 9.21*b* and to move downward and toward the well system in the direction of ground-water flow. Fortunately the drought was short in duration, and fresh water once more returned to the canal forcing out the tongue of salt water, and the underground supplies began to be recharged with fresh water. However, the salt-water mass continued its progress toward wells and contaminated the water supply as shown in Fig. 9.21*c*. By prolonged pumping the salinity of the water finally was reduced to tolerable limits, and normal conditions were resumed.

WATER IN CARBONATE ROCKS

Solution and Precipitation of Calcium Carbonate

Limestone and dolomite are subject to rapid solution by rain water only because the reaction involved depends primarily upon the amount of dissolved carbon dioxide. Carbon dioxide from the atmosphere is soluble in rain water, the degree of solubility being controlled by temperature. There exists, therefore, an equilibrium between car-

[12] The example selected is described by Gerald G. Parker, "The Encroachment of Salt Water into Fresh," *Water, The Yearbook of Agriculture* 1955, U.S. Department of Agriculture, p. 619.

bon dioxide, water, and carbonic acid in which the amount of carbonic acid formed varies inversely as the temperature. In turn, carbonic acid dissociates into its component ions, the principal relationship being illustrated by the following equations:

$$H_2O + CO_2 \rightleftharpoons H_2CO_3 \rightleftharpoons [H^+] + [HCO_3^-] \qquad (17)$$

The reaction proceeds to the right as more carbon dioxide is made available and to the left as carbon dioxide is withdrawn. Addition or removal of CO_2 from rain water is inversely proportional to temperature, the percentage of CO_2 decreasing as temperature increases. Solution of limestone depends principally upon formation of the HCO_3 ion formed through dissociation of carbonic acid. Indirectly, the concentration of HCO_3 ion affects solution or precipitation of calcium carbonate. The ionic equilibria concerned follow:

$$\underset{\text{(solid)}}{CaCO_3} \rightleftharpoons \underset{\text{(solution)}}{CaCO_3} \rightleftharpoons [Ca^{++}] + [CO_3^{=}] \qquad (18)$$

$$[Ca^{++}][CO_3^{=}] = K_{sp}(CaCO_3) \qquad (19)$$

$$\frac{[H^+][CO_3^{=}]}{[HCO_3^-]} = K_{HCO_3^-} \qquad (20)$$

$$\frac{[H^+]\,[HCO_3^-]}{H_2CO_3} = K_{H_2CO_3} \qquad (21)$$

The general behavior of these equilibria can best be understood by some simplification of the actual conditions. For purposes of the consideration here let us assume that temperature and pressure remain constant and that the only ions involved in the equilibria are those listed above. Suppose that initially the calcite in the limestone was in equilibrium with fresh water at sufficiently high temperature that, in effect, no carbon dioxide was present in solution. This water is stated as having an H ion concentration of 10^{-7} mole per liter (i.e., $p\text{H} = 7$). As the water cools to 25°C the relationship in existence between solid calcite and Ca and CO_3 ions is that of eq. 18, and the equilibrium controlling the concentration is that of eq. 19 as follows:

$$[Ca^{++}][CO_3^{=}] = K_{sp\,(CaCO_3)} = 4.8 \times 10^{-9} \qquad (19)$$

Concentration of the Ca and CO_3 ions is equal and has a value of approximately 7×10^{-5} (i.e., the square root of 4.8×10^{-9}).

Rain water saturated with CO_2 is permitted to enter, mixes with the solution in contact with calcite, and establishes the equilibrium of eq. 21. Assume, also, that sufficient dissociation of H_2CO_3 takes place to increase the H ion concentration from 10^{-7} to 10^{-6} (i.e., by a factor of 10). In order to increase the H ion concentration from 10^{-7} to 10^{-6} enough H_2CO_3 dissociates to provide 9×10^{-7} moles per liter of HCO_3 ion. (1×10^{-7} original H ion + 9×10^{-7} added H ion = 1×10^{-6} H ion). The equilibrium as specified by eq. 21 is now represented by substituting the proper values, namely $H^+ = 10^{-6}$, $HCO_3^- = 9 \times 10^{-7}$, and $K_{H_2CO_3} = 4.3 \times 10^{-7}$

$$\frac{[1 \times 10^{-6}][9 \times 10^{-7}]}{[H_2CO_3]} = 4.3 \times 10^{-7}$$

$$[H_2CO_3] = \frac{9 \times 10^{-13}}{4.3 \times 10^{-7}} = 2.1 \times 10^{-6} \tag{22}$$

Equation 20 is expressed as:

$$\frac{[1 \times 10^{-6}][CO_3^=]}{9 \times 10^{-7}} = 4.7 \times 10^{-11} = K_{HCO_3}$$

$$[CO_3^=] = \frac{[9 \times 10^{-7}][4.7 \times 10^{-11}]}{10^{-6}} = \frac{4.2 \times 10^{-17}}{10^{-6}} = 4.2 \times 10^{-11} \tag{23}$$

and, at saturation:

$$[Ca^{++}][CO_3^=] = K_{sp} = 4.8 \times 10^{-9} \tag{19}$$

$$[Ca^{++}] = \frac{4.8 \times 10^{-9}}{4.2 \times 10^{-11}} = 10^2 \text{ (approximately)} \tag{24}$$

Ca ion concentration increases from 7×10^{-5} to 10^2 when the pH of the solution changes from 7 to 6. Indeed, approximately 10 million times as much Ca ion is required to saturate a solution when the pH is 6 as when it is 7. For this reason large amounts of calcite are dissolved by the average rain water of H ion concentration of 10^{-6} as opposed to the very slight amount of solution of calcium carbonate which occurs when the water is neutral (i.e., $p\text{H} = 7$).

The reader must also remember that in the weathering environment the system discussed is not closed. With each successive rainfall new water unsaturated with Ca ion is brought into contact with the calcite of the limestone, and spectacular solution of limestone proceeds. Elsewhere, wherever these waters carrying large amounts of Ca ion in solution are neutralized, principally by removal of carbon dioxide from solution through increase of temperature, precipitation of $CaCO_3$ must occur. As the H ion concentration decreases to 10^{-7} or less the equilibria discussed are reversed and Ca ion must be lowered from the order of 10^2 to 10^{-5}. This change demands removal of

Ca ion from solution and precipitation of the carbonate as that solubility product is exceeded.

Cave Development

Percolation of ground water through limestone must take place principally along joints and bedding planes because, normally, the rock itself is low in pore permeability. Solution along the walls of such fissures results in their enlargement, and eventually they become prominent channelways. These carry water from the surface through an elaborate subterranean system to its eventual return and discharge into some stream system. (See distribution of large springs in Fig. 9.17. Many of these are loci of important flow from the Ocala limestone.) As the water moves along certain preferred bedding planes solution enlarges favorably situated cavities. Eventually these become caverns, sometimes of enormous proportions, interconnected by hundreds of miles of channelways. Most complete development of caverns is to be found in the vadose zone. Here, infiltrating ground waters are, as yet, undersaturated with Ca and CO_3 ions, and the H ion concentration is relatively high. Beneath the water table where openings are filled with water solution progresses at a much slower pace inasmuch as saturation of the solution has been attained, and the system in this sense is closed. At still greater depths actual precipitation of $CaCO_3$ may be in progress in small openings and intergranular space. Cave formation, therefore, is a surficial phenomenon. Some ancient caverns currently filled with water are found hundreds or even thousands of feet below the present surface. The geologist interprets this condition as indicating that at the time of cave formation such rocks were much closer to the earth's surface and within the vadose zone.

In certain areas cavern development is particularly rapid. These are localities of high rainfall in generally cool or cold climates where dissolved CO_2 in rain water is maintained at a high level. In tropical regions much less CO_2 can be held in solution, and caves are not formed at such a rapid rate although given sufficient time and heavy rainfall very large caverns do develop.

As caves increase in dimension the roof may weaken and collapse, sometimes breaking through to the surface. Such collapsed structures are circular depressions and are designated by the term *sink*. Sinks form important funnel accumulators for surface waters fed directly into caverns. During times of rain the cavern system locally may be filled with water moving under strong velocity, and active mechanical erosion of cave walls ensues. In many areas this process is considered to have been extremely important in enlarging the interconnecting channelways between caves.

Limestones which contain an important percentage of clay as part of their composition yield this clay as an insoluble-residue fraction as solution progresses. The clay residue (*cave earth*) gradually accumulates along the cave walls, floors, and channelways. The low permeability of this material tends to reduce the rate of solution, but in those localities where surface waters stream through such channels the insoluble fraction is swept away, and the walls are left bare and smooth.

Depositional Features in Caves

A spectacular aspect of some caves is the *dripstone,* consisting of stalactites (suspended from the cave roof) and their counterparts, stalagmites, along the cave floor. Such dripstones along with drapery-like deposits lining cave walls are the principal features representing calcium carbonate precipitation inside the cave. The deposits result when the solubility product of $CaCO_3$ is exceeded and it is forced out of solution.

Presence of dripstone in caves is an important indicator of depth to the ground-water table inasmuch as dripstone does not accumulate in those openings which are filled with water. Caves now filled with water but which also contain dripstone indicate that there has been a relative rise of ground-water level since the dripstone was deposited.

Limestone Landscapes

Well-drained uplands of limestone or dolomite often are barren of soil, and bare rock exposures display fluted and pitted weathered surfaces typical of carbonate rock. Sometimes the fluting is strongly pronounced, and a very irregular or serrated surface exists. Differences of 10 to 15 feet between trough and pinnacles are not uncommon in areas of heavy rainfall. Crevices with overhanging sides expand into caves, and collapsed structures leave steep-wall depressions. Large-scale failure of cave roofs produces a "pock-marked" surface very typical of many limestone areas. This sink-hole topography, often also catalogued as *karst,* is distinguished on air photographs or topographic maps from areas of glacial out-wash by the occurrence of streams which terminate in sinks and also by the large size of some of the circular depressions.

Collapse of adjacent caves frequently leaves an arch structure between the two portions and a feature recognized as a natural bridge. These are particularly spectacular where underground stream development has been pronounced and a "graded" system has been reached by the interconnected caves.

Where the water table is near the surface sinks contain small ponds whose level may rise and fall with seasonal changes and after rainfall. Others are plugged with clay, and the water of the pond is perched above the mean water-table level. Interconnection of subterranean and surface systems of drainage is typical, and large springs often are associated with the emergence of an integrated cavern drainage. Some of the springs are enormous and may actually mark the beginning of a river. Elsewhere, a sink is the terminus of the exposed part of a stream drainage, and the entire flow is directed underground. These are conditions which lead to wholesale bacteriologic contamination of water. Disease-bearing organisms such as typhoid bacilli may be carried long distances from their source to their eventual introduction into a community water supply.

Karst areas are characterized by a minimum of surface drainage. Streams are few, usually short, and there is no semblance of the well-integrated system so typical of areas of heavy runoff. Where rainfall is heavy this water must rapidly become part of the subsurface drainage where its influence is exerted in eroding channelways and walls of caves. Although erosion is manifested principally by solution the abrasive effects of sediment-loaded water are known to aid in the removal of carbonate rock and cave earth. Through these processes the near-surface zone is, in a sense, being eroded from the interior toward the surface, and ultimately the extent of this erosion is reflected by the surface topography.

Selected Supplementary Readings

Bretz, J Harlen, "Caves of Missouri," *Missouri Geological Survey and Water Resources,* Vol. 39, 2nd Series, 1956, pp. 1–37. A description of cave-forming processes and features.

Hendry, C. W., Jr., and Lavender, J. A., "The Progress of an Inventory of Artesian Wells in Florida," *Florida Geological Survey, Information Circular No. 10,* 1957, pp. 1–12. Illustrates the conditions of the Floridan aquifer.

Meinzer, O. E., "The Occurrcnce of Ground Water in the United States," *U.S. Geological Survey, Water Supply Paper* 489, Government Printing Office, 1923. An old but valuable discussion of the principles of ground-water occurrence.

Muller, S. W., *Permafrost or Permanently Frozen Ground and Related Engineering Problems,* J. W. Edwards, Inc., 1947. A condensation of the conditions of permafrost and the engineering problems imposed.

Tolman, C. F., *Ground Water,* McGraw-Hill Book Co., Chapter XV. A summary of conditions controlling springs.

Veatch, A. C., et al., "Underground Water Resources of Long Island, N.Y." *U.S. Geological Survey, Prof. Paper* 44, 1906, Chapter II. An excellent example of a simple artesian system.

10

Wind deposits

FACTORS CONTROLLING WIND DEPOSITION

Movement within the envelope of air which surrounds the earth simulates that of water in the great oceans. Strong planetary wind systems exist as analogues to the great oceanic currents. Important wind streams encircle the globe confined to certain latitudes as do their fluid counterparts the great oceanic currents. Violent local masses of whirling air are similar to strong tidal currents emerging through narrow-necked bays. Turbulence in the atmosphere carries particles far above the land surface and transports them elsewhere.

Dune scene, Cape Cod, Massachusetts.

Finally, such small particles settle at a rate approximately determined by Stokes' law.

Inasmuch as water and air are so similar in behavior how do they differ, and to what extent do their erosive and depositional aspects vary? The most obvious distinctions between them as transporting media are their relative densities and viscosities. The density ratio of air to quartz is approximately 1/2,000, whereas for water and quartz the ratio is about 1/2.6. This great difference in ratio of densities is responsible for marked differences in behavior of particle transport. The momentum of a sand grain moving at the same speed as the surrounding water is about 1250 times greater than when moved by air whose velocity is the same as that of water. A current of air, in order to impart to a solid particle a velocity equal to that delivered by a fluid moving with specified velocity, must lose momentum equivalent to more than 1000 times that lost by the fluid. At the base of the air column reduction in the velocity of air by virtue of the required momentum exchange needed to move the solid particle is sufficiently large to produce a very strong drag.

Drag Effect on Wind Velocity[1]

The nature of the irregularity of the surface of the air-land interface is important in controlling the height at which the drag becomes marked. Figure 10.1 is a plot showing the relationship of the logarithm of height above the air-land interface and the wind velocity. The data from which these relationships were developed indicate that for any selected surface the height above the air-solid interface at which drag is very effective is independent of the wind velocity. This is a very important observation as it reveals that no matter how strong the wind velocity a threshold height exists below which its velocity approaches zero.

Qualitative observations do indicate that winds are more effective transporting media over some surface than others. Every schoolboy knows that wind velocity is less in the lee of a large boulder or cliff. The same holds for small particles which armor the earth's surface. A diagrammatic representation appears in Fig. 10.2 in which is plotted a wind of uniform velocity moving across three surfaces of different roughness. As the roughness increases so likewise does the height

[1] Much of the following consideration of drag effect on wind velocity is based upon the concepts presented by R. A. Bagnold, *The Physics of Blown Sand and Desert Dunes,* Methuen and Co., 1941.

above the air-solid interface at which wind velocity approaches zero. A good approximation of the height above this interface at which the

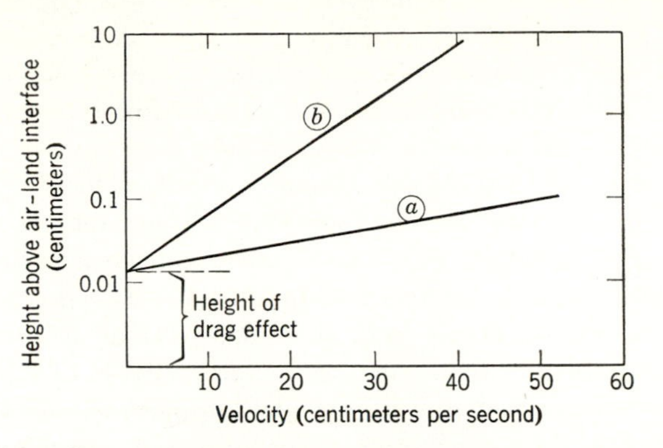

Fig. 10.1 Relationship of ideal conditions of two winds, *a* and *b*, of two different velocities, showing that over the same surface the height is fixed at which the drag reduces wind velocity to approximately zero. (Modified from Bagnold, Methuen and Co., Ltd., 1941.)

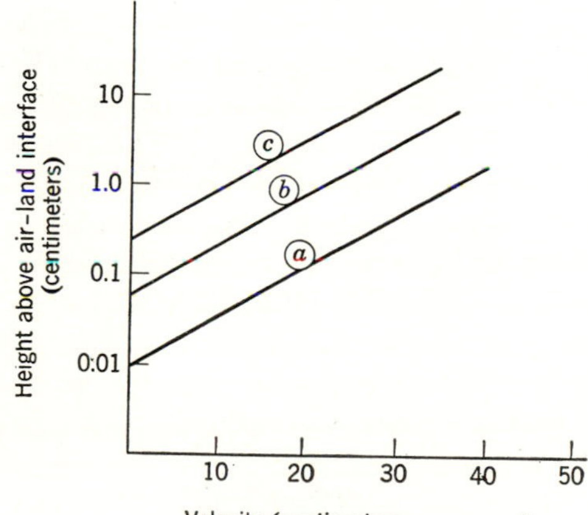

Fig. 10.2 Diagrammatic relationship of ideal conditions of wind of the same velocity moving over surfaces of three different magnitudes of surface roughness. Wind *a* is moving across the surface of minimum roughness, wind c across the surface of maximum roughness. Note that the height at which zero velocity is attained is highest above the interface in the case of the surface of greatest roughness. (Modified from Bagnold, Methuen and Co., Ltd., 1941.)

wind velocity diminishes to approximately zero is obtained by dividing the average diameter of the solid particles by 30. For example, if the average particle diameter coating a surface is about 3 mm (i.e., granules) wind velocity approaches zero at 0.1 mm above the interface. If, however, the surface is armored with boulders whose average diameter is 300 mm wind velocity is ineffective at a height up to 1 cm above the interface.

Conversely, if the wind velocity (centimeters per second) is plotted against the log height (centimeters) a straight-line relationship exists. Extension of the line of such a plot to its intersection with the ordinate of zero velocity indicates the height of the zero-velocity layer. This value multiplied by 30 is the approximate diameter of the average particle coating the surface at the locality of measurement.

The geologic significance of the relationship between surface roughness and zero wind velocity should become increasingly well defined as the student reads further in this chapter. It should be apparent already that removal of fine debris such as particles of silt and clay is possible only from very smooth surfaces. On gravels removal of sand is possible, but silt and clay particles tend to remain. Extensive surfaces coated by weathered rubble of cobble and boulder size are very likely to accumulate small particles. As sand and finer sizes are moved into the site they become trapped in the zone of zero wind velocity. Such an area of boulders can become gradually buried by fine debris until an equilibrium condition is reached. Exceptions appear when small but violently turbulent masses of air (dust devils) sweep across a surface coated with fine debris. Such eddies have a strong upward gradient, and fine silt or clay-sized particles are removed when laterally moving wind is unable to accomplish the same result.

The shear force, or drag per unit area of ground surface, as the wind moves across this surface can be expressed as:

$$\tau = \rho \bar{V}^2 \tag{1}$$

where τ = shear force per unit area.
ρ = density of air.
$\bar{V}$ = a velocity directly proportional to the rate of increase of wind velocity with the logarithm of the height above the air-solid interface.

Indirectly, $\bar{V}$ is a measure of the amount of drag which the land surface exerts against the wind, and, hence, its value as a velocity varies. Figure 10.1 illustrates an example of two winds whose rate of velocity increase differs with an increase in log height above the ground surface. For wind *a* the velocity for any selected height above the ground is not only considerably greater than for wind *b*, but the rate of velocity increase with each unit of increase in height is greater than wind *b*. $\bar{V}$ for wind *a* is greater than $\bar{V}$ for wind *b*. $\bar{V}$ is appropriately called *drag velocity,* and its value is determined by the tangent of the angle of slope of the line such as a or b times a factor of proportionality which is 1/5.75.

To find the value for τ (i.e., surface drag per unit area) determine the wind velocity at each of any two heights. Plot these velocities on semi-logarithmic graph paper on which the logarithm scale is used for heights and the arithmetic scale is used for velocities. Connect the two points with a straight line. Select velocities at two heights, one of which is 10 times the other.

Then,
$$\bar{V} = \frac{V_2 - V_1}{5.75\,(\log h_2 - \log h_1)} \tag{2}$$

and
$$\log h_2 - \log h_1 = \log 10 = 1$$

and
$$\tau = \rho\left(\frac{V_2 - V_1}{5.75}\right)^2 \tag{3}$$

where V_2 = higher wind velocity at height h_2.
V_1 = lower wind velocity at height h_1.

Sand-Grain Motion

Once sand is activated and begins to move through the air it does so by saltation. Wind velocities for which $\bar{V}$ is large (and consequently so is τ) are those in which the length of the leap is great. As a grain of sand moves forward the path described is asymmetric. Most of the grain's forward motion is accomplished when it is traveling near its greatest height above the ground, whereas its initial rise is nearly vertical. Such differences in direction of motion indicate that the grain acquires its maximum forward momentum after its initial rise. During the forward motion the grain is exerting its greatest drag upon the wind. Moving sand, therefore, alters the state of the wind and tends to increase the height above the ground where zero wind velocity exists. As an illustration suppose a wind to be blowing across a smooth dried lake bottom surfaced with hard clay. The level of zero wind velocity is very close to the air-clay interface. The same wind blowing across a wet sandy surface somewhat rougher than the clay would show a somewhat greater height above the surface where zero wind velocity exists. Over dry sand of the same average particle diameter the limit of zero velocity would be even higher by virtue of the drag effect of the moving sand.

So far we have considered the average sand grain as representing grains of uniform diameter. Now suppose the average diameter to be kept constant but the range of size distribution increased. Small grains will upon falling strike large grains and rebound from their surface rather than to form a crater in the neighboring grain cluster. The average upward velocity is increased thereby, and the height and distance of saltation also are increased. Addition of grains in the air stream tends to increase the total drag on the wind, and the level of zero velocity is raised somewhat higher than when grains are uniform in diameter.

Height of Transportation of Solid Particles

Unlike stream transportation particles moved as a part of the traction load are not large in dimension. Reports exist of stones 2 inches or so in diameter being moved by violent storms in Turkestan. A wind of measured velocity of 40 miles per hour 5 feet above ground level was observed to roll granules 5 to 10 mm in diameter and to cause granules 2 to 3 mm in diameter to move by saltation. Coarse material, therefore, tends to remain behind as fine debris is carried away. In a qualitative way this is illustrated in Fig. 10.3 in which are shown terminal velocities of particles settling through air and water. For small particles, i.e., diameter less than approximately 0.1 mm, Stokes' law controls velocity of fall, but for larger particles the impact law governs the rate of fall. (See p. 125.) This is illustrated by the departure from the straight line of the equation of Stokes' law. Impact of the "fluid" against the settling particle involves its inertia as well as its viscosity, and the inertial forces of the fluid play an increasingly important part on the speed of settling as the particle increases in size. Note that the slope of the settling curve in air is similar to that of water; hence, the same forces can be considered as being active in both media.

Primarily because of the density differences settling velocities through air are greater than through water. For particles smaller than 0.5 mm the ratio of settling, air to water is approximately constant. (See Fig. 10.3.) For larger sizes the two curves have different slopes, and the rate of settling through air rapidly increases over that of water becoming more than 1000 times as great. This departure of the two curves is ascribed to the increased importance of the inertial difference between the two media as the impact law controls the rate of fall.

An additional interpretation can be drawn from the data of Fig. 10.3. The great increase in settling rate through air of particles of coarse sand or larger indicates that grains exceeding 1 mm fall so rapidly that they constantly tend to remain near the earth's surface even though they may be lifted into the air by exceptional storms.

Height of travel of solid particles is shown, in a general way, in Fig. 10.4. Coarse sand and granules hug the earth's surface moving

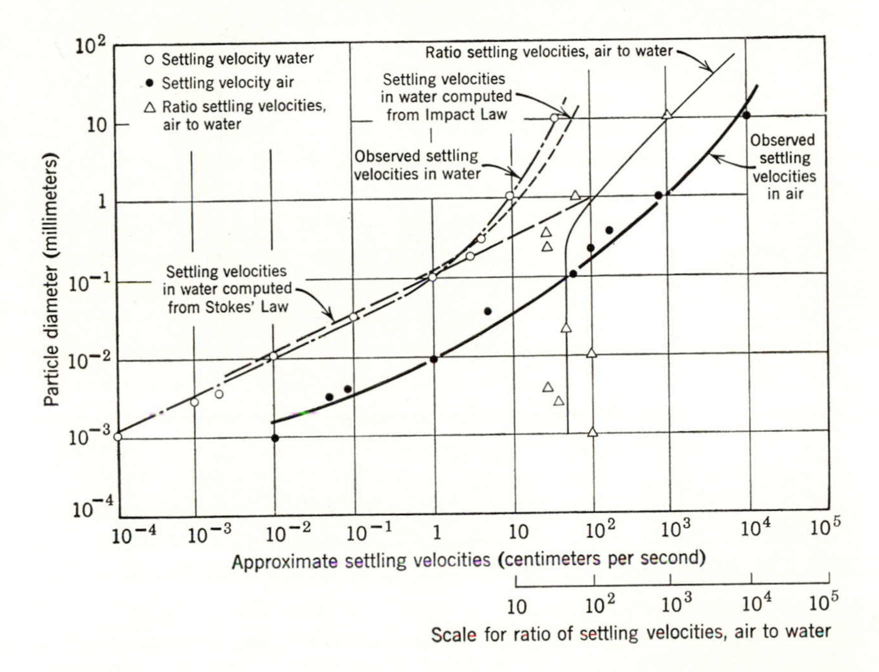

Fig. 10.3 Comparison of settling velocities of particles through air and water. Note the similarity of the slopes of the two curves. Stokes' law holds for particles with diameters less than 0.1 mm, but for larger particles the settling velocities are controlled by the impact law. For particles up to about 0.5 mm diameter the ratio of air-water settling velocities remains uniform, but for larger sizes the ratio increases logarithmically.

in a narrowly restricted zone. Fine sand is swept higher into the air but tends to remain concentrated within a few feet of the ground. Very fine sand and silt is carried to progressively greater heights of several thousands of feet. For particles of these diameters note that the logarithm of the height of transport bears a straight-line relationship to the log of particle diameter. As the colloidal dimensions of clay particles are reached this relationship no longer holds, and minute particles are carried by turbulent currents of a thunderhead into the stratosphere. Here, under the driving force of earth-circling wind streams such particles are carried on a global journey and eventually settle downward into the lower atmosphere.

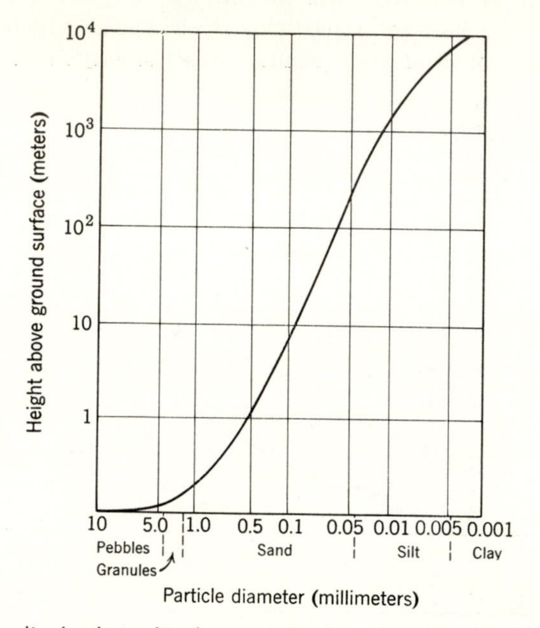

Fig. 10.4 Generalized relationship between the size of a particle and the height at which it can be carried in the atmosphere.

Distance of Particle Transport

Limitations to the influence of the wind on the earth's surface reach a minimum in those regions where there are few obstructions to air movements. In forested belts the damping effect of trees reduces wind velocity at the ground level to minimum values. Also, in such areas the land surface is likely to be covered by a blanket of decaying vegetation and abundant small plants whose roots are effective in retaining mineral grains in position. Similarly, in grassland areas the mat of grass prevents removal of fine debris. Except for local spots where grass is destroyed or forests removed wind action is ineffective. In these local areas eventual equilibrium is attained between removal of fine debris and vegetation growth.

Within the belts of Trade Winds are extensive areas where rainfall is so low or concentrated within such a short period of the year that vegetation virtually is unable to mantle the soil. Such desert areas are the principal sites of wind erosion. From these areas dust is being removed and carried great distances, or sand is being driven into mounds of dunes.

Large particles such as granules and sand tend to linger near the source of supply. Occurrences of dunes, therefore, are to be associated with areas of local sand supply. Exceptionally long transportation may be anticipated in some of the great sand deserts, but even

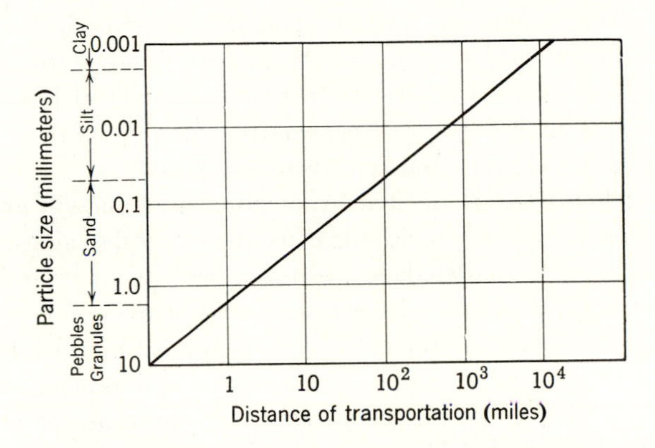

Fig. 10.5 Conceptual relationship between particle size and the distance which it may be transported from source. Selective transportation tends to segregate size groups such as sand or silt.

in such regions it is doubtful that dune migration has exceeded 100 miles. Regions of extensive occurrence are to be interpreted as resulting from the existence of many local source centers. Coalescence of local sand belts produces the impression of long distance sand migration.

Silt may be carried at great heights and, hence, can be expected to be transported far from the sources of supply. This is illustrated in Fig. 10.5 which presents a first approximation of the relationship between particle size and distance of transport from a single source area.

GRAVEL DEPOSITS

Lag Gravel

In most all dry regions local areas exist where wind turbulence is strong. Despite the retarding influence of drag, silt and clay particles are carried off. The effect is to produce a concentrate of coarse particles which wherever pebbles and cobbles occur results in a surface veneer known as *lag gravel.* Certain lag gravels are found where residual soils are thin, and the concentrated fraction represents the broken subsoil which rests upon the bedrock.

Lag gravels are common in regions of alluvial fans where by previous water deposition pebbles and cobbles are abundant in layers. As wind action removes the sand, silt, and clay-sized particles among which the larger fragments are embedded the latter becomes a concentrate which effectively retards additional removal of fine debris. A common condition, therefore, is to find the lower slopes of alluvial fans armored with lag gravels, especially when the fan is no longer receiving additional contributions from streams.

Pebbles and cobbles are blasted by swirling sand and silt and abraded to characteristic shapes. Sand blasting develops flat sides and sharp edges. Often such sharp-edged pebbles are cut on more than one side, and pyramidal shapes (*dreikanter*) are not uncommon. Coarsely crystalline rocks often are pitted and individual mineral grains etched as the rock surface is worn according to the differential hardness of the minerals. Monomineralic rocks such as limestone often are fluted by directed sand blast. The total effect is to produce fragments of very low roundness and to polish the surface oftentimes to a glazed finish. Such fragments are known collectively as *ventifacts.*

SAND DEPOSITS

Particle Distribution

Sand removed during concentration of lag gravels is swept along near the earth's surface periodically coming to rest behind a larger object or wherever the wind velocity is lowered below its competency to transport the particle. This sporadic movement concentrates particles of similar dimension and weight and permits winnowing of silt. Dune sand tends to be uniform in particle mean size, and the standard deviation of this size is low. (See Fig. 2.8.)

As sand shifts away from the source of supply gradual dissipation must result. Dunes tend to become smaller as the area of sand distribution is increased. Inability of the wind to carry sand much above the rock-air interface forces sand to move into and across areas which may be unfavorable for dune accumulation.

Very extensive areas of sand accumulation are restricted to desert areas of long duration where sand supply is the result of weathering of broad expanses of outcropping sandstones. Wherever sand is being supplied from playas, beaches, or ephemeral streams which periodically freshen their sand supply large. although not necessarily widespread, dunes are to be expected. (See Fig. 7.28.)

Accumulation of Sand

In areas of very active sand accretion particles lodge between and behind pebbles coating the surface until the eventual condition is one in which the pebbles are buried completely. Additional sand is added, and the patch of dune increases in dimension splaying out as an apron beyond the position of prime accumulation. Some of these deposits are built by obstructions in the wind path. Bushes, boulders, and cliffs are such obstacles, and the accumulated sand is dependent upon their existence. Position of the sand is controlled by the obstacle, the removal of which will result in dissipation of the accumulated sand. Such sand deposits of fixed position are known as *sand shadows.* (See Fig. 10.6.)

A true dune is a mound usually separated from its neighbors and possessing windward and leeward slopes of fundamentally different inclination. Dunes exist as lone individuals, as colonies, or in chain-like ridges and are not dependent upon fixed obstacles for their development. As they shift their position from place to place they maintain their characterizing slopes although their over-all outline may change with vagaries in wind velocity.

Normally, the windward slope is gentle, rarely exceeding 10 degrees of inclination from the horizontal. In contrast the leeward slope is approximately 30 degrees from the level surface and varies only slightly from dune to dune. Lee slopes tend toward uniformity because they represent slip faces maintained as the angle of shear dictates the slope of a pile of sand. Sand swept up the windward slope is blown or rolled beyond the crest. Here it gathers as a mound whose steepness is maintained by the angle at which sand will shear under its own weight and slide downward toward the base.[2]

Deposits of Fixed Position

Distinction is to be drawn between a sand shadow which is an accumulation of sand piled immediately downwind of an obstacle and

[2] The angle of slope which is maintained by loose granular rock debris is known also as the *angle of repose* and varies according to the particle size, particle shape, and moisture content of the accumulation.

a *drift* formed in the lee of a gap in a cliff. (See Fig. 10.6.) Sand blowing toward such a gap may gather in piles on either side windward of the rock mass. The sand is forced into the gap inasmuch as it is unable to accumulate at the vertical contact between the rock

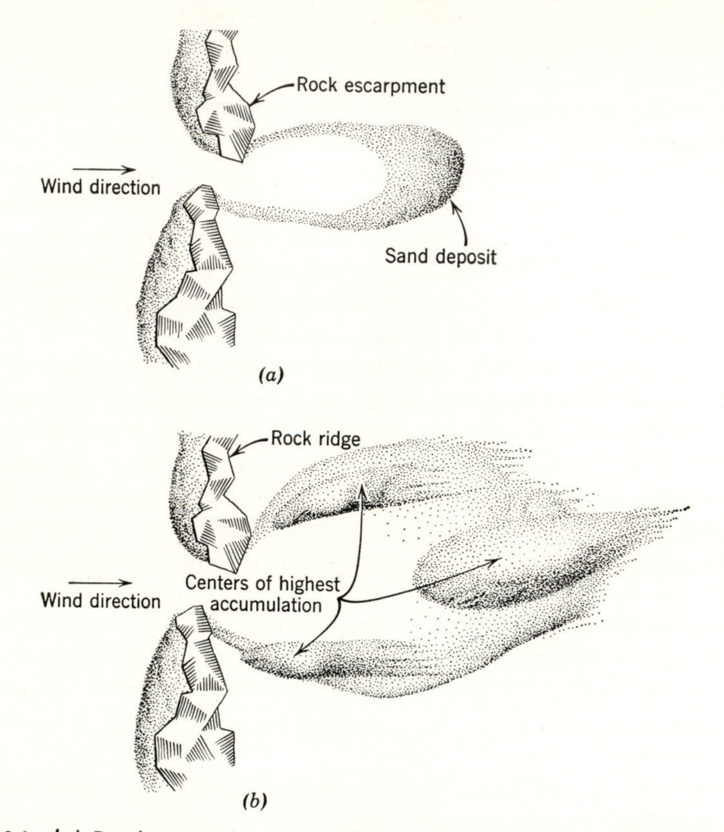

Fig. 10.6 (*a*) Development of a sand shadow. The wind drives the sand along a rocky barrier to a gap where wind velocities are accelerated by restriction. Once through the gap, wind velocity falls and the sand is dropped. (*b*) A steady-state condition of sand drift where loss of sand from a large surface equals the amount furnished through the gap.

wall and the opening. Sand which arrives over a wide frontage drifts toward the gap through which it is concentrated. By this process more sand per unit area is delivered to the windward side than over the adjacent level surface. Also, winds are stronger through the gap than in the neighboring area because of the funneling and confining influence of the rock barrier. Once through the gap the restricted air current spreads and loses velocity. Its transportational energy declines and sand is dropped.

Dunes

Range in velocity of the prevailing wind permits accretion of excess sand blown from sand shadows, drifts, or other sources of more or less fixed sand. Once the small mound is established distribution of drag velocity along the windward surface is important in controlling the growth of the embryo dune. Wind velocity is highest on the windward side near the crest of the dune and is lowest near the ground on the lee side. If wind velocity is low sand is removed from the highest section of the windward side near the crest of the dune and carried to the lee side where it is dropped. Near the dune margins the sand has no movement. Under such conditions the dune mound tends to flatten to a sheet inasmuch as no sand is furnished to the windward side, and no general accretion occurs.

Increase in wind velocity creates movement of sand in front of the windward side. Sand is added to the dune, and if the total sand left behind on the lee side exceeds that carried off dune growth continues. Air drag is strong along the sand surface because large amounts of sand in motion have the same effect as a rough surface, and air speed is reduced near the base of the upward rise of the dune front. Some sand is added to this sector, but much is carried over the dune crest toward the lee side. During strong winds of uniform direction the dune is built upward and correspondingly increased in surface area.

Development of the Lee Slope (Steep Face)

At the very summit of the dune deposition approaches zero as sand is in transit across this section. Also an effective zero value of deposition must be reached somewhat beyond the lee margin of the dune. Between these two extremes sand deposition must reach a maximum. In fact, this point is nearer the summit than the base. At this position sand accumulation results in prominent build-up and steepening of the lee slope. Eventually, the slope becomes unstable, a shear plane develops of nearly 30 degrees, and excess sand slides to the base of the dune. The shear surface appears to disrupt the former air streamline pattern, and eddies form. In the zone of eddies lowering of forward velocity prevails, and much sand falls on arrival from the windward side. Additional sand accumulates, and a new shear plane forms in a position farther downwind. By this mechanism the dune is advanced and is started on its migratory journey.

Dune Shapes

Dune shapes are responses to wind velocity, constancy of wind direction, and sand supply. The principal control, however, is uniformity of wind direction. In regions of steady prevailing winds dune forms are recognized as belonging to three major types. Elsewhere, where winds are more erratic in their directional distribution dune shapes are modified in various ways. The fundamental forms are:

a. Transverse.
b. Barchan.
c. Longitundinal.

Modified varieties are:

a. Blowout shapes.
b. Irregular or complex amoeboid outlines.

Despite the pattern of outline each is asymmetric in cross section showing a gently inclined windward slope and a steeper lee slope generally characterized by a slip face. Slopes may be modified and rendered much more symmetrical particularly when the dune becomes fixed by grass or forest cover. The slip face gradually disappears under the influence of creep and soil-forming processes.

The transverse dune. Low-lying sandy shorelines constitute line sources of sand blown inland under prevailing winds. It is commonplace that this sand gathers into dunes in a series of parallel ridges arrayed with long dimension perpendicular to the wind direction, hence their name *transverse*. Where these dunes are not restricted to a narrow belt they have symmetry and form akin to their smaller counterparts, the ripples on dune surfaces. (See Fig. 10.7.) Wind requirements for the development of a transverse dune are gentle to moderate (i.e., less than 25 miles per hour) blowing with strong prevailing tendencies across a narrow direction band. Over extensive areas of loose sand this is the primary dune form. Such a shape does not appear to be stable, and all indications suggest that its occurrence is limited in areal distribution. So long as an adequate sand supply is available, winds remain low, and vegetation absent, the transverse dune can persist, but under most circumstances a transformation in shape occurs. This is especially true when the dune ridge reaches an unbroken length of several miles, whereupon it no longer can remain as a single individual.

An unstable condition is established when wind velocities fluctuate between strong and gentle such that at times a wind of low sand-carrying capacity near the windward margin of the dune has sufficient velocity to cause erosion at a low wind gap. The result is a dune feature known as a *blowout* in which a segment of the transverse dune is moved downwind (see Fig. 10.8). As the gap becomes enlarged additional sand drifts into the blowout area from the margins of the

Fig. 10.7 Belts of transverse dunes along the coast line of Point Sal, Santa Barbara County, California. Note that in the foreground, under the strong wind drive, the sand is currently accumulating as sand shadows downwind from small vegetation-fixed dune masses, but further inland the transverse shape is assumed as the sand is dissipated from the "shadow." (By permission from Spence Air Photos, Los Angeles.)

transverse dune, and growth of the blowout is sufficiently accelerated to cause its lateral movement to exceed that of the parent. A new and larger dune is developed, and separation from the parent becomes an eventuality. When sufficiently removed from its original site excess sand no longer is supplied from the parent transverse dune, and the height and forward motion of the blowout are decreased.

The barchan dune. As the blowout dune advances beyond the divided transverse form its outlines no longer correspond to the transverse pattern. At first its shape is that of a sand hummock of uniform dimension. Later, when isolation is completed there appears a tendency to assume a crescent shape. The latter is a stable form and when clearly developed characterizes a dune known as a *barchan*. The barchan displays crescent tips, or wings, whose terminal positions point in the direction of wind flow (see Fig. 10.9).

Within the shadow of the slip face wind velocities are effectively zero, whereas at the margins of sand accumulation wind velocities increase, and sand is driven downwind. The effect is to cause appearance of the wing tips marking the most advanced forward position of the barchan. From these extreme positions sand is dissipated and lost.

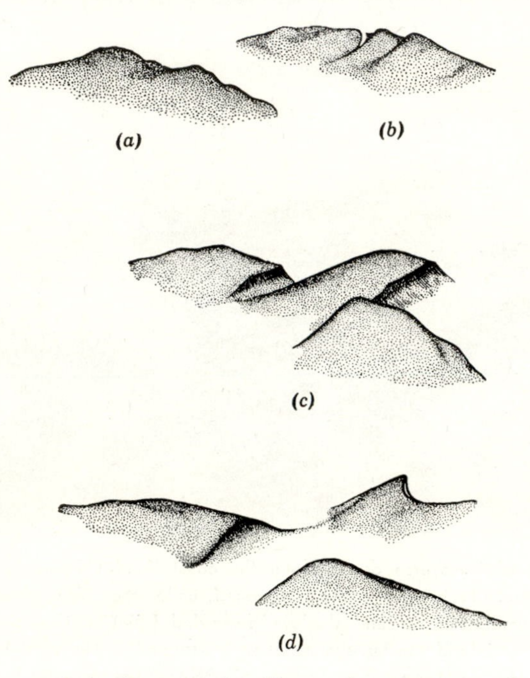

Fig. 10.8 Sequence of development of a blowout in a transverse dune. (*a*) Initial shape; (*b*) gap forms due to increase in velocity of underloaded wind; (c) a blowout forms and moves downwind of the remainder of the dune; (*d*) the isolated dune loses height as sand supplied from the former parent is diminished.

Barchans advance across the desert floor as individuals or in groups depending upon the quantity of available sand. Their best development is in localities where they move across a nonsand floor and derive their sand supply from some line source. A population of barchans is most numerous in the locality of greatest sand supply. Here, they tend to overtake one another often developing composite or double forms. Small dunes move faster than large ones and upon overtaking these add their bulk to form first a composite and later a single variety. For this reason the area of primary barchan development often is a dune complex of transverse and barchan shapes, and only at a distance from the source area do individuals become isolated. Separation between barchans increases with distance from sand supply, and also there is a gradual reduction in size.

Relative heights of barchans vary widely, and their average is low. Large dunes tend to spread out reaching lengths of one-quarter mile, but their height does not appear to exceed 30 m. In some localities this height is not reduced greatly as the sand supply dwindles. Rather,

Fig. 10.9 Air photograph of barchan dunes near Ancón, Peru. Note that the symmetrical lunate form is developed when the dune is isolated from its neighbors. (From Johnson, *American Geographical Society Sp. Publ.* 12.)

the height remains reasonably constant, but the spacing between individuals becomes progressively greater, and they may become separated by several miles. Far less commonly, the height dwindles, and eventually the dune is dissipated.

Relationship between forward motion of the dune and dune height is as expected, i.e., downwind motion is greater for small dunes. An example of available data on dune height and rate of travel is presented in Fig. 10.10. Here are shown rates of annual movement for dunes in several different desert areas. A line has been drawn through the distribution of points as a very rough approximate measure of the average relationship. Extension of this line toward the intersection with the horizontal coordinate axis suggests that as a dune height diminishes to 2 or 3 m its forward motion has reached an average value of about 22 m per year. In like manner, extension of the dashed line toward its intersection with the vertical coordinate indicates that a tendency to stagnate appears as dune height exceeds 30 m. This may provide the explanation for the occurrence of great piles of dunes 400 to 500 feet high which appear to be fixed in position and whose shapes appear to be composites of many dunes.

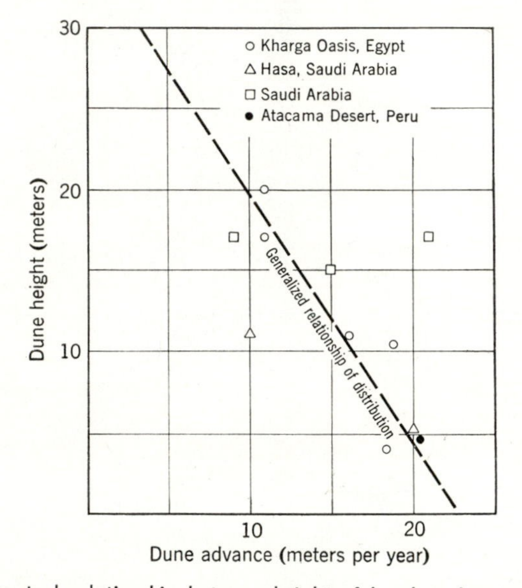

Fig. 10.10 Conceptual relationship between height of barchan dune and rate of movement. Dunes moving 25 meters per year tend to dissipate, whereas dunes over 30 meters high move only a few meters per year, tend to stagnate, and become great, high, dune complexes. (Data on movement from Beadwell, Kerr and Nigra, Melton.)

The longitudinal dune. A third major form is known as the *longitudinal dune.* In outline it is not dissimilar to the transverse dune, but its orientation is with its long dimension in the wind direction. In the Arab States such dunes because of their elongation and oftentimes pointed terminus are known by the term *seif* meaning sword. This shape is stable under rather extreme wind conditions when wind velocities are high, and there is a strongly preferred direction of prevailing winds within an angular distance of 30 degrees. Ideally, two strong wind directions should prevail, separated from one another by the 30-degree angle. One wind tends to establish the barchan shape, whereas the other causes one wing tip to move downwind well in advance of its counterpart on the other side of the dune.

Shifting of wind direction unbalances the sand supply so that the position of the slip face is altered. Sand is blown along one wing tip, and general development of a new, but attached, barchan appears. (See Fig. 10.11.) Under special conditions where storm winds have a persistent and narrow range of direction composites of longitudinal dunes developed from original barchan shapes result in a continuous chain extending many miles in length.

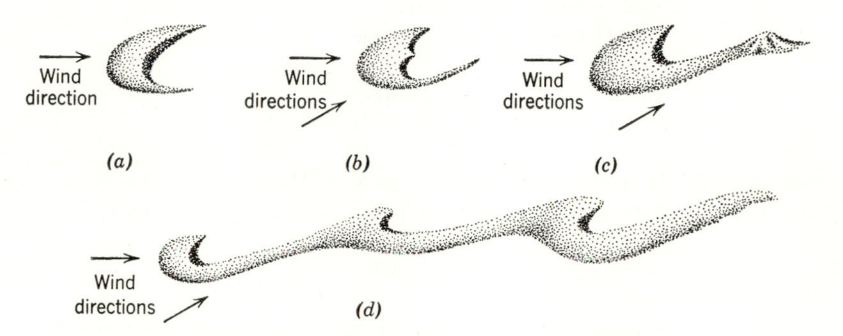

Fig. 10.11 Stages in development of longitudinal or seif dune from barchan. (*a*) Barchan dune with uniform wind direction; (*b*) modification due to swing in wind directions; (*c*) advanced stages of (*b*) and beginning of second barchan tied to its parent; (*d*) continuation of process to produce second and third barchan linked in long chain or seif. (Modified from Bagnold, Methuen and Co., Ltd., 1941.)

Dune shape and wind characteristics. Air photos of extensive dune belts reveal shapes which do not readily classify under any of the three fundamental types. Some have shapes which have characteristics of more than one of the major types, whereas others are strongly modified by blowouts. Such generally irregular shapes result principally from wide ranges in wind direction. In some areas storm winds blow repeatedly from opposite directions. Slip faces tend to appear on both sides of a dune and develop a general dome shape. Others display long tongues of sand extending from the dune to produce an irregular or amoeboid outline. Such tongues are upwind slopes produced by strong winds of differing directions of approach.

Uniformity in wind direction, sand supply, and absence of rainfall provide conditions for dunes of ideal outline. Within a belt of regular dunes irregular shapes are unusual and result from some anomalous local condition. Once a dune belt is recognized regularity of outline of individual dunes provides the basis for interpretation of general climatic conditions which prevail (see Table 10.1 and Fig. 10.12). Provided adequate sand supply and restricted wind direction exist, the principal controls of dune form are wind velocity and the nature of the surface over which the dune moves.

Table 10.1. Dominant Factors Controlling Ideal Dune Shapes

Dune Shape	Wind Direction	Wind Velocity	Sand Supply	Character of Floor
Transverse	In one dominant direction, but range of direction may be high.	Up to 25 miles per hour. Become irregular (blowouts) at higher velocities.	Large beaches, alluvial valleys.	Loose sand.
Longitudinal	In one or two dominant directions, but range is less than 30 degrees, generally only a few.	Greater than 25 miles per hour. Stable strong wind form.	Moderate to large.	Loose sand or hard floor. Often continuous with rocky ridge parallel to wind direction.
Barchan	In one narrowly restricted direction.	15 to 25 miles per hour.	Moderate to low, limited.	Firm floor often a rock surface.
Irregular	Variable and unsteady.	Less than 25 miles per hour.	Low to high.	Not significant.

SILT AND CLAY DEPOSITS

Occurrence

Dust storms distribute silt and clay debris throughout an enormous volume of air through which the particles settle as prescribed by Stokes' law. (See Fig. 10.3.) As these particles shower to the earth's surface some accumulate on land and others in lakes, rivers, and oceans. The contribution to the latter two is small when compared with the total fine debris transported by rivers, but on land surfaces, in small lakes, and swamps the total silt and clay supplied from the air may constitute important deposits. Silts and clays deposited on land must bear the hallmarks of subaerial deposition. Principally, these are absence of lamination and the presence of land-dwelling

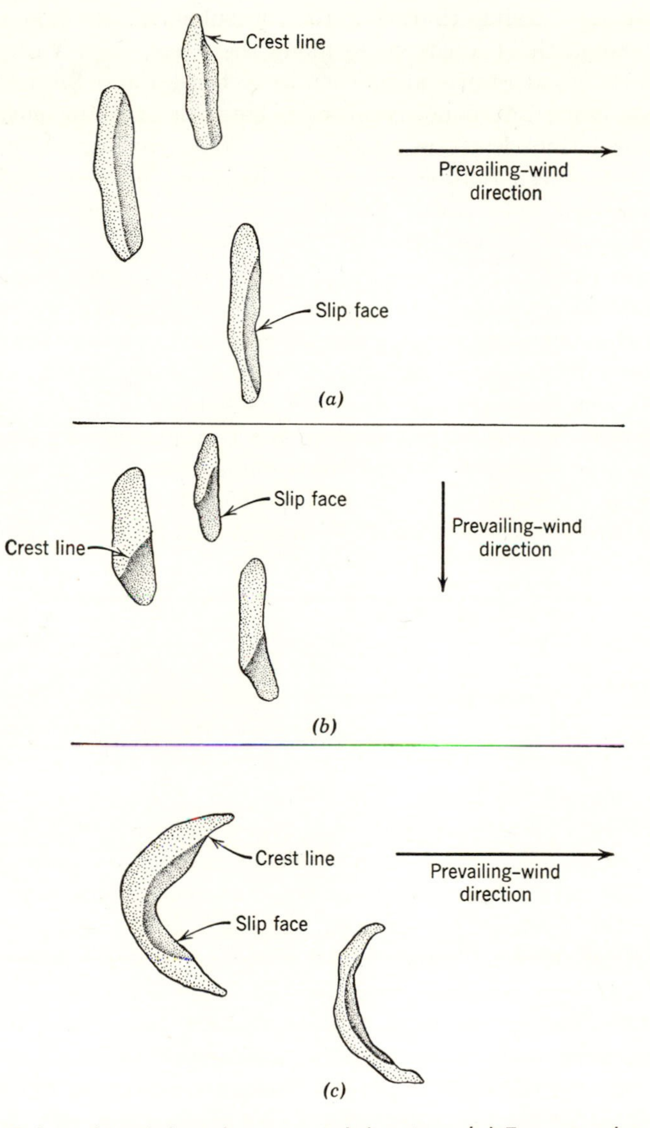

Fig. 10.12 Relationship of dune shapes to wind direction. (*a*) Transverse dune orientation of slip face is perpendicular to wind direction; (*b*) longitudinal dune orientation of slip face is parallel to wind direction; (c) barchan with crescent oriented in wind direction.

organisms which grew upon or lived within the accumulating deposits. Similar material deposited in water is bedded and carries remains of aquatic organisms.

Sediments showing characteristics of subaerial accumulation are widespread in the United States along the Mississippi Valley and in some of the great prairie areas such as Nebraska and Iowa. In these localities important deposits of clay silts blanket the surface and attain local thicknesses as much as 90 feet. Similar sediments are known in northern Europe, northern China, and scattered localities in South America and Africa. Everywhere they show certain characteristics suggestive of some common mode of origin.

On close examination these silts display special characteristics whose origin has been variously interpreted. Some investigators are of the opinion that such features are responses to deposition in water, others to soil-forming processes, and still others to wind deposition. Inasmuch as these deposits constitute the surface sediments over very extensive areas they must be handled in construction of dams, highways, canals, water reservoirs, soil terraces, foundations, and in agricultural production. It is, therefore, to be expected that considerable interest be shown in connection with the physical and chemical properties of these silts and indirectly in their mode of origin.

Loess

Definition

Without doubt the most striking aspect of certain of the silts in question is the tendency to stand with nearly vertical walls along stream channels and road cuts. This characteristic in addition to the primary requisite of particle size distribution (silt and clay) is often the only one employed to catalogue the deposit under the term *loess.* The term, of Germanic origin, was used in the Rhine Valley to indicate yellowish, fine-grained, slightly loamy clay generally unstratified but showing some vertical jointing. Material satisfying this description is now known to cover enormous areas of the northern hemisphere and in particular to be distributed in a band paralleling the maximum southern limits of the Pleistocene ice.

As with all other earth materials research on the nature of loess has shown it to be very complex, and no complete agreement exists concerning which characteristics are to be used to define what is to be called loess. For example, a definition which from an engineering viewpoint excludes certain silty clays does not coincide with a definition satisfactory to agronomists. However, certain characteristics are accepted generally as typial of loess but not of all silty clays. Table 10.2 lists these properties to provide the basis for general recognition of the material in question.

General Properties

Color does not appear to be a particularly diagnostic indicator of unaltered loess, but yellow buff is typical. Weathering increases the brown and red tones, and such colors are commonplace along present-day erosion surfaces or old ones of the Pleistocene. Soil colors vary principally with the amount of incorporated vegetation, and as loess is a parent-soil material colors of the soil horizons alter the color of initially deposited loess.

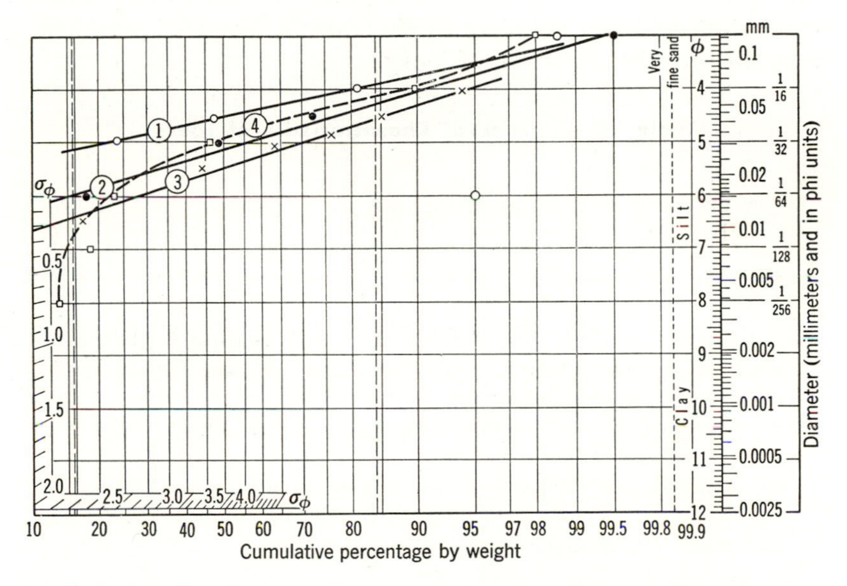

Fig. 10.13 Particle size distribution of loess and dust plotted on Otto graph base. (1) Loess, Frontier County, Nebraska (av. of 4 samples). (2) Loess from northwestern Kansas (av. of 42). (3) Loess, Washington County, Mississippi. (4) Dust from storm, Mead, Kansas. (Analyses (1), (2), and (4) by Swineford and Frye, (3) by Russell.)

Analysis of grain size distribution (Figs. 2.11, 2.12, and 10.13) indicates that it tends to be log normal; i.e., the logarithm of the grain size plotted against weight per cent frequency is a symmetrical bell-shaped probability curve, or "normal curve of error." All log normal distributions will plot as straight lines on the base (Otto base) of Fig. 10.13 and, in general, are typical of very good sorting. Well-sorted deposits are those which have been winnowed by currents of air or water until there remains a concentrate of restricted size range and distribution. Sorting typical of loess is to be expected in fine silts and clays which have settled out of the atmosphere. Curve 4, Fig. 10.13, represents the size distribution of dust which accumulated on a window sill in Meade, Kansas, after a dust storm. Note that the bulk of the size distribution of the dust approximates that of loess, but important differences are noted particularly in the very fine sizes. The departure from a straight line indicates that the distribution is not precisely log normal and, hence, not exactly like the distributions of loess. The departure from identical shapes of the curves of the dust and loess particle size distribution is suggestive of the basis for some of the controversy which exists concerning the origin of loess. Is there sufficient similarity between the curves to claim they have been deposited by a common transporting medium, or is there reason to doubt that wind is primarily responsible for the deposition of loess?

Table 10.2. General Characteristics of Loess

Color:

Yellow or yellow buff characteristic, also ash gray, gray buff, or brown. "A" horizon often very dark.

Texture:

a. Friable.

b. Dominantly silt but also contains clay- and sand-sized particles (see Fig. 10.13). Ordinarily the fraction .01 to .05 mm contains at least 50% of the material by weight.

c. An aggregate of clay particles appears to surround each large silt or sand grain.

d. Silt and sand particles are angular to subangular.

e. Microscopically shows an open texture with little interlocking of grains and many intergranular voids.

Composition:

a. Silt fraction consists of quartz (±50%); *feldspar often as abundant as ±20%;* calcite, dolomite, micas less than 10%; locally volcanic-ash shards; chert, hornblende, chlorite, and pyroxene are minor.

b. Clay fraction is somewhat variable, but montmorillonite and illite are most abundant.

c. Chemical composition. (See Table 10.3.)
Calcite is an important constituent although not always present. Occurs as silt grains, concretions, tubes, snail shells, and cement.

Structures:

a. Stands with vertical faces principally in artificial cuts and banks undercut by streams. (See Figs. 10.14 and 10.15.)

b. Vertical jointing is noted in most exposures. (See Figs. 10.14 and 10.15.)

c. Tubular structures may reach a maximum of 30 to 40 feet below the surface. They are coatings on roots and often are associated with small rootlet tubes 1 mm or less in diameter.

Fig. 10.14 Loess exposed at Sioux City, Iowa, showing typical smooth-weathering face, absence of bedding, and tendency to erode in columns because of incipient vertical joint system. (Height of face about 10 feet.)

Table 10.2. General Characteristics of Loess (*Continued*)

d. Incrustations of carbonate appear on fracture faces, around exposed roots, and below overhanging walls, often as stalactitic growths.

e. Concretions, particularly of calcite, are very common, and the depth from the surface to their first occurrence is directly proportional to rainfall. (See Fig. 10.16.)

Engineering Properties:

a. Vertical permeability to water is greater than horizontal.

b. Saturation of loess prior to loading results in rapid consolidation after loading. Dry undisturbed loess will withstand loads which it is incapable of supporting after saturation by rain or canal water.

c. Liquid limit of five samples from Iowa: 38, 36, 46, 39, 30.

d. Plastic limit of five samples from Iowa: 24, 25, 26, 24, 21.

e. Plastic index of five samples from Iowa:* 14, 11, 20, 15, 9.

Topographic Forms:

a. Chimneys or pipes occur along valleys, narrow ravines, and sharp divides. (See Fig. 10.15.)

b. Blankets the underlying topography. (See Fig. 10.17.)

c. Intricate drainage pattern especially in the tributary system. Slopes show steplike terraces ("cat-steps") due to minor landsliding or creep.

* Liquid limits and plasticity indices from C. S. Gwinne, "Terraced Highway Side Slopes in Loess, Southwestern Iowa," *Geological Soc. Amer. Bull.*, Vol. 61, 1950, p. 1350.

Fig. 10.15 Loess near Unionville, Tennessee, showing typical "chimneys" developed by erosion. The exposure is 53 feet, of which (as shown) horizon 1 is 11 feet of Loveland loess, horizon 2 is 4 feet of Farmdale loess, and horizon 3 is 32 feet of Peorian loess. A good example of superposition of loess sheets outside the limits of glaciation. (Courtesy of Leighton and Willman, *Jour. Geology*, 1950.)

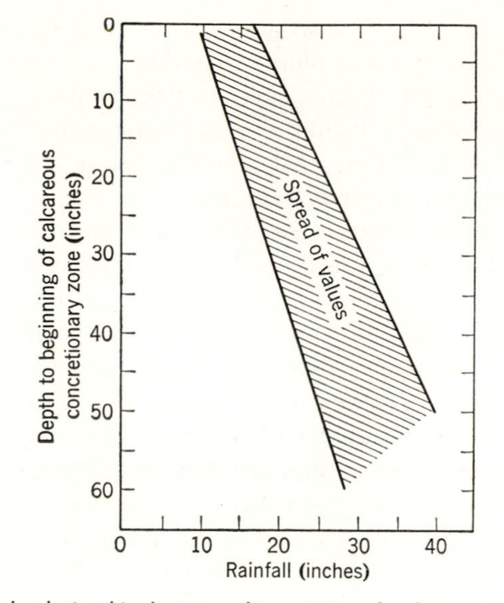

Fig. 10.16 General relationship between the position of calcareous concretions in loess and the average rainfall in an area. The tendency for concretions to develop near the surface in dry regions is the effect of marked capillary potential developed in the sediment by evaporation. In wet regions carbonates are leached by the continued progress of downward-moving waters following rains. (Modified from *Factors of Soil Formation*, by Jenny, copyright 1941, McGraw-Hill Book Co., Inc.)

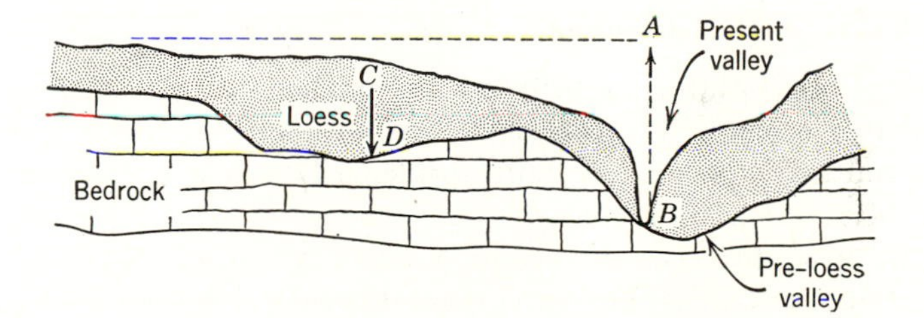

Fig. 10.17 Blanket effect of loess burying underlying topography. Note the measured thickness *A–B* is exaggerated beyond the limits of the true maximum thickness *C–D*, known through drilling.

Coarse sand and gravel occur with some frequency in association with loess. These deposits often are well bedded and sometimes are interlayered with laminated silts whose texture is identical with loess. Such anomalous associations have been explained as representing concentrates upon which the loess accumulated or deposits in lakes in which the dust fell. Their occurrence is considered coincidental to deposition of loess by wind action.

Terrestrial snails are overwhelmingly the most abundant organism occurring in loess. Many are species still living today and are typical of those which inhabit flood plains of rivers and dwell in calcareous soil. Fragments of vertebrates such as fish, rabbits, gophers, otters, beavers, squirrels, deer, elk, buffalo, and mastodons have been found in loess of several localities. All of these animals are characteristic land dwellers or inhabitants of lake and river margins.

Scattered large calcareous fragments are commonplace on loess surfaces or piles. These are secondary in development and display typical concentric structures or hollow tubular outlines of concretionary growths. As they are not original deposits they must be excluded from particle size distribution determinations. They are so common, however, that they may appear as zones which to the uninitiated could be confused with gravel. Some of the concretions are covered with excrescences and are very irregular in outline. Others form around roots and are cylindrical in shape, whereas still others are smooth, rounded, and often crudely suggestive of the outline of human form. In dry regions concretions appear in the "A" soil horizon, but as rainfall increases $CaCO_3$ is leached from the upper soil and is concentrated in the "B" horizon. In areas where rainfall is 40 inches annually $CaCO_3$ may be more than 50 inches below the surface, whereas in desert areas of 10 inches of annual rainfall $CaCO_3$ is concentrated at the surface. (See Fig. 10.16.)

Thickness and Distribution

In the United States the principal occurrence of loess is in the north central states where it is interstratified with glacial tills. Loess also occurs south of the glacial limits principally along the eastern bluff of the Mississippi Valley.

In the glaciated section several distinct sheets of loess are recognized. Of these there are two, the Loveland or Sangamon and the Peorian, which are widespread. The period of Loveland and Sangamon loess deposition followed melting of the Illinoian ice (see p. 434), whereas the Peorian loess marks a time of ice withdrawal during early Wisconsin glaciation, but outside the limit of ice loess deposition continued into late Wisconsin time. Each of these loess formations is continuous over an impressive area and contains a distinctive fauna which clearly indicates amelioration of climate during its accumulation.

Within the glaciated section these individual loess sheets can be identified by their relative position to overlying or underlying glacial clays (tills). Outside the boundaries of continental glaciation distinction between beds of loess becomes difficult inasmuch as individual

Fig. 10.18 Excellent example of superposition of loess sheets near the limits of continental glaciation. Exposure near New Harmony, Indiana. showing: (1) 5 feet of leached glacial clay (till), (2) 5 feet of pink-brown, calcareous Farmdale loess, and (3) 15 feet of yellow-brown, calcareous, fossiliferous Peorian loess. Color distinctions are prominent. (Courtesy of Leighton and Willman, *Jour. Geology*, 1950.)

loess sheets are not separated but often occur superimposed one upon the other (Fig. 10.18).

Determination of the extent of individual sheets and their thickness becomes extremely uncertain in unglaciated areas. Nevertheless, careful field work has provided important information sufficient to construct an isopach (thickness) map of one of the important loess formations. Figure 10.19 represents such a map of Peorian loess showing it to be thickest in Nebraska and thinning eastward to less than 2 feet in the Great Lakes region where it becomes generally unrecognizable. Progressive thinning in an easterly direction has led some observers to consider the source of the silt to have been prin-

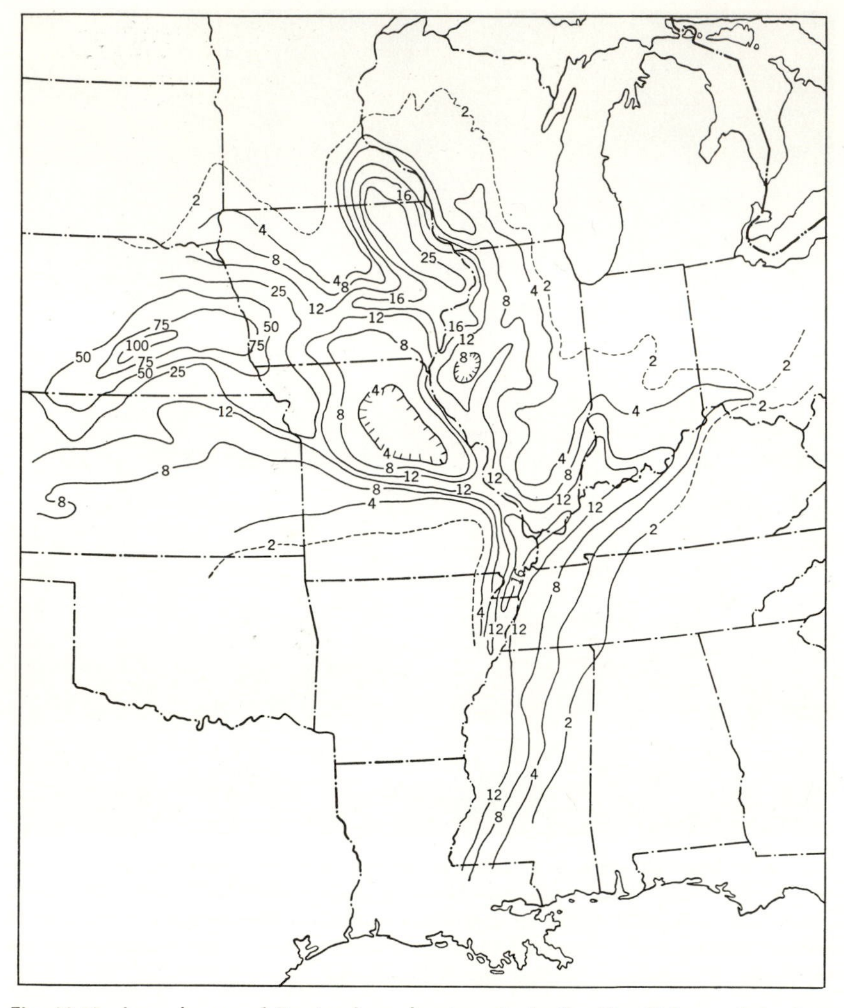

Fig. 10.19 Isopach map of Peorian loess (contours in feet). The thickness of the loess increases systematically from Illinois into Nebraska, but local increases in thickness are to be noted along each of the major river valleys, e.g., Ohio, Mississippi, Missouri, Illinois. (Data from various state publications and map of *Pleistocene Eolian Deposits* of the United States by Com. Nat. Research Council, published by Geological Soc. Amer., 1952.)

cipally from the semi-arid Great Plains. Silt and clay gathered in that area could readily be transported by the wind to the depositional sites.

The blanketing effect of loess deposition filling valleys and uplands alike tends to give the field observer a somewhat exaggerated measurement of loess thickness (see Fig. 10.17). Hence, early estimates were

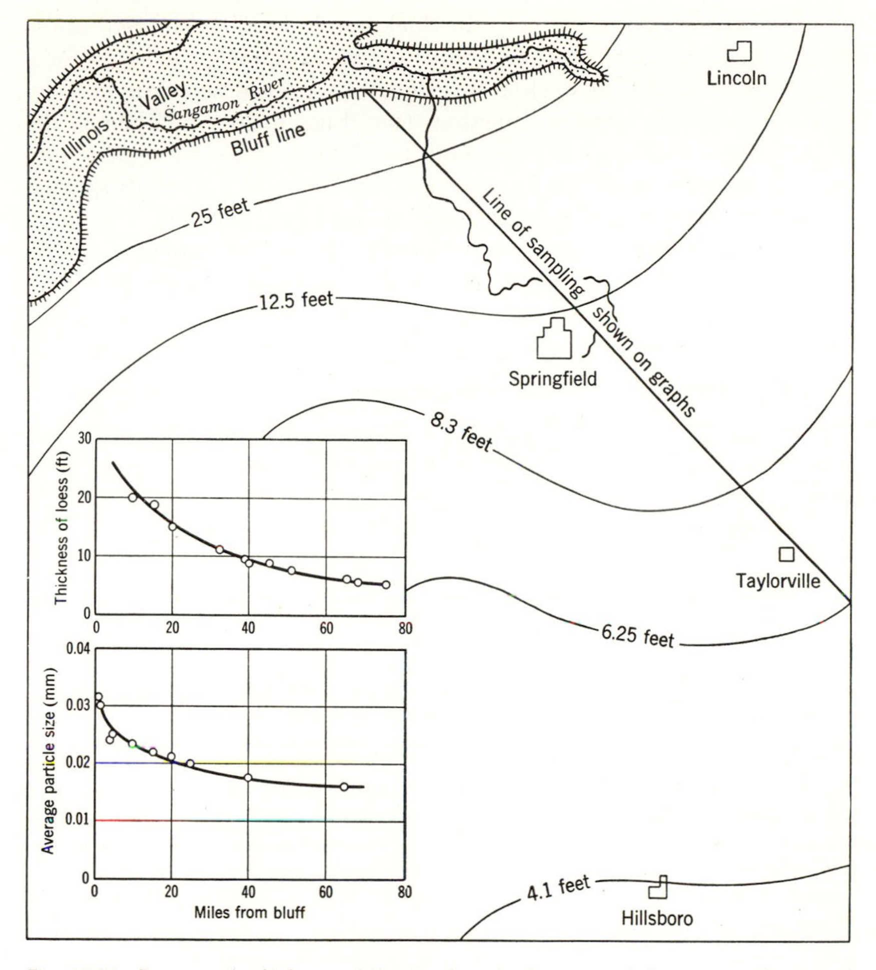

Fig. 10.20 Decrease in thickness of Peorian loess with increased distance southeastward from Illinois River bluffs. Inset graphs show decrease in thickness and average particle size along the line of sampling (see also Fig. 10.21). (From Krumbein and Sloss, W. H. Freeman Co., 1951. Data from Smith, *Ill. Agric. Exp. Sta.*, 1942.)

much greater than values now made available through drilling. This information reveals that decrease in thickness from areas of maximum values is not uniform with distance but reaches low values near the central portions of Iowa and Missouri and thickens eastward toward the Mississippi Valley. This anomalous direction of increase is not peculiar to the Mississippi Valley but is typical toward all major river systems of the area. Note in Fig. 10.19 that the thickness values increase along the valleys of the Missouri, Illinois, Wabash, and Ohio rivers. A study of the decrease in thickness of Peorian and Sangamon loess with progressive distance from the Mississippi River bluffs indicates that the rate of decline is geometric and can be expressed by means of an exponential function (see Fig. 10.20). Although the values listed in Fig. 10.20 refer only to Illinois, a similar relationship can be expected to obtain elsewhere. Note that in Fig. 10.19 spacing of individual contours is approximately uniform, yet with a few exceptions their values increase geometrically. The decline in loess thickness, therefore, is not arithmetical with distance from the river valleys.

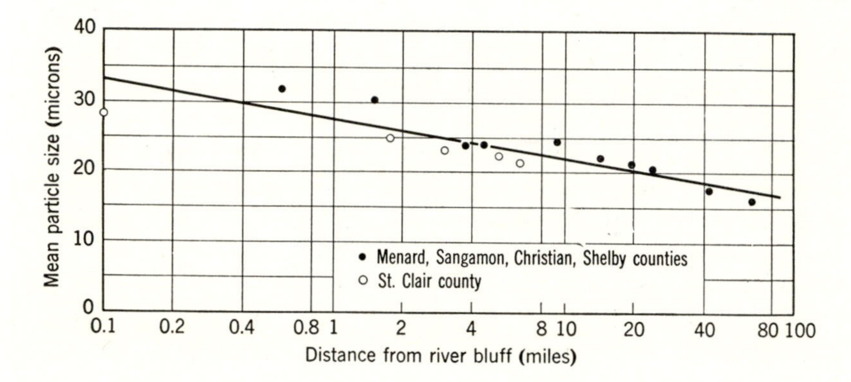

Fig. 10.21 Decrease in mean particle size of Peorian loess with increased distance from Illinois River bluffs in Menard, Sangamon, Christian, Shelby, and St. Clair counties, Illinois (same data as in Fig. 10.20). The distribution is plotted to show that the relationship between distance and mean size can be expressed by an exponential equation. The relationship suggests very strongly that the local source of some loess was from valleys of large streams with broad flood plains of debris supplied by melting glaciers. (Data from Smith, *Ill. Agric. Exp. Sta.*, 1942.)

In Nebraska, northwest of the area of maximum thickness, the loess becomes increasingly sandy and merges with beds of sand. The average particle size, therefore, increases in that direction suggesting that the loess was derived from a western source. More precise information is known concerning particle size relationships in Illinois. Here as shown in Fig. 10.21 mean size decreases uniformly with the logarithm of the distance from the river valleys. A well-founded geological rule is that sediments coarsen in the direction of their source. In the case of loess increase in particle size toward river valleys would indicate that such areas were local sources for the silt and clay.

Composition

Loess differs from most silts in its composition. Most silts are dominated by quartz and clay minerals and minor amounts of feldspar. Other minerals appear as traces. Loess, however, contains considerable amounts of feldspar and more than traces of other silicate minerals such as hornblende. Feldspar, in particular, is a very important constituent. Certain river silts and very fine sands such as those carried by the Mississippi also contain high percentages of feldspar, and compositionally they are not unlike loess. (See Table 10.4.) This similarity precludes positive identification of loess as a wind-blown deposit on the basis of composition alone. The high carbonate content of loess is also a strong identifying feature. In this respect most loess differs from the composition of river silt (see percentage of CaO, Table 10.3). Where soil profiles are being developed in loess, concentration of carbonate in the lower horizons, through leaching of the overlying portion, affects the analyses. Nevertheless, widespread distribution of high carbonate in the unleached part has not as yet been satisfactorily explained.

Table 10.3. Chemical Composition of Loess and River Silts*

Locality	SiO_2	Al_2O_3 and TiO_2	Fe_2O_3	CaO	MgO	K_2O	Na_2O	H_2O
Loess, average of 17 from Kansas†	74.10	13.04	3.25	3.05	1.68	2.78	1.60	0.51
Loess, near Galena, Ill.‡	69.2	11.80	3.34	5.79	3.95	2.20	1.44	2.20
Loess, Vicksburg, Miss.‡	68.0	9.50	3.68	10.0	5.10	1.21	1.31	1.27
Loess, Kansu, China§	65.1	13.20	4.25	10.70	2.50	2.36	1.97	1.05
Mississippi R. silt, composite of 235 samples‖	74.3	11.81	3.71	2.30	1.51	2.45	1.61	2.08
Rhine silt from delta in Lake Constance‖	83.2	7.85	4.42	1.27	0.56	0.91	0.89	1.63

* All analyses have been recomputed to a comparable basis.

† Analysis by Russell Runnels, State Geol. Survey of Kansas.

‡ Analysis by F. W. Clarke, "Data of Geochemistry," **U.S. Geol. Survey Bull.** 770, p. 514.

§ Analysis by G. B. Barbour, **Ann. Rpt. Smithsonian Inst.,** 1926–1927, p. 283.

‖ From F. W. Clarke, "Data of Geochemistry," **U.S. Geol. Survey Bull.** 770, p. 508.

Table 10.4. Per Cent Mineral Composition of Silt Fraction of Loess

	1. Peorian Loess, Kansas	2. Peorian Loess, Tennessee	3. Fine Sand Mississippi R., Cairo, Ill.	4. Atmospheric Dust, Baton Rouge, Louisiana
Quartz	65	78	62	53
Feldspar	13	21	25	8
Rock grains	—	—	3	—
Volcanic glass	13	not reported	T	24
Mica	4	"	T	2
Chert	2	"	8	—
Clay aggregates	—	—	—	4
Hornblende	minor	.8	T	T
Hematite	—	—	—	2

1. Analysis approximated from data reported by Swineford and Frye, *Jour. Geol.*, vol. 59, pp. 312–314.

2. Analysis recalculated from Wascher, Humbert, and Cady, "Loess in Southern Mississippi Valley," *Soil Sci. Soc. Amer. Proc.* 1947, vol. 12, 1948, p. 391.

3. Analysis generalized from R. D. Russell, "The Mineral Composition of Mississippi River Sands," *Amer. Jour. Sci.*, vol. 48, 1937, p. 1318.

4. Analysis generalized from R. D. Russell, "The Mineral Composition of Atmospheric Dust," *Amer. Jour. Sci.*, vol. 31, 1936, p. 58.

Selected Supplementary Readings

Bagnold, R. A., *The Physics of Blown Sand and Desert Dunes,* Methuen and Co., 1941. A most comprehensive treatment of the movement of sand particles.

Flint, R. F., *Glacial and Pleistocene Geology,* John Wiley and Sons, 1957, Chapter X (eolian features). A good summary of the characteristic features of loess.

Holmes, Arthur, *Principles of Physical Geology,* Thomas Nelson and Sons, 1944, Chapter XIII. A good description of wind action in the development of desert landscapes.

Lobeck, A. K., *Geomorphology,* McGraw-Hill Book Co., 1939, Chapter XI. A summary of distribution of loess throughout the world.

Wright, W. B., *The Quaternary Ice Age,* The Macmillan Co., 1914, Chapter X. A presentation of the controversial origin of loess.

11 Processes associated with glaciation

INTRODUCTION

Definition of Glaciers

World distribution of solid water is far more restricted than either the gaseous or liquid forms, but in areas of its occurrence it is an impressive agent of landscape alteration. This results indirectly from the transformation of snow into ice which under the application of

Northernmost Labrador. (Photograph by Forbes, courtesy American Geographical Society.)

stress forms a slow-moving stream. On the land surface extensive masses of ice are known, and where these are caused to move as a result of gravitational stress they are identified as *glaciers*.

Glaciers exist only in areas where much of the annual supply of moisture is in the form of snow and where the average temperature is low enough to permit ice to remain throughout the year. The limit of such a sheet of ice must be reached where the annual heat input from solar radiation is sufficient to melt the ice as rapidly as it advances, and the terminal zone must necessarily fluctuate in position with long-term variations in temperature and rates of ice flow. For this reason during certain time intervals large parts of frontal areas of ice sheets have tended to remain stagnant. Still other glaciers have become detached from the main body under conditions of excessive melting and have wasted away without any motion to the ice.

Accumulation of Snow

The ratio of the amount of moisture which falls as rain or as snow is broadly controlled by a combination of two factors, the latitude of the area and the altitude. If some snow is to be preserved throughout

Table 11.1. Relationship between Latitude, Elevation of Snow Line, and Annual Precipitation*

Latitude	Region	Elevation of Snow Line (Feet)	Approximate Annual Rainfall (Inches)
0°	Ecuador	19,000	60
3° N	East Africa	17,000	80
4° S	New Guinea	16,000	60
11° N	Colombia	15,000–16,000	80
29° N	Tibet	18,000–20,000	60
41° N	Pyrenees	8,800–9,200	80
45° N	Wyoming, U.S.A.	11,000	20
47° N	Alps	8,000–9,500	80
71° N	Scandinavia	3,000	10
80° N	Franz Josef Land	±1,000	10

* Data on elevation of snow line principally from R. F. Flint, *Pleistocene Geology and the Ice Age,* John Wiley and Sons, 1947.

the year the ratio of the total amount of water falling as snow to that which falls as rain must be of the order of at least ten. Near the equator this ratio prevails only at great heights (above 19,000 feet), but with an increase in latitude the altitude at which the necessary ratio occurs becomes progressively lower. Table 11.1 shows the observed lower elevation where permanent snow exists with a change in latitude. This inverse relationship has only general significance in view of other factors which contribute to the accumulation of snow. Under the optimum combinations of altitude, latitude, temperature, and amount of precipitation snow will accumulate and remain from one year to another, and this situation establishes the condition of glacier development.

Varieties of Glaciers

The snow line establishes a physical limit below which snow does not accumulate and below which a glacier cannot originate. At or above the snow line the accumulating snow is best protected from dissipation in some valley particularly at the base of a cliff where the steep walls shut off much of the daily sunshine. Snow fields which develop under such circumstances will continuously grow in size from year to year as the snow supply is increased. Gradually, under the weight of the accumulated snow ice may develop and initiate the condition of a glacier. Once the ice begins to flow, primarily under the influence of gravity, it is restricted by the enclosing valley walls and forced to move along a predetermined route. Such glaciers have been appropriately named *valley glaciers*.

Where the snow line lies below a plateau or intersects the earth's surface in areas of low relief accumulation and preservation of snow is not limited to valleys, but an extensive snow field may mantle a large area covering both lowlands and divides. Under such circumstances the supply of snow will become cumulative, transformation into ice will occur, and glacier conditions are established within a disc-shaped mass of ice and snow. This mass has no limits defined by the walls of a valley, but the boundaries are established by other physical conditions such as the amount of snowfall, the rate of melting, the periphery of the plateau, or the shores of an ocean. Such sheets of ice are known as *ice caps* or *continental glaciers*.

In low-latitude regions valley glaciers are isolated units, each controlled by the local physical environment; hence, some are larger than others and extend farther down the mountain slopes. With increase in latitude such glaciers extend into the lower mountain valleys, and many tributary units coalesce to form large master glacial streams. Where these emerge from the mountains onto plains the valley no longer confines the ice, and a general outward spreading occurs. Coalescence of several major glacial tongues will produce a broad lobate mass whose limits are reached when the amount of ice locally melted is balanced by the amount which is received locally. Glaciers of this type are called *piedmont* because of their distribution along mountain fronts.

Ice caps develop where for a long time the rate of ice accumulation exceeds the rate of outward flow from that region. Under this condition ice thickness increases until little, if any, land surface is exposed above the glacier. Development of an extensive ice cap must be restricted to regions where the snow line is low in elevation, and the precipitation is high. These necessary climatic conditions can center in a region of low mountains in which all but the highest portions become engulfed in the mass of outward moving ice. The ice cap spreads beyond the limits of the mountain area until the equilibrium-bounding limits are reached. In regions of low relief the developing ice cap rapidly covers all the land surface and spreads from the area of accumulation in all directions into the region where melting predominates. This condition where melting and evaporation loss exceeds the amount of ice supplied is identified as *ablation,* and all glaciers can be subdivided into the section of accumulation and the section of ablation.

PROPERTIES OF ICE CONCERNED WITH GLACIER MOVEMENT

Transformation of Snow into Glacial Ice

Individual flakes of freshly fallen snow consist of skeletal or partly developed crystals of ice. As such they represent an unstable crystalline state, and there is strong tendency for continued growth of the space lattice to complete the crystal form. The existing equilibrium between the vapor and solid phases of water permits sublimation of additional solid material to certain snowflakes and removal of other flakes by passing into the vapor phase. Large snowflakes, therefore, tend to fill out their crystal form at the expense of smaller ones.

Transfer relationships between the solid and vapor phase represent only one of the conditions in the growth of ice inasmuch as liquid-solid transformation probably is more important. Melting and refreezing result in the loss of smaller snowflakes and the growth of others. This process alters the skeleton outline of snowflakes into individual crystals of ice of irregular outline. In the early stages of its development such granular ice known as *firn* or *névé* is a loose aggregate of ice grains and represents an intermediate stage in the development of glacial ice.

Alteration from snow to firn is associated with attendant decrease in porosity and an increase in density (see Fig. 11.1 and Table 11.2).

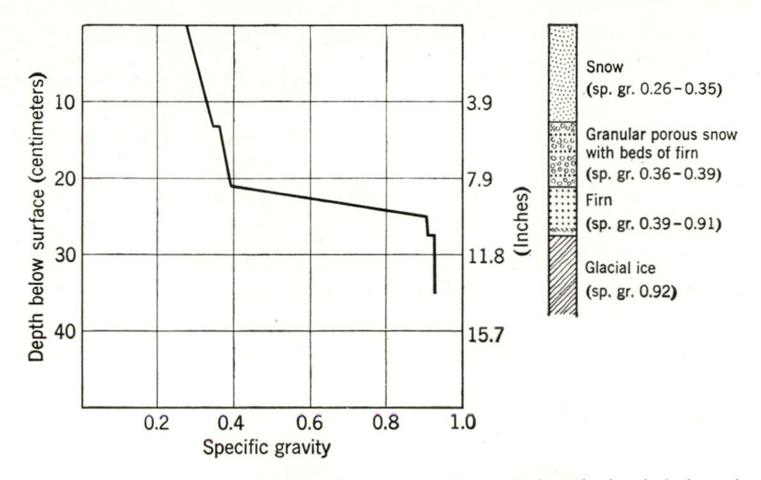

Fig. 11.1 A plot of the specific gravity of frozen material with depth below the surface of the glacier at Kings Bay, Spitzbergen. At right is the profile from the surface of the snow to the glacial ice with increased depth as shown, at left is the change in specific gravity with depth. Note that in the snow and granular firn the increase in specific gravity is attendant principally upon compaction, whereas in the zone of firn, rapid transition to ice occurs. Such a rapid change is very unusual, and the thickness of firn generally observed may exceed 100 feet. (Data from Ritter, and Son Ahlmann, *Geografiska Annaler*, 1933.)

As layers of snow are piled one upon the other the specific gravity increases gradually under the superincumbent load from average values of 0.26 to as much as 0.40 particularly where thin layers of firn are present. Firn has a much greater specific gravity ranging between 0.3 and 0.8, the higher value being that of slightly porous ice. Some localities are reported where with increased depth below the surface there is a gradual rise in the proportion of firn to snow, and the over-all specific gravity increases. Still other occurrences are known such as illustrated in Fig. 11.1 where transition from alternating beds of snow and firn terminates abruptly and thick layers of firn lie upon dense ice. This condition can be recognized by the sharp rise in values of specific gravity from 0.4 to 0.8 or 0.9 and by the existence of glacial ice. Transition from closely packed firn to glacial ice must be recognized as a matter of increased interlocking of crystals

Table 11.2. Physical Properties of Ice*

Crystallography	Form Ice 1—common form, hexagonal. Changes to form Ice 2 (denser) only at 30,000 lbs per sq. inch and −22°C.
Hardness	$1\frac{1}{2}$ to 2—normal temperatures 4 — −44° C ±6 — −50 to −78°C
Strength	(Depends on crystallographic orientation, temperature, etc.) Shearing—95 to 110 lbs per sq. inch = 7 to 8 kg per sq. cm. Crushing—350 to 1000 lbs per sq. inch = 25 to 70 kg per sq. cm. (Granite = 1500 kg per sq. cm.)
Specific gravity	Newly fallen snow—0.1 to 0.3 Old snow —0.3 to 0.5 Firn —0.3 to 0.82 (0.82 = 13% porosity) Glacier ice —0.82 to 0.92 (av. = 0.90) Pure Ice 1 —0.92 (blue ice—no air bubbles)
Thermal properties	Latent heat of fusion—80 cal per g. Latent heat of vaporization (sublimation)—600 cal per g. Pressure-melting relations—(assume hydrostatic pressure) Lowering of melting point 7.5×10^{-3}°C per atm, i.e., about 1.92×10^{-2}°C lowering for each 100 feet of ice.

* These data are from a syllabus of a course entitled *Geologic Principles* (Northwestern University) offered by L. H. Nobles.

and attendant decrease in porosity until pore space is eliminated completely.

Change from snow to firn is accomplished over a considerable range in depth below the surface of the snow and is controlled in part by the season and the rate and the locality of snow accumulation. For each glacial tongue there exists a position where the amount of snow furnished each year is equal to that amount lost through melting. This position of material balance is known as the *firn line,* and as shown in Fig. 11.2 its location above sea level may vary even in the same general region. Above the firn line accumulation exceeds ablation, and glacial ice is generated. At elevations below the firn line ablation loss exceeds accumulation, and glacial ice is dissipated in part.

Recrystallization of firn into glacial ice is accomplished by only minor increases in pressure supplied by the total weight of the superincumbent snow and firn inasmuch as the density of closely packed firn approaches that of glacial ice. The latter is the end product of the ice-crystal growth, and the interlocked and welded crystals develop a texture identical to that of an igneous rock. The nonporous mass consists of individual crystals some of which are large in size, more than an inch in length, but normally the crystals are of uniform dimension of about ¼ inch across. Where flow is important orienta-

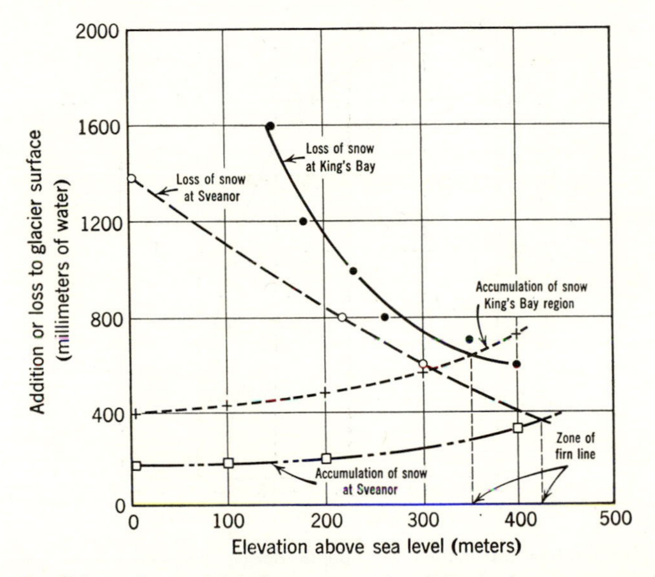

Fig. 11.2 Conditions of material balance at two localities in the same general latitude (±80° N.), Sveanor and Kings Bay, Spitzbergen showing accumulation of snow and loss of snow with change in height above sea level in each of the two localities. Balance is attained where the curves cross, and this is the position of the firn line. Note that the elevation of the firn line in the region ranges between 350 and 425 m above sea level. (From Son Ahlmann, *Geografiska Annaler*, 1933.)

tion of individual crystals is brought about, and textures similar to metamorphic rocks are indicated.

The welded crystalline mass is unstable physically under certain conditions which are known to exist below the glacier surface. Such conditions of instability can be enumerated as resulting from:

a. Ice pressure–temperature–melting-point relationships.

b. Rotation of individual crystals into crystallographic orientation under applied stress.

c. Formation of new ice in proper crystallographic orientation on the existing crystals.

d. Plastic flow.

Each of these processes is considered to be partially responsible for movement within the glacier and its attendant flow beyond the limit of centers of ice accumulation. Glacier movement, however, is an extremely complex process which as yet is known only approximately. Despite this limitation the following paragraphs are introduced to acquaint the reader with the extent of current knowledge and some of the questions which remain as yet unanswered.

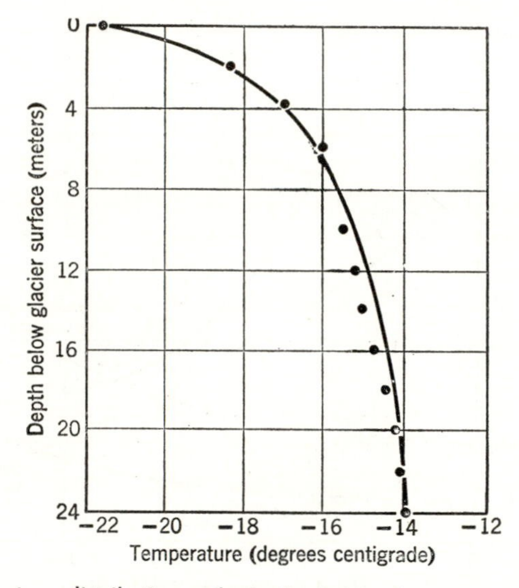

Fig. 11.3 Temperature distribution with depth during the winter in a polar glacier, the Borg glacier in eastern Greenland. Observe that the value at the position of greatest depth of reading indicates that the temperature curve is becoming asymptotic to a value between −13 and −14°C, i.e., far below the value of the pressure-melting temperature. (Data from Koch and Wegener, cited by Fjeldstad, *Geografiska Annaler*, 1933.)

Ice Temperatures

Records of temperatures prevailing within existing glaciers are scanty and somewhat inconsistent. However, those which are available show curves of similar shape when temperature is plotted against depth below the ice surface. Figure 11.3 is one such distribution based on data obtained from the Borg glacier in East Greenland. The temperatures indicated represent average winter conditions which are necessarily lower than must exist during the warmer season. Near the surface the rate of increase in temperature with depth is considerably more than that deeper within the ice, whereas observations near the base of the bore hole suggest that the rate of increase approaches zero.

Analogous temperature records in glaciers of mid-latitudes indicate the ice is much warmer and near the melting point, but as in the Borg glacier the temperature approaches uniformity with depth. On this basis two types of temperature regimes are recognized. *Polar glaciers* are those characterized by extremely low temperatures, whereas *temperate glaciers* are those whose temperature at depth hovers near the melting point throughout much of the year. In polar glaciers the temperature distribution in a thick ice sheet approaches uniformity at a value considerably lower than the freezing point. Temperate glaciers appear to be in a state of delicate temperature balance as maintained by pressure–melting-point relationships. In either case during the warmer season the extreme surface layer reaches the melting point, but in temperate glaciers the thickness of this layer is much greater and large quantities of melt water are furnished to the underlying firn and ice. When such a condition exists water from the melting snow moves downward through the available pore space in the firn until the existing temperatures permit freezing in crystallographic orientation to local ice crystals. By this process the density of the firn gradually increases until no more pore space remains, and solid ice is the product. The general case is one where the gradual transition from snow to nonporous ice occurs through a thickness of more than a hundred feet, particularly in temperate glaciers, but in polar glaciers there are some areas where solid ice is developed very close to the surface. One such example is illustrated by Fig. 11.1 in which in the King's Bay region of Spitzbergen glacial ice forms within one foot below the snow surface. In this locality the processes leading to the development of solid ice are identical to those of the temperate glacier but are telescoped within an exceptionally short interval, and the changes in specific gravity from snow to ice are sharply defined. To the surficial skin of snow enough heat is supplied from solar energy to produce melting, but the low temperatures of the ice immediately underlying cause freezing of the water and elimination of a thick zone of porous firn.

Lowering of Freezing Point

Depression of the freezing point with increasing pressure is illustrated in Fig. 11.4 in which freezing temperatures are plotted with changing pressures. A load of 1000 m of ice (0.92×10^5 g per sq. cm) will lower the freezing point from 0°C at 1 atm to −1°C. Under such a load if the ice temperature of a temperate glacier rises to −1°C, even momentarily, this solid phase begins to transform to the liquid through the stage of a mushy mass of ice. Providing the latent heat of melting is furnished more and more ice will be transformed to the liquid until much melt water will be present. When the temperature of the ice-water mixture is lowered below the melting-point temperature for the corresponding pressure the liquid no longer is a stable phase and refreezing begins. As the water refreezes much of its bulk can be added to the already existing ice in crystallographic orientation; ice crystals grow in size and develop a texture of interlocking crystals.

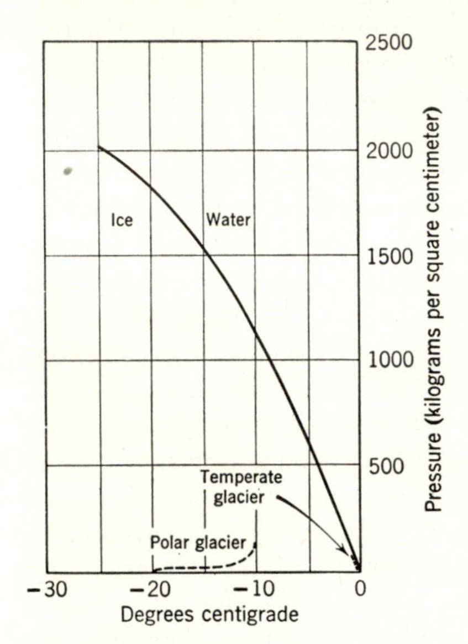

Fig. 11.4 Pressure-melting temperature curve of ice and distributions of temperatures observed in polar and temperate glaciers.

The influence of pressure on lowering of the freezing point has been considered important at the base of thick temperate glaciers which near their terminal margins may attain temperatures approaching 0°C. The concept proposed is that, ideally, the lower part of the ice would be rendered mushy and forward motion would be accelerated with the partly melted ice as a lubricant. Obviously the process does not apply in those parts of temperate glaciers which are colder than the pressure-melting-point gradient of approximately 1°C for each kilometer of depth. For example, at the base of a glacier 2 km thick the pressure is approximately 180 kg per sq. cm, a value which would lower the melting point of ice to about −2°C. However, some observed temperatures with increased depth in the ice suggest that a constant value is reached which is somewhat lower than the calculated melting temperature, and so the ice must remain solid. Equilibrium conditions between the liquid and solid phases can not be established at the observed temperatures, and under such circumstances only the solid phase can exist.

Viscosity of Ice

Photographs such as Fig. 11.5 illustrate clearly the similarity in appearance between valley glaciers and a river system. The distribu-

Fig. 11.5 Air photo of Muldrow Glacier and Mt. McKinley, Alaska, showing the glacier in the steady-state condition. Dark stripes are moraines of rock debris carried toward the glacier terminus, and serve to emphasize the fluid aspect of the ice. The ice flows past a selected position with relatively small change in elevation with the passage of years. (Photograph by Bradford Washburn.)

tion of debris on the surface of the ice emphasizes the pattern of flow, and there is no difficulty in understanding why they are often called streams of ice. Thus at first glance the major difference between the behavior of glacial movement and stream flow would appear to be primarily one of viscosity. These aspects led the great naturalists of the nineteenth century to conclude that glacial movement could be described in terms of a highly viscous fluid, and, therefore, relationships between ice thickness and flow velocity could be calculated, such as

$$t = \sqrt{\frac{2\mu V_0}{\rho g \sin \alpha}} \qquad (1)$$

where t = thickness of the glacier.

μ = coefficient of viscosity of ice.

V_0 = surface velocity along the mid-point of the glacier between valley walls.

ρ = density of ice.

g = acceleration of gravity.

α = inclination of the rock bed.

or surface velocity should vary as the square of the ice thickness.

Accordingly, if eq. 1 were valid the calculated surface velocity should agree with the observational data in those cases where the thickness and surface velocity of ice are known. However, information gathered in the area of ice accumulation of certain Alps glaciers demonstrates that the surface velocity is not uniform and its value is about one-half as great in the summer as in the winter. Yet, the additional winter snow layer approximates less than 1 to 2 per cent of the total weight

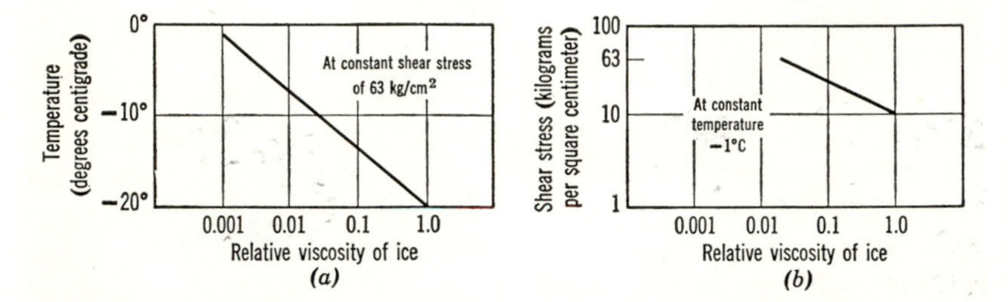

Fig. 11.6 Relative viscosity of ice under (*a*) a condition of change in temperature at constant shear stress of 63 kg per sq. cm, i.e., sufficient to produce rupture under compression and far in excess of rupturing strength under shear; and (*b*) a condition of change in shear stress at constant temperature of −1°C. (From Höppler as cited by Perutz, *Jour. Glaciology,* 1947.)

of the ice. The great difference between the summer and winter velocities of the same glacier of approximately the same thickness indicates clearly that equations based upon viscosity and ice thickness cannot be applied even for very rough approximations. Rather, the surface velocity appears to be largely independent of ice thickness and is not a response to simple laminar flow of a liquid material of high viscosity.

Recent experimental evidence indicates that the viscosity of ice varies with applied stress and change in temperature. Figure 11.6 shows variation in the relative viscosity of ice with change in temperature at constant stress and with change in stress at constant temperature. Note that increases in temperature and stress act together to lower viscosity. These are properties of plastic flow within a solid rather than simple liquid flow.

Orientation of ice crystals into parallel planes exerts much influence on the viscosity of the ice. Thus, in a sample of frozen snow showing random crystal orientation the apparent viscosity is 70 times greater than in a multicrystalline block with crystals oriented such that all glide planes in the ice are parallel to the direction of shear.[1]

Still one more control which effectively reduces viscosity is the amount of pore water which can form when ice at melting temperature approaches the terminus of the glacier. Here, where the ice is still thick the effect of pressure on the freezing point can be important in transforming ice into pore water and lower the bulk viscosity of a unit volume.

Movement of Ice[2]

Statements made in the preceding paragraphs have implied that the movement of ice most nearly simulates plastic flow in a solid. This type of motion appears to be a combination of viscous flow and intracrystalline rearrangement within the space lattice of individual oriented crystals. Together they result in plastic yielding to applied shear stresses. Hydrostatic stress appears to be important only when this produces some melting or after a steady state of flow is achieved. Much remains to be understood about the movement of glacier ice but some clarification of the nature of the processes is important to an understanding of glacial erosion and deposition. There follows, therefore, a standard but admittedly oversimplified statement of the development of stress within the ice and its relationship to other properties common to glaciers.

Ice movement is best understood and illustrated in the valley glacier where inclination of the valley floor establishes a horizontal component of stress applied in the direction toward the mouth of the valley. Hence, a unit of the glacier can be selected as a typical segment representative of movement throughout the theoretical glacier. Movement of the ice is laminar, i.e., a theoretical ice layer moves directly forward and does not exchange position with overlying layers of ice as in the case of turbulent flowing waters. Suppose that the segment is considered to be a slab of ice of rectangular dimension frozen to an inclined smooth rock surface (see Fig. 11.7). The ice is then in a position to move down the inclined surface only by virtue of plastic flow under the application of a shear stress great enough to cause yielding; the process is identified as *gravity flow*.[3]

[1] M. F. Perutz, "Report on Problems Relating to the Flow of Glaciers," *Jour. Glaciology,* Vol. 1, 1947, p. 47.

[2] An excellent summary of the subject as well as a bibliography is presented in R. P. Sharp, "Glacier Flow—A Review," *Geological Soc. Amer. Bull.,* Vol. 65, 1954, pp. 821–838.

[3] The concepts expressed are principally those of Max Demorest, "Glacier Thinning during Deglaciation," *Amer. Jour. Sci.,* Vol. 240, 1942, pp. 29–66.

Assuming that such should occur then,

$$Y_s = t\rho(g \sin \alpha) \qquad (2)$$

where Y_s = yield strength of ice under shear stress.[4]
α = angle of inclination from horizontal.
g = acceleration of gravity.
ρ = density of ice.
t = thickness of ice.

For a given slope angle, ρ and $g \sin \alpha$ are constant, and Y_s varies directly as t.

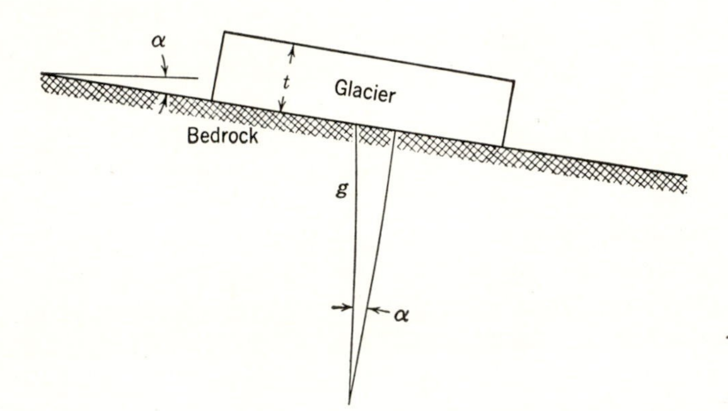

Fig. 11.7 Conceptual representation of conditions of gravity flow. The ice block is frozen to a rock surface and inclined downhill. Movement occurs by plastic flow when the shear stress exceeds strength. The stress applied is proportional to the thickness of the ice and to the slope. When the slope is low flow is brought about by a thickness which exceeds the strength of ice under compression. (From Demorest, *Amer. Jour. Sci.*, 1942.)

The surface over which the glacier moves is not of constant slope and varies locally over short distances in which the ice thickness, t, remains essentially uniform. Under such circumstances Y_s must vary directly as the increase in slope. As the value of Y_s varies from point to point along the length of a glacier a longitudinal pressure gradient is established between the points of high and low values of Y_s, and plastic deformation can occur. If a zone of high gradient is followed by an extensive distance of low gradient the additional horizontal force required for pushing the glacier along may be so great that it exceeds the resistance of the ice to longitudinal compression. Thick-

[4] Rupture under shear stress is considered independent of temperature and ranges between 7 and 8 kg per sq. cm (see Table 11.2).

ening of the ice results, and the cross-sectional area is increased until it can transmit the horizontal force without the yield strength being exceeded. Distribution of this longitudinal pressure gradient permits glacial ice to move uphill for local distances and probably is the major factor which enables a glacier to override rocky obstructions such as local hills or low mountain masses in the case of ice caps.

Plastic deformation occurs also under the direct application of static stress due to the thickness of the accumulated ice. Assuming a thin column of ice resting on a horizontal base

$$Y_c = g\rho t \tag{3}$$

where Y_c equals the yield strength of ice under compressional stress.[5] When t reaches a value which results in the critical compressional yield stress, Y_c, being exceeded plastic flow occurs at the base of the ice. If the values for Y_c between 25 and 62 kg per sq. cm are substituted in eq. 3 the critical thickness, t, above which plastic flow should occur approximately ranges between 25 and 70 m.

Where the areal dimensions of the ice block are very large in comparison to the thickness of the ice as is the case with ice caps the above equation requires modification. Plastic extrusion is then opposed by shear stresses acting between the layer near the horizontal base and the extreme upper part of the block above the limit of plastic deformation. In order to overcome these opposing stresses the pressure exerted by the upper rigid part of the ice must need to exceed both the shear and compressional yield strengths of the ice. But inasmuch as the great range in values of ice strength is associated with its behavior under compression the effect of shear strength should be considered of secondary significance in controlling the thickness of the ice necessary to produce plastic flow. Rather, temperature as a control upon compressional strength appears to assume greater importance.

On the basis of eq. 3 the height at which an ice cap will develop equilibrium conditions of plastic flow from its center outward can be approximated. In a lens-shaped ice cap the stress applied in the area of accumulation is essentially vertically directed and static, and

$$Y_c = g\rho t \tag{3}$$

But the total horizontal force across the glacier base per unit distance outward from the center is

$$F = \tfrac{1}{2} g\rho t^2 \tag{4}$$

[5] The strength at rupture under compression varies with temperature; at 0°C the value is about 25 kg per sq. cm, and at —17°C about 62 kg per sq. cm.

where F = total horizontal force across the glacier base per unit distance outward.

t = maximum ice thickness.

Inasmuch as the cross section of the ice represents equilibrium conditions the curved profile of the surface must be the result of the application of forces which cancel one another. The force F must be opposed by some other horizontal force, namely the shear stress which must act along the glacier surface.

$$Y_s = Y_c w, \qquad \text{by substitution} \tag{5}$$

where $w = \sin \alpha$ = distance along glacier base from center to margin.

In equilibrium

$$\tfrac{1}{2} g \rho t^2 = Y_c w \tag{6}$$

and

$$t = \sqrt{\frac{Y_c w}{\tfrac{1}{2} g \rho}} \tag{7}$$

Thus in the case of the Greenland ice cap a first approximation is w equals about 450 km and t equals about 3 km.

Velocity Distribution in the Glacier

Our discussion so far actually describes two distinctly different theoretical varieties of glacial motion inherent in the major classes of glaciers. The valley glacier represents an accumulation of ice along a sloping floor in which the major source of energy of movement is supplied by a horizontal component of gravity. In such a case the cross-sectional outline of ice is controlled by the valley shape, and the ice accumulates to the necessary thickness, t, sufficient to exceed the shear-yield stress of ice for the given slope angle. As ice continues to accumulate the rate of flow down the valley must increase until a steady-state condition is reached when the rate of ice accumulation is balanced by the rate of ice flow. This equilibrium condition approaches the case of the theoretical glacier of uniform thickness resting on a surface of uniform slope (see Fig. 11.7). A maximum of differential motion occurs near the base of the ice of which in the ideal case a thin layer can be considered as frozen to the bedrock. The minimum differential motion is at the glacier's surface where the shear-yield strength of the ice has not been exceeded. Within the ice, movement can be considered as the shearing of one minute layer over another, in the process of gravity flow, with an increase in total displacement toward the upper surface. Figure 11.8 illustrates the ideal relationships between relative displacement by shear and the distribution of velocity. In this connection it should be noted that per unit of time the upper surface of the ice must move farther than successively lower layers as there is a summation of all the movement beginning with the theoretical zero value at the rock-ice boundary and reaching a maximum at the ice-air interface.

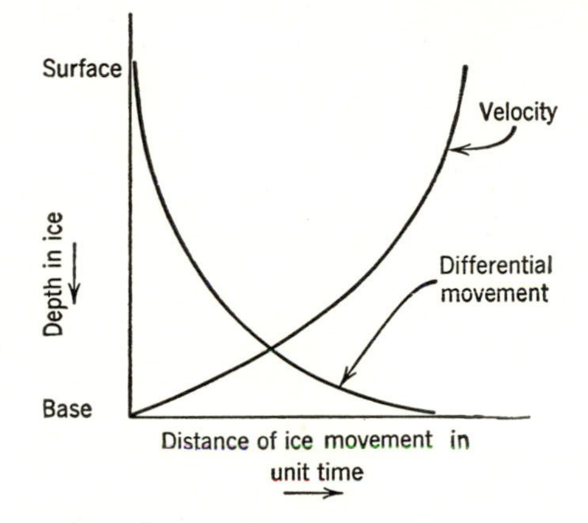

Fig. 11.8 Schematic relationship between over-all velocity and differential movement of ice in the ideal condition of gravity flow. Differential movement between adjacent laminas of ice is least near the surface and greatest near, but not at, the base. The total movement in unit time is greatest at the surface, inasmuch as that position represents the summation of all forward motion.

The established ice cap represents a steady-state condition because the addition of ice in the accumulating area is balanced by the rate of outward flow. The ice cap is not confined between valley walls, and there is no over-all general slope to the floor over which it moves. Ice flow, therefore, is controlled by movement outward from a center of greatest accumulation. The primary energy source for such motion is the thickness of the ice as indicated by eqs. 3 and 7. This is the case of plastic flow unrelated to an inclined surface and is called *extrusion flow*. In the frictionless state the velocity of extrusion flow should be greatest at the base of the ice, but, actually, the velocity of the basal ice is retarded by friction with the rock surface. The theoretical velocity distribution should be as illustrated in Fig. 11.9, and the zone of most rapid movement should lie close to, but not at, the glacier's base.

Within the valley glacier flow cannot be attributed entirely to gravity flow inasmuch as the ice pressure–melting-point relationship, the weight of incorporated rock material, and local areas of horizontal rock floor must contribute to the existence of some extrusion flow. Local observation on velocity distributions appear to substantiate such a conclusion (see Figs. 11.10 and 11.11). The two profiles illustrated are schematic but are based upon data gathered from glaciers in Switzerland.[6] Note that the zones of maximum velocity shift

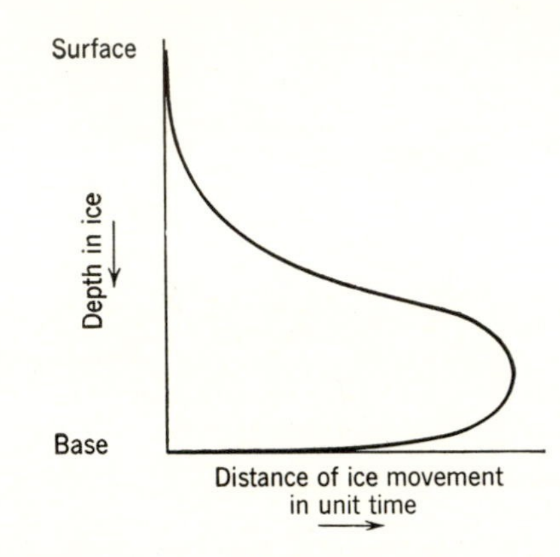

Fig. 11.9 An interpretation of the relative velocity distribution between laminas of ice in conditions of extrusion flow typical of the ice cap.

position, sometimes being concentrated near the upper surface and not necessarily confined to the glacier base. At present the velocity distribution is not subject to precise mathematical expression, but a qualitative understanding of the shift in velocities within the ice brings us closer to the solution of glacier movement.

Zones within the Glacier

The velocity distribution illustrated by Fig. 11.11 shows a zone of maximum value generally about two-thirds of the distance from the ice surface to the base. Locally, however, this zone shifts position and even appears at the ice surface. Where the maximum velocity reaches the surface the ice thickness is insufficient to cause plastic flow, and movement must be accomplished by shearing layers of ice over one another. At the site of the intersection of successive shear planes and the ice surface a series of pressure ridges and crevasses will cross the ice in a transverse direction.

Crevasses representing failure of the ice under tension also are common on the glacier surface. These appear particularly where there is a sudden expansion of the ice tongue such as when a valley glacier

[6] G. Seligman, "Extrusion Flow in Glaciers," *Jour. Glaciology,* Vol. 1, 1947, pp. 14–16.

passes from a narrow tributary into a master valley or when the confines of the valley are left behind and the glacier becomes a piedmont type. Soundings of the depths of these crevasses show that they rarely exceed 100 feet (30 m). This depth delimits the upper zone or shell

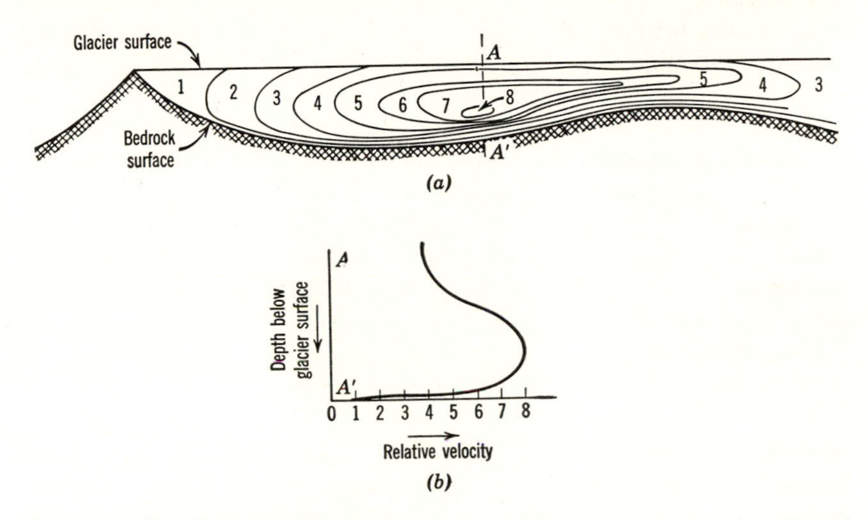

Fig. 11.10 (*a*) Part of the cross section of a valley glacier showing hypothetical distribution of velocity zones. (*b*) A section perpendicular to (*a*) and in the plane of direction of flow showing velocity distribution. (Modified from Streiff-Becker in Seligman, *Jour. Glaciology,* 1947.)

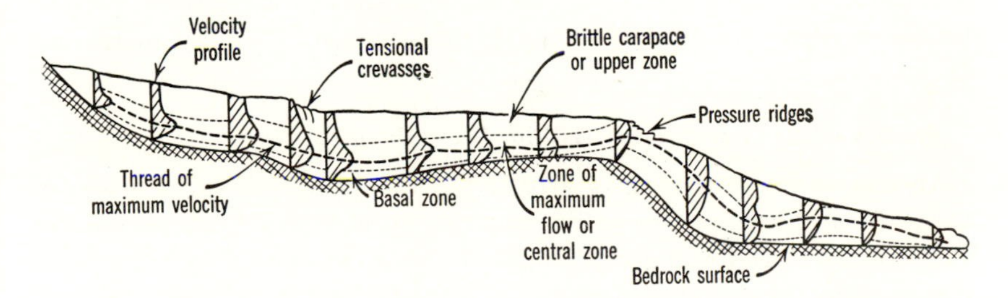

Fig. 11.11 Sketch of a long profile of a valley glacier showing the position of the thread of maximum velocity and velocity profiles at selected intervals. Note relationship of crevasses and pressure ridges to velocity changes. (Modified from Streiff-Becker in Seligman, *Jour. Glaciology,* 1947.)

which in the case of valley glaciers is carried forward by gravity and extrusion flow of the underlying ice. Differential forward motion in this upper zone establishes stresses, primarily tensional, which exceed the strength of the ice, and rupture or crevassing results. The upper 100 feet of ice, therefore, can be described as the brittle carapace of the glacier. From theoretical relationships as expressed by eq. 3 the thickness of the brittle layer below which plastic flow must occur is computed to be at a maximum depth of between 25 and 70 m which, considering the uncertainties involved, is a good agreement with the observations.

Below the brittle zone both extrusion and gravity flow operate in the same direction to produce the zone of maximum velocity. In this zone the movement primarily is in a horizontal direction, but in the region of ice accumulation the downward component of motion must be significant. Locally, the upper surface of the zone of flow may be sharply defined from the carapace wherever the yield strength of the ice is exceeded, but the lower surface is believed to be poorly delimited from the underlying zone in which motion is less. Normally, the upper boundary of the basal zone can be placed where the velocity drop is rapid, but such statements can be only generally applied as shown in Fig. 11.11.

The basal zone of the ice is of primary importance inasmuch as it is in contact with the rock surface. Considering valley glaciers the cross-sectional profile is not unlike that of the wetted perimeter of streams and may be considered a semicylindrical shell marking the rock-ice interface. In ice caps the basal-zone position is approximately horizontal. The lower zone is the part of the ice which effects active erosion of rock walls and floor bounding the glacial ice. It must also be the zone in which is carried the greatest amount of rock debris eroded at the rock-ice interface. In turn, the incorporation of much rock debris into the lower ice must inhibit plastic flow by increasing the over-all strength of the ice-rock mixture.

Steady-State Conditions within the Glacier

In mid-latitudes the valley glacier advancing from its area of source toward the mouth of the valley must eventually reach an altitude of the firn line. If for a given length of time the annual conditions of precipitation and temperature remain fixed the firn line is stabilized. Flow of ice from the area of accumulation reaches equilibrium with the amount moving past the firn line, and a steady-state condition prevails. Except for the extreme upper limits of the glacier the elevation of the surface must become relatively fixed so long as the climate remains uniform, and an aspect such as Fig. 11.5 must be developed.

Below the firn line the flow of ice from the zone of accumulation is stabilized, and loss through melting and evaporation is approximately uniform from year to year for any given elevation. At successively lower elevations the loss through ablation increases, and the ice mass becomes progressively thinner. Despite such thinning of the ice the velocity of flow does not appear to change markedly although in this section of the glacier somewhat greater velocity is observed during the warm season than in winter, and this may be due to the decrease in viscosity as the water-ice ratio increases. Continued forward motion can be expected to a position near the ice terminus where thinning is accelerated by melting and forward motion is progressively slowed down. In other words the velocity declines as less and less ice volume must flow between equally spaced distances along the ice tongue. Finally, at the terminus the ice is effectively stagnant and scarcely has any forward motion.

If the glacier were completely free from rock debris the ice tongue would represent a condition of equilibrium in which the position of the ice terminus would remain fixed by balance between melting and ice supply. The presence of rock debris incorporated within the ice modifies this condition inasmuch as at the ice terminus the rock debris must accumulate on the valley floor. As the valley floor receives more and more debris the cross-sectional area of the ice is reduced, and some increase in velocity occurs. This permits the glacier locally to override some of the previously deposited debris. The ice front, therefore, is characterized by areas of stagnant ice, slowly moving debris-laden ice, and actively moving ice shearing over debris-laden stagnant ice.

Steady-state conditions are best established under a fixed climate, and with each change in climate positions of the ice front are responses to the new temperature and moisture environment. Pleistocene glaciation was characterized by four major climatic cycles. Each cycle began as glacial climatic conditions were imposed upon regions formerly of moderate climate. The close of each cycle was marked by progressive deglaciation brought about by the onset of a warmer climate. During each such cycle the factors which were required to be placed in equilibrium were constantly being altered. For example, as the glacial cycle progressed through the frigid portion the firn line must have been lowered steadily in elevation, and the glacial mass increased in dimensions. The advancing ice must have overridden the debris formerly deposited at its terminus and progressively invaded formerly unglaciated territory. During the last half of the cycle the opposite conditions must have prevailed, and the ice front must have melted gradually leaving its debris to mark its former extent.

EROSION IN THE AREA OF THE VALLEY GLACIER

Erosion above the Valley Glacier

When conditions of steady state are approximated in the valley glacier much of the mountain mass rises above the level of the ice. This unglaciated or locally glaciated zone is subject to considerable snowfall much of which avalanches to the ice surface (see Fig. 11.12).

Fig. 11.12 Schematic drawing of a region experiencing alpine glaciation devised to illustrate terminology. *Glacier types:* 1. valley, 2. hanging, 3. regenerated. *Sections of glacier:* 4. firn basin (accumulator above firn line, F.L.), 5. glacier tongue (ablator), 6. glacier cap. *Crevasses:* 7. Bergschrund, 8 and 13. marginal, 9. transverse, 10. ice fall and seracs, 11. pressure arches (ogives, Forbes bands). *Ice features:* 12. Ice cave, 14. hummocks, 15. glacier tables, 16. glacier mills, 17. dead ice beneath moraine. *Moraines:* 18 and 21. lateral, 19. terminal, 20. medial. *Special features:* 22. glacial stream, 23 incipient avalanche glacier, 24. ice apron. (Redrawn from original by Streiff-Becker, *Jour. Glaciology,* 1947.)

Soil mantle is removed, and bare rock exposed to the wedge action of freezing water. Water freezing in fractures (joints) in the rock results in spalling off large and small slabs which at some later date are carried to the glacier surface by an avalanche. Constant prying action of the freezing water and removal of the rock to the glacier surface tends to increase the inclination of the mountain slope. With increase in steepness wedge action and avalanching are accelerated, and the mountain generally becomes a bare rock mass with a serrated and peaked outline. Many characteristic shapes are produced, most of which are identified by terms familiar to the mountaineer.

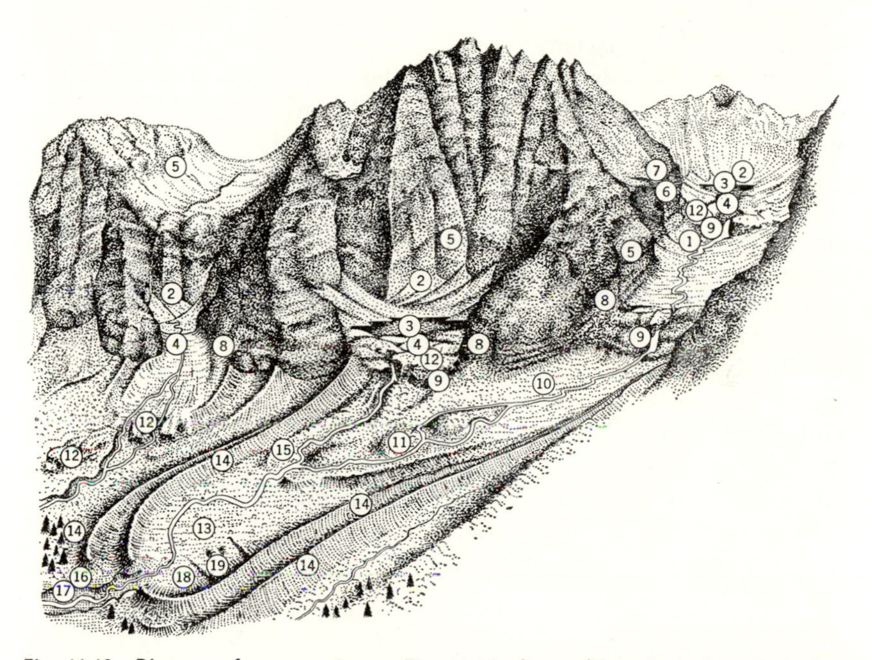

Fig. 11.13 Diagram of same region as Fig. 11.12 after melting of glaciers, devised to illustrate terminology applied to landscape features. 1. Glaciated valley, 2. cirque, 3. cirque lake, 4. rock barrier, 5. upper limit of ice erosion, 6. rock wall, 7. rock shoulder, 8. glacial grooves, 9. valley steps, 10. glacier bed, 11. drumlin, 12. roches moutonées, 13. ground moraine, 14. terminal moraine, 15. recessional moraine, 16. glacio-fluvial gravels, 17. river channel, 18. eskers, 19. kames. Redrawn from original by Streiff-Becker, *Jour. Glaciology,* 1947.)

Debris which is carried to the ice surface may be concentrated in zones along the rock-ice contact, and as the ice moves forward the loose aggregate is carried as by an endless belt toward the terminus of the ice. These ridges (*moraines*) serve to emphasize the flow lines of the glacier. (See Figs. 11.5 and 11.13.)

The Cirque and Cirque Lake

Erosion below the ice surface consists of two different processes: one is abrasion caused by rock surfaces grinding together, and the other is quarrying of blocks from the bedrock. Of the two processes the more important is the removal of fragments of the bedrock by freezing to the glacial ice of a rock isolated by fissures. With forward motion of the ice the block is pulled from its bed and incorporated within the ice. The space formerly occupied by rock is now filled by ice, and the process is repeated. Such glacial quarrying is particularly important at the head of the glacier where undercutting tends to produce perpendicular walls. Headward erosion by the ice is analogous to that produced by streams, and the effect is to develop oversteepened slopes where the unsupported rock has sheared from its position of attachment and fallen to the glacier surface. When valley glaciers disappear under the impress of a warm climate the steep walls at the source of the old glacial basin are exposed to view. These walls surrounding the basin contribute much to the impressive scenery of landscapes of mountain glaciation and are known by the French term *cirque.*[7]

The rock floor in the area particularly at the head of the glacier also is extensively quarried and leaves a water-filled depression when the glacier has melted away. Such lakes (*cirque lakes*) at the foot of the cirque wall add to the spectacular grandeur of the glaciated mountain and often are the sources of hydroelectric power. Junctions of several cirques may leave a more or less pyramidal shaped peak known as a *horn,* a term famous to mountaineers.

The U-Shaped Valley

Among other erosional features characteristic of valley glaciation is the general cross-sectional profile of the valley. Prior to glaciation mountain streams show the normal V-notch cross profile of the youthful stream. After glaciation, however, the cross profile is U- or semicircular shaped, and the valley floor is often broad and flat. The pattern of glacial erosion has altered the valley cross section by steepening the walls as well as eroding the floor. The explanation for such a change in profile appears to be found in the velocity distribution pattern of the ice tongue.

As ice begins to flow down the initial V-shaped valley the velocity and energy distributions are not symmetrically arrayed although the center moves more rapidly than the margins and the top faster than the bottom. Erosion is concentrated more along the valley walls than at the bottom as in the case of the stream valley of symmetrical cross section (see Fig. 7.7) and finally the semicircular cross section is

[7] See Figs. 11.12 and 11.13 for identification of terms used in glaciated terrain.

attained. Erosive energy is distributed equally along the rock-ice interface, widening and deepening of the valley proceeds, and the U-shaped cross section is accentuated by the height of the valley walls.

Development of the ideal cross section of the glaciated valley often is modified by certain conditions omitted from the theoretical consideration above. These include variation in the resistance to erosion by the rocks of the valley. If easily eroded rocks constitute the floor and the valley walls are resistant, valley deepening will proceed despite the further unbalancing of energy distribution within the ice. Excessive deepening of this type is to be expected in all areas where glaciers plowed across constricted areas of easily eroded rocks. Some, but not all, of the excessive depths observed in fjords appear to have resulted from conditions where large tongues of ice constricted between valley walls continued to deepen the valley some distance beyond the limits of the present shoreline.

Deepening of the valley by glacial erosion creates a topography in which tributaries to the former stream valley no longer join the master stream at accordant levels but intersect the main valley at elevations above the floor of the master stream. These *hanging valleys* are typical of mountain glaciation and often are marked by spectacular waterfalls.

Long Profile of the Glacial Valley

Typical long profiles of the glaciated valley show marked departure from a uniform gradient. Zones of very steep slope are separated by zones of low slope. The entire aspect is one of a series of steps with short steep risers between the longer treads. The "tread" zones often are basined and contain small ponds or swamp areas, whereas the "risers" are zones of bare rock exposures and water cascades. Explanations of the development of this profile utilize evidence such as multiple advances of the ice, the presence of extensive zones of fissures (joints) in some parts of the exposed rock and general absence of fissures in others. For example, localities where rocks are cut by zones of closely spaced intersecting joints are areas where glacial quarrying is likely to be active and the bed of the valley can be deepened locally, whereas zones where the joints are widely spaced are rock areas which tend to resist erosion. The importance of joint spacing probably is overemphasized, and the development of the "treads" and "risers" may be dependent more upon velocity distribution or shear planes within the ice than any attribute of the bedrock. (See Figs. 11.11, 11.12, and 11.13.)

EROSION BY THE ICE CAP

Nature of Erosion by the Ice Cap

The magnitude of the ice cap establishes conditions which are somewhat at variance with those of the valley glacier. The ice cap is not restricted by enclosing rock walls; hence, as has been described its outline is a function primarily of the forces established under extrusion flow. Data gathered from existing caps indicate that the surface slope of ice from the center outward is less than one degree and probably averages less than one-half degree. This is a remarkably flat surface and indicates establishment of equilibrium between accumulation and lateral flow. Where the entire landscape is buried ice sculpture is not restricted to general reduction of the land surface, but preferential plowing of areas of readily eroded rock, deepening large valleys, and shaping rock hills has precedence over grooving and scratching the bedrock surface.[8] Much of this erosion is on a major scale and extensive areas of rock-scoured lake basins occur in belts formerly occupied by ice. Erosion of rock hills produces one gently sloping (*stoss*) and one steeply sloping (*lee*) side. This asymmetric profile results from extensive quarrying of rock blocks on the lee side as the ice advances from the stoss to the lee side.

Debris held near the base of the ice forms an armored surface which effectively abrades the bedrock as the glacier advances. Abrasion is considered to be much less important as an erosional process than quarrying and primarily is effective in grooving, scratching, and polishing the rock surface over which the ice moves. Theoretical reconstruction of pre-glacial surfaces in New England suggests that in the development of lee and stoss hills quarrying has been several times more effective than abrasion. Nevertheless, striated surfaces and glacial grooving are important indicators of the former presence of glaciation, and their orientation is useful in determining the direction of local glacial movement.

Zones of Erosion in the Ice Cap

Evidence currently available indicates that except for important valleys general reduction of the surface by Pleistocene continental

[8] Scratches on the surfaces of soft rocks such as limestones are very characteristic and are called *striae*.

glacial action was on the whole small, averaging somewhere between 50 and 100 feet. In the areas of the former ice centers erosion was particularly slight and, in some localities, was insufficient to remove the pre-glacial weathered rock. Locally, where ice moved over belts of readily eroded rocks or followed pre-glacial stream valleys cut in soft strata erosion of the magnitude of 1000 feet is reported. Examples are the Great Lakes and the Finger Lakes of New York. Other areas

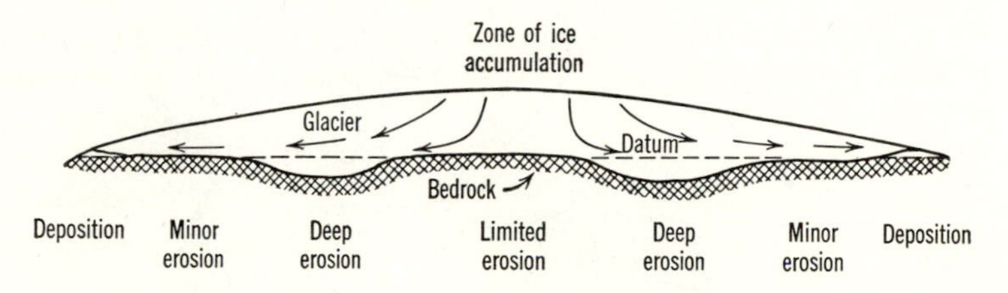

Fig. 11.14 Conceptual illustration of zones of accumulation, deep erosion, minor erosion, and deposition in the ice cap (see Fig. 11.15).

of intensive erosion are marked by the presence of rock-basin lakes and deeply plowed valleys some of which later have been filled with glacial debris.

On the basis of such information the interpretation is drawn that when the expanding ice cap finally has reached its maximum extent and attained equilibrium erosion appears to be concentrated in a belt midway between the center of accumulation and the peripheral zone. This may be due in part to the distribution and predominant direction of extrusion flow, schematic representation of which is shown in Fig. 11.14. Streamlines in the central zone of accumulation show downward-moving ice gradually spreading laterally near the base of the glacier. Outward from the central zone lateral motion is prominent, and in this zone erosion is concentrated. Still farther from the center, in the peripheral zone, lateral motion is reduced by the volume of ice lost through melting and evaporation, the energy source for erosion is reduced, and much ice remains effectively stagnant.

The concept of concentric zones of erosional intensity applied to the area in North America subjected to continental glaciation is illustrated in Fig. 11.15. This map must be interpreted as presenting only the general aspect rather than being specifically applicable to local conditions. The effect of easily eroded rocks is also important inasmuch as there is a very rough coincidence between the belt of maximum erosion and a boundary between hard (igneous) and soft (sedimentary) rocks.

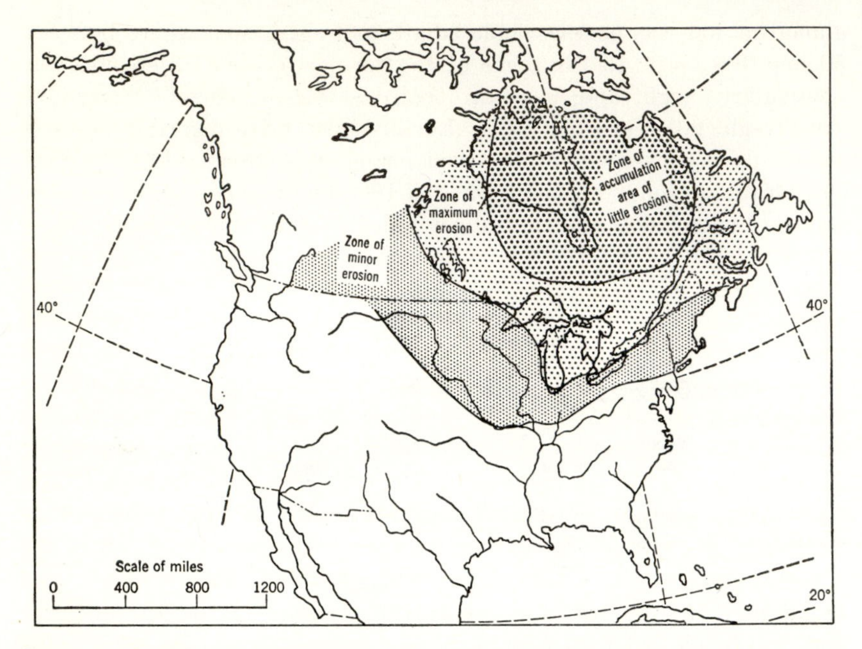

Fig. 11.15 Highly generalized limits of zones of erosion in the North American ice cap. The belt of maximum erosion lies outside the zone of major ice accumulation.

DEPOSITS OF THE GLACIER ENVIRONMENT

Transportation of Rock Debris

Valley glaciers receive much rock debris from avalanches which cascade to their surfaces. This mass of fragmented rock locally accumulates in piles on the ice surface and forms the medial and lateral moraines. (See Figs. 11.12 and 11.13.) Ideally the density difference between this debris and ice should permit some of the material to sink downward and eventually reach the base of the ice, but almost all the broken rock is carried along on the ice surface as a ridge. The bulk of the rock debris, however, is gathered either by quarrying or abrasion along the rock-ice contact and remains in that position. Rock loads characteristically are composed of large blocks, boulders, pebbles, and all size grades including much of clay dimension, and as a heterogeneous mixture become incorporated within a matrix of ice. As large particles scour across bedrock surfaces they are striated, grooved, or abraded into somewhat characteristic shapes. Locally, during the journey to the ice margin this assortment is deposited beneath the ice.

Although the mechanism controlling this deposition is imperfectly understood the deposit is a pebbly sediment called *till,* which generally but not necessarily is very high in clay content. Debris of similar sizes and composition is carried to the ice terminus where it is deposited in concentric ridges known as *terminal moraines.* Some of this material is sorted partially by melt water into gravel, sand, silt, and clay and deposited in ponds, lakes, or the flood plains of streams adjacent to the glacier.

Observations of existing valley glaciers indicate that the debris is carried primarily at the base of the ice or on the surface, whereas the central part of the ice is relatively clean. Noteworthy exception is immediately below medial and lateral moraines where the debris penetrates deep into the ice. The presence of rock debris beneath the lateral and medial moraines is the result of: (1) gradual increase in addition of ice to the glacier burying the avalanched rock debris, (2) plastic flow within the ice carrying debris downward as well as laterally, (3) theoretical settling of rock particles through ice of reduced viscosity when the ice temperature is at the melting point. The importance of the first two conditions requires no further analysis, but the extent to which particles may settle through the ice has not been evaluated.

Data on the rates of forward motion of glaciers show much variability. Some of the highest rates of motion are reported from Greenland where part of the outlet ice moves at the rate of 5600 feet per year (17×10^4 cm). Low rates appear to be of the order of less than 1 inch per day (9×10^2 cm). For a glacier 100 km (10^7 cm) long, a particle moving 17×10^4 cm per year would require 60 years to travel the length of the ice, whereas if the ice were moving 9×10^2 cm per year 10^4 years would be needed to reach the terminus. Reference to the hypothetical calculations of Table 11.3 shows that a boulder of 100-cm radius could sink 1 cm in the 60 years allowed to move 100 km (at rate 17×10^4 cm per year) or would sink 170 cm during its travel of 100 km (at rate 9×10^2 cm per year). Smaller particles would scarcely settle at all through the ice during the course of travel.

The preceding analysis suffers from oversimplification of glacial conditions and properties and the likelihood that Stokes' law does not properly apply, but it is intended to provide a rough approximation of settling rates which doubtless would be much slower than calculated above. For this reason our results are conservative, and we may infer that the majority of rock debris of the medial and lateral moraines fails to reach the base of the ice before it is carried to the ice

Table 11.3. Hypothetical Rates of Settling of Spheres of Rock Through Glacial Ice According to Stokes' Law

(Average viscosity of ice 10^{14} poises, average particle density 2.6)

	Radius of Particle in Centimeters	Settling Velocity Centimeters per Second	Number of Years Required to Sink 1 cm	Number of Years Required to Sink 1 km (10^5 cm)
Clay	0.0001	5×10^{-16}	6×10^{7}	6×10^{12}
Sand	0.01	5×10^{-14}	6×10^{5}	6×10^{10}
Pebble	1.0	5×10^{-12}	6×10^{3}	6×10^{8}
Boulder	100.0	5×10^{-10}	60	6×10^{6}

terminus. Debris that is present at the base of the ice, therefore, must be gathered at the ice-rock interface and remain more or less fixed in that position until deposition.

The glacier's regime includes ice transporting a load of rock fragments ranging between clay and boulders, stagnant but loaded ice, and clean ice. Primarily near the ice terminus but also locally throughout the glacier melt water is active in carrying and sorting part of the rock load. It is to be expected, therefore, that an intimate relationship should exist between sediments deposited directly by the ice at the glacier's base and sediments deposited by melt waters moving through and from the ice. Sediments deposited by water action normally are stratified, whereas ice-deposited debris primarily is massive and heterogeneous. This difference in structure has led to subdivision of glacial sediment into two classes, namely stratified and nonstratified deposits. Long before the glacial origin of these materials was recognized they were thought to have been carried by water or rafted into position by floating ice. All such sediment was named *drift*. This term became deeply engrained in geologic literature and still remains in common use to connote stratified and unstratified glacial material.

Unstratified Drift (Till)

Unstratified drift consists of such an intimate mixture of large and small fragments that direct deposition by ice must be acknowledged. Fragments too large to be moved by wind or water action are in direct contact with and surrounded by grains of clay of colloidal dimension. Absence of layers of particles of approximately the same size such as is dictated by settling through water or air is noteworthy.

Unstratified drift predominantly is a bouldery clay deposit, and the term *boulder clay* often is applied to such material. However, in many localities sand or silt constitutes the matrix material, and only small quantities of clay are present. Elsewhere, boulders are absent, and the deposit consists of a very tough clay with a few scattered granules or small pebbles. These varieties of unstratified drift are not suited to the term "boulder clay" and the more general term of "till" is applied. Although till is restricted to unstratified drift local pockets of stratified material often are incorporated within larger masses of till. Commonplace also are masses of till within much stratified debris, and distinction between till and stratified drift is not always made easily. The concept of intergradation should be kept clearly in mind. Many occurrences are reported where till grades upward into stratified drift, the latter representing debris which was carried in the upper part of the ice and deposited as the ice melted.

Properties of Till

Till is an extremely variable sediment. It may consist of as much as 99 per cent clay or virtually all boulders or any combination of these and intervening size grades. Pebble varieties are numerous, and it is not uncommon to note 10 to 15 different major rock types occurring in different proportions. Distribution of rock types in Table 11.4 shows a strong predominance of sedimentary rocks over other varieties. Similar counts made on tills in New England, northern Wisconsin, and other localities where igneous and metamorphic rocks are abundant show sedimentary rocks to be infrequent. This difference in percentage distribution of rock types is not random but undergoes progressive changes depending upon the character of the bedrock over which the ice has moved. Pebbles from tills listed in Table 11.4 are chiefly dolomite and limestone. The underlying bedrock is of similar composition, whereas igneous and metamorphic rock pebbles whose percentage is small represent fragments from outcrops far to the north. The local source of the preponderant pebbles is illustrated by Fig. 11.16 which shows the location of till samples in the Valparaiso moraine near Chicago and the character of the underlying bedrock. The predominance of dolomite and limestone pebbles rapidly declines and is replaced by shale and siltstone as the bedrock changes. The changeover is not confined precisely to the boundary between bedrock types, but appears to be accomplished within 50 miles of the bed-rock boundary change. The local nature of the predominating pebbles is useful in providing samples of the nature of the underlying bedrock. Of more general use is the determination of the pebble character to

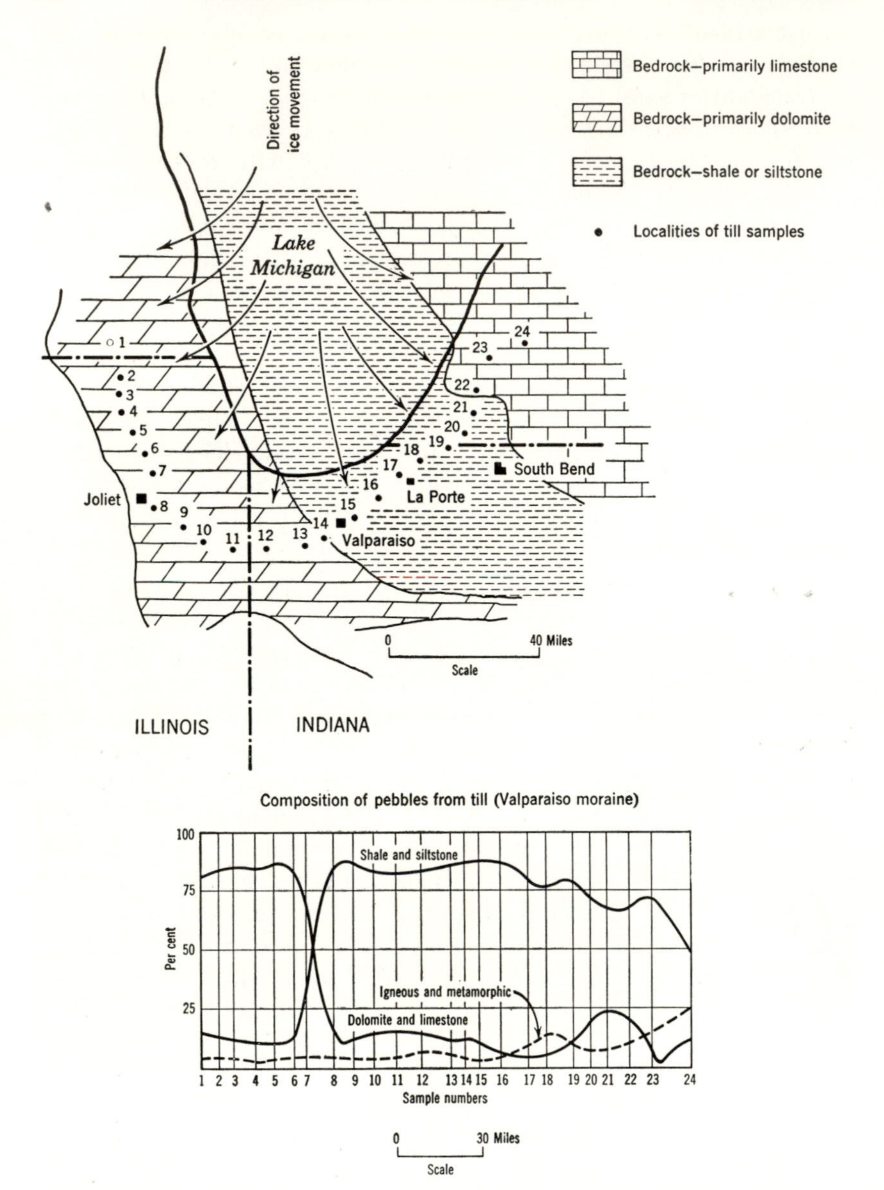

Fig. 11.16 Map of the distribution of bedrock types beneath and surrounding the southern end of Lake Michigan, of the dominant direction of ice flow, and of the localities where counts were made of pebbles in the glacial till (Valparaiso). Below is a plot of the distribution of pebble types with respect to the sample sites. Note the strong relationship between the type of predominating pebble and the proximity to bedrock of the same composition. (Modified from Krumbein, *Jour. Geology,* 1933.)

Table 11.4. Distribution of Rock Types among Pebbles From Several Moraines* In Northern Illinois

Kind of Rock	Moraine			
	Marengo	Gilberts	Marseilles	Valparaiso
	Distribution in Per Cent			
Chert	4.3	3.6	5.8	1.8
Dolomite	70.3	56.7	76.4	75.4
Limestone	9.7	26.8	4.8	14.2
Sandstone	2.1	4.5	2.2	3.9
Shale	6.8	3.1	6.2	0.8
Basalt	2.2	1.5	1.1	0.7
Diorite	0.5	0.6	0.1	0.5
Felsite	0.1	0.5	1.0	0.0
Granite	1.7	2.2	2.3	0.7
Quartzite	1.1	0.3	0.1	1.2
Totals	98.5	99.8	100.0	99.2

* Data from Jessie B. Stark and Evelyn M. Turpin, *An Analysis of Some Physical Characteristics of the Glacial Tills from Moraines in North-Eastern Illinois.* Unpublished Thesis, M. A., Northwestern University, 1945.

be expected in tills by use of a geologic map of the underlying bedrock surface.

Some tills are characterized by the presence of boulders (often very large) completely foreign to the local bedrock. These boulders large or small are known as *erratics* and sometimes are useful as a means of determining the direction of former glacial movement. Special rock types known to crop out in restricted areas supplied fragments which have been carried to new sites. Erratics of this kind are very helpful in determining the ice centers from which certain drifts were derived as well as providing information on ice paths. The small percentage of igneous and metamorphic erratics in the pebble analysis of the moraines of northern Illinois (Table 11.4) contains some types which can be traced to outcrops north of Lake Huron.

Bedrock control also is reflected in the fine-grained fraction of the till. Fragments from dolomite and limestone bedrocks are ground to a calcareous clay and shales to noncalcareous clays, whereas granites, quartzites, and sandstones produce sandy tills with very little material in the clay-size fraction.

Some concept of the great size range of till as well as proportionate distribution of sizes can be obtained from Fig. 11.17*a*. This diagram represents a composite analysis of ten samples of till from the Valparaiso moraine in northern Illinois (see Fig. 11.25). In this case the

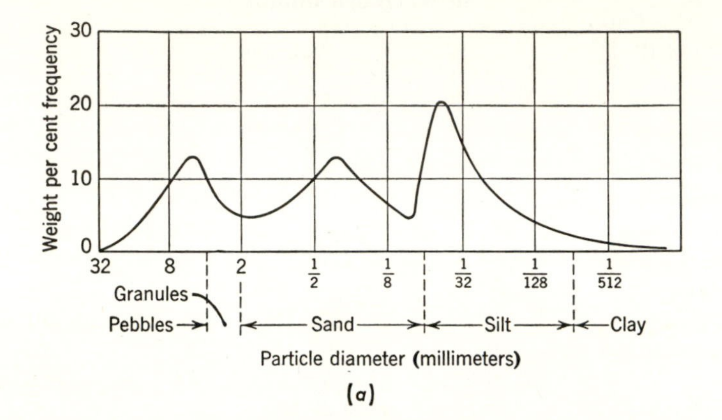

(a)

(b)

Fig. 11.17 (a) Particle size frequency distribution of composite of 10 samples of till from Valparaiso moraine in northern Illinois. Note the polymodal distribution which characterizes some tills, a departure from the law of simple crushing. (Data from Lundahl, cited by Pettijohn, Harper and Brothers, 1949.) (b) Exposure of typical coarse boulder till, Bull Lake moraine, near Bull Lake, Wyoming.

composite sample is high in the silt sizes, whereas the clay fraction is low. Other samples show a concentration in the clay sizes. Concentration of several sizes is important in producing the conglomeratic texture (coarse particle surrounded by matrix of fine particles) typical of till. Certain analyses indicate the size distribution to be in accordance with that predicted by Rosin and Remmler's law of crushing (see p. 121). This is to be expected in view of the crushinglike processes involved in glacial erosion and the absence of tendency to segregate material into beds of similar sizes.

Unweathered till usually is dominated by a gray color in shades ranging between light and dark. Often the gray is modified by different hues of blue, green, buff, brown, and red depending upon the color of bedrock from which the till is derived. Weathering alters the color appreciably and develops gray buffs, buff yellows, and several shades of browns.

Fresh exposures reveal the general massive character of till which at first glance appears to lack any orderly character. (See Fig. 11.17*b*.) Further inspection of a clayey till reveals a fracture pattern of more or less irregular outline. These fractures control the appearance of individual clay fragments producing blocky, laminated, and irregular pieces which rapidly become shapeless on wetting or weathering.

Pebbles and larger rock fragments are variable in shape although there is a strong tendency to be equidimensional. Corners and edges are slightly rounded, and the general appearance is subangular. Ranges in pebble shape are extreme inasmuch as strongly laminated rocks such as slate, schist, or shale break along the cleavage surfaces and form bladed or disc-shaped fragments. Homogeneous rocks such as limestone, dolomite, granites, etc., tend toward equidimensional form. If the latter varieties dominate a shape distribution illustrated in Fig. 11.18 is observed. This shape analysis is based on the ratio of three axial lengths in a theoretical triaxial ellipsoidal pebble.[9] Axial ratios of intermediate to long are plotted as ordinates against ratios of short to intermediate which are plotted as abscissae. Four general shape categories are obtained: namely, discs or low prisms, blades, rods or high prisms, and spheres or cubes. Pebbles from moraines of northern Illinois appear to be preferentially in the shape of modified discs or cubes. Tills over schist or slate bedrocks should contain pebbles which are chiefly discs or blades.

Surfaces of the pebbles of soft fine-grained rock (shale, dolomite, or limestone) very commonly are striated, whereas hard or coarse-grained rocks rarely show any sign of surface marking. Striae are

[9] This technique developed by Zingg is described by W. C. Krumbein and L. L. Sloss, *Stratigraphy and Sedimentation,* W. H. Freeman and Co., 1951, pp. 78–84.

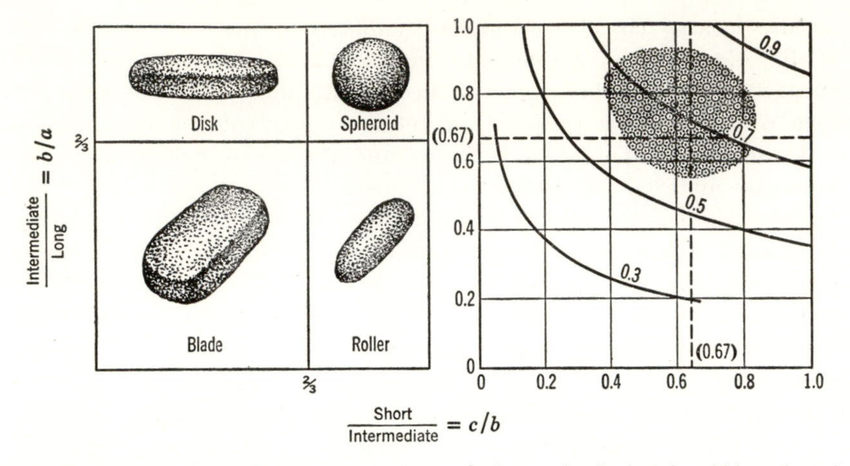

Fig. 11.18 Generalized distribution of shapes (sphericity) of glacial pebbles plotted on chart originally developed by Zingg. Ratios of axial lengths of intermediate to long and short to intermediate characterize a certain over-all sphericity whose value is indicated by curved lines in right-hand illustration. Area of pattern is general sphericity of glacial pebbles from limestone bedrock. (Modified from Krumbein and Sloss, W. H. Freeman Co., 1951.)

very typically developed during glaciation and should be used as important guides in recognizing glacial till despite the presence of similar markings which appear on fragments in landslide debris unrelated to glaciation.

Drumlins

Certain extensive but widely scattered areas are characterized by the presence of elongate ovaloid hills, each separate from the other but clearly related in size, shape, and direction of elongation. These hills known as *drumlins* are constructed primarily of till, often very bouldery, but some contain stratified deposits to a greater or lesser degree. The ideal drumlin appears to be plastered to the surface of low-lying morainic debris (*ground moraine*) above which it rises as a discrete unit and is associated always with other neighboring drumlins. Normally, the individual drumlin is a mile or less long, about a $\frac{1}{4}$ mile wide, and averages about 100 feet high although the range is between 20 and 200 feet. A single ovaloid outline with one gentle and one more steeply inclined side along the long axis is typical although round outlines, domed shapes, and doublets of two joined together are not uncommon. The gentle slope continues with gradually diminished elevation until it becomes part of the ground moraine and is oriented in the direction of ice flow, the streamlined shape being attributed to movement of the ice over the hills. The reader will note that this is a profile similar to the lee and stoss rock hill but reversed in orientation 180 degrees, i.e., with the gentle side indicating the direction of ice movement.

The internal structure of the drumlin is variable but often possesses a core of bouldery clay which when examined displays greater strength under unconfined compression than the peripheral material. Other drumlins are known to have bedrock cores against which till has been plastered (Fig. 11.19*b*). Cores of tough well-compacted, pebbly or bouldery clay appear to represent till carried in the basal part of the

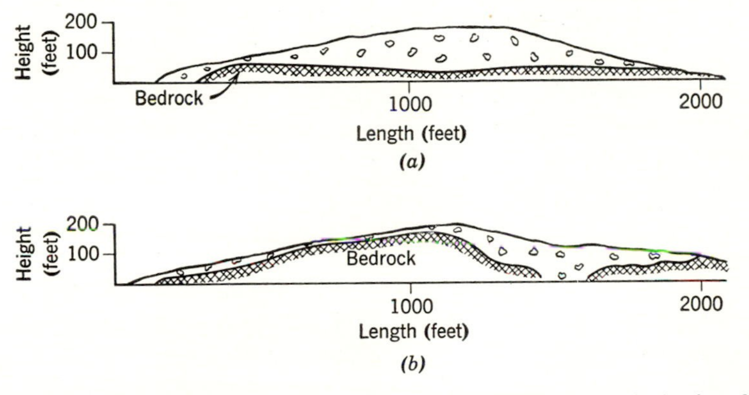

Fig. 11.19 Cross sections through two drumlins to show relationship to bedrock surface. (*a*) Mt. Ida, Newton, Massachusetts, somewhat oblique to long axis. (*b*) Transverse section through Parker Hill, Roxbury, Massachusetts. (From Crosby and Lougee, *Geological* Soc. *Amer. Bull.*, 1934.)

ice and plastered in rude successive layers against some local irregularity of the former surface. Overriding by the ice is believed to be responsible for compaction of the till and elongation of the form in the direction of ice flow. As the ice stagnates and melts additional stratified debris may become associated or superposed upon the older core but is not fundamental to the feature.

Drumlins are not distributed uniformly but are concentrated in certain areas for reasons which as yet are not clearly understood. The regional pattern is impressive inasmuch as they appear to radiate outward in clusters from the centers of concentration. This radial pattern of orientation has been interpreted as related to actively spreading sections of the glacial tongues in areas near the ice margins. Their sporadic distribution, however, suggests they must be the product of certain special conditions as yet unrecognized which are prevalent near the margins of ice caps. They are not considered to be a feature associated with valley glaciation although there appears to be no reason why they could not be formed under such conditions.

Moraines

Originally the term *moraine* was applied by peasants living in the French Alps to elongate ridges composed of loose boulders and smaller-sized sediments in heterogeneous mixture which parallel the margins of existing glaciers of the region. Later, after the origin of the sediment composing the moraine was recognized the name was applied to the material irrespective of its topographic form. Terms such as *ground moraine* were applied generally to signify till and other debris deposited beneath a glacier but lacking any topographic lineation. Within recent years there has been a general return to the original concept of particular topographic expression as part of the definition. As for example, ground moraine is distinct from *end moraine* which is a ridgelike accumulation constructed along any part of the continental glacier margin.

Primarily, the term moraine has place significance with respect to the ice tongue. *Lateral* and *medial* moraines refer to respective positions on the surface of the valley glacier. (See Figs. 11.12 and 11.13.) Ground moraine refers to a widely distributed veneer of drift which was built beneath the ice. End or terminal moraines are concentric bands of drift with elevated topographic expression marking the accumulation during stages of steady-state conditions at the ice front.[10] Recognizable end moraines are associated with maximum advances of individual ice sheets (i.e., from ice caps) or with local ice stands during the last stage of deglaciation as the ice front successively melted toward the accumulating center.

End moraines of the most recent ice advance are preserved in much of their original appearance. They are recognized by a series of parallel ridges constituting an arcuate belt of the shape of the ice lobe or composites of several ice lobes. Figures 11.20 to 11.23 show the limits of end moraines for individual glacial stages, but the composite of all the end moraines marking the positions of the ice-invaded portions of the United States is shown in Fig. 11.24. In detail the arcuate pattern is complex as illustrated by the distribution of end moraines associated with the Lake Michigan ice tongue, a local expansion of a

[10] The term *terminal moraine* is best applied to a rock-debris ridge marking the outer margin of an ice tongue in a valley but is applied also to similar ridges marking important stands of ice during continental glaciation of the Pleistocene.

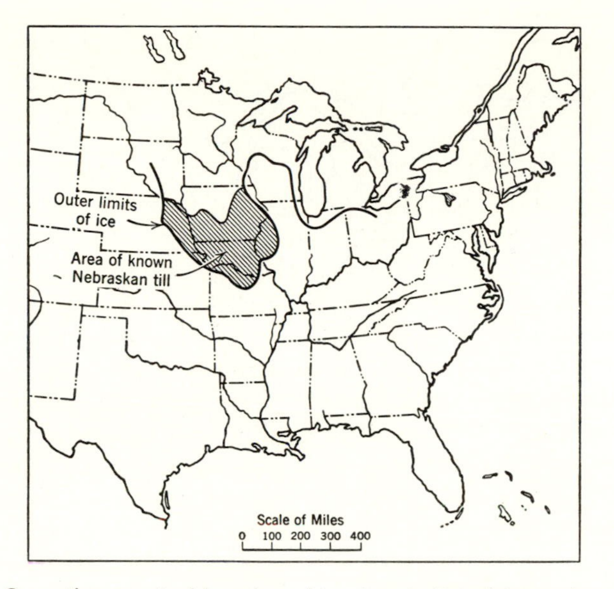

Fig. 11.20 Currently recognized boundary of ice sheet in United States during Nebraskan glacial stage. (Data from Flint, John Wiley and Sons, 1947.)

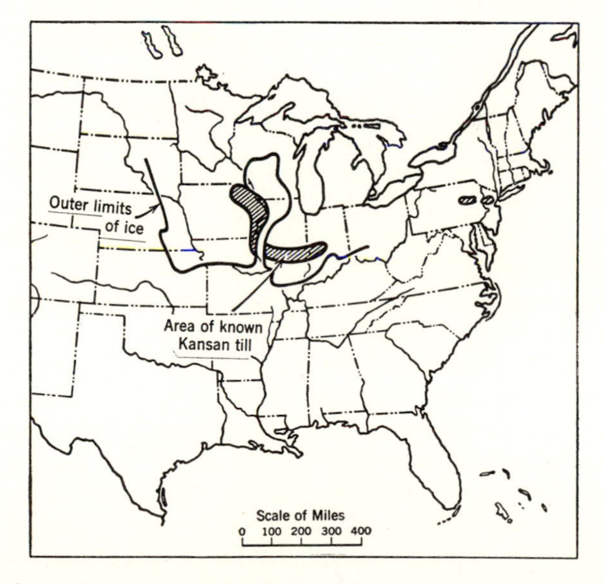

Fig. 11.21 Currently recognized boundary of ice in the midwestern United States during Kansan glacial stage. Note the deep indentation of ice-free land in southwestern Wisconsin. This is part of the so-called Driftless Area. (Data from Flint, John Wiley and Sons, 1947.)

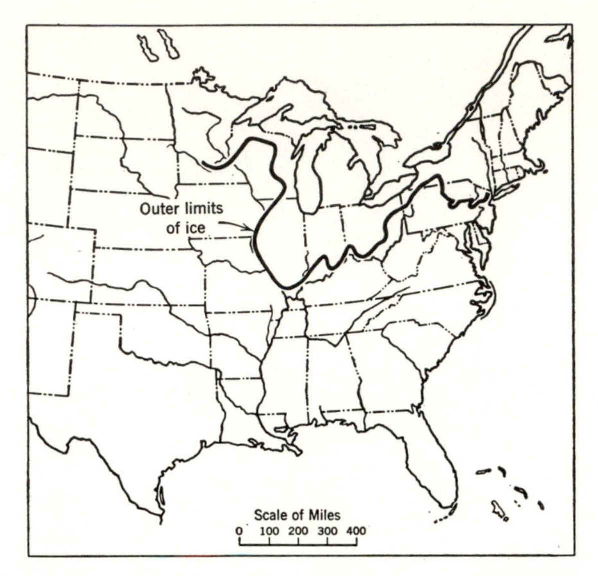

Fig. 11.22 Currently recognized boundary of ice in the United States during the Illinoian glacial stage. Note the positions of the Ohio and Mississippi rivers with respect to the ice limits. (Data from Flint, John Wiley and Sons, 1947.)

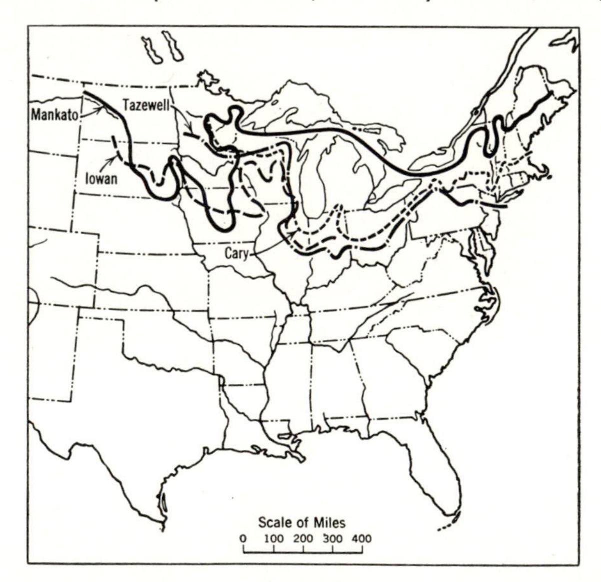

Fig. 11.23 Currently recognized boundaries of ice during various substages of the Wisconsin stage of glaciation. (See Table 11.5.) Note the general northward retreat and preservation of the unglaciated Driftless Area in southern Wisconsin. Note also the correspondence of position of the western reaches of the Missouri River and the limits of the Mankato ice. (Data from Flint, John Wiley and Sons, 1947.)

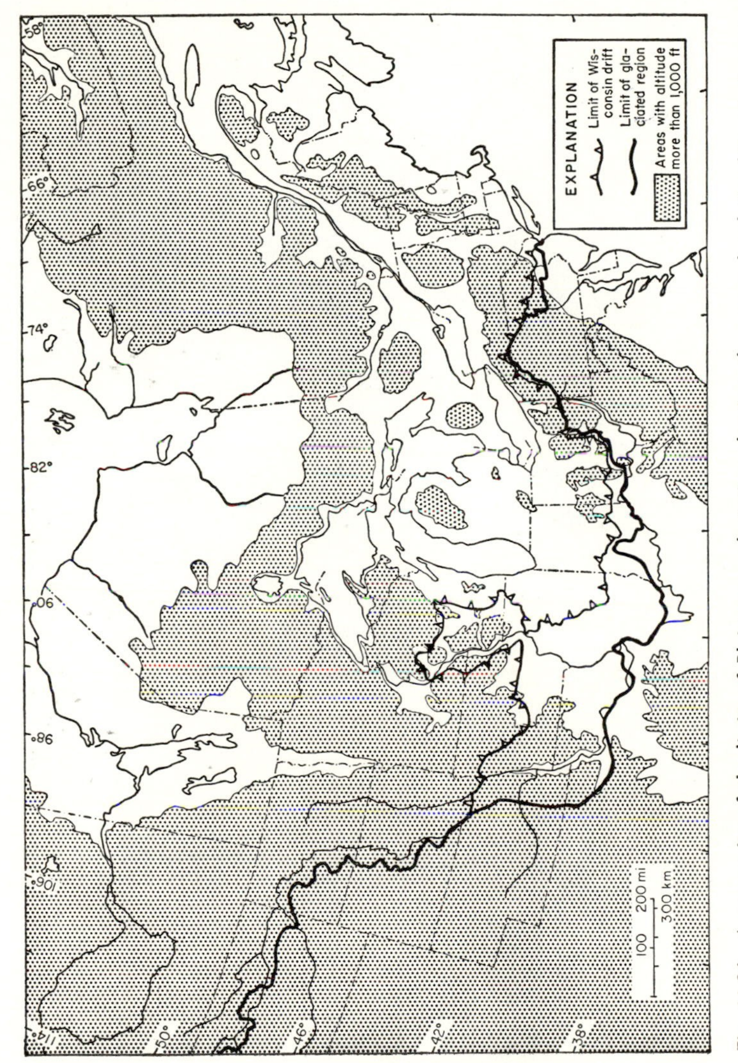

Fig. 11.24 A composite of the limits of Pleistocene glaciation showing the general relationship of deepest southerly invasion in the region of lowest land elevation and isolation of the Driftless Area. Note also the position of the Missouri and Ohio river systems as trunk streams marginal to the ice. (Reprinted with permission from Flint, *Glacial and Pleistocene Geology*, copyright 1957, John Wiley and Sons, Inc.)

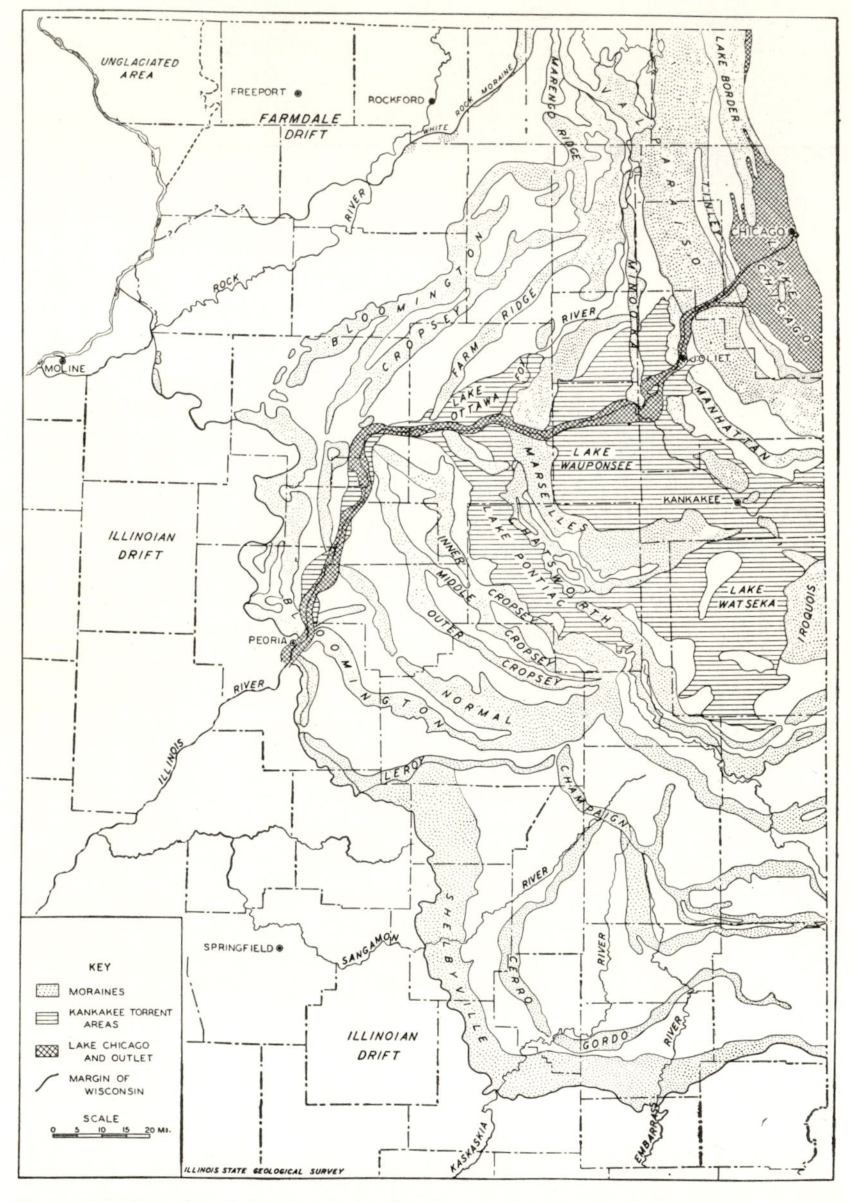

Fig. 11.25 Example of details of the distribution of moraines of the Wisconsin stage and associated outwash deposits in northeastern Illinois. Note the general lobate outline associated with the south end of Lake Michigan. The merging of moraines and the overlap of older deposits by younger ones can be noted in several localities. Observe also the distribution of lake deposits between moraines. The Illinois River drainage is of a later origin (see Fig. 11.33). West of the limits of the moraines of

much larger mass. The system of roughly parallel ridges some of which have poorly developed topographic form accords with the present lake outline. (See Fig. 11.25.)

End moraines of continental glaciers rarely have a topographic relief which is more than 100 feet. Those in mountain valleys commonly exceed several hundred feet and have been reported as much as 1000 feet high. Except for the general ridged aspect the local details of the end moraine are variable. Some are smoothly rolling, whereas others are confusions of more or less semicircular depressions and individual knobs characterized by an absence of any well-defined stream drainage. Small lakes, ponds, swamps, and undrained depressions are scattered between higher parts of the moraine and serve to accentuate the irregularity of topography. As a general rule old moraines are subdued by subsequent stream erosion and downward creep of soil. Two compositionally similar moraines but of different ages often can be distinguished on the basis of the stage of integration of a drainage pattern or the degree to which the respective topographies are subdued.

End moraines contain much stratified drift along with till. The stratified material is the product of deposition from melt waters at the ice front, and large aprons of stream-transported gravels, sands, and silts mantle the unstratified debris. Local readvance of the ice results in deposition of till on the stratified deposits, and the end-moraine zone becomes a disordered complex of stratified and unstratified sediment.

Surfaces of ground moraines show low relief often accentuated by swampy patches and local concentrations of erratic boulders scattered on the present surface. Ground-moraine areas are extensive and constitute the zones between end moraines which often are widely separated. Drumlins, local patches of stratified drift, and post-glacial wind, stream, and lake deposits often are laid down on the surface of the ground moraine and locally alter its topographic expression.

Stratified Drift

Stratification is the single criterion by which some drift is distinguished from till. If the stratification is clearly visible there is no

---

the Wisconsin stage are deposits of the Illinoian ice. These deposits can be observed to be overlain by deposits of Wisconsin ice. In the extreme northwestern corner of the map is a part of the boundary between the Illinoian drift and the unglaciated section known as the Driftless Area. (Map by G. E. Ekblaw, courtesy of Illinois Geol. Survey, 1957.)

difficulty in its recognition, but some debris is very poorly layered and as such is only arbitrarily differentiated from till. The stratifying agent primarily is running water, and the aspect of individual layers is the result of concentrations of particles of more or less uniform size. Layers of similarly sized particles were deposited in a given locality during a certain energy regime of the stream. Inasmuch as these regimes change abruptly with local increase or decrease in quantity of water, beds of coarse gravels may be deposited on fine sand, or silt may be deposited on coarse sand. The result is a parallel set of neatly defined layers. Each layer is characterized by a particle size distribution which is independent of the layer above or below, but all the layers have a size range which is more narrowly restricted than that of the corresponding till.

Outwash

The great bulk of stratified drift is called *outwash* and generally is intimately associated with end moraines. As the name suggests outwash is the product of redistribution by water of rock fragments carried to the end-moraine zone. Ideally, outwash forms a large apron spreading outward from the glacial end zone and decreasing in over-all particle size grade with increased distance from the ice terminus until fine sand and silt dominate the deposits. The outwash apron begins with the establishment of equilibrium conditions at an ice front, gradually is extended throughout the glacier's growth, and reaches maximum development during the time of melting. Outwash surfaces generally are level, but the surface often is pitted by depressions left by melting of blocks of ice rafted into position and buried by stratified drift.

In valley glaciers the outwash is confined by the valley walls, whereas in continental ice sheets the distribution of an individual outwash apron is dependent upon the localities of major streams draining the ice. Fingers of outwash in the broad channels of the glacial streams may extend many miles downstream from the ice boundary, and the areas between the old stream channels may be locally free from such deposits.

Unsystematic variability in distribution, quantity, particle size, and clay matrix is the rule rather than the exception in outwash deposits. Tongues and pockets of till may be interlayered with coarse gravel. Some outwash is deposited rapidly and still contains a high clay-size fraction, whereas immediately adjacent may occur a deposit of uniformly sized sand or gravel. Yet, the over-all aspect reveals a concentration of gravel near the moraine and a diminution in amount and

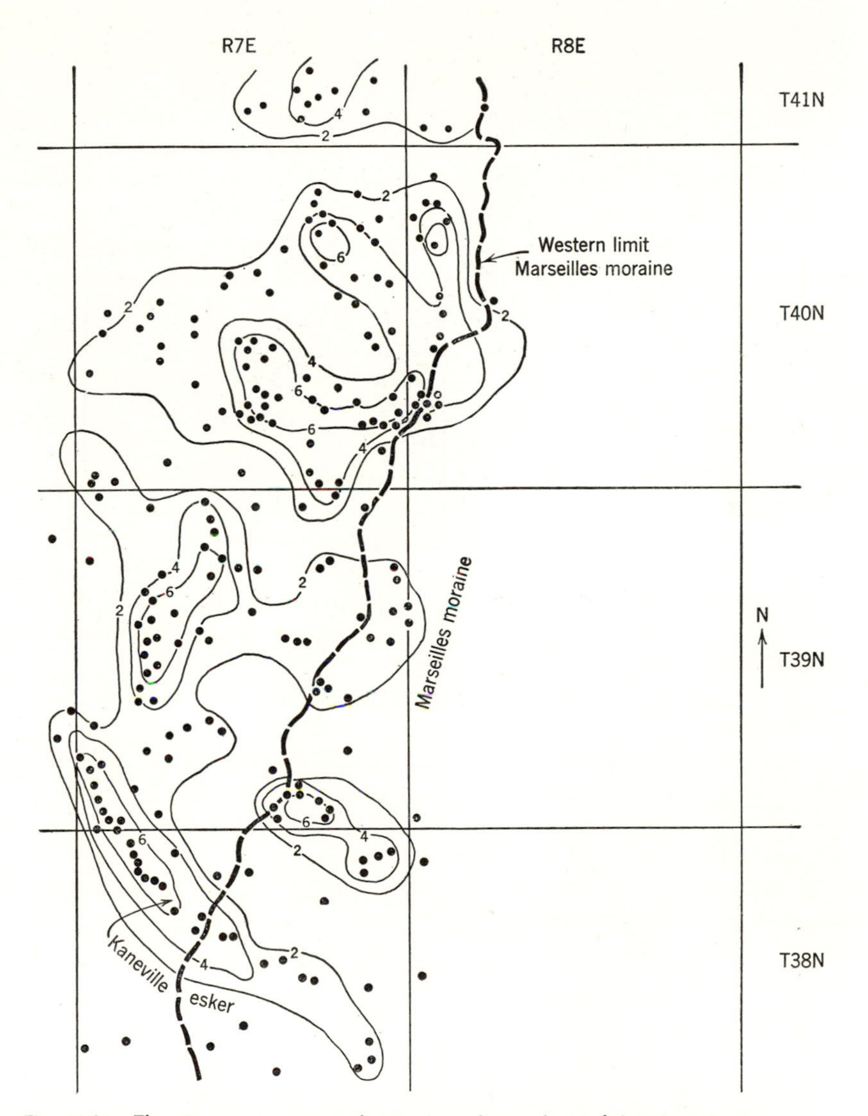

Fig. 11.26 The contours represent the number of gravel pits (closed circles) per square mile in the outwash of the Marseilles moraine in a local area of northeastern Illinois, north of the Illinois River (see Fig. 11.25). Concentration of gravel is within a few miles of the moraine margin and diminishes to the west. Gravel of the Kaneville esker seems to be distributed within a single stream channel, whereas other deposits are considered as those of braided streams. (Modified from Powers, *Pleistocene Geology of the Geneva, Illinois, Quadrangle*, Ph.D. thesis, Harvard University, 1931.)

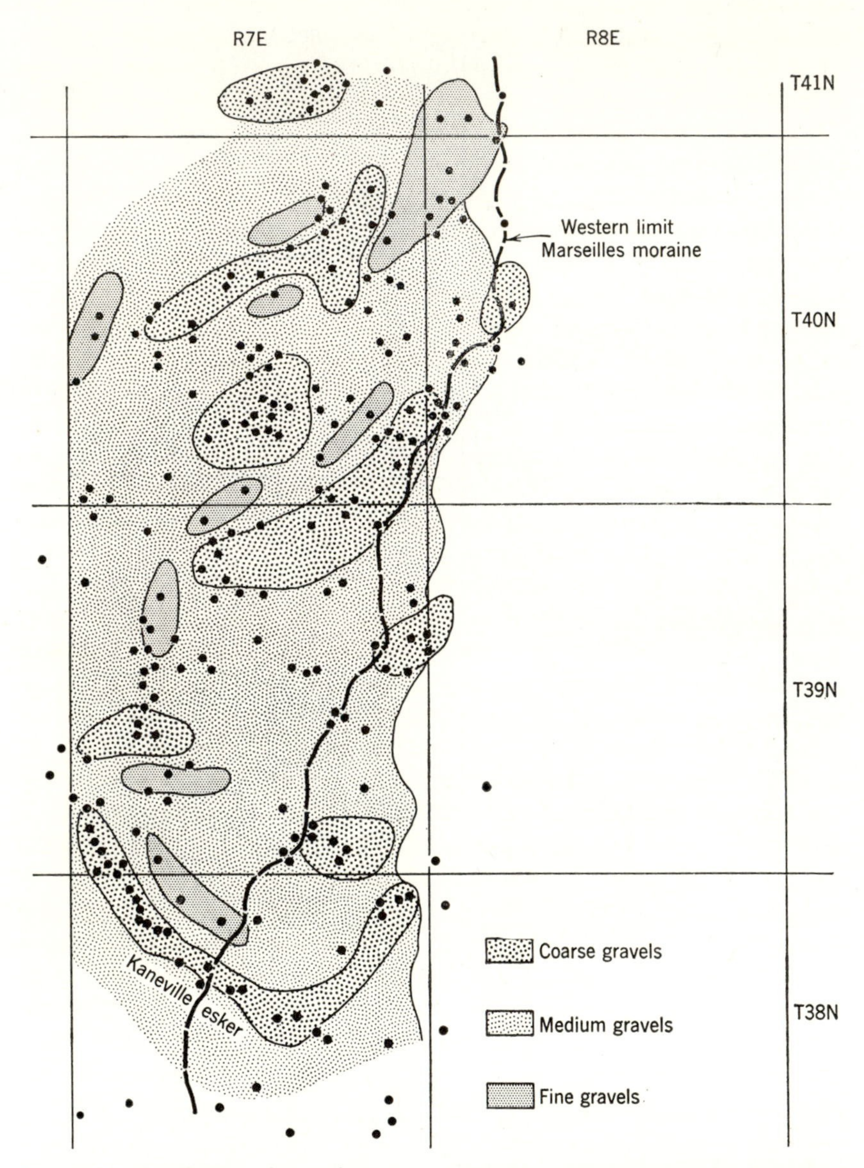

Fig. 11.27 Distribution of gravel sizes in gravel pits (closed circles) in the outwash of the Marseilles moraine in northeastern Illinois (see Fig. 11.26). Note the tendency for coarse gravel to occur near the moraine boundary. (Data from Powers, *Pleistocene Geology of the Geneva, Illinois, Quadrangle*, Ph.D. thesis, Harvard University, 1931.)

size with increased distance downstream from the ice. Figure 11.26 shows the concentrations of gravel pits in a limited area adjacent to the Marseilles moraine in northern Illinois. (See also Fig. 11.25.) This locality has been thoroughly exploited for gravel, and one can assume that local absence of pits indicates general absence of gravel. Contours on the number of gravel pits per unit area (1 sq. mile) reveal the positions where clean gravel is concentrated with respect to the moraine margin. To the west gravel deposits diminish in number indicating that the particle size of the outwash is below that of desirable gravel. The deposits also are distributed in a very irregular pattern although the general parallelism to the moraine border is demonstrated as is concentration of good gravel several miles west of the moraine. This distribution is attributed to the general admixture of till and gravel at the moraine border, whereas deposition of clean gravel occurred several miles from the ice stand. Observe, also, that certain systems of stream channels appear to have been more favorably disposed to carry gravel as indicated by the paucity of gravel pits in several narrow and sinuous strips near the upper, middle, and lower parts of the map.

Figure 11.27 is a map of outwash from the Marseilles moraine but is a generalized representation of the distribution of gravel sizes. There exists a rather crude correlation between average coarseness of particles and concentration of gravel pits, but medium-sized gravels are the most abundant. As a whole coarse gravels are concentrated within 3 to 4 miles of the moraine where systematic reworking of formerly deposited material separated the coarse fraction. However, the sinuous trend to some of the occurrences of coarse gravel is interpreted as a response to deposition in the channel of a localized stream rather than a systematic gradation of sheet wash in well-defined belts outward from the moraine. Pockets of fine gravel and sand are distributed close to the moraine margin and may occur immediately adjacent to coarse deposits. This is in accordance with the expected behavior of streams which are heavily laden with coarse and fine debris. Bulk sediment dumping occurs when flood stages fluctuate and are followed by low-water stages, i.e., such as the physical condition leading to the development of the braided stream.

Stratified Ice-Contact Deposits

Kame terrace. Certain stratified drift is related specifically to contact with ice. One of these deposits is the *kame terrace*. This is an accumulation of outwash which is developed between a wasting tongue of ice and local valley walls. (See Fig. 11.28.) Ideally kame terraces are developed in regions of valley glaciation, but they are common marginal deposits of ice caps where local tongues of ice invade pre-existing valleys which during the stage of ice melting confine the ice.

Stratified drift is built layer upon layer against the valley wall and the edge of the ice as a narrow flat-topped strip bordering the ice

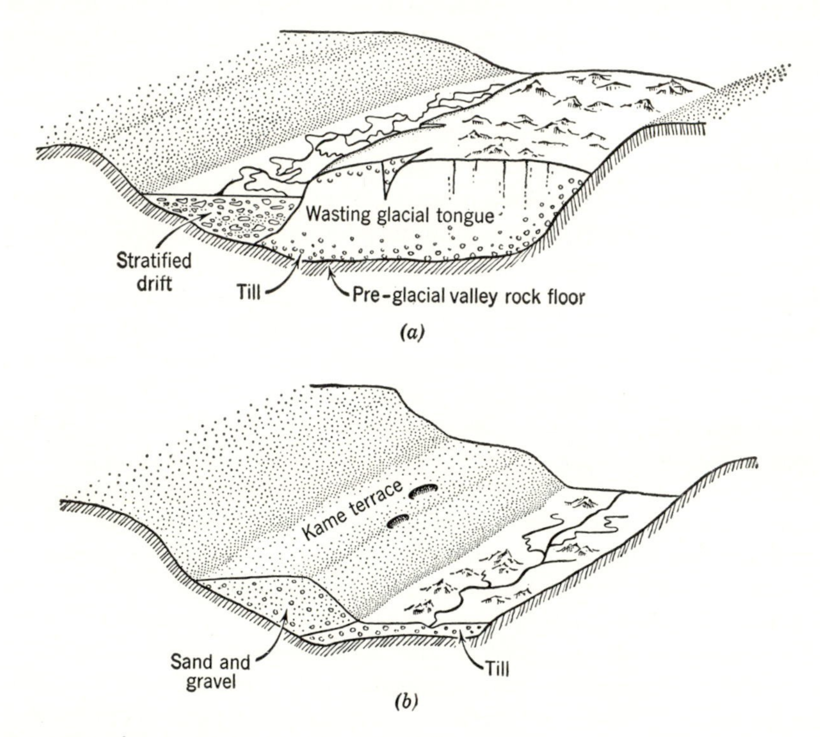

Fig. 11.28 (*a*) A sketch of conditions leading to the development of a kame terrace. The wasting ice tongue which formerly occupied the valley furnishes outwash to the zone between the valley wall and the ice margin. (*b*) After the melting of the ice the belt of gravel is left as a terrace on the valley floor and locally rests upon till.

margin. When the ice wastes away the deposit remains as a terrace along the valley wall and has the general land form of a stream-cut terrace. However, there are certain distinguishing differences among which are the following:

a. Kame terraces lack the regularity of stream terraces. Often they occur only on one side of the valley or are irregular patches with variable surface elevation.

b. They can be distinguished by the aspect of the valley floor. As, for example, in a region of stream terraces the valley fill will consist of stream alluvium, whereas the floor of the valley containing kame terraces will be an irregular surface of glacial drift much of which is till.

Kame. Low steep-sided conical or dome-shaped hills occur near the moraine margin often on the outwash plain. Such a hill known as a *kame* often stands alone on the landscape surface. The hill is rarely more than 100 feet high but because of its isolated position may form a conspicuous feature of the glacial landscape. Kames are gravel deposits which have accumulated in a crevasse near the ice border or where a conical-shaped gravel alluvial cone has been constructed by a stream cascading from the ice front. Their prominent position above the general land surface makes them ideal locations for gravel pits. Gravel is easily removed by small equipment, and the position of the gravel above the general water table makes the deposit self-draining.

Kame distribution is extremely variable. Individuals occur scattered over several miles from the end moraine on the outwash plain, but statistically they are concentrated within, or peripheral to, the moraine belt. A frequent locality of occurrence is on the outwash plain immediately in front of well-defined gaps in the moraine. Such interruptions in a moraine ridge are not uncommon and appear to mark the position occupied by glacial streams actively dumping a cone of outwash during construction of the moraine.

Esker. *Eskers* are low sinuous narrow ridges of stratified drift found primarily on the ground-moraine surface. Usually, the ridge is less than 50 feet high, but, locally, it may exceed 100 feet, and 300 feet in breadth rarely is exceeded. The length is variable and often interrupted, but an extent of several miles is not uncommon, and several have been reported to have been traced for as much as 300 miles.

Inasmuch as most eskers occur singly and on low surfaces of the ground moraine they stand out with topographic expression. They will persist, however, through an end-moraine belt and sometimes terminate as a delta or kame on the outwash plain. Some of these relationships can be seen to hold for the Kaneville esker (Fig. 11.27) which extends through the Marseilles moraine and continues as a prominent ridge on the outwash surface.

Eskers rest upon other glacial material and appear to represent deposition during stages of ice wasting. The sinuous shape and internal stratification suggest deposition by streams traveling in tunnels and channels in the ice. These tunnels or channels are oriented with the long dimension of the ice tongue and general flow of the ice as reconstructed from directions of striae and grooves on bedrock surfaces.

Pebbles in the esker gravels are of local derivation and coincide with variation in bedrock types. Figure 11.29 is a generalized representa-

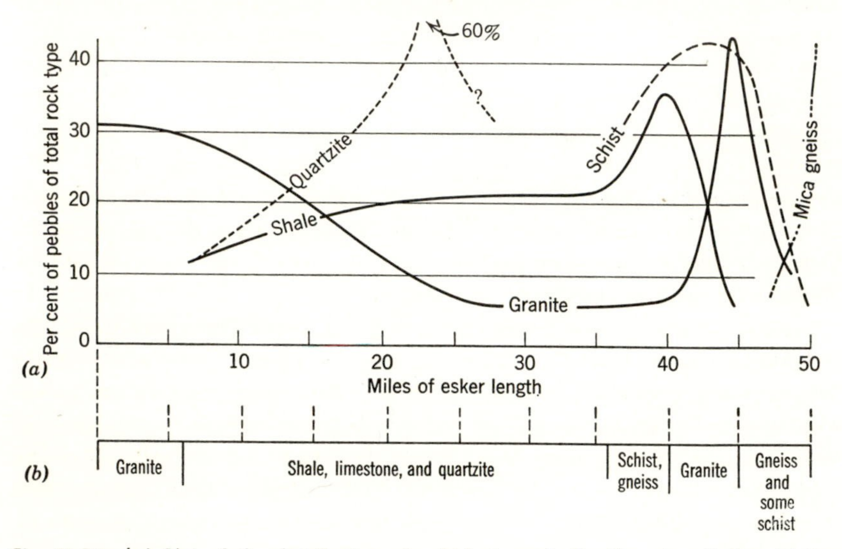

Fig. 11.29 (*a*) Plot of the distribution of pebble types in the Kennebec Valley, Maine, esker compared to distribution of bedrock (*b*) along the esker length. Note the general sensitivity of the esker composition to the bedrock control. (Data from Trefethen and Trefethen, *Amer. Jour. Sci.*, 1944.)

tion of the distribution of pebble types in the Kennebec, Maine, esker gravel, and the nature of the bedrock beneath the locality of sampling.[11] The reflection of bedrock type is striking. This is particularly true of hard rocks such as granite the fragments of which resist reduction in size, but also shale becomes an important constituent within the belt of shale, limestone, and quartzite. In the schist belt shale locally increases in proportion because it forms harder pebbles than the schist, but a few miles farther from the area of shale bedrock shale pebbles become a minor constituent. Correlation of esker pebbles and local bedrock, therefore, follows the pattern described earlier as characteristic of till.

Eskers are common sources of gravel, and their distribution and

[11] The study was reported by J. M. Trefethen and H. B. Trefethen, "Lithology of the Kennebec Valley Esker," *Amer. Jour. Sci.*, Vol. 242, 1945, pp. 521–527.

location generally are noted on local geologic maps. Maps of recent publication may indicate the quality of such gravels as construction material, but in others only the position of the esker is noted. However, general prediction of the quality of an individual esker gravel can be made on the basis of the character of the local bedrock. If this is shale or schist the esker gravels will be poor, whereas if the bedrock is limestone, quartzite, or granite the gravels are likely to be of good quality.

Glacial-Lake Deposits

Ponding of waters in large or small lakes is an associated characteristic of glaciation. Some of these lakes develop as the ice melts from its former basin, and in such cases the end moraines often are the dams.

Lakes which subsequently have been drained have modified the landscape by erosional and depositional features generally readily recognizable. (See Fig. 11.25.) Primary erosional features are wave-cut benches and cliffs marking the former shoreline. Depositional features are the extremely level bottom surface and bottom and shore deposits. These consist of beaches, bars, spits, hooks, deltas, and quiet-water sediments. Of these the last merit attention not because they are any more abundant but because they are somewhat unique in development.

Varves

Quiet-water bottom deposits of glacial lakes often are remarkably stratified in thin bands which are sharply separated one from the other by a prominent difference in color tone. Close inspection of individual layers reveals alternations of color tones in which one dark layer is overlain by a light layer which in turn is succeeded by a dark layer, the succession being repeated manyfold. Some gradual transition exists between a lower light and an upper dark band, but a sharply defined line marks the appearance of the succeeding light band. Gradation between the light and dark layers has led to the recognition that the two form a closely related pair distinct from a similar overlying or underlying unit. Each couplet of light and dark is known as a *varve* and is believed to represent an annual deposit. The light layer represents the summer accumulation and the dark layer the addition during the winter season.

Varves are developed only in clays and silts in which the grain size is small enough to develop the thin laminations which are characteristic of the deposit. The light-toned layer is the coarser of the two and contains more silt in particles of uniform dimension. Most of the silt is quartz which accounts for the light color of the layer. Dark layers are the clay fraction, and generally there is enough incorporated organic matter to develop the dark tone in the clay lamina. Individual laminas range in thickness from several inches to as little as $\frac{1}{16}$ inch, with $\frac{1}{8}$ to $\frac{1}{4}$ inch being the most common thickness. Uniformity in thickness of individual units is striking and persists over extensive areas covering many miles. Persistence in thickness is more typical of the dark layers representing the winter accumulation than the lighter or summer deposit.

Accumulation of one of the semi-annual layers constituting a varve requires that fine sediment be distributed in about equal amounts over an extensive portion of the lake. The particle must sink through the quiet waters at approximately the same rate such that a layer of sediment approaching uniform thickness is developed during one season of the year. Agitation of water must remain at a minimum so that no serious interruptions in settling velocities develop. These conditions are established in rather extensive glacial lakes where quiet water exists some miles from the inflowing glacial streams. Ideally, still water remains at the temperature of maximum density except for the epilimnion, and particles of identical sizes and densities settle downward with velocities controlled by Stokes' law. Glacial streams carrying a mixed-size load from the melting ice enter the lake at a temperature only slightly above the freezing point. Particles of sand size or larger rapidly settle out near the mouth of the entering stream leaving a turbid mass of fine-sediment-laden water, which floats upon the somewhat warmer, and therefore heavier, clear lake water simulating conditions of plane jet flow (see p. 216). The light water spreads rapidly over the surface of the lake and distributes its fine-grained sedimentary load more or less uniformly over a large area. Particles begin to sink with velocities generally in direct proportion to the square of their radii. Silt particles reach the bottom first, but clay particles remain in suspension gradually flocculating into aggregates sufficiently large to settle to the lake bottom. During the summer season the supply of sediment is greater and somewhat coarser than during the winter season. Silt, therefore, is the summer-season deposit. With the approach of the winter season the lake freezes over, and the supply of debris from the stream is reduced greatly or cut off. During this season the clay particles settle, and the dark layer accumulates with its incorporated organic debris which grew in the surface waters during the summer.

Some varved clays or silts contain isolated boulders or cobbles which were rafted frozen to cakes of ice into quiet lake waters. Infrequent lenses of sand also reflect temporary shifting in shore currents or stream outlets.

STRATIGRAPHIC SUCCESSION IN GLACIAL DEPOSITS[12]

The Law of Superposition

Near the margins of end moraines of Pleistocene continental glaciers particularly in Iowa, Illinois, and Indiana there is unquestioned evidence of a vertical succession of deposits. Red-colored till may lie beneath outwash which in turn is covered by gray till. In other cases two tills of very distinctly different physical properties (i.e., clay content, pebble counts, stage of weathering, etc.) can be observed one upon the other. The nature of these associations has led students of glacial sediments to unravel a complex history of glacial advances separated by long periods of warm climate. In this connection much of the establishment of chronologic sequence has been based upon a fundamental geologic concept known as the *law of superposition.*

A general statement of the law of superposition defines a layer of rock or sediment as younger than the one upon which it lies. In the case of the two tills mentioned above a young till lies upon an older till. The law does not delimit or help to determine the length of elapsed time between the deposition of the two deposits, but we do know that some short or long time interval is involved. Again using the example of the two tills above if the interval of elapsed time between their respective deposition was short they may have been laid down during the same glacial stage. On the other hand if the interval between the deposition of the two tills was very long the second ice advance marks the advent of a new glacial stage following a period of deglaciation. Thus determination of the length of elapsed time between the deposition of the two tills assumes very real importance. Solution of the problem resolves in assembling geologic data, the interpretation of which leads to measures of time lengths.

[12] An excellent treatment of the subject is to be found in R. F. Flint, *Glacial and Pleistocene Geology,* John Wiley and Sons, 1957, Chapters 16–20.

Estimates of Time Involved in Interglacial Stages

On the basis of certain observations to be discussed in later paragraphs four major glacial and three interglacial stages have been recognized. (See Table 11.5 and Figs. 11.20–11.23.)

Table 11.5. Major Glacial and Interglacial Stages Recognized in the United States

Glacial Stage		Interglacial Stage	Order of Magnitude of Duration in Years*†
Wisconsin	Mankato		25,000
	Cary		10,000
	Tazewell		10,000
		Peorian	-----
	Iowan		10,000
		Sangamon	135,000
Illinoian			100,000
		Yarmouth	310,000
Kansan			100,000
		Aftonian	200,000
Nebraskan			100,000

* The values listed in years are highly speculative and may vary particularly in the glacial stages by 100% inasmuch as no satisfactory methods exist to determine the length of time required for a glacier to reach its maximum extent.

† Data from C. O. Dunbar, *Historical Geology*, John Wiley and Sons, 1949, p. 451.

Each ice-sheet advance extended tongues into the north-central part of the United States, but the localities of penetration were not exactly coincidental, and certain areas were covered with ice during one or two ice advances only. Other areas escaped glaciation altogether although the combined advances have surrounded them with glacial drift. The most extensive of these unglaciated "islands" is the *Driftless Area* centered chiefly in southwestern Wisconsin. (Fig. 11.24.) Approximate limits of the drift assigned to each stage of glaciation are shown in Figs. 11.20–11.23. On the basis of the position of ice front at the time of maximum ice advance for each stage the Driftless Area can be demonstrated to have been surrounded by drift but never completely enclosed by ice.

Estimates bearing on the elapsed time between deposition of glacial tills have been based upon observations, the classification of which can be assembled under the several headings which are given below. Each technique employed to provide a measure of time is an important contributor to the evidence that a glacial condition followed by milder climate has been repeated in cyclical order several times in the recent geologic past.

Measuring Time by Varved Clays

The evidence provided by the deposition of the glacial-lake sediments, varved clays, is based upon counting the annual layers accumulated in one locality. Spacing and individual thickness of each varve layer in a sequence is matched with its counterpart in a nearby locality where the counting is continued. By painstaking effort of matching each locality of varved clay with its neighboring correlative sequence information has been assembled which demonstrates the progressive appearance of extensive glacial lakes which existed for thousands of years between ice advances. Also, the stages of melting and final disappearance of ice from the glaciated section have been reconstructed with magnificent detail.

Significance of Forest Beds

Local forest beds buried between two tills contain plants which are identified as associated with mild climates. At one such locality in central Iowa a peat bed has been examined for its plant succession. At the base the pollen is of northern conifers typical of glacial climate, but upward the pollen grades into that of grasses and oaks of climate generally like that of the present. This is strong evidence to support the concept that more or less complete destruction of ice sheets occurred during the accumulation of the forest bed. Later, as demonstrated by the presence of overlying till the glacial climate reappeared, and ice once again moved outward from centers of accumulation.

Position of Loess Sheets

Extensive deposits of loess are often associated with the buried forest beds and intimately interlayered between tills. At some localities fresh loess overlies a till sheet which has been very deeply and thoroughly weathered, but the loess is relatively unaltered and contains the remains of organisms which lived in a mild climate. Elsewhere a loess deposit shows a well-developed soil profile. Together these observations are to be interpreted as indicating long periods of weathering and soil formation between glacial advances.

Development of Gumbotil

Intensive weathering in till sheets has been observed in many widely scattered localities and has led to the understanding that some of the older tills were exposed to long periods of soil development before a layer of loess or new till was laid down. The weathered till has been given the name of *gumbotil,* a descriptive term derived from the popular name *gumbo.* The latter has long been used to connote a clay which is extremely sticky when wet.

Gumbotil usually is light dull gray in color, but in places it is mottled with brownish or reddish tints. Dry surfaces are light gray and often cracked by shrinkage on drying. Throughout the clay a few small pebbles are scattered. These are predominantly quartz, chert, quartzite and a very few resistant crystalline rocks. Surfaces of the quartz and chert pebbles usually are smooth. Downward the gumbotil grades into yellowish or chocolate-colored till in which pebbles are more numerous although still minor in importance. With increased depth this zone merges into yellow-colored till which contains calcareous pebbles and still lower into unaltered dark-gray till which may be strongly calcareous. Some of these gradations are more rapid than others, but the net effect is to produce a series of distinct zones characterized by special properties.

The general case of alteration is illustrated in Fig. 6.9 in which is illustrated the complete gradation from gumbotil to its unaltered parent. Downward-moving water tends to move colloidal iron and aluminum hydroxides into the "B" zone where they are flocculated by Ca ions and gradually develop a zone high in colloids. The iron hydroxides generally are not precipitated until the transition zone between the "B" and "C" horizon is reached. Here, small concretionary pellets of limonite appear. Somewhat higher in the "B" horizon the aluminum hydroxides and silica are arranged in the space lattice of illite clay which dominates this zone of gumbotil.

The effect of concentration of illite in the "B" horizon is to produce a clay of high plasticity and swelling properties when wet and a hard mass of reduced volume when dry, i.e., typical hardpan. This zone also acts as a membrane of low permeability to water. During the wet season surface water ponding is common, and recharge of ground water is reduced, whereas runoff is increased over that of other clay soils under the same conditions.

A measure of the long period of leaching (weathering) required for gumbotil development is reflected by the reduction in number and volume of pebbles compared to the unaltered till. In Fig. 11.30 are

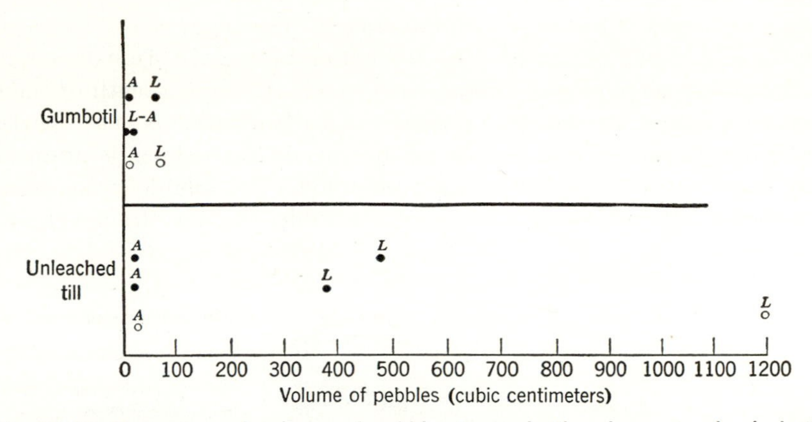

Fig. 11.30 Comparison of volumes of pebbles in gumbotil and parent unleached till in Taylor Co., Iowa. *A* is the volume of an average pebble, *L* is the volume of the largest pebble. A closed circle indicates till of Nebraskan age, an open circle Kansan. Position along the ordinate has no significance. Observe the small differences in volume of the average pebbles, but the very great difference between the largest pebbles in the weathered and unweathered deposits (see also Fig. 11.31). (Data from Kay and Apfel, *Iowa Geol. Survey*, 1929.)

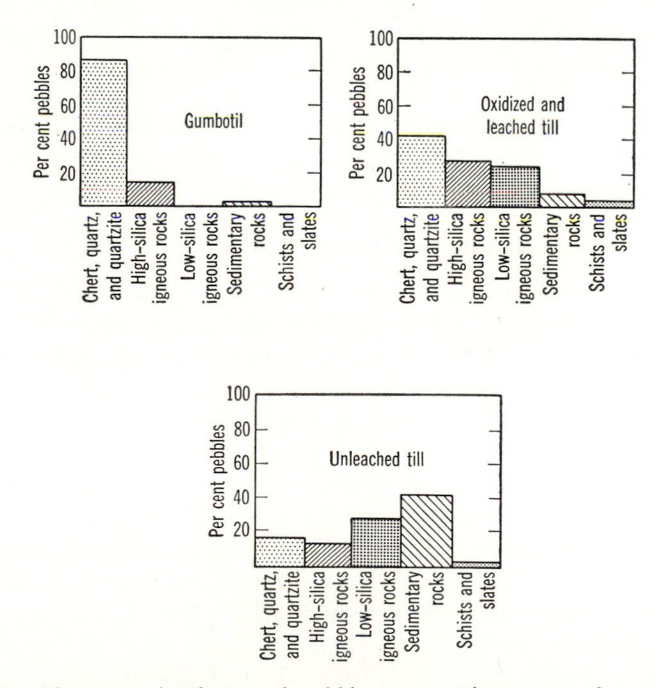

Fig. 11.31 Change in distribution of pebble types with progress of weathering of Kansan till in Iowa. Observe that the insoluble fraction is concentrated in the gumbotil. (Data from Kay and Apfel, *Iowa Geol. Survey* Spec. Rpt., 1943.)

shown volumes of the largest and average pebbles in gumbotil and also in the underlying unaltered till. Reduction in average size is not significant, but large pebbles are reduced to as little as one-tenth of their former volume. By the same process the predominating pebble variety is changed also. This is illustrated by Fig. 11.31 in which is graphed the gradual transition between a predominance of soluble rocks in an unleached till and concentrates of insoluble siliceous fragments in gumbotil.

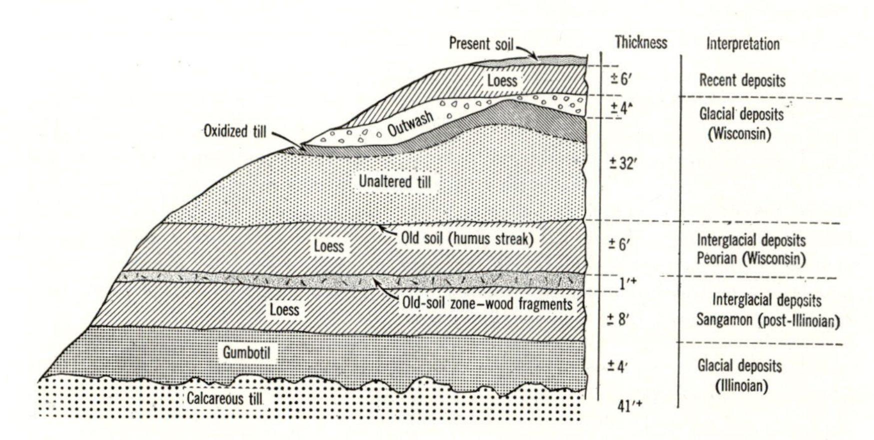

Fig. 11.32 Generalized interpretation of exposure of glacial and interglacial deposits near Peoria, Illinois (Farm Creek Section). (Simplified from Leighton, *Illinois Geol. Survey*, 1926.)

Interpreting the Sequence of Glacial Deposits

An example of the interpretation which is applied to a layered sequence of glacial and related deposits can be shown by reference to the classic exposure at Farm Creek, Illinois. (See Fig. 11.32.) Here at the base of the valley cut is to be seen a calcareous till which grades upward into 4 feet of gumbotil. This gumbotil displays the typical characteristics of the "B" soil zone of Fig. 6.9, and we deduce that the lowermost (Illinoian) till was exposed to intensive weathering after deposition. The thickness of the gumbotil also is a measure of the length of the weathering period. We are, therefore, led to conclude that the time of gumbotil development was long particularly if due allowance is made to possible reduction in its thickness by contemporaneous stream erosion.

Loess lies upon the gumbotil from which it is sharply defined. This loess contains feldspars and other minerals which decompose into clay minerals under the weathering process. The presence of the unstable (i.e., to weathering) minerals indicates that the loess must have accumulated rapidly and was buried before it could be altered to a gumbotillike deposit. Deposition of loess was interrupted long enough to develop a soil profile at its upper surface, and identification of incorporated wood fragments reveals that trees similar to the present local assemblage grew on this soil. Another period of loess accumulation interrupted the soil-forming process, and the old soil was buried and preserved. At some later date deposition of the uppermost loess also came to an end as indicated by a poorly developed soil zone.

The presence of the soil zone between the two loess layers dictates that there must have been two distinct periods of loess deposition separated by sufficient length of time to develop a soil profile but not long enough to produce gumbotil. Hence, the distribution of time since deposition of the till at the base of the cut must be: (1) a very long portion for the development of gumbotil, (2) a short period for the deposition of the lowermost loess, (3) a longer period for the development of a soil, (4) the shortest period for the development of the overlying loess and its soil.

Upon the second (from the base of the cut) layer of loess is found calcareous till. The presence of this material indicates that conditions of glaciation returned to the region and an interglacial stage was brought to an end. Unweathered specimens of each of the two tills are much alike, and the pebble count distribution suggests that the ice moved over the same bedrock path in each case. From this observation the interpretation is drawn that both sheets advanced into the area from a common center of ice accumulation.

Disappearance of the glacial conditions is indicated by the presence of loess overlying the till and outwash. We interpret the outwash as reflecting melting of the ice, and following its deposition oxidation began to affect both the outwash and till. Weathering, however, was not permitted to continue long enough to develop a substantial soil profile before loess once again buried the surface. These last relationships supply a measure of the relative length of elapsed time since deposition of the last till compared to that represented by the interval between the deposition of the two tills. Sangamon interglacial time is, therefore, recognized as of much longer duration than post-Wisconsin time. (See Table 11.5.)

Changes in Stream Drainage

Major glacier advances had a profound effect upon stream drainage patterns already developed in the invaded areas. Some channels favorably oriented to the ice advance were deeply plowed and modified;

others crossing the path of advancing ice were buried, and the streams were forced to develop new channels. Many of the new courses were aligned marginal to an ice sheet at the time of its farthest advance. Some of these marginal streams became important drainage routes and established patterns which in gross measure are preserved today. Of these perhaps the two most important in the United States are the Ohio and Missouri river valleys. The position of the present valleys with respect to the limits of farthest ice advance is striking (see Fig. 11.24). These two streams carried melt waters from the ice sheets to a junction into the Mississippi drainage.

Detailed inspection of the stream courses and information from many borings outside the valleys reveal that ice marginal control of the Ohio and Missouri valleys is only general and that important divergences from the present courses existed. As more information is assembled the old drainage systems can be reconstructed. (See Fig. 11.33.) Some of the old drainage is pre-glacial as indicated by very old till filling bedrock channels. Other channels are found which have been cut in old till and filled with loess or younger till. These represent the drainage systems of an interglacial stage. The cross sections of some of these valleys reveal profiles typical of maturity in the stream erosion cycle. Mature stream valleys took time to develop and serve to emphasize the length of duration of interglacial time.

Dating by Radioactive Carbon Analysis

One of the most promising techniques of measurement of the time involved in the various glacial and interglacial stages appears from the determination of the per cent distribution of certain isotopes of elements contained in organic material. One of these elements, radioactive carbon (C^{14}) is created when a neutron from cosmic rays bombards nitrogen in the earth's atmosphere. The product of this bombardment is C^{14} and a proton (H^+). Radioactive carbon is unstable, emits beta rays and disintegrates with a half life of 5,570 years, and returns to nitrogen. Through this process a small but fixed amount of C^{14} is generated and oxidized into carbon dioxide. Each unit volume of carbon dioxide in the atmosphere and hydrosphere, therefore, contains a fixed ratio between C^{12} and C^{14} in equilibrium, established by the rate of C^{14} decay. Organisms consuming the CO_2 directly or indirectly also maintain an equilibrium ratio between C^{12} and C^{14}, a situation which is continued so long as metabolic processes remain in effect. Upon death and cessation of metabolism the equilibrium no longer is maintained and the percentage of C^{14} in the organic matter is reduced by its characteristic rate of decay. Determination of the C^{14}

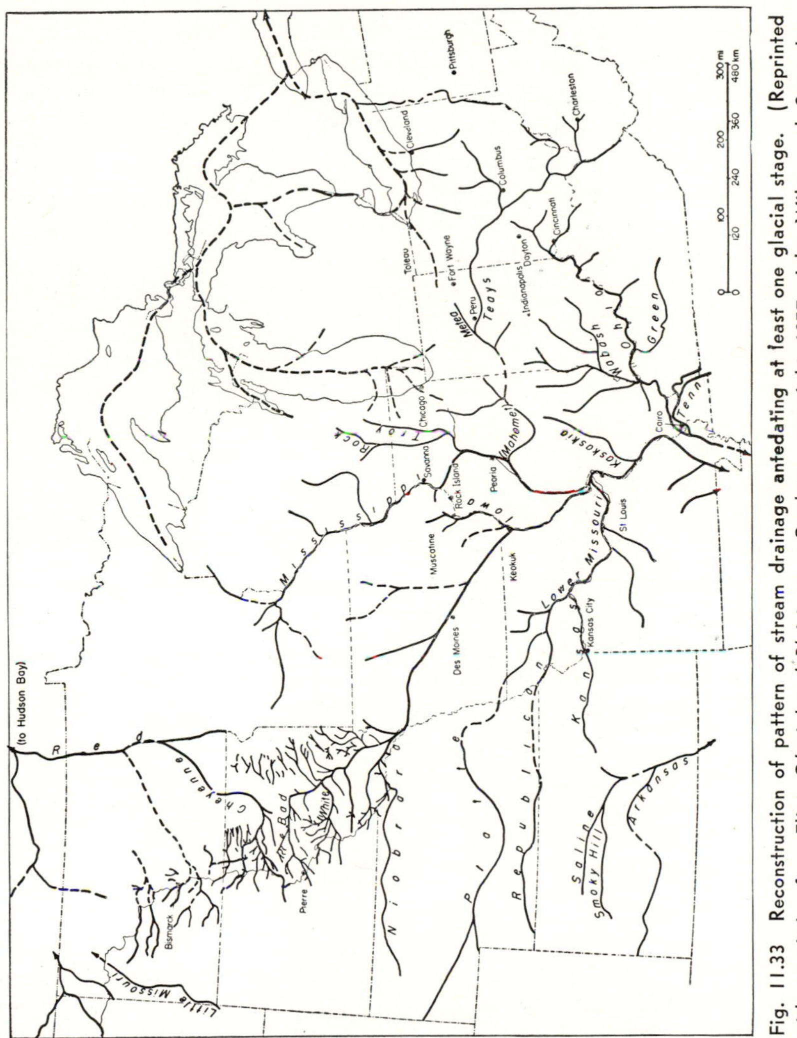

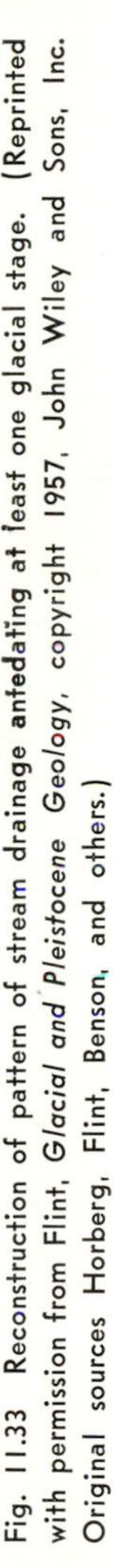
Fig. 11.33 Reconstruction of pattern of stream drainage antedating at least one glacial stage. (Reprinted with permission from Flint, *Glacial and Pleistocene Geology*, copyright 1957, John Wiley and Sons, Inc. Original sources Horberg, Flint, Benson, and others.)

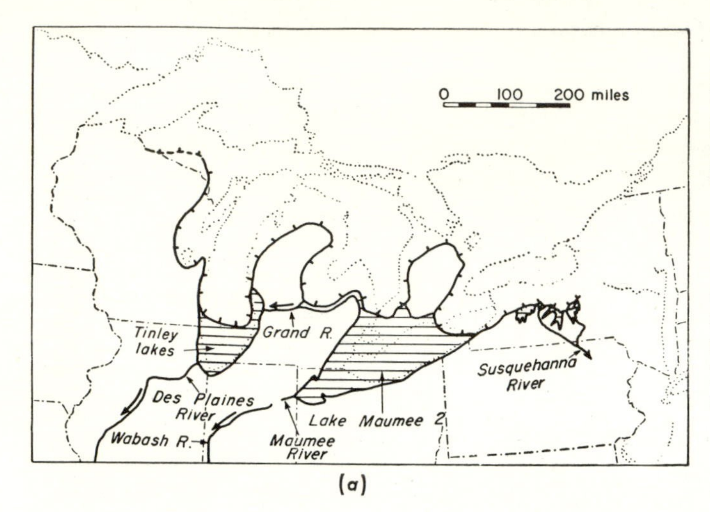

(a)

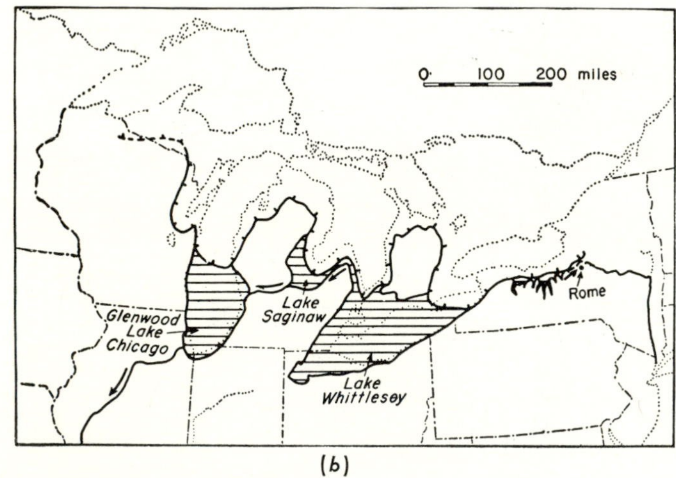

(b)

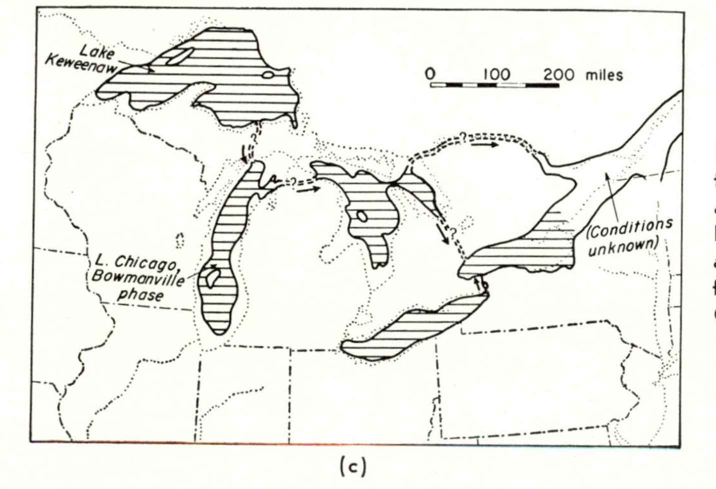

(c)

Fig. 11.34 Current interpretation of stages in the history of the Glacial Great Lakes.

(*a*) Small lakes bordering ice front draining principally into Desplaines-Illinois and Wabash river systems.

(*b*) Expansion of condition of (*a*).

(c) Important retreat of ice front and appearance of large lakes. In Chicago area the lake levels fell below that of the Glenwood stage.

(*d*) Readvance of ice and raising of lake levels.

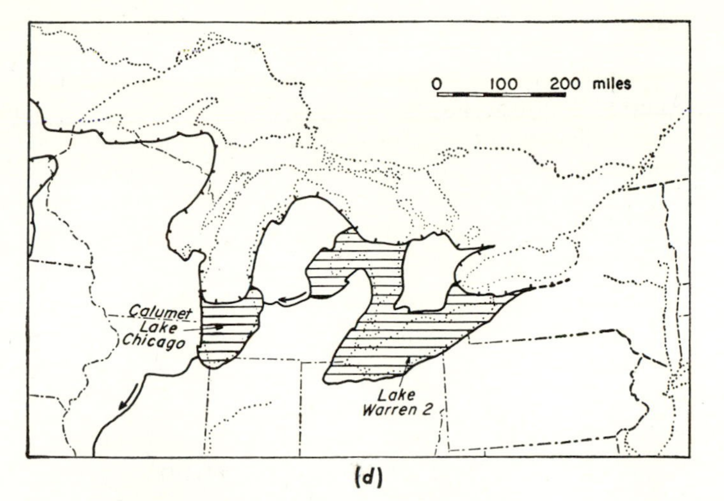

(*d*)

(*e*) Gradual retreat of ice front and drainage of lakes via the Mississippi, Illinois, and Mohawk rivers.

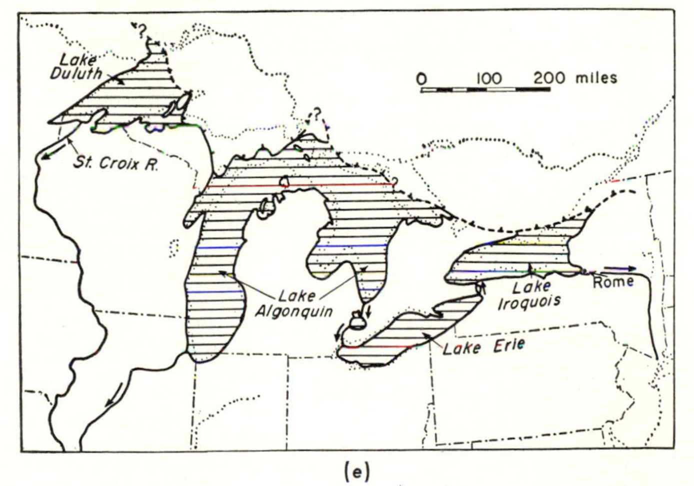

(*e*)

(*f*) Shift of drainage to eastern outlet and invasion of the St. Lawrence and Lake Ontario basin by the sea. The current condition is the result of recent elevation of the land.

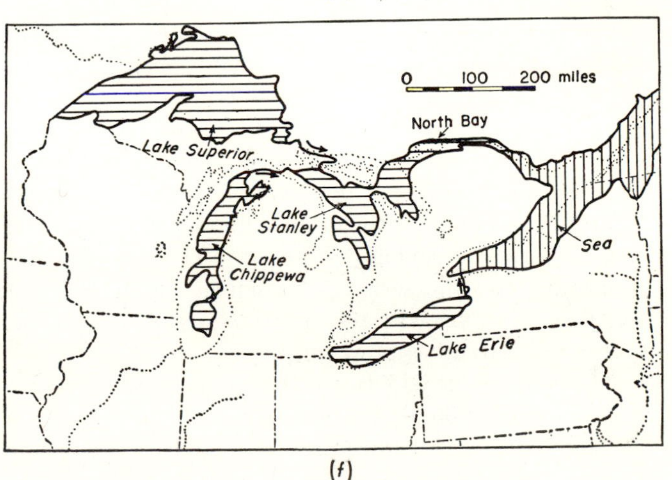

(*f*)

(Reprinted with permission from Flint, *Glacial and Pleistocene Geology*. Copyright 1957, John Wiley and Sons, Inc.)

remaining in the organic matter and its comparison to the amount in normal equilibrium under living conditions can be equated into terms of the length of the number of half-life periods involved and a measure of the length of time elapsed since death. Unfortunately, the method is restricted by the short half-life period of C^{14} to lengths of time not exceeding 45,000 years inasmuch as the errors involved become excessive. Very small errors are observed in the dates obtained from wood from Egyptian tombs whose age is known from cryptographic deciphering, but for organic debris collected from deposits which antedate Wisconsin glaciation the values obtained lack significance.

Major Features of Deglaciation

Progressive amelioration of the glacial climate is apparent from the successive limits of end moraines of the Wisconsin substages. As the ice margins retreated northward extensive bodies of water were impounded between the ice fronts and higher morainal land. Some of these lakes were extremely shallow; others occupied glacially plowed channels. All of these lakes originally stood at high levels which progressively were reduced as new outlets appeared or old outlets were lowered. During the time they occupied such levels many of the erosional and depositional features associated with bodies of standing water were developed, and these remain today to mark the former sites of the post-glacial lakes. Beach lines, flat sandy bottoms, lake clays, deltas, spits, and bars are to be found locally lying upon Wisconsin drift and forming the parent material of the present soil. (See Fig. 11.25.)

In North America the most impressive of all of these lakes is a precursor to the present Great Lakes. At one time this ancestor, Lake Algonquin, was a single body of water occupying somewhat more area than the present individual bodies but of the same general outline. This enormous lake developed from the coalescence of smaller independent ice-restricted lakes whose general water elevation stood 50 or more feet above the present lake levels. As individual lakes grew larger they passed through an elaborate history of growth the outlining of which is beyond our scope. The bare essentials, however, are illustrated by the series of drawings Fig. 11.34. Note the positions of the small lakes with respect to the ice sheet and the various drainage outlets as the lakes grew larger.

Other lakes developed north and northwest of the ancestral Great Lakes (Fig. 11.35). Although some of these were very extensive such as Lake Agassiz they were also very shallow and were drained when the ice sheet melted. The old bed of Lake Agassiz, however, is well defined, and varved clays deposited on its bottom are widespread in North Dakota and Saskatchewan.

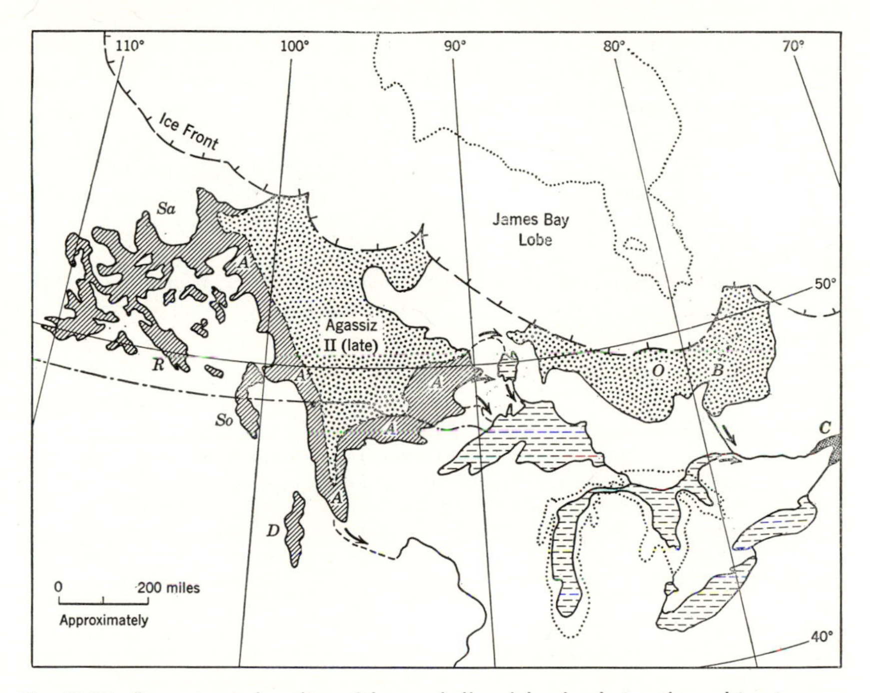

Fig. 11.35 Reconstructed outline of large, shallow lakes bordering the melting ice cap. Central North America in very late Wisconsin time (about 7,000 years to 8,000 years ago), showing sites of earlier glacial lakes (inclined ruling), the glacial lakes then extant (dots), correlative former glacial lakes (horizontal dashed lines) and the Champlain Sea (cross-hatched). Lakes indicated by letters: A, early phases of Lake Agassiz; R, Lake Regina; Sa, Lake Saskatchewan; So, Lake Souris; D, Lake Dakota; O–B, Lake Ojibway–Barlow; C, Champlain Sea. Arrows indicate outlets. (Compiled by Prof. J. A. Elson from Geological Survey of Canada and other sources.)

Deposits accumulated during the span of deglaciation are distributed widely on the present earth surface. Their greatest abundance is to be expected in the regions which were invaded by ice. Nevertheless, such material has found its way by means of large trunk streams to valley terraces and other alluvial fills far beyond the boundaries of the ice. The influence of the discharge of enormous quantities of glacial debris currently is being recognized in such great deltas as that of the Mississippi; and in cores from the depths of the oceans layers of sediment associated with the period of deglaciation are being recognized and studied. Slowly the geography of the times is being reconstructed, and there is much published information available to the interested student on the properties of sediments and the organisms which were contemporaries of early man.

Selected Supplementary Readings

Flint, R. F., *Glacial and Pleistocene Geology,* John Wiley and Sons, 1957, Chapters 2, 3, 4, 5, 6, 7, 8, 9, and 10. This book supplants the earlier one entitled *Glacial Geology and the Pleistocene Epoch* (1947) and is a most comprehensive treatment of glaciation.

Gilbert, C. K., "Lake Bonneville," *U. S. Geological Survey Monograph* 1, 1890. A classic presentation of the reconstruction of the history of a Pleistocene lake outside glacial limits.

Leverett, Frank, and Taylor, F. B., "The Pleistocene of Indiana and Michigan and the History of the Great Lakes," *U. S. Geological Survey Monograph* 53, 1915. A classic description of the stages of development of the Glacial Great Lakes.

Thwaites, F. T., *Outline of Glacial Geology,* Privately Published, 1946, Part III. An excellent description of the glacial stratigraphic succession of North America.

12

Sedimentary rocks

PRIMARY GROUPS

Preservation of Strata

In a general way rocks forming on the earth's surface are ephemeral and except for those being produced in the ocean basins are continually being eroded to form new sediments. In contrast, deep-seated igneous rocks, such as batholiths, and intensely metamorphosed strata are long persistent and remain as important markers of past events in the rock column. The great size of these masses, particularly of batholiths, and the depths to which they are known to be found cause such rock to remain preserved despite the enormous erosion which takes place in the areas of their occurrence. Yet most of the uppermost rocks of the earth are sedimentary in nature. This seemingly anomalous situation is the result of three processes fundamental to the accumulation and preservation of sedimentary rocks. The first of these processes is stream erosion which eventually results in the transportation of rock debris to the seas. Here the second process begins. In the region of the shoreline and in the shallow offshore waters there occurs remixing, resorting, retransportation, and redeposi-

Thin-bedded sandstone.

tion. The third process is the one which enables these sedimentary particles to remain preserved. This is the process identified as *diastrophism,* i.e., the relative rise and fall of localized parts of the earth's surface. Subsidence of an area adjacent to the coast line results in marked alteration in the distribution of the continental shelf, and extensive areas, formerly land, are placed in a condition to receive sediments moved by shore agencies.

Large-scale diastrophic subsidence can be classified into three different conditions of behavior:

1. Some regions are known to have been depressed at exceptionally slow rates with remarkable uniformity over enormous areas. Their present-day counterpart is considered to be the areas of broad continental shelves. As these subside slowly sediment is piled upon sediment, and a thin veneer is accumulated over a long period of subsidence. Regions having experienced such conditions are called *shelf.*
2. Other areas within the present-day continental margins have in the past undergone subsidence at a rate somewhat faster than the surrounding region. Some of these depressed segments are trough-like in outline, whereas others are ovaloid, but collectively they are identified as *basins.* Generally these are sites of impressive accumulation of sediments in amounts ranging in thickness in excess of 20,000 feet.
3. Still other regions of pronounced subsidence are linear troughs of spectacular dimension. These are the great *geosynclines* into which are poured thousands of feet of sediment eroded from an adjacent rising land mass. Sediments of the geosyncline are the sources of metamorphic rocks inasmuch as these depressed areas eventually become the sites of great folded mountain chains.

Subsidence of a local area whether rapid or slow is the single process which permits thick accumulation of sediments. Only in these downwarps is the sedimentary record preserved. Although these sites of subsidence are chiefly submarine there are important inland localities where basins have developed and where thousands of feet of sediment did accumulate in the nonmarine environment. In the geologic record, however, such occurrences are not as abundant in total as the marine deposits.

Diastrophism also operates to cause withdrawal of the sea from subsiding areas as the process is reversed and uplift takes place. Uplift can be gradual and minor such as has been typical of much of the central part of the United States, or it may be rapid and of great magnitude. In the former case where upward rise has been halted at elevations not much higher than sea level deeply buried sediments are preserved from erosion. Wherever the uplift has been great, however, sediments have been stripped away, and no record remains of their former existence. Still other occurrences are known where by random preservation small patches of sedimentary rocks have escaped erosion or are deeply buried in troughs. By the never-ending processes of depression and uplift local segments of sedimentary rocks are preserved to establish the geologic record over much of the continental blocks.

Clastic and Nonclastic Rocks

Ideally all sediments can be classified as belonging to one of two major groups. That group recognized as clastic consists of individual mineral particles separate from one another. Nonclastic sediments are precipitates, either chemical or organic, in which individual crystals are held together by chemical bonds or are interlocked one within the other. Clastic sediments are primarily detrital and represent loose aggregates which can be shifted by sediment transport and ultimately deposited. Nonclastic sediments are also defined as *chemical deposits* because they represent products precipitated at the site of their accumulation. Most clastic sediments consist of land-derived detritus representing the materials of weathering and erosion. Other clastic rocks are not nearly so neatly defined. For example, volcanic ash blown from a volcano and deposited in adjacent marine waters is a clastic sediment, yet it is not strictly land-derived detritus. Fragments broken from coral reefs or comminuted shell sand are not land-derived detritus, yet they constitute clastic sediment.

Chemical rocks often are important parts of the rock column and may be very extensive in distribution. Coal, gypsum, common salt (NaCl), and certain limestones and dolomites are the principal types. Oftentimes they are pure deposits without any incorporated land-derived detritus. More commonly there is contained clastic debris. Whenever the land-derived clastic is small in quantity it is an adulterant to the chemical sediment. As the quantity of clastic material is increased to more than half, the chemical fraction assumes an increasingly minor role and dilutes the clastic debris. This relationship indicates that a complete gradation exists between clastic and nonclastic sediments, each of which is an ideal end member of a continuous series of sedimentary types.

The presence of the chemical fraction is the important control in altering a sediment to a sedimentary rock. This is the substance which when precipitated in the interstitial pores between clastic grains cements them together. The degree of this cementation identifies the extent to which the rock has been consolidated. Friable sedimentary rocks are loosely cemented, and individual grains are easily separated. Quartzites are thoroughly cemented, and individual mineral grains are complexly interlocked by the addition of quartz to these grains. Indeed, well-cemented rocks are those in which the chemical end member is an important fraction of the rock.

Clastic Rocks

A typical clastic sedimentary rock consists of a principal fraction consisting of particles and a minor fraction which is chemical. Ideally the chemical fraction is restricted to filling pore space and serves to interlock individual detrital particles together. In this capacity it is acting as a cement and is so identified. The principal fraction is derived through some process of erosion and is carried to the depositional site by an agent of transportation. During transportation and deposition certain grains are sorted from the original mixture and others are added. The total effect is to produce a mixture of particles whose dimensions may range widely or may be narrowly restricted. When the particle size distribution is more or less uniform the sediment is said to be well sorted. (See Figs. 2.8 and 2.9.) This characterizing feature indicates that currents of uniform velocity were active in preventing larger particles from reaching the site and smaller particles from being deposited. Sediments consisting of important coarse and fine fractions have particle size distributions covering a broad range. They are regarded as poorly sorted and are to be interpreted as indicating wide variations in current velocities which permit deposition of large and small particles at the same site.

Relationships of mineral grains to one another and their relative sizes collectively constitute a very important fundamental property of sediments. This property is defined as *texture*. In those sediments where the fine fraction predominates over the coarse a texture is produced in which an individual large fragment is embedded in a mass of fine debris.

Conglomerates, pebbly siltstones, and sandy shales consist of large particles, a matrix of small particles, and a cement which binds them together. Particle and matrix relationships reflect the influence of sorting, whereas the amount of cement is a measure of the extent to which chemical action has affected the rock. In other rocks, for example some sandstones, sorting has reduced the amount of matrix material to a small fraction of the total sediment. In such cases the matrix occurs primarily as filling of the interstices between the large particles and reduces the space available for cement.

Sedimentary rocks also vary in composition within wide limits. Obviously the composition of a sediment varies according to the percentage distribution of the fragments which it contains. For example, some sands consist almost exclusively of quartz, others of calcite, and still others of quartz, clay, and feldspar. The composition of each is different, and any added cement will likewise alter the composition further. Composition, like texture, is a fundamental property. Most sediments can be classified on the basis of composition and texture alone, and most of the systems of classification which have been proposed are based upon these two fundamental attributes. Only indirectly do composition and texture reflect the environment in which each sediment was deposited. In a general way the sedimentary environment controls the nature of sorting which also partly controls the composition by affecting the distribution of mineral species. Nevertheless, the environment produces certain conditions which affect such mass properties as variety and degree of fossil content, bedding, ripple mark, and mud crack. (See Fig. 12.1.) Such features identified as *structures* are very important in reconstruction of the depositional environment and, hence, constitute the third principal basis for classification.

Textural distinction between clastic sediments is based principally upon differences in the diameter of particles. Gravels contain particles of pebble or cobble dimension, sands consist of grains of sand dimension, and clays of clay-sized particles. Gravels and sands are described as coarse, fine, or medium, but such adjectival endings convey different meanings to individual observers. Even the trained observer finds his descriptive terminology affected by the over-all coarseness of rocks which he is examining. For this reason certain proposals of standardization in nomenclature have been made of which Table 12.1 is an example. As yet no clearly defined limits have been established of the proportionate amounts required of any given size. The general assumption is made that a majority of the particles must fall within one class in order to carry that name. However, certain limitations to class names have been proposed an example of which is shown in Fig. 12.2.

By virtue of the differences in mineral composition of each of the

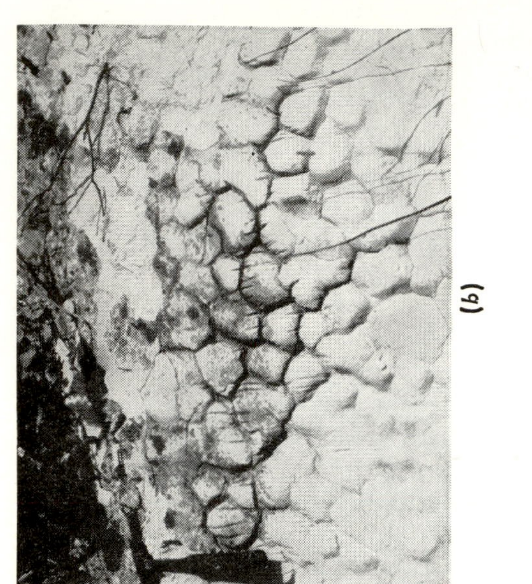

(b)

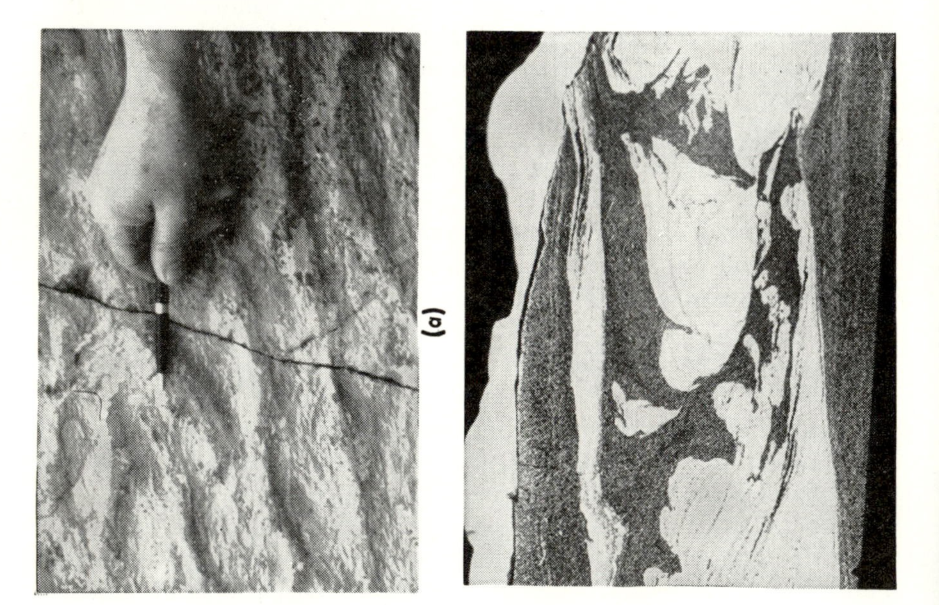

(a)

Fig. 12.1 (a) Ripplemark in sandy dolomite. (b) Mudcrack and linear shrinkage cracks in silty dolomite. (c) Loadcasts in siltstone and sandstone layers. The dark layers of sand penetrated the silt which was deformed by the load applied by deposition of the sand. Structures of this type are developed when the sediments are water-saturated and not cemented. (Photograph (a) courtesy of L. H. Nobles, photograph (c) courtesy of J. E. Hemingway.)

Table 12.1. Terminology Applied to Clastic Sediments and Rocks Ranged according to Size Limits*

Diameter of Particle (millimeters)	Sedimentary Source				Volcanic Source	
	Shape of Particle				Shape of Particle Rounded to Angular	
	Rounded to Subangular		Angular			
	Sediment	Rock	Sediment	Rock	Pyroclastic Fragments	Rock
>256	Boulder Gravel	Boulder Conglomerate	Block Rubble	Block Breccia	Volcanic Blocks	Volcanic Breccia
256 to 64	Cobble Gravel	Cobble Conglomerate	Rubble	Breccia	Bomb	Breccia Agglomerate, or Agglomerate
64 to 4	Pebble Gravel	Pebble Conglomerate	Angular fragments pebble size	Breccia	Lapilli	Lapilli Tuff
4 to 2	Granule Gravel	Granule Conglomerate	Granules	Granule Breccia	Coarse Ash	Coarse Tuff
2 to $\frac{1}{16}$	Sand	Sandstone	Sand	Grit or Sandstone		
$\frac{1}{16}$ to $\frac{1}{256}$	Silt	Siltstone	Silt	Siltstone	Fine Ash	Fine Tuff
$<\frac{1}{256}$	Clay	Shale	Clay	Shale		

* Modified slightly from V. T. Allen, "Terminology of Medium Grained Sediments," Report of Committee on Sedimentation 1932–1934, *National Research Council Bull.* 98, 1935.

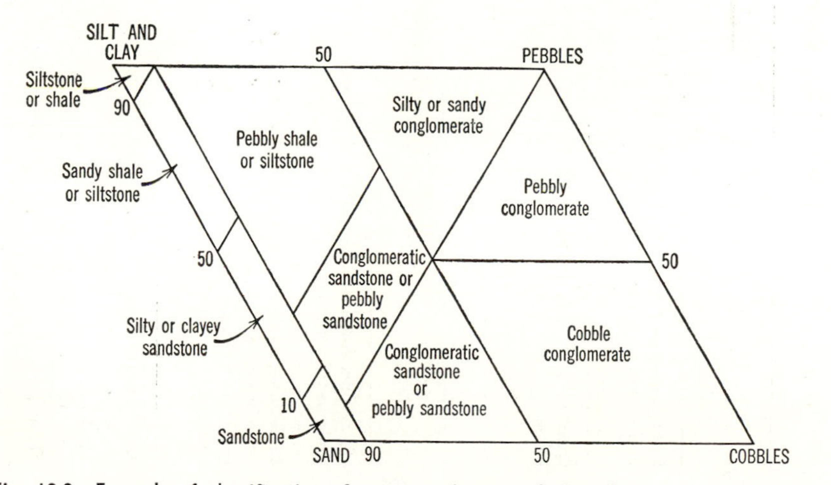

Fig. 12.2 Example of classification of common clastic rocks based upon per cent occurrence of end members of cobbles, pebbles, sand, silt, and clay.

Table 12.2. Classification of Conglomerates and Breccias*

	Composition			
	Rock Particles	Matrix	Cement	Examples
↑ PRIMARILY ONE VARIETY ↓	Quartz, quartzite, chert, flint and some carbonate	Generally large amounts, quartz sand dominant, also calcareous sand	Siliceous common, calcareous less common	Quartz pebble conglomerate or breccia Quartzite pebble conglomerate or breccia Chert pebble conglomerate or breccia
	Quartz, chert, flint, iron oxides	Fine particles of iron oxide, quartz, and clay	Ferruginous or siliceous	Ferruginous-chert-pebble conglomerate or breccia Iron oxide-pebble conglomerate or breccia Quartz pebble conglomerate or breccia
	Limestone, dolomite, siderite	Calcareous or argillaceous, generally in small amounts	Generally calcareous, but may be siliceous	Limestone conglomerate or breccia Dolomite conglomerate or breccia Siderite conglomerate or breccia
	Limestone, dolomite, siderite, shale in thin fragments	Calcareous or argillaceous in large amounts		Flat pebble or edgewise conglomerate or breccia
	Limestone, dolomite, often large in size, associated with much fossiliferous-fragmental material	Calcareous or argillaceous generally in small amounts, often a calcareous sand	Generally indistinguishable from matrix	Reef-flank conglomerate or breccia Fossiliferous-fragmental conglomerate or breccia (These may be better classified as carbonates. See Table 12.6.)
	Limestone or dolomite, may have gypsum or anhydrite, some red shale	Calcareous		Solution breccia
	Shale	Argillaceous or sandy	Siliceous, ferruginous or slightly calcareous	Shale pebble conglomerate or breccia

Few varieties	Plutonic igneous and metamorphic	Ranges between small and large amounts; feldspathic sand dominant	Siliceous, ferruginous, rarely calcareous	Granite-schist conglomerate or breccia Granite-slate conglomerate or breccia Diorite-gneiss conglomerate or breccia Gabbro-schist-calcareous conglomerate or breccia
	Dominantly effusive, and basic igneous rocks	Large amounts of sand size, eruptive rock particles, mica, chlorite and clay		Basalt-andesite agglomerate Basalt-scoria agglomerate Rhyolite-andesite agglomerate
	Chiefly granite, some feldspar (alkaline) crystals	Feldspar, mica, quartz, moderate amounts of clay	Generally siliceous, calcareous less common	Arkose conglomerate
Many varieties	Igneous and metamorphic more abundant than sedimentary	Quartz dominant, feldspar, mica, rock fragments, chlorite; chiefly sand grain size, some clay	Siliceous, ferruginous, carbonate	Granite-schist-dolomite graywacke conglomerate or breccia Diorite-slate-sandstone graywacke conglomerate or breccia Granite-gneiss-schist graywacke conglomerate or breccia
	Sedimentary more abundant than igneous or metamorphic	Quartz dominant, rock particles, some feldspar, very small quantities of clay	Carbonate common, siliceous or ferruginous	Limestone-quartzite-granite quartzose conglomerate Sandstone-basalt-granite quartzose conglomerate
	Wide range of varieties, wide range in particle size	Dominantly clay dimension but wide mineral range; may also be sandy with wide mineral range	Carbonate common, also siliceous	Glacial till Fanglomerate Mudflow

* Condensed and modified from E. C. Dapples, W. C. Krumbein, and L. L. Sloss, "The Organization of Sedimentary Rocks," *Jour. Sedimentary Petrology*, Vol. 20, 1950, p. 11.

textural classes there must also be a difference in the basis of classification of each group. The schemes which follow are examples of classifications which are employed.

Conglomerates

Most conglomerates (see Table 12.2) can be classified upon the basis of size of individual fragments and upon the ranges in their composition. Thus, there are conglomerates in which the individual pebbles or cobbles consist of many rock varieties, those in which the pebbles represent few rock varieties, and those in which the fragments are essentially of one rock variety. Conglomerates consisting of many rock types are typical deposits of very intense erosion in regions undergoing rapid elevation. Local areas experience mountain uplift and shed large amounts of coarse debris by means of rivers to some depositional site. The mixture of pebbles, cobbles, or boulders consisting of different rock types indicates the source area to be geologically complex in order to expose to erosion rocks of many kinds. Usually such deposits are associated with general diastrophic instability in both the source area and the depositional site. The rising mountain section was undergoing faulting or folding during the time of conglomerate accumulation. Oftentimes this structural instability extends into the zone of accumulation, and the conglomerates are themselves folded and faulted. Exceptions to this geologic condition are to be made in the case of conglomerates known to be associated with glacial deposits. These contain many rock types, but as has been previously stated fragments from the local bedrock dominate in numbers. Ancient glacial outwash and till are extremely rare among the rocks of the geologic column; hence, except for Pleistocene deposits they need not be considered.

Conglomerates of few rock varieties are subject to different geologic interpretations depending upon the rock types present. If the pebbles consist of rock varieties generally considered to be unstable in the weathering environment (e.g., limestones or gabbro) the deposit is considered to have resulted from rapid uplift, rapid erosion, and rapid burial all in a local area. This interpretation of origin implies that a condition of structural instability existed and that local faulting or folding prevailed during the time of accumulation. Oftentimes a conglomerate consisting of granitic rock fragments is to be found. Much of this granite can be observed to be weathered and fragmented so that the fine sizes contain abundant partly altered feldspar as individual grains. Material of this type is called *arkose* when in the sand sizes, and when in larger fragments it is identified as arkosic conglomerate. Arkose is typically associated with local uplifts where granitic rocks are the source, and their partially weathered debris is carried to some local subsiding area to be deposited. Such local subsiding sites often are separated from the uplifted adjacent source area by important faults.

Conglomerates of a few rock types, or one rock type, which are demonstrably stable to weathering are common in regions of gently inclined or essentially horizontal rocks. Pebbles are quartz, quartzite, chert, hematite, or limonite. Usually they consist of well-rounded pebbles indicating important abrasion. Such fragments are clearly a residue of extensive weathering and erosion and represent a concentrate of the coarse fraction which may have already passed through one cycle of deposition. Chert pebbles indicate that limestones or dolomites have undergone extensive weathering in order to supply the required amount of chert. Quartzite pebbles and vein quartz indicate derivation from a metamorphic terrain, whereas hematite and limonite fragments generally are derived from concretions of iron carbonate (siderite) which have altered to iron oxide on weathering. Thin zones of conglomerate composed of such stable products often are indicative of accumulation on an old erosion surface commonly associated with a sink-hole topography in limestones.

Conglomerates consisting of pebbles of a single, but unstable, rock type such as limestone indicate rapid local erosion of such a rock terrain and rapid burial in an adjoining sedimentary basin. In some limestone areas, particularly in flat-lying strata, thin platy fragments of limestone are embedded within a limestone bed. Frequently such fragments are identical in composition to the incorporating rock. Such conglomerates are called *edgewise* or *flat pebble conglomerates* and are interpreted as fragments torn from the limestone by wave action during its accumulation. They should not be considered as indicative of important diastrophic conditions.

Sandstones

Rocks classified as sandstones (see Table 12.3) range into conglomerates when they contain a few pebbles and into shales when they contain much clay. They are, therefore, described on a textural basis as pebbly, silty, or clayey sandstones. (See Fig. 12.2.) They are classified also on the basis of their mineralogy. Pebbly sandstones are related to conglomerates, and although such pebbles can range widely in rock composition they are very often chert, quartz, or granite. Pebbles of limestone, coal, shale, schist, slate, basalt, etc., are known but are not frequent in occurrence. Clayey sandstones also

Table 12.3.

Appearance		Composition	
Bulk	Special	Particles	Matrix
Monomineralic (quartz > 90%) +detrital chert	Very light colors predominate	Quartz, light chert	Generally none, minor amounts of silt or clay
Quartz + detrital chert dominant	Red or brown colors	Quartz with iron-oxide coating	Finely divided iron oxide
	Green colors	Quartz, glauconite	Minor, finely divided quartz and glauconite
	Gray colors, muscovite flakes	Quartz, muscovite concentration on bedding surfaces	Minor amounts, chiefly quartz silt, clay minor
	Gray colors, small amounts feldspar	Quartz and mica dominant, feldspar minor, coal fragments common	Moderate amounts, chiefly quartz silt, clay moderate, some carbonaceous matter.
	Gray colors, moderate amounts feldspar	Quartz, feldspar, minor biotite and muscovite	Moderate amounts, chiefly quartz silt; clay and iron oxide minor
			Minor amounts of matrix
Other mineral species abundant (except carbonates or sulphates)	"Salt and pepper" gray color, black chert prominent	Quartz, black chert predominate; mica, glauconite less prominent	Moderate amounts; quartz, mica, silt, and clay in equal proportions
	Tuffaceous, laminated (porcellanite)	Feldspar, glass fragments and quartz prominent; mica minor	Moderate amounts, fine felt of glass and feldspar
	Pink gray or red gray colors, generally coarse grained	Feldspar >25%, quartz, mica less prominent	Minor to moderate amounts, chiefly clay, some quartz silt, mica, and iron oxides
	Medium gray colors	Quartz, rock fragments, mica, small amounts of feldspar	Moderate to large amounts, chiefly clay, mica, chlorite
	Medium gray colors "salt and pepper"	Plagioclase, pyribole, mica, and black chert dominant; moderate amounts quartz, chlorite, and rock fragments	
	Dark gray colors	Quartz, mica, pyribole, rock fragments and chlorite	
Carbonate particles dominant (See Table 12.6)	Crystals, fossil fragments or particles, with or without noticeable matrix	Calcite, dolomite, rarely gypsum	Calcite, dolomite, clay, etc.

Classification of Sandstones*

Cement	Examples	Group Name
Siliceous, rarely calcareous, often poorly cemented	Pure quartz sandstone	
Ferruginous, siliceous, gypsiferous	Quartz-iron-oxide sandstone	
Siliceous, rarely calcareous	Quartz-glauconite sandstone	Quartzose sandstones
Siliceous to slightly calcareous, often poorly cemented	Quartz-muscovite sandstone	
Siliceous to slightly calcareous, generally moderate	Quartz-mica-feldspar subgraywacke sandstones Quartz-mica-carbonaceous subgraywacke sandstone	Subgraywacke sandstones
	Quartz-feldspar subgraywacke sandstone (<25% feldspar)	
	Arkosic sandstone (<25% feldspar) Quartz-feldspar sandstone (<25% feldspar)	Arkosic sandstone Quartzose sandstone
Calcareous common, also siliceous, moderate to well cemented	Quartz-black chert subgraywacke sandstone	Subgraywacke sandstone
Siliceous	Sodic-feldspar–glass subgraywacke sandstone	
Calcareous common, also siliceous	Gray arkose Pink arkose Granite wash arkose (thin sheets)	Arkosic sandstone
	Quartz-mica graywacke sandstone	
Siliceous to slightly calcareous	Calcic-feldspar–pyribole graywacke sandstone	Graywacke sandstone
Siliceous	Quartz-chlorite-pyribole graywacke sandstone	
Usually not distinguished from matrix; if present, mainly calcareous	Dolomite and calcite sandstones Foraminiferal sandstones and marls Gypsum sandstone Oolitic sandstones (See Table 12.6)	

* Condensed and modified from E. C. Dapples, W. C. Krumbein, and L. L. Sloss, "The Organization of Sedimentary Rocks," *Jour. Sedimentary Petrology*, Vol. 20, 1950, p. 12.

called "muddy" sandstones often are poorly cemented and are associated with stream, delta, and lagoonal deposition.

Diastrophism also exerts an important influence on the composition of sandstones by affecting the extent of weathering in the source region and the rapidity of burial in the depositional site. Rapid and intense uplift of the source area supplies mixtures of incompletely weathered rock to the streams. Conversely, a low-lying immobile region such as has been typical of much of the mid-western section of the United States supplies rather completely weathered debris to streams.

Quartzose sandstones. (See Table 12.3.) Weathering in the topographically low area concentrates quartz and clay in particular, and the mineralogy of the sediment has been rendered less complex than the source rock. Relative stability of the depositional site results in very slow burial. Much of the debris which is deposited is only temporarily at rest, and in shallow-water marine environments wave and current action keeps the upper layer of sediment in constant agitation. During this stage of the depositional history the clay fraction is removed by winnowing, and there remains an accumulation of almost pure quartz sand devoid of interstitial clay. Gradually as slow subsidence occurs these sands are buried, and the particles are cemented to a greater or lesser degree by mineral matter precipitated from percolating ground-water solutions. Such is the origin of the ideal quartzose sandstone.

Quartzose sandstones consist of well-sorted grains of quartz generally of exceptional roundness. Individual grains approach spheres particularly those which are near 1 mm in diameter. Under the microscope a thin section of rock reveals the round outline of every detrital grain. (See Fig. 12.3*a*.) Individual grains touch one another at points of short contact. This open arrangement permits much interstitial space and consequently a high porosity. Quartzose sandstones are characteristically almost free from clay. If the sand is accumulating in a marine environment in which calcium carbonate is being precipitated this interstitial space becomes completely or partially filled with calcite. The rock is well cemented or poorly cemented depending upon the extent to which calcite fills the pore space. Sometimes the sand is buried without any cement. Later, as ground water moves through the buried sand silica, carbonate, or iron oxide may be deposited in the interstices. In those cases where silica is deposited it is precipitated on the individual quartz grains in crystallographic continuity. Addition of quartz to round detrital grains results in development of crystal faces and eventual transformation into a sharply outlined crystal. Individual crystals become mutually interlocked, and the rock is transformed into a tightly cemented tough aggregate called *quartzite*. (See Fig. 12.3*a*.)

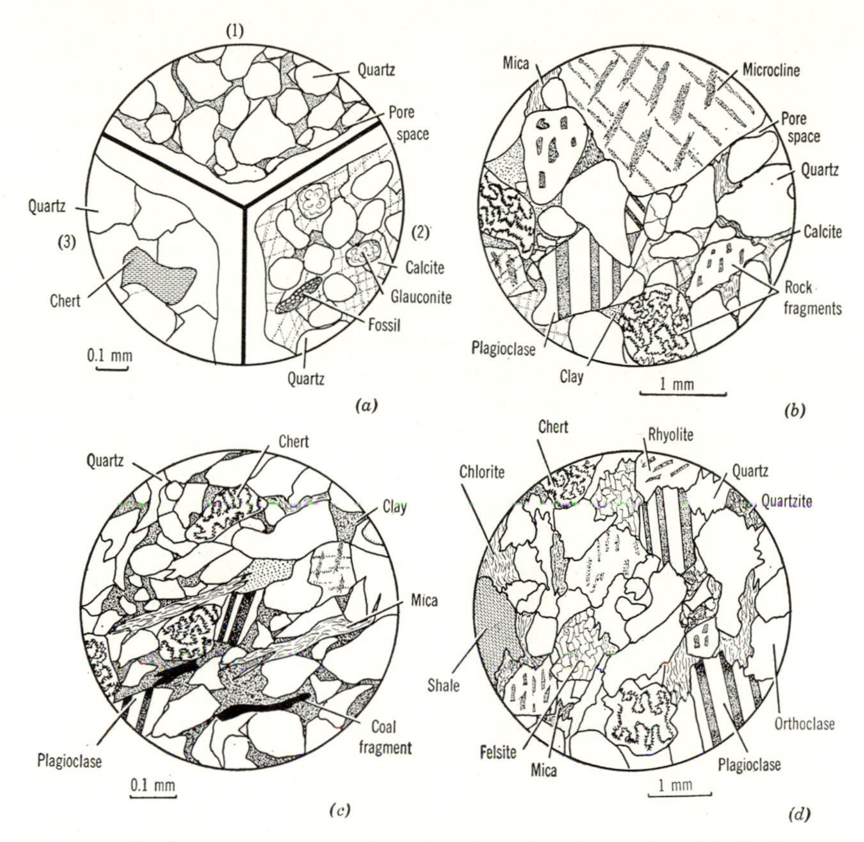

Fig. 12.3 Sketches of textures and mineral composition of typical clastic sedimentary rocks in thin section under the microscope. (*a*) *Quartzose sandstone:* (1) poorly cemented, Cambrian, Iron Mountain, Michigan, (2) cemented by calcite, Davis shale (Cambrian), French Village, Missouri, (3) cemented into quartzite by overgrowths of quartz on quartz grains, Cambrian, Jackson, Wisconsin; (*b*) *Arkose,* Nonesuch shale, Precambrian, Hancock, Michigan; (*c*) *Subgraywacke,* Pennsylvanian, Pleasantview sandstone, Fulton Co., Illinois; (*d*) *Graywacke,* Precambrian, Saganaga Lake Ontario.

Arkose. (See Table 12.3.) Strong uplift or faulting in an area where granitic rocks can be exposed through erosion and in proximity to a rapidly subsiding area nearby results in accumulation of the richly feldspathic sandstone called *arkose*. Thick arkose deposits generally are wedge shaped and often attain thicknesses of several hundred to several thousands of feet near the source but taper rapidly with increased distance outward. Wholesale disintegration of the granitic rock supplies a coarse rubble of granite much of which has been disaggregated into fragments of sand or granule size. Light pink to red colors generally characterize this sand which consists of abundant orthoclase or microcline feldspar. Intense weathering darkens the pink color to red which becomes the characteristic tint of arkose[1] Arkose is also of gray color when the feldspar is dominated by plagioclase. Such arkoses are not easily recognized except by careful identification of feldspar. Clay interstitial matter is not a prominent feature of the typical arkose, rather it appears to be a mixture held by a cement. Absence of clay interstitial matter is to be expected because arkose is like a mass of crushed granite reconstituted by cementing individual grains of quartz, feldspar, and rock.

Arkose consists of a poorly sorted sediment in which individual grains range between fine and coarse. (See Fig. 12.3*b*.) This permits packing of individual grains so that much of the interstitial space is filled with fines, or large fragments are buried by small grains. Also these grains tend to be of a low order of roundness, and in the packing process a crudely interlocking texture is produced. Introduction of cement binds this mass together into a generally tough rock, often nearly as strong and massive as its original parent.

Graywacke. (See Table 12.3.) A third important group of sandstone is designated graywacke. These are gray or greenish gray colored rocks, usually very well cemented, and occur typically in belts of closely folded strata. Characteristically they contain fragments of sand size derived from a large variety of rocks. Slate, schist, quartzite, basalt, felsite, rhyolite, sandstone, and limestone are abundant constituents and are intimately mixed with quartz and feldspar. A clay-sized fraction occupying interstitial space between grains or when abundant as a matrix consists principally of clay minerals, chlorite, mica, and chert. Quartz is usually but not necessarily the most abundant constituent, followed by rock fragments, matrix mixture (clay, chert, micas, chlorite), feldspar, and scattered minerals unstable to weathering.

Graywacke has a special texture which consists of large, splintery or irregular grains buried in a mass of very poorly sorted fine debris. (See Fig. 12.3*d*.) Some grains have well-defined boundaries, but others are extremely irregular and deeply embayed. This corroded aspect is characteristic of small as well as large grains, and the embayed areas are occupied by matrix minerals which have recrystal-

[1] The red color in potash feldspar is due to oxidation of the minute amounts of iron which are occupying voids or random positions in the crystal lattice. In some cases a strong red color is produced.

lized and have penetrated larger grains. Such an interlocking produces a rock which is a welded mass, and cement is not an important part of its makeup. The characteristic intergrown texture is the basis of the extreme toughness of the rock and the tendency to break with conchoidal fracture.

Graywackes are rocks of singular occurrence. Their distribution is very clearly restricted to areas which were undergoing strong deformation during the course of deposition and which later became the site of great mountain chains. The sediments have been accumulated in great linear geosynclinal troughs whose rate of subsidence kept easy pace with the dumping of enormous quantities of rapidly eroded detritus. Diastrophic instability in the depositional site is indicated by the common association of graywackes interbedded with basalt lava flows. Many of these show pillow structure, a characteristic attributed to solidification under water.[2] Such display of volcanism is a clear indication of crustal unrest. Rapid subsidence of the geosynclinal trough appears to establish deep-water conditions adjacent to the shore area supplying the sediments. The graywacke-producing topography must, therefore, have been one of mountainous aspect in order to supply the required enormous quantity of incompletely weathered multirock-type debris.

Bedding in graywackes is not uniformly spaced except in the sequences of black shales which are intimately associated with the sandstones. Generally, individual beds range between a few inches and a few feet in thickness. Thin beds dominate, and within these small units there is gradation of grain size from coarse to fine upward. Certain structures such as cross-bedding and minor contortion of bedding planes are notably of small scale, usually restricted to units a few inches thick.

Subgraywacke. (See Table 12.3.) A rock intermediate in aspect between graywacke and quartzose sandstone is called subgraywacke. Generally this is a moderately cemented rock characterized by a relatively high percentage of clay or chert interstitial matter. Its composition is dominated by angular or splintery quartz grains packed in a loosely interlocked framework into the pores of which has accumulated clay. Rock fragments, feldspar, muscovite, and coaly material are minor but common and characteristic constituents. Small, rounded grains of dark chert often are so abundant as to produce a speckled appearance, and the rock is described as a "salt and pepper" sandstone. Another common aspect is that in which muscovite is

[2] Pillow structure appears as a series of individual ellipsoidal-shaped masses, approximately 1 to 2 feet long, piled one upon another.

scattered throughout the rock but principally concentrated on bedding surfaces. Such micaceous sandstone is very common among coal-bearing strata.

Subgraywackes range widely in textural fabric. (See Fig. 12.3*c*.) Some resemble quartzose rocks and consist of rather well-rounded quartz grains and interstitial clay. With the interstitial space filled with clay the permeability of the sediment to water is low. Solutions from which cement is precipitated, therefore, do not move freely through the rock, and poor or irregular cementation is a characteristic property. In some subgraywackes interlocking of splintery and irregular grains produces a reasonably strong bond, but those which are held by clay are remarkably weak. Very muddy or clayey subgraywackes often disintegrate when placed in water. Irregularity of cementation in subgraywackes makes prediction of their behavior under stress unreliable. They must be regarded as weak rocks and very undesirable construction materials.

Subgraywackes have certain associations which ally them to graywackes. For example, they are typical sandstones of the geosynclinal margins. That is, they occur in belts which have undergone important subsidence in regions where the limits of the great geosynclines are becoming nebulous and where the subsiding areas appear to have consisted of individual basins surrounded by more stable areas. Within such subsiding basins subgraywacke sandstone accumulates with its associated shales, coals, and frequently limestones. In contrast to the arkose and in particular to the graywacke which may attain thousands of feet in thickness of repeated units the subgraywacke is seldom more than 200 feet thick as a single unit without being interrupted by shale. Within a unit it may occur in massive beds as much as 25 feet thick or in 1 to 2 inch thin slabs. As a general rule thin bedding is a counterpart of significant clay content, whereas the more massive sandstones tend to be low in clay. Also subgraywackes are part of an association of sediments which are cyclical or repetitive in occurrence. These are beds of sandstone, shale, coal, and shale occurring in order and repeated with impressive frequency.

Subgraywackes are characteristic "coal basin" rocks. In this occurrence they are frequently slabby, carbonaceous, and contain much clay. Usually the rock is a fine- to medium-grained sandstone with uniform and often smooth bedding surfaces. The under part of bedding surfaces shows casts of flow and interpenetration due to differential loading of water-saturated sediments before consolidation. (See Fig. 12.1.) Thin-bedded units often are micaceous particularly upon the surfaces of bedding planes, and such surfaces also are marked with imprints or well-preserved remains of plants.

Other subgraywacke sandstones occur in associations of shales and limestones, often in cyclical sequence, and in which the preserved fossils are recognized as nearshore marine- or brackish-water dwellers. Many of the marine sandstones are massive and stand in bold outcrop revealing rather poorly defined bedding. These marine rocks also tend to contain less clay, and the interstitial space is occupied by cement, principally calcite and chert.

Shales

(See Tables 12.4 and 12.5.) The term *shale* is applied to a variety of rocks ranging between clays and silts. Their normal aspect is thinly bedded in well-displayed layers up to several inches thick. Surfaces of these layers are planes of actual parting or incipient separation and are planar and smooth. The plane of separation is a cleavage parallel to the general surface of deposition; in some cases it is so well developed that actual sheets of sediment can be separated as from a tablet of paper. Shales of this description are appropriately called *paper shales,* and the attribute of cleavage is termed *fissility.*

Although to the unaided eye shales appear to be composed of nearly uniformly sized material the actual range of particle size is very large. Some shales consist exclusively of clay-sized particles, whereas others are mixtures of clay, silt, and even sand. This textural difference often endows the shale with distinct properties. Differences in shales, however, appear to be responses to the contained minerals. For example, shales composed primarily of illite or montmorillonite tend to be saturated with water and are soft and greasy to the touch. They lack hardness and strength, and they usually disintegrate when placed in water. Shales dominated by quartz are much harder, more brittle, and remain unchanged in water.

Color is a very conspicuous attribute of shale, and much reliance is placed upon this property as a broad index of environmental conditions. Black shales usually are rich in organic matter which has contributed the dark color. Their environment of development is stagnant water where oxygen is low and chemically reducing conditions prevail. Gray colors indicate less contained organic matter and are to be interpreted as having been deposited in a more aerated environment where oxygen reaches the bottom and bacteria remove the excess organic waste. Green colors indicate the existence of mildly reducing conditions during deposition and usually the presence of illite or minerals such as glauconite containing ferrous iron. Red colors indicate strongly oxidizing conditions and are not typical of marine environments. This is the characteristic color developed through oxidation in the continental environment, particularly in the absence of vegetation. For this reason red shales are generally assumed to have been deposited on land and in arid climates where very little organic waste could be deposited with the accumulating clays.

Shales are considered to be the most abundant sedimentary rocks, but they are among the least understood, and a uniformly acceptable classification has not been proposed. This is due to the inability to determine their mineral composition except by X-ray techniques. Attempts to classify shales by color or by bulk physical attributes have not proved satisfactory. Another approach has been to classify them according to some characterizing mineral and general association with rocks typical of accumulation under certain diastrophic conditions. This is the classification shown in Tables 12.4 and 12.5. Accordingly quartzose shales are recognized as those accumulating in association with quartzose sandstones.

Feldspathic shales are associated typically with arkose, micaceous shales with subgraywackes, and chloritic shales with graywackes. However, all shales are not easily placed in these groups despite their association with certain well-defined sandstones or limestones. For this reason resort must be made to some system of identification similar to that of Table 12.5.

Environments of Clastic Deposition

Marine Deposition

Marine transgression of the continent can be understood by hypothecation of a gradual subsidence of the Mississippi Valley region much of which lies at an elevation less than 600 feet above sea level. Suppose that subsidence took place at the rate of 1 foot in 1000 years. Then the entire area would be below sea level in 600,000 years, and the North American continent would be divided in two, with a seaway from Hudson Bay to the Gulf of Mexico. If this region continued to subside at approximately the same rate and deposition kept pace with subsidence a belt of marine sediments would be accumulated within the continental margins. Uplift of this area sufficient to expel the sea would preserve some of the marine deposits inasmuch as some sediment would remain at elevations below sea level.

It is important, also, to visualize the nature of the marine invasion and regression. As subsidence was initiated the present shoreline would gradually invade the coastal plain, the beach gradually migrating inland leaving a blanket of sand representing coalescence of individual beaches. The delta of the Mississippi would gradually become submerged, but delta sediments would migrate upstream as the great Mississippi lowland gradually subsided below sea level. The pattern of sediment distribution in the nearshore zone of the Gulf of Mexico would gradually shift inland, assuming a new pattern as the topography dictated. Sand bars surrounded by silt areas of the Texas coast would be shifted inland as would the great belt of limestone deposition around the peninsula of Florida. The new pattern of limestone deposition, delta deposition, and sand and silt deposition would not be the same as the present, but such belts would be continuous with present deposits as the marine transgression took place. Regression of the sea would cause the depositional belts to shift once more toward the Gulf of Mexico, but the positions of the belts would not necessarily coincide with their respective positions during the stage of transgression. By this mechanism sediments representative of different depositional conditions lie one upon the other in sedimentary sequence, and their individual thickness is dependent upon the local rate of subsidence and the amount of sediment which is furnished.

Sandstones represented in this transgression would reveal the characteristics of the immediate nearshore environment. Along the barrier beaches sands of the shore would be well sorted and generally free from clay, whereas the subaerial portion would be composed of remarkably well-sorted dune sand. Such sands would consist of little other than quartz and remain spread as a thin sheet at the margin of the advancing sea. This is the environment of deposition of most quartzose sandstones which often occur as continuous sheets for hundreds of miles. Features typical of the littoral zone such as swash mark, small cross-bedded dunes, and fossils of shore-dwelling organisms living in this zone should be preserved in the sediments. In turn, these deposits would lie upon muddy and organic sediments of the brackish-water lagoon across which the sea was encroaching.

Seaward the conditions of the neritic zone would be encountered. This zone is limited by water depth extending from the littoral zone (low-tide limit) to depths of 100 fathoms (600 feet). Such a depth range spans differing marine conditions, and, normally, the quieter and deeper water receive the finest sediments. As a rule shale deposition characterizes the deeper-water areas, whereas sandstones are more typical of the shoreward margins. In this zone sands grade into silts and clays which eventually become siltstones and shales.

Table 12.4. Classification of Shales and Siltstones*

Bulk Appearance		Cement and Other Features	Examples	Group Name
Texture	Color			
Claystones predominate, also silty claystone, sandy claystone, and clayey siltstones Bedding blocky to fissile	Light shades of green, gray, and blue predominate	Cement poor, often calcareous, may contain calcareous pellets, "slip" fracture common	Glauconite-quartz shale Glauconite-quartz siltstone Glauconite-quartz clayey siltstone	Quartzose shales
			Quartz shale Quartz siltstone Quartz silty shale "Fireclay"	
	Shades of red and brown predominate		Iron-oxide–quartz shale Iron-oxide–quartz siltstone "Red-bed" shale	
		Swells markedly when wet	Bentonite	?
	Dark shades of gray to black	Cement poor to moderate, often calcareous, may contain calcareous pellets	Pyrite-quartz shale, carbonaceous-quartz shale	
		Cement moderate, often siliceous, may contain plankton, spores	Quartz-pyrite shale	Quartzose shales

Siltstones predominate, also clayey siltstones, sandy siltstones, and silty claystones Bedding, thin to less commonly blocky, fissile, fracture splintery or "pencil"	Dark shades of gray and green	Cement moderate, generally slightly calcareous, also siliceous	Glauconite-mica clay siltstone	Subgraywacke shales
	Dark shades of gray; red, brown, to black	Cement moderate, siliceous to slightly calcareous	Quartz-mica siltstone Feldspar-mica clayey siltstone Carbonaceous-mica siltstone "Fireclay"	
			"Red-bed" shale Iron oxide-mica siltstone Gypsum-mica siltstone etc.	
			Carbonaceous-mica siltstone or shale	
	Dark shades predominate, commonly gray	Cement moderate to good, generally calcareous	Phosphatic-mica siltstone Phosphatic, calcareous, mica siltstone or shale	
		Cement good, generally siliceous	Tuffaceous-mica siltstone or shale	
	Dark shades of red, brown, variegated, also grays and greens	Cement poor to good; commonly calcareous; also siliceous	Arkosic siltstone Feldspathic siltstone or shale Iron-oxide feldspathic-clay siltstone	Arkosic shales
			Gypsiferous arkosic siltstone Iron oxide, gypsiferous arkosic silty shale	
'Dense" rocks predominate, siltstones and claystones	Dark shades predominate commonly gray; black	Cement good, generally siliceous, associated with graywackes	Chloritic siltstone Siliceous-chloritic clayey siltstone Pyrite-chlorite silty shale Tuffaceous-chlorite siltstone Pyritic-carbonaceous siltstone	Graywacke shales; grade into subgraywacke shales
Bedding blocky to splintery	White to light gray		Porcellanite	Graywacke shales

* Condensed and modified from E. C. Dapples, W. C. Krumbein, and L. L. Sloss, "The Organization of Sedimentary Rocks," *Jour. Sedimentary Petrology*, Vol. 20, 1950, p. 14.

Table 12.5. Characterizing Aspects of Shales*

Type	Thickness	Other Rock Associations	Condition in Depositional Site	Inferred Environment f Deposition
Quartzose	Uniform over wide areas. Usually less than several hundred feet thick.	Quartzose sandstones, cherty limestones and dolomites.	Stable, very slow subsidence.	Broad shallow seas, or wide flood plains, marginal lagoons. Black shales indicate restricted circulation. Stable shelf deposition.
Arkosic or feldspathic	Generally uniform over wide areas but with abrupt local differences. Thickness as above.	Arkosic sandstones, nodular limestones.	Locally unstable and subsiding more than surrounding regions.	Similar to above. Unstable shelf and basin deposition.
Micaceous	Thickness shows abrupt changes or transitions, regional changes pronounced.	Subgraywacke sandstone, argillaceous or "clean" limestones, special associations are gypsum and other salts.	Slowly or actively sinking basins.	Shallow seas, alluvial plains, deltas, marginal lagoons, restricted marine evaporite basins, or stagnant black-shale basins. Unstable shelf or basin deposition.
Chloritic	Thickness shows abrupt changes locally and regionally.	Graywacke sandstones, siliceous, dark limestones.	Strongly subsiding areas.	Rapid burial, range between terrestrial and marine. Geosynclinal deposition.

* Modified from W. C. Krumbein, "Shales and Their Environmental Significance," *Jour. Sedimentary Petrology*, Vol. 17, 1947, pp. 101–109.

Gradation may be laterally abrupt or gradual as the current conditions dictate.

Clastic sediments of the neritic zone are characterized by uniform bedding and general continuity of a stratum. They contain fossils of organisms, particularly benthos which inhabit this zone and whose remains are buried quickly. Bedding as well as continuity of the shales is more uniform than sands which tend to be more irregular in distribution and thickness. Cross-bedding typical of strong current action along bars is common in the sands.

Certain areas, for example the neritic zone stretching gulfward from the Florida peninsula, receive very little land-derived clastic debris. These are the sites where in warm waters carbonate sediments accumulate through organic activity. In cold waters thin deposits of clay and organic residue slowly accumulate, eventually yielding black shales. Carbonate deposition as a whole is considered chemically controlled, and discussion of these sediments is reserved for special consideration.

In the vicinity of the delta the neritic zone is most variable in types of deposits. Lenses of sand sometimes covering many square miles in area accumulate off the mouths of distributaries. Here current action is sufficiently intense to concentrate much coarse debris along with some clay. This is the environment of deposition of the typical subgraywacke sand. Nearby in protected areas behind the extended natural levees silts and clays will accumulate sometimes as mixtures or as end members. These deposits enclose the sand lenses as the mouths of the distributaries shift in position. Variability of sediment type with rapid lateral change is the characteristic aspect, and deposition typical of the long straight coast line receiving its sediments from longshore current activity does not exist. In the deeper neritic waters abrupt changes of sediment become less and less typical, and such areas are dominated by silty clays which upon compaction and cementation become shales.

The conditions described above produce sediments of the common and normal varieties. These are subgraywacke and quartzose sandstones and gray shales. Arkose and graywacke are much less common and require special conditions of accumulation. These are localities of mountainous relief adjacent to deep-water shores. In the case of arkose, granitic rocks must be disintegrating in the mountain belts, and the weathered products must be transported by streams to be dumped into deep nearshore waters. Graywacke requires much the same conditions of accumulation, but the source area terrain must consist of a geologic complex where rocks of many kinds are exposed.

In each case, however, the sediment is dumped into the depositional site and receives only minor sorting by currents. The principal depositional controls are the rate of settling and rapid subsidence within the burial site.

Marine and Continental Transitional Deposition

Certain environmental conditions straddle the marine and continental realms. Principal among these are the delta, the lagoon and lagoonal swamp, the estuary, and the littoral zone of the beach. Sediments of the delta, the lagoon, and the littoral beach are not uncommon in the rocks of the geologic column, and identification of their environment of deposition is important in reconstruction of the geologic history of an area.

The delta. This is not a single environment but consists of composites of several independent environments such as beach, alluvial channel, swamp, and lagoon. As distributaries shift position coarse-sand deposition occurs in localities formerly receiving silts and clays. Lenses of sand become incorporated in the body of silts and clay. During periods of exceptionally high sea level marine conditions invade the mouths of distributaries, and biologic activity is correspondingly changed. Brackish-water organisms such as oysters, clams, and certain marsh vegetation shift positions along the delta front and become incorporated in the lenticular deposits of silts, sands, and clays. Locally muds are exposed above sea level, dry out, and develop regular or irregular patterns of cracks (mud crack). Elsewhere currents produce cross-bedded sands or erode channels in previously accumulated deposits. Later such channels become filled once more with sediments, often coarse sand. Channel fills of this type are characterizing features as are ripple marks, mud cracks, buried vegetation, laminated silts, and slump structures. Most of the sediments are poorly sorted, and the subgraywacke sands are characteristically high in clay. Such sands usually are fine grained and often contain small pods of "clean" sand $\frac{1}{4}$ to $\frac{1}{2}$ inch in thickness and a few inches in length in an otherwise very argillaceous sand.

The lagoon. Lagoons are of various shapes but are typically elongate, marshy, and filled slowly with silts and clays. Much of the water is brackish and inhabited principally by bottom dwellers such as clams which burrow in the muds and feed upon the accumulating organic debris. The sediments are characteristically colored dark gray to black from the organic residue, and thin layers of accumulated vegetation become buried in the silts and clays. The lagoonal environment is ephemeral and is frequently destroyed by a barrier beach advancing inland. Vertical sequences of clays, silts, organic muds, and sands, each as lenticular or irregularly shaped bodies, are a characteristic of ancient lagoonal sediments.

The littoral beach. This is characteristically a gravel or sand deposit although silty and muddy tidal flats are within this zone. Except for ancient beaches which are recognized by their association with wave-cut cliffs and marginal strips of conglomerate this environment is not readily recognized. When the ancient sediments contain organisms typical of the tidal flat (worms, crabs, clams) such fossils prove to be the best means of recognizing the environment.

Continental Deposition

Stream deposits. Principal among the deposits identified as continental are those deposited by streams. These deposits are accumulations along broad alluvial flats or aprons of fans or in marginal swamps.

Among the most prominent of the alluvial-flat deposits are those classified as red beds. Some red beds are massive units of sandstone or siltstone (sometimes arkose), whereas others consist of beds of red-colored shale, sandstone, or siltstone separated by similar rocks of buff, gray, or green-gray color. Except for the massive units which may have continuity over many square miles, individual layers are distinctly discontinuous within short distances. Beds a few feet thick abruptly taper and intertongue with other red-bed units of different color. In addition to the color, channel fillings are characterizing features. These are conglomerate or sand lenses representing deposits within former stream channels. Such lenses of coarsely granular sediment are concave in their under portion and can be observed to cut across bedding of the enclosing rocks. Among the common fragments of the conglomerates are pieces of petrified wood, coal chips, rounded shale or clay pellets, and bones of land-dwelling vertebrates. Surfaces of shale layers irrespective of color often are marked by abundant remains of plants. The fossils are clearly defined imprints of vegetation bordering the former stream courses. Locally such deposits appear to have accumulated in shallow lakes as indicated by limy layers containing abundant fossils of snails and algal spheroids.

Other red-bed deposits prticularly those containing arkose appear to have been accumulated as alluvial fans mantling granitic mountain uplifts near the sea coast. Locally the arkose beds fill channels in the red feldspathic shales, whereas closer to the source they occur as more continuous beds. Continental arkosic red beds merge into those which accumulated under marine conditions, and in turn these grade into quartzose sandstones typical of shallow water and strong wave action.

Other alluvial-flat deposits are accumulated under high-moisture conditions and are associated with much evidence of former vegetation. These deposits are intergradational with the lagoon, marginal swamp, and deltaic environment and are typically illustrated by strata containing coal beds. The associated strata consist of subgraywacke sandstones, gray and black shales, and clays. Typical exposures consist of orderly repetitive sequences of these beds in units of a few feet each through many hundreds of feet of strata. Some of the sandstones are continuous, but more commonly they are fillings of sinuous outline occupying channels cut in the underlying beds by streams. The sandstones range between coarse and fine grained and are locally conglomeratic particularly near the base. Pebbles of the conglomerates consist of quartz, shale, coal, and broken "shells" of concretions of siderite (iron carbonate). Overlying the sandstone there is to be found frequently a light to dark gray structureless clay which on the weathered surface becomes light gray, soft, and sticky. Such beds are rather continuous units extending over many square miles but seldom are more than 2 feet in thickness. The clays consist almost entirely of kaolinite, illite, and quartz and are often very refractory. For this reason they are called "*fireclays*" and are the principal source of much high-grade brick and tile for furnace linings. Even when dry, fireclays are soft rocks, but when wet they swell and heave under loads. There is virtually no structure within the fireclay, but there are many occurrences of fossilized tree stumps and roots which have led to the interpretation that the clays are the former soils upon which grew a swamp forest vegetation.

Almost without exception a coal bed overlies the fireclay. Such coals range in thickness from a mere streak of less than 1 inch to the exceptional bed of 10 feet or more. Coals are surprisingly continuous despite their relative thinness and are extremely useful in tracing the position of strata from one locality to another. Usually the coal rests upon the fireclay with sharp definition, but the boundary can be gradational, and fossilized tree stumps with roots in the fireclay are known to continue upward into the coal. Deposition of coal is considered to be typical of the marginal swamp or bog where vegetation may accumulate with only minor admixture of silt and clay.

Overlying the coal a gray or black shale frequently is observed. This deposit brought an end to coal deposition by invasion of the swamp by fine silts and clays. Shales which overlie the coal beds generally are very fossiliferous, and well-preserved remains of the coal-producing plants are plentiful. Such shales also contain concretions of siderite (iron carbonate) as nodules, spherical masses, and discontinuous layers an inch or so thick. The concretions are secondary in origin but their development appears to be partly controlled by the environment of deposition. For example, the presence of such carbonate nodules associated with fossils of brackish-water organisms indicates that some of the gray shales accumulated in environments marginal to the sea. Oftentimes this interpretation is strengthened by the appearance of thin limestone beds bearing marine fossils interbedded with the gray plant-bearing shales. Local marine invasions of the alluvial flats and marginal lagoons testify to the low-lying character of the topography under which the coal-bearing strata must have accumulated.

Dune deposits. Climates of aridity appear to be surprisingly persistent throughout geologic time. Desert regions appear to remain geographically fixed for extended periods, and the sediments accumulated in such areas have been preserved. Some red beds appear to belong to this category and are associated with extensive beds of sandstone considered to have been dunes. Some of the dune sand is red or pink in color, but generally it is light gray or light buff. The rock is unusually well sorted, a characteristic typical of dunes, and individual grain sizes tend to be medium to coarsely grained. (See Figs. 2.7, 2.8, and 2.9.) Outcrops are typically massive exposures with very poorly preserved normal bedding but cross-bedding is a common structure. Foreset bedding inclined in many different directions and cut-and-fill structure so common in modern dunes are to be observed on the exposed faces of some outcropping beds. Some geologic formations in various parts of the world are characterized by dune sandstones and are known to extend over very large areas.

Hand specimens of the dune sandstone contain scarcely any interstitial silt or clay. Individual grains of quartz are nearly spherical in shape, and large grains have surfaces which are frosted and pitted. The rock is variably cemented even in the same outcrop. Some portions are thoroughly bonded with quartz or calcite cement, whereas nearby the sandstone may be essentially uncemented. For this reason dune sandstones are good aquifers wherever they can gather enough water, and in some regions they are important producers of artesian supply.

Nonclastic (Chemical) Rocks

The typical nonclastic rock (see Tables 12.6 and 12.7) is a finely crystalline marine limestone. A thin section of this rock examined

Table 12.6.

Bulk Textural Appearance	Particles or Crystals		Matrix
	Appearance	Composition	Composition
Coarse to medium particles, crystals, or fragments in matrix	Tests or parts of tests of macroscopic organisms	Calcareous, less commonly siliceous or pyritic	Calcite or dolomite, rarely argillaceous
			Dolomite more common than calcite
	Particles and matrix similar	Limestone, dolomite	Dolomite or calcite; anhydrite
		Limestone, dolomite, chert, etc.	Dolomite, calcite, anyhdrite; red clay common
	Particles and matrix similar, flat pebbles	Limestone or dolomite	Calcite, dolomite, commonly argillaceous and glauconitic
	Pisolitic or oolitic	Calcite, dolomite, silica, siderite, hematite, phosphate, etc.	Calcite, dolomite, silica, hematite, phosphate, clay
	Interlocking or cemented crystals	Calcite or dolomite	Dolomite, calcite, anhydrite
		Gypsum, anhydrite, halite, etc.	Gypsum, anhydrite, halite, etc.
	Nodules; dense, irregular	Limestone, dolomite, siderite	Limestone, dolomite, commonly argillaceous
Fine particles or fragmental in matrix	Tests of microscopic organisms; fragments of macroscopic organisms	Calcite or dolomite; less commonly silica	Calcite, dolomite, silica; commonly argillaceous
Dense to finely crystalline	Subvitreous or porcellaneous, with scattered crystals of organic remains	Calcite or dolomite, rarely siderite, anhydrite, etc.	
	Subvitreous to vitreous	Silica	

* Condensed and modified from E. C. Dapples, W. C. Krumbein, and L. L. Sloss, "The Organiza-

under the microscope reveals a texture not unlike a fine-grained igneous rock. (See Fig. 12.4*a*.) Crystals of calcite can be seen to be mutually interlocked in a texture similar to that developed by the solidifying igneous melt. Chemical analyses of some of these limestones demonstrate they are practically pure $CaCO_3$ with only traces of silica and alumina. They are interpreted as having been precipitated from water which upon being warmed lost carbon dioxide and from which $CaCO_3$ was forced out of solution. Wherever this condition exists near to areas receiving typical land-derived clastics particles of quartz and clay are carried to the site of limestone

Classification of Chemical Rocks*

Other Features	Examples	Group Name
May have structures of detrital rocks	Fusulinid limestone Coquina Crinoidal dolomite	Fossiliferous fragmental
Poorly bedded, rubbly	Reef-core dolomite	
Typical structures of detrital rocks	Dolomite "sand" Sugary limestone	Fragmental
Poorly sorted, angular fragments; massive; obscure bedding and lateral continuity	Solution breccia Collapse breccia	
Shingled and edgewise pebbles commonly cross-bedded, ripple-marked, etc.	Edgewise conglomerate Flat-pebble conglomerate	
Fossils aberrant; many structures of detrital rocks	Oolitic limestone Pisolitic phosphate rock	Oolitic
May have fossils and primary structures obscured by recrystallization	Sugary dolomite Calcite "sand" Crystalline dolomite	Crystalline
White, pink, translucent, soft	Bedded anhydrite Bedded salt	Evaporite (crystalline)
Conchoidal fracture or breaking along nodule contacts	Nodular limestone	Nodular
Well bedded to massive, well-preserved organisms	Foraminiferal limestone Chalk, Diatomite	Normal marine (finely crystalline)
Scratched with steel; conchoidal fracture; often with blebs and mottling of different color and coarser texture	"Lithographic" limestone Dense dolomite Siliceous limestone	Dense (crystalline)
Scratches steel; conchoidal fracture; translucent; beds, nodules, lentils, etc.	Chert	Chert

tion of Sedimentary Rocks," *Jour. Sedimentary Petrology*, Vol. 20, 1950, p. 16.

deposition. These detrital particles become incorporated in the sediment as the insoluble fraction. This is a case where the clastic fraction is minor and the chemical fraction dominates. As the distance toward the source of land-derived detritus is decreased the amount of clastic debris will gradually increase to become the major fraction, and the carbonate will become the cement.

Carbonates of Organic Origin

Another form of calcium carbonate precipitation is that accomplished by those organisms which construct a shell during their life

Table 12.7. General Classification of Carbonate Rocks*

1. Fragmental.
   A. Fossiliferous fragmental—composed principally of fossil fragments.
   B. Oolitic—composed principally of oolites.
2. Nodular—composed of nodules of limestone usually segregated from neighboring ones by films or masses of clay.
   A. Fossiliferous nodule—consists of fossil fragments aggregated together as a nodular mass and cemented by calcite.
   B. Algal nodule—consists of individual balls or spheroids of calcareous algal deposits.
   C. Lithographic nodule—consists of individual nodules of very finely crystalline limestone. They show no internal structure.
3. Crystalline—composed of crystals without obvious organic structures often produced as a result of recrystallization or dolomitization of fragmental limestones.
   A. Megacrystalline—crystals > 10 mm in length.
   B. Coarsely crystalline—crystals 2 to 10 mm in length.
   C. Mediocrystalline—crystals 1 to 0.1 mm in length.
   D. Finely crystalline—crystals 0.1 to .02 mm in length.
   E. Cryptocrystalline—crystals $< .02$ mm in length.
   F. Sugary—crystals are individual rhombs of calcite or dolomite, more or less loosely packed and loosely cemented.
   G. Lithographic or dense—a cryptocrystalline texture which breaks with conchoidal fracture. The rock is generally unfossiliferous but scattered remains are well preserved.

* Modified from L. L. Sloss, "Environments of Limestone Deposition," *Jour. Sedimentary Petrology,* Vol. 17, 1947, pp. 109–113.

processes. When the organism dies the shell becomes part of the inorganic debris. It is now a clastic particle and under wave attack is transported, broken, and abraded into small grains many of which are of sand dimension. In tropical marine waters extensive beaches of calcium carbonate sand are produced through such a procedure. This sand is a clastic sediment, yet upon consolidation and slight recrystallization following burial the rock is transformed into a crystalline limestone with a texture much like an igneous rock. (See Figs. 12.4*b* and *c*.)

Still another carbonate rock is produced by colonial organisms such as algae and corals. These organisms precipitate small amounts of calcium carbonate about their body wall which is attached to that of its neighbor. In this manner the growing mass attains characteristic and complex shapes. Some are branching and treelike, others are large serrated forms, whereas still others are generally ovaloid with

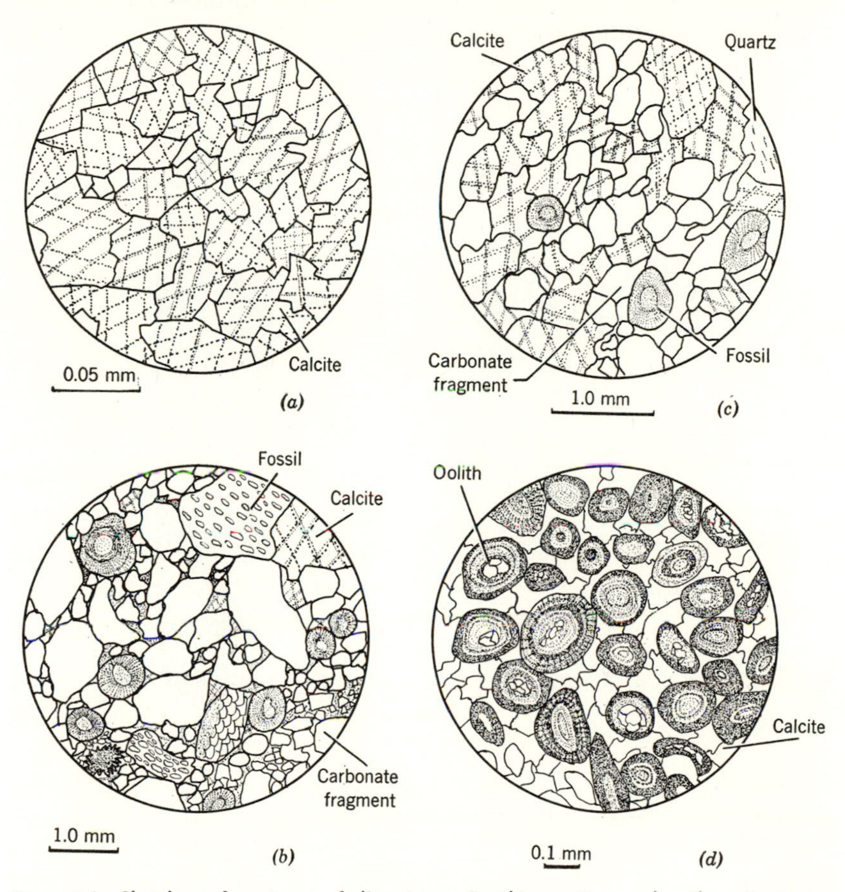

Fig. 12.4 Sketches of textures of limestones in thin section under the microscope. (*a*) *Finely crystalline*, Kokomo limestone, Silurian, Kokomo, Indiana. (*b*) Poorly cemented *fragmental*, Mississippian, Grassy Cove, Tennessee. (c) Partly recrystallized *fossiliferous fragmental*, Mississipian, Grassy Cove, Tennessee. (*d*) *Oolitic* texture showing ooliths with radiating fibers as well as concentric shells, Fredonia oolite, Mississippian, Anna, Illinois.

crenulated surfaces. Within and upon the surfaces of the growing organisms calcium carbonate is precipitated so as to form internal wall structures, and the organism can be identified by these characterizing features which are preserved by the calcium carbonate. Individual colonies grow together to form a reef, extending their dimensions laterally as well as vertically and growing to the water surface where is to be found an abundance of light, aerated water, and food. By such growth the reef develops a shallow-water environment of its own, and wave attack is directed toward the disintegration of the mass. Particles broken loose are clastic fragments, and as these are

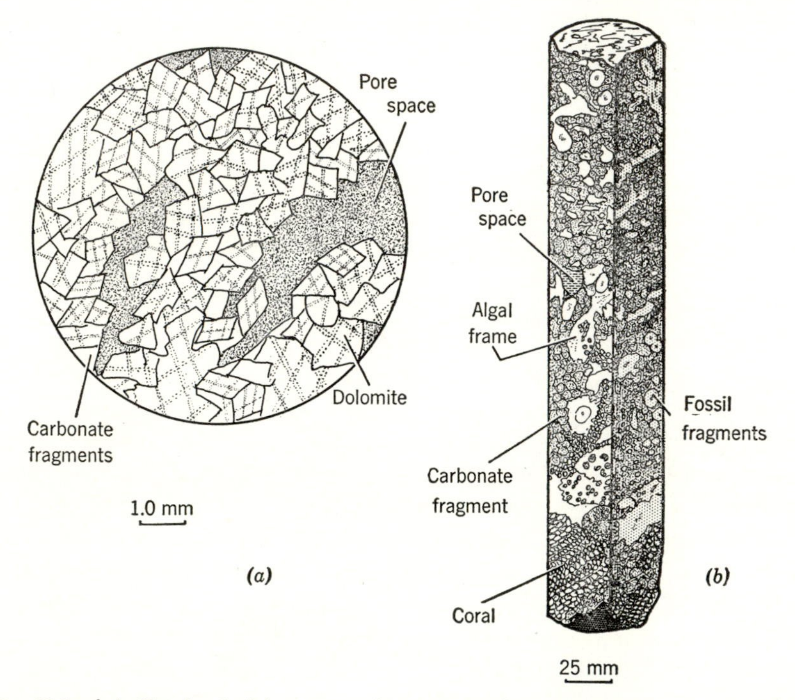

Fig. 12.5 (*a*) Sketch of dolomite in thin section showing "*sugary*" texture, Racine dolomite, Silurian, Thornton, Illinois. (*b*) Sketch of core from drill hole in ancient coral reef. Texture shows typical *reef-core* texture of framework of algal growth binding carbonate fragments. Silurian strata, Tilden, Illinois.

comminuted by wave and current action they become lodged in cavities of the organic structure. This procedure continues and the solid structure of the reef is constructed by two processes: (1) shells of reef-dwelling organisms and reef builders are broken to make an abundance of clastic sediment; (2) the clastic sediment fills space between the growing parts of the reef-building organisms, and these bind the sediment into solid rock by precipitation of calcium carbonate as a part of their growth processes. The resulting rock is both clastic and nonclastic and, hence, displays the characteristics of both (see Figs. 12.4*c* and 12.5*b*).

Classification of Carbonates

Because of such difficulties in differentiation no uniformly satisfactory system of classifying limestones has been proposed, and the one presented in Table 12.7 is no exception. This table is preferred because it is simple and useful for hand-specimen identification. The essence of this classification is the recognition of three basic textures: (1) fragmental, (2) nodular, and (3) crystalline. (See Table 12.6 for identification of these rocks.) Fragmental texture is the product of clastic-carbonate sediments derived principally from the shells of organisms. Nodular texture appears to be developed in shallow-water environments where abnormally high salinity causes precipitation of calcium carbonate about some center or in shallow brackish water in association with calcareous mud flats. Crystalline texture may be due to direct precipitation of calcium carbonate or it may represent recrystallization of some previously existent texture.

Fragmental texture. Fragmental textures convey information regarding the condition of the depositional environment. Some limestones are very argillaceous and consist of partially broken and whole fossil shells embedded in a matrix of land-derived clay. This fragmental texture is to be interpreted as representing the probable growth environment of the organisms represented. Whole shells of clams and other mollusks indicate they have not been transported far, and half shells when oriented with their convex surfaces upward are in a stable position to resist movement. In the case being illustrated the environment was one of muds, and the calcareous fraction which dominates is present because of the ability of this environment to support molluscan life. The presence of mud is interpreted as an indication that bottom currents were not strong, but they were sufficient to roll a half shell into its stable position.

Another fragmental texture consists of sand and granule grains recognized to be pieces derived from the shells of organisms. This rock is composed entirely of rounded particles held by calcite cement, and there is definite absence of very fine-grained material. Outcrops reveal cross-bedding and ripple marks. The granular texture and such associated structures lead to the interpretation that the depositional environment was one of considerable wave and current activity. Shells were broken into sand which was built into bars as indicated by the cross-bedding. Directional-current activity is indicated also by ripple marks. Sediment of silt and clay dimension must have been removed by winnowing, and the concentrated product was a well-sorted fossiliferous-fragmental sand. By cementation and partial recrystallization this sediment was transformed into rock, and its texture is described as "clean" fragmental. (See Figs. 12.4*b* and *c*.) Such rocks are the calcareous counterparts of quartzose sandstones, and the two are common associates. (See Fig. 12.3*a2*.)

Oolitic texture is fragmental (see p. 288). It is a product of concentration of oolites moved by current action as sand grains. Individual ooliths are swept along to their depositional site where in total they appear as cross-bedded sediments and are intergradational with fossiliferous-fragmental beds. Cementation by calcite binds individual ooliths together into a poorly cemented or well-cemented rock depending upon the amount of calcite precipitated. In this respect fossiliferous-fragmental and oolitic sediments are much alike, and frequently they are very permeable to water when cementation of individual fragments is poor.

The environment of oolite development is warm shallow salt water in which salinity is slightly higher than normal. Here, oolites appear in the zone where active wave action creates strong turbulence. Typical occurrence is on the surface of the coral reef or the shallow aerated banks where fossiliferous-fragmental sands are accumulating. In these rough-water sites certain objects become nuclei for the oolith. A nucleus may be a shell fragment, an entire but microscopic shell, a quartz grain, or the fæcal pellet of some small organism. Gradually the nucleus becomes encased by calcite added as concentric layers or as fibers radiating outward (see Fig. 12.4*d*). Concentric internal structure is by far the most characteristic feature, and additional layers are added until the oolith becomes buried after which time growth no longer occurs. Some ooliths grow to dimensions of granule, or larger, size whereupon they are called *pisolites,* but there is no distinction in their composition. Oolites of calcium phosphate are associated with some oolitic limestone and are believed to have been deposited under conditions of chemical reduction rather than the oxidizing environment typical of the calcite oolite. Siliceous or dolomitized oolites also are reasonably common. They are not original deposits but are developed through replacement of calcite by silica or dolomite. The process whereby this replacement occurs is not well understood despite its common occurrence, but there has been mineral substitution in such a fashion that original microscopic structures are preserved faithfully.

Textures associated with organic reefs. An organic reef is described as a mass of carbonate rock which has been constructed through the growth processes of organisms. These processes have resulted in vertical building of the reef and the development of a wave-resistant structure. Reef-constructing organisms produce a framework which binds fragmental material broken by heavy wave action against the reef mass. As the organically erected part of the reef grows upward eroded material is spread as an apron of clastic debris around its base. The structural entity of the reef, therefore, consists of two parts: a wave-resistant organically developed center and a detrital apron. In its active part the reef grows upward to the water surface where

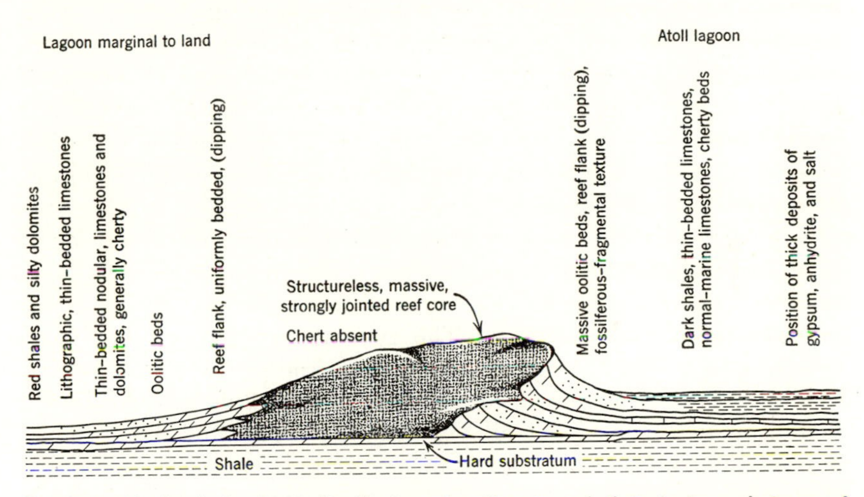

Fig. 12.6 Idealized sketch of distribution of sediments and their textures during reef growths. The reef core is the wave-resistant structure built by organisms to the water surface and maintains topographic relief above the apron of the carbonate detritus over which it grows. Land-derived clastic sediments intertongue with the carbonate deposits.

the proper conditions for organic growth are to be found. From this position growth continues laterally over its own detrital debris. Colonies of reef-building organisms erect the framework which holds the fragmental debris, and this structure is characterized by a massive aspect lacking bedding. For this reason it is identified as the *reef core* and is distinguished from the flanking portion by the well-bedded structure and prominent dip of the latter. *Reef-flank* beds are inclined in all directions from the core, beginning with dips approximating 30 degrees but decreasing rapidly in magnitude within short distances. (See Fig. 12.6.)

Rock from the structureless reef core is porous. It consists of coarse pieces of shells, corals, and algae partially replaced by calcite or dolomite and held by well-preserved coralline or algal growths. The effect is to produce a rock consisting of fossil debris often containing openings $\frac{1}{4}$ to $\frac{1}{2}$ inch across. (See Fig. 12.5*b*.) Reef-flank texture also is fossiliferous fragmental, but individual grains usually are of sand or granule size and are held by calcite or dolomite cement. These strata are much less porous than their reef-core counterparts, and their permeability depends to a larger degree upon the extent of cementation or recrystallization.

Outward from the immediate reef flank several textural types commonly are developed. Oolite zones intertongue with coarse reef-flank debris, whereas thin and nodular beds, often bearing chert, are to be observed somewhat farther distant. Textures typical of the quiet-water lagoons are cryptocrystalline (lithographic) and characterized by uniform and thin beds which break with conchoidal fracture. In scattered localities some of these strata are gradational laterally into gypsum, anhydrite, and halite (NaCl) which under certain conditions are in beds totaling hundreds of feet in thickness.

Some ancient reefs are extremely large extending over several square miles in area; others are small and only a few feet in diameter. Some are thick attaining several thousands of feet, whereas the small ones may be as thin as one foot. The general structure of thick reefs indicates that they have grown upward, expanding their flank as they did so. However, from bottom to top they contain the remains of organisms which must inhabit very shallow water and require much oxygen and sunlight.

Nodular texture. Nodular textures are to be observed principally in outcrop. In surface exposures individual nodules can be seen embedded in more evenly bedded rock. The aspect is enhanced by weathering which accentuates boundaries between the nodules. Perhaps the most clearly defined texture of this type consists of a nodular mass of fragmental limestone, consisting principally of whole shells and broken fragments, surrounded by calcareous clay containing much fossil debris. Surfaces of the nodules are encrusted with fossils suggesting that such organisms were attached to these fragments during their growth period and that the nodule grew by addition of shell matter. No such surface was available in the adjacent soft clay, and this environment was occupied by "mud" dwellers.

Some carbonate rocks are nodular by virtue of the enclosed algal "balls." These may be scattered along bedding planes as small individual colonies, or they may form small reeflike mounds of larger colonies. Individual algal balls are recognized by a concentric structure much like that of concretions; hence, they are often called pisolitic algae (see Fig. 12.7). Nodular algal limestones are considered to be developed primarily in fresh or very saline water. Algae inhabit waters of normal marine salinity also, but their colonies are large, and their growth forms are branching. In this environment they are associated with corals with which they make important reefs. Ap-

(*a*)

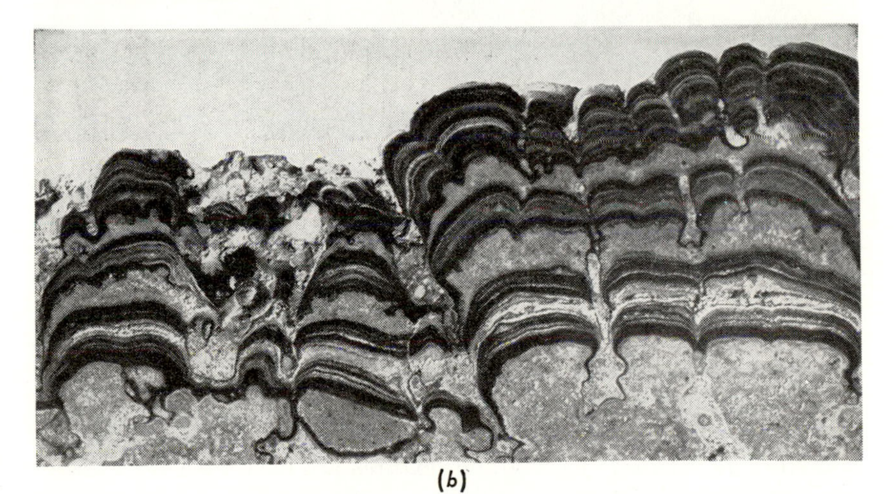

(*b*)

Fig. 12.7 (*a*) Upper surface of a compound reef which consists of alternating zones of algal and inorganic limestone, Green River formation, Eocene, Sweetwater Co., Wyoming. (*b*) Transverse section of (*a*) cut and polished to show alternating layers of organic and inorganic limestone; scale approximately full size. (From Bradley, *U. S. Geological Survey Prof. Pap.* 154-G.)

parently rigorous conditions established by extremes in salinity are required to keep colonies small and, hence, to form the nodules which become incorporated in the limy sediments.

Certain nodular limestones consist of a finely crystalline rock enclosing small rounded masses of lithographic or cryptocrystalline texture. Upon weathering these nodules are more resistant than the enclosing rock, and they may form a loose rubble on the weathered surface. Individual nodules are without structure and consist of the same cryptocrystalline texture throughout. They tend to contain rather high amounts of clay which concentrates on the weathered surface of the nodule to give it a silty appearance. Such rock is sparingly fossiliferous and not uncommonly contains spheroids of

Fig 12.8 Spheroids of chert in limestone. St. Louis limestone, Mississippian, White Co., Tennessee. Note that the nodules are individuals scattered throughout the stratum.

chert. (See Fig. 12.8.) Strata of this description are not widespread areally and, hence, are believed to be developed in some special local environment. As yet this environment is not clearly established, but such textures have been observed in association with the lagoonal portions of large coral reef complexes, particularly in that portion which occupied very shallow water near land.

Crystalline texture. Crystalline textures are developed through recrystallization of fossil fragments and other carbonate material and also are the product of original crystallization. Recrystallization may completely obliterate pre-existing textural aspects, and as a result many carbonates are not easily identified as representing original fragmental material. (See Fig. 12.4*a*.) With the exception of the texture described as "sugary" individual crystals are interlocked closely, and a generally strong and impermeable rock has been produced.

"Sugary" texture consists of individual rhombs, principally of dolomite, piled one upon the other in such a fashion as to leave conspicuous openings, often ¼ inch in diameter, and arranged in a complicated network separated by generally interconnected channels. (See Fig. 12.5*a*.) Such textures are extremely permeable. Individual rhombs are strikingly uniform in size, averaging about 1 mm, and may be so poorly interlocked as to be held by small quantities of cement only. Under such circumstances the rock is usually weak and crushes at very low values for carbonates. Other crystalline textures are characterized by well-interlocked crystals and in these circumstances produce rock of very great strength in all directions of orientation. (See Fig. 12.4*a*.)

As a general rule coarse crystallinity is associated with metamorphosed carbonates (marble), but certain rocks composed of broken thick shells produce a very coarse fragmental texture. With slight recrystallization this texture becomes megacrystalline.[3] Cryptocrystalline textures are typical quiet-water lagoon deposits and are particularly characteristic of dolomites interbedded with thick sections of gypsum and salt.

Common Saline Deposits

There is general agreement among geologists that deposits of gypsum, anhydrite, and common salt are precipitates from highly concentrated waters resulting from excessive evaporation; hence the deposits are classified by the bulk term *"evaporite."* Some evaporites are associated with red beds and are known to have originated through concentration of waters similar to those of Great Salt Lake, Utah, but such lakes are incapable of supplying sufficient salt to accumulate thicknesses of hundreds of feet of these deposits. Rather, these must accumulate in areas where evaporation pans are connected with seas which can supply a steady flow of salt-laden water to the evaporating site. Principal among evaporites are gypsum ($CaSO_4 \cdot 2H_2O$), anhydrite ($CaSO_4$), and halite ($NaCl$). Only rarely are these found with a high degree of purity, but they are usually interbedded with shales and carbonates, principally dolomite. Locally layers of other salts are found with halite, but their common occurrence is as individual crystal impurities.

The texture of evaporite deposits ranges between megacrystalline

[3] Coarse crystals are produced largely from the stems of an invertebrate animal identified as a crinoid. This organism has a cup, or calyx, of frondlike tentacles mounted upon a segmented stem. The stem is fastened to solid rock and imparts a plantlike appearance. Upon death the stem becomes fragmented into cylindrical pieces and constitutes an important part of most fossiliferous-fragmental textures (see Fig. 12.4*c*).

and cryptocrystalline, but coarse crystalline varieties are more common than those of small crystal size. Special textures such as fibrous are found in gypsum. This is an arrangement of thin but elongate crystals in parallel orientation to produce a texture resembling a fibrous mass. Halite is usually very coarsely crystalline and occurs in massive units without a break of any sort. When gypsum and halite form in shales individual crystals may reach dimensions several

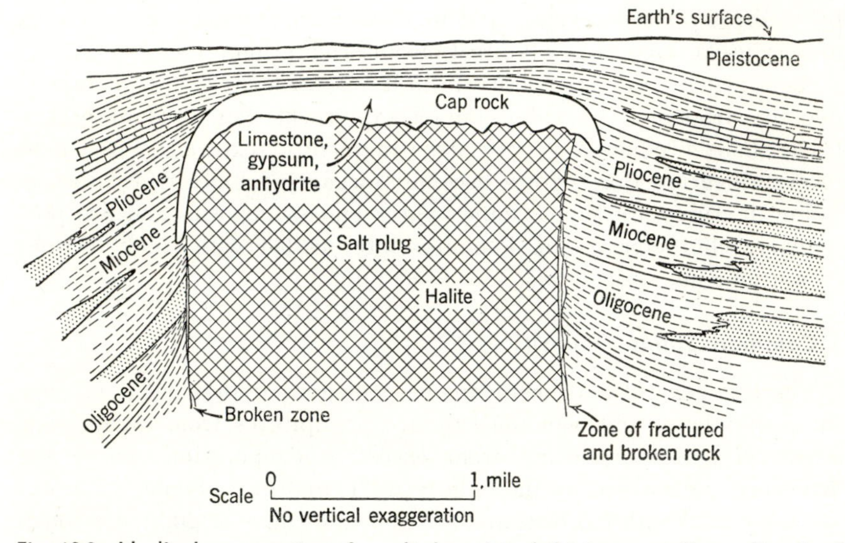

Fig. 12.9 Idealized cross section of a salt dome intruded into surrounding sediments of Tertiary age. Barker's Hill dome, Chambers County, Texas. (Simplified from *Houston Geol. Soc. Guidebook*, 1941.)

inches in length. Gypsum has a habit of large flat clear crystals of selenite, and halite of large cubes. Anhydrite does not appear to develop in shale but does form in dolomite generally as veins.

Bedding in evaporites is not readily recognized and usually is indicated by a change in color from light to dark. However, thin shale separations can be observed in some deposits. Because these materials tend to flow rather than fracture when under stress jointing is not a common feature, rather they deform more or less as a plastic mass as indicated by contortion of the bedding. The nature of plasticity under stress is akin to that of ice deformed by glacial movement, and evaporite deposits are known to yield readily under stresses applied near the earth's surface. Even minor folding tends to produce conditions of instability within the salt which may be forced through enclosing strata in much the same manner as an igneous intrusion.

In the essentially horizontal strata of the Gulf Coastal Plain deeply buried layers of salt have been irregularly loaded by sediments. Unequal distribution of these loads has caused the salt to yield and intrude the superincumbent strata. Such intrusions, known as salt domes, corklike in shape and frequently exceeding a mile in diameter, have broken their way through the overlying limestones, sandstones, and shales until equilibrium is attained. (See Fig. 12.9.)

BASIC CONCEPTS OF STRATIGRAPHY

Source of Stratigraphic Information

The occurrence of sedimentary rocks in layers separated one from the other by bedding planes and differences in composition presents a unique aspect among the three great rock groups (i.e., igneous, metamorphic, and sedimentary). Some of these layers contain special materials such as the evaporites mentioned above or deposits of ores or oil. There is more than academic reason, therefore, to map the position of such layers and to trace them as discrete units from place to place. In some instances this can be done with ease inasmuch as such strata can be followed along their outcrops, but eventually they pass beneath the surface and their position is a matter of speculation. Commonly, where they emerge at the surface once again their aspect is changed, and the question must be raised as to whether they represent the same group of strata or whether they are different layers. The technique of recognizing the relative position of such beds and the interpretation of the environments of deposition which they represent are studied in a specialized field of geology known as *stratigraphy*. As in many other fields of specialization stratigraphy is still largely in its descriptive stage because of the gigantic task of defining and cataloguing the relative positions of sedimentary layers throughout the world. Initially this work was accomplished entirely by means of mapping surface exposures on foot, but for the past half century an increasing amount of information has become available from bore holes. The data compiled each year have assumed spectacular proportions principally because of the intensive search for oil. Some formations now described never crop out at the surface anywhere, and their characteristics are known only through the information of the borings. Others can be shown to change drastically in texture and composition as they are traced from the surface into the subsurface and could not otherwise be recognized.

Law of Faunal Succession

Inasmuch as the special contribution of stratigraphy involves arrangement of strata into units the basis of this cataloguing was early recognized to be the relative age of deposition of the sedimentary layers involved. A time chart was erected which in an early form was based upon local relationships observed in Great Britain (see Table 12.8). With the progress of exploration there was need to compare the ages of strata with those recognized in continental Europe, and eventually on a world-wide scope. In part, the sequence listed in Table 12.8 (Lyell's) was developed through the law of superposition, and in Wales this was done with considerable success. However, the most valuable basis of dating the relative ages of individual formations was soon to be recognized as being the characterizing enclosed fossils. Through painstaking examination of many thousands of fossils a clearly defined vertical sequence of organisms was demonstrated to exist. A regular succession of fauna and flora was observed in strata over the explored world, and nearly identical organisms were being found in strata of the separate continents. From such observations was enunciated the *law of faunal succession*. This law stipulates that with the progress of time organisms follow specific patterns of evolution and become increasingly complex irrespective of their geographic distribution. Some organisms followed such a pattern of development and then became extinct. Others changed only slightly through long periods of time, but the direction of change was always toward increasing specialization. Gradually, by much careful labor on the part of paleontologists and stratigraphers a succession of organisms and their order of change has been established, and strata can be dated on the basis of the enclosed fossilized remains. There exist, however, certain fundamental assumptions which have been incorporated into the law of faunal succession. Principal among these is the concept that organisms will spread over the world very rapidly and with very little differentiation. Moreover, the rapidity of spread of life is assumed to be much greater than the rate of geologic change. In large measure these assumptions have been demonstrated to be correct, and strong reliance may be placed upon this method of dating strata.

Division of Geologic Time

Early understanding of the vertical distribution of fossil remains led to the establishment of groups of rocks classified as belonging to

Table 12.8. Subdivision of Geologic Time as Recognized in North America

As Recognized in 1833*		As Recognized in 1949†		
Major Divisions	Subdivisions	Eras	Periods	Epochs‡
Recent	Newer Pliocene	Cenozoic	Quaternary	Recent Pleistocene
Tertiary	Older Pliocene Miocene Eocene		Tertiary	Pliocene Miocene Oligocene Eocene Paleocene
Secondary	Cretaceous Wealdian	Mesozoic	Cretaceous	Upper Lower
	Oolite or Jura		Jurassic	Upper Middle Lower
	Lias New Red Sandstone		Triassic	Upper Middle Lower
		Paleozoic	Permian	Ochoan Guadalupian Leonardian Wolfcampian
	Coal Measures		Pennsylvanian	Virgilian Missourian Desmoinesian Atokan Morrowan Springeran
	Mountain Limestone		Mississippian	Chesterian Meramecian Osagian Kinderhookian
	Old Red Sandstone		Devonian	Senecan Chautauquan Erian Ulsterian
			Silurian	Cayugan Niagaran Medinan
	Transition Group		Ordovician	Cincinnatian Mohawkian Canadian
			Cambrian	Croixian Albertan Waucobian
Primary		Precambrian	Keweenawan Huronian Temiskamian Keewatinian	

* Classification of Charles Lyell, *Principles of Geology*, John Murray, London, 1833.
† Classification as listed by R. C. Moore, *Introduction to Historical Geology*, McGraw-Hill Book Co., 1949.
‡ Names of epochs are locally applied and change geographically as required by local differences in organisms and changes from marine to continental strata.

specific episodes of time. This classification was based upon the association of certain organisms which persisted through long intervals of geologic time before becoming extinct. Groups of strata overlying one another contain an association of organisms which change abruptly between certain rock layers. On this basis all strata containing certain organisms are classified as belonging to one major interval of geologic time distinct from that of the underlying and the overlying group. Rocks are thus arranged into great time systems.

Each system was believed to be separated from another by world-wide catastrophic events and the annihilation of certain organisms. Support to this concept appeared in the evidence supplied by groups of strata separated by sinuous surfaces of erosion. Below each such surface the rocks contain certain organisms which do not occur in the rocks above. Moreover, fossils in the younger strata are of higher evolutionary rank. The surface of erosion known as an *unconformity* is to be interpreted as representing an interval of time of unknown length spent in erosion of the underlying strata before deposition of the overlying rocks. (See Fig. 7.33.) Some unconformities are of spectacular extent separating strata which can be demonstrated to be of very radically different ages, and use of the unconformity to bound groups of rocks has been most effectively employed. By such methods major subdivision of the rocks of the geologic column has been essentially completed for Europe and North America. Much additional work requiring special techniques of dating remains to be done in order to catalog the minor subdivisions not separated by significant differences in fossil remains. As additional information becomes available the stratigraphic position of local groups of strata is constantly undergoing revision.

The most significant pioneer work in stratigraphic subdivision was done in the nineteenth century in Great Britain, where following recognition of major groups identified as Primary, Secondary and Tertiary (already so-called on the Continent) subsidiary units were established. To each such sequence of strata an appropriate name was applied customarily after some distinct physical property. (See Table 12.8.) Strata belonging to the group called the Old Red Sandstone were distinguished from those of the Mountain Limestone and these in turn from the overlying Coal Measures. Some of these names are still employed today but not precisely in the same manner as they were first used as subdivisions of the Secondary System of strata. Rapid strides in paleontology in England as well as on the Continent demonstrated that strata belonging to such groups of rocks were rich in distinctive fossil remains. On the basis of the occurrence of certain fossils as well as superposition and unconformities a new classification was erected employing names such as Cambrian, Ordovician, Silurian, and Devonian but not in the order currently accepted. Strata low in stratigraphic position were observed to be dominated by invertebrates (brachiopods) and fish. Higher in the succession of strata, but separated by an unconformity from those underlying, molluscan invertebrates (cephalopods) and reptiles were found to occur. Overlying strata separated by an unconformity were recognized as belonging to the Tertiary sequence and were identified by certain mollusks, particularly oysters and clams, and the presence of mammal remains. Recognition of such distinct assemblages of organisms typical of very large sequences of strata led to the concept of the classification of geologic time in terms of major developments in organisms and to the definition of large time units, identified as *eras*. Each of the eras, Paleozoic, Mesozoic, and Cenozoic, bracket certain unequal lengths of geologic time recognized in the rocks by the evolutionary development of particular organisms.

Subdivision of the eras into periods was based upon separation by organisms and unconformities. These are surprisingly consistent over very large areas, but the boundaries are not precisely the same for all of the periods; hence, some difference in intercontinental classification is recognized. Further subdivision of periods into epochs is even more local in nature. Within the borders of a continent strata identified as belonging to a single geologic period are subdivided into different epochs which are not synchronous between widely spaced regions. This situation has arisen because of uncertainty in time equivalence between widely separated marine and nonmarine beds. For these reasons names applied to epochs vary from place to place, and those listed in Table 12.8 are not uniformly appropriate to all parts of North America. The student must, therefore, become acquainted with the terminology in vogue locally. Uncertainty in time correlation is bound to prevail until some method of precise dating of sedimentary rocks becomes available.

Stratigraphic Correlation

Perhaps the primary task of the stratigrapher is to measure, describe, and trace strata from one locality into their correlative units elsewhere. He does this by more or less standard procedures and eventually assigns such beds to some epoch, period, and era of geologic time. In this form they are catalogued in chronologic order without regard to lithology. At one time the ultimate stratigraphic achievement was considered to be the completion of a catalog of all strata arranged in their respective chronologic order and grouped into epochs, periods, and eras. In this connection each unit was to be arranged in position on the basis of its contained fauna and flora and in accordance with the law of faunal succession. This arrangement appeared to indicate some rather profound differences between the organic remains of strata grouped into different eras. Moreover, the nature of the organic remains indicated a periodicity in the behavior of the earth during the course of each era. Every era appeared to be initiated by gradual invasion of the continents by shallow marine waters. This invasion was culminated when the continents were very extensively submerged and was followed by withdrawal of the seas and initiation of an episode of mountain building of world-wide scope. Withdrawal of the seas and eruption of the great mountain chains were considered to have exerted a profound influence upon the existent organisms and resulted in extinction of an impressive list of organisms. The end of each era was, therefore, considered to be a great "age of dying," and with the dawn of a new era more advanced and different species of organisms appeared. To a lesser extent each period was considered to have patterned the cycle of the era. A period of geologic time was considered to be separated from the one to follow by extensive withdrawal of seas from the continent and a time of sufficient erosion to produce an unconformity between the strata.

Time-Rock Units and Rock Units

Present-day concepts have modified such rigid definitions of eras and periods, and increasing use is made of assemblages of strata called *time-rock units*. A time-rock unit consists of a sequence of strata recognized as representing deposition during a distinct interval of time. Time-rock units are considered without recourse to unconformities or invasions and retreats of the seas. Rather, they are rock bodies possessing well-defined properties but recognized to have been deposited in a designated interval of time irrespective of the thickness of rocks concerned.

Certain bodies of rock of uniform composition can be traced over extensive areas and can be demonstrated to transgress time "planes." Others are rock layers of nearly identical areal aspect which appear to parallel time "planes." In each case the strata in question can be shown to be continuous and traceable over many square miles. The strata concerned can be precisely identified as being limited by well-defined boundaries which separate them from underlying and over-

Fig. 12.10 Winsor formation, Jurassic, in wall of Yellow Creek canyon, Garfield County, Utah. The capping layer (cliff-forming) is part of an overlying formation, the Dakota sandstone, Cretaceous. The two formations are distinguished because they show different lithologic aspects and contain distinctive fossils. (From Gregory, *U. S. Geological Survey Prof. Pap.* 226.)

lying rocks of different aspect. These are *rock units* which are distinguished as geological formations or groups. Such formations are assigned names after the geographic locality where they are typically represented. Formation names such as St. Peter sandstone (named after St. Peter, Minnesota) and Galena dolomite (named after Galena, Illinois) are examples of rock units in strata of the Ordovician Period and Paleozoic Era. Practical use of such formations can be made without reference to the position in the time scale. They are considered in terms of their properties only and not with regard to their age. Only when their relative age equivalents are to be considered in widely separated geographic sites does the question arise of their position in the geologic time scale.

The formation. The *formation* is regarded as the fundamental stratigraphic unit. It is considered to represent a body of rock with areal continuity and which having been deposited in continuous sequence is thick enough to be represented on the ordinary geologic map. Often it is a single lithologic unit such as sandstone or conglomerate, but it may be also an assemblage of lithologies such as alternating beds of limestone and shale. (See Fig. 12.10.) Ideally the upper and lower boundaries should be clearly distinct from the overlying and underlying rock units, but occasionally a formation of variable lithology is encountered where upper and lower boundaries are vaguely defined. This complexity increases as formational units are traced areally. Very often there is sufficient lithologic change to make uncertain the precise position of the bounding limits of the formation. Also, with lateral extent the over-all rock aspect eventually changes, or the terminal limit of the particular deposit is reached. Beyond these limits the formation can no longer be recognized as originally defined, and a new formation name is used to apply to the different strata. Identical formational names should not be applied to unlike lithologies even though the two can be shown to be intergradational and chronologically identical. Rather, it is necessary to apply a new name which has been selected from some appropriate geographic locality where the typical lithology of the new formation is displayed.

From time to time there is need to recognize certain individual parts of a formation. This may be a coal, thin limestone, or bentonite bed or some other distinct lithologic unit with considerable areal extent, such a unit is designated a *member*. The terms *"lentil"* and *"tongue"* are also used; the former to indicate a lithologic unit of very local distribution, and the latter to describe a bed which is known to terminate as a wedge.

Use of time-rock chart. The dual system employing time-rock and rock-unit classification is proving to be an effective device of examining strata because it permits consideration of rock units as separate entities. With few exceptions the engineer is not concerned with time-rock groupings unless there is need to know the relative ages of strata between widely separate regions. Use of the time-rock chart is best understood by reference to Fig. 12.11 which is an example using certain Cambrian strata in the Northern Rocky Mountains and Great Plains region of the United States. This chart is arranged so that individual formations as recognized in the indicated geographic localities are listed in their relative positions with respect to geologic time. The time span represented by each formation is indicated, but except in general terms the chart does not disclose the nature of the rocks. For example in southeastern Idaho the Langston limestone is the time equivalent of the Wolsey shale in southwestern and central Montana, the Gordon shale in northwestern Montana, and the lower shale member of the Gros Ventre formation of northern Wyoming. However, in the Black Hills of South Dakota and the Williston Basin of North Dakota strata of this age are absent. In central Montana strata from the top of the Park shale to the base of the Flathead sandstone are considered to be of Middle Cambrian Age and distinct from the Pilgrim formation which is assigned to the Upper Cambrian. Charts of this type are very useful in stratigraphic work because they are part of the never-ending program of cataloguing all geologic formations of the earth. From an engineering point of view, however, they are not of particular importance inasmuch as they do not disclose the actual relationships of the rock bodies concerned, nor do they indicate the nature and thickness of the units named.

Age		S.E. Idaho	Central Idaho	S.W. Montana	Central Montana	N.W. Montana	N. Wyoming	Black Hills	Williston basin
Cambrian	Upper	St. Charles ls	?	absent / Red lion fm	Pilgrim fm	---?---	absent	Deadwood fm	absent
						Devils Glen dol	Boysen fm		
		Nounan ls ---?---		Hasmark fm					
	Middle	Bloomington fm	Bayhorse dol	Park sh	Park sh	Switchback sh	Gros Ventre fm: Upper shale member	absent	absent
				Meagher ls	Meagher ls	Steamboat ls	Death Canyon member		
						Pentagon sh			
		Blacksmith ls				Pagoda ls			
			---?---			Dearborn ls			
		Ute ls	Garden Creek phyllite			Damnation ls			
		Langston ls		Wolsey sh	Wolsey sh	Gorden sh	Lower shale member		
		Brigham qtzite		Flathead ss	Flathead ss	Flathead ss	Flathead ss		

Fig. 12.11 Time-rock correlation chart of the Cambrian system in the Northern Rocky Mountains and Great Plains of the United States. Formations and their subdivisions are arranged according to their relative times of deposition. (From Sloss and Moritz, *Amer. Assoc. Pet. Geologists Bull.*, 1951.)

Use of rock-unit chart. Figure 12.12 is a chart of rock-unit correlation. As shown it represents a series of stratigraphic columns of the distribution and thickness of Cambrian strata in two counties of southwestern Montana. The strata in question are the same as those listed in one column (S.W. Montana) of Fig. 12.11, and no attempt is made to show their relationship to other formations outside the local area. Each formation is identified by symbol. By custom this symbol begins with the geologic period of its deposition (Ꞓ for Cambrian) followed by the first letter or other suitable abbreviation of the formation name in lower-case letters. At the specific locality where the section was measured lithologies of each formation are shown by conventional patterns as indicated in the legend. By custom, sandstone is shown by dots, shale by broken horizontal lines, limestone by blocks,

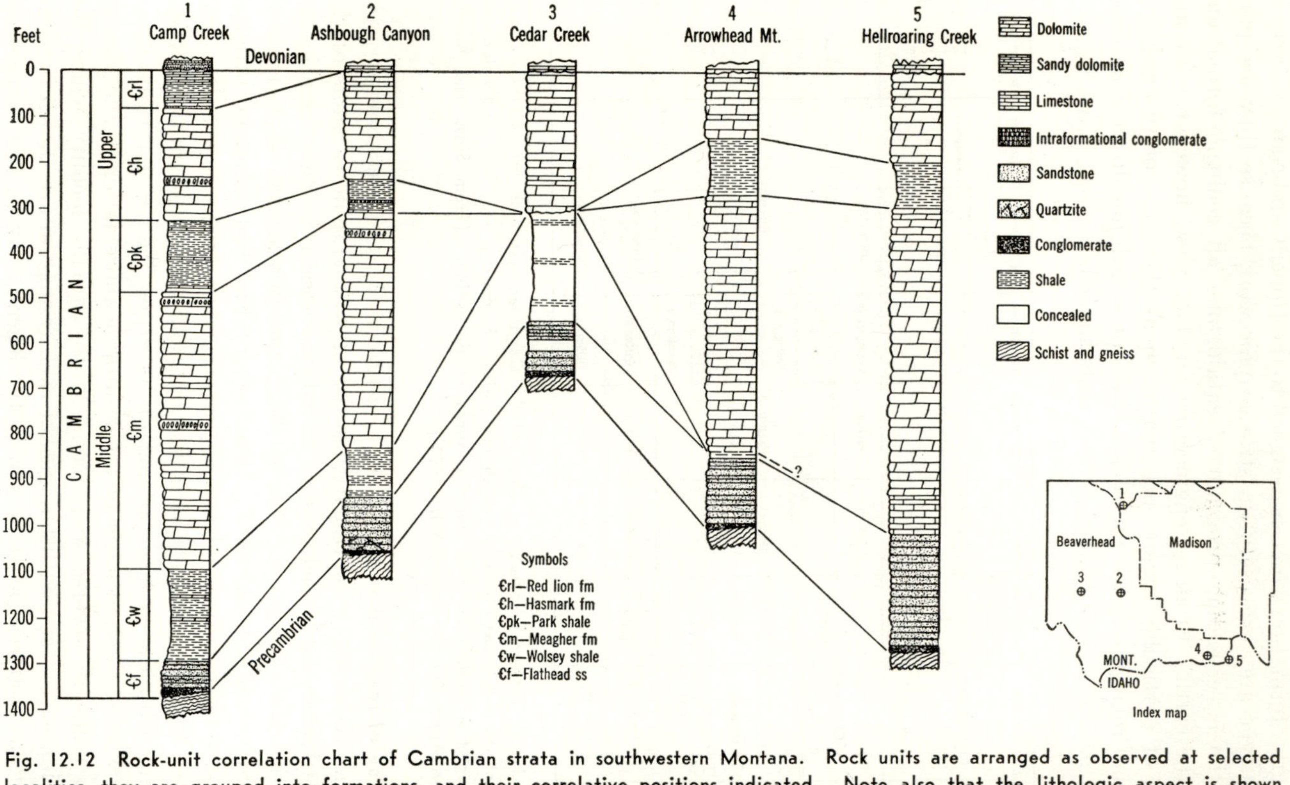

Fig. 12.12 Rock-unit correlation chart of Cambrian strata in southwestern Montana. Rock units are arranged as observed at selected localities, they are grouped into formations, and their correlative positions indicated. Note also that the lithologic aspect is shown by appropriate symbols. (Compare with Fig. 12.11.) (From Sloss and Moritz, *Amer. Assoc. Pet. Geologists Bull.*, 1951.)

and dolomite by blocks with inclined ruling. Along the left-hand margin is indicated the vertical scale suitable for all columns, and the recognized ages of strata are in the adjacent column. Note that in contrast to the time-rock chart (Fig. 12.11) the rock-unit chart does not indicate relative time of deposition along the vertical axis. The rock-unit chart shows by lines connecting the columns the correlative strata and localities where certain formations are absent. For example, at Cedar Creek (locality 3) the Park shale and Meagher formation are missing, and the dolomite which overlies the Wolsey shale is not the Meagher as in the other localities but is in fact the Hasmark formation.

Rock-unit charts are based upon some selected datum plane. In Fig. 12.12 this is the base of Devonian strata which are separated from those of the Cambrian by an unconformity. Erosion which preceded deposition of the Devonian strata removed the Red Lion formation except in the Camp Creek region and cut deeply into the Hasmark formation in the Arrowhead Mountain area. However, were this not the case the presence or absence of an unconformity would have no influence upon the selection of the datum. Any rock unit which underlies the entire area under consideration can be used. In this connection the surface of the Precambrian at the base would serve the purpose of a datum equally well, except it is customary to establish the upper surface as a straight line and allow the irregularities of thickness to be reflected at the bottom.

East of Arrowhead Mountain the Wolsey shale is no longer present as a recognizable lithologic unit, and its formational limits have been reached. Despite the fact that the lower part of the Meagher formation may have been deposited at the same time as some of the Wolsey shale no attempt should be made to recognize the latter beyond its occurrence as a shale unit. The section at Hellroaring Creek (locality 5) indicates the presence of a limestone at the base of the Meagher formation. Should this prove to be a locally continuous unit to the east of this locality it could be designated as a limestone member of the Meagher formation. If Hellroaring Creek proved to be the best exposure of this limestone and the name was not already used the unit could be designated the Hellroaring Creek limestone member of the Meagher formation.

Through custom and by nature of occurrence strata are indicated on the rock-unit chart with oldest at the base as though representing a record of a bore hole in horizontal layers. In actual outcrop the beds may be inclined and may have been measured on a mountain slope, in a canyon, or along a road cut where the rock sequence was pieced together along an exposure continued over a distance of a mile or more. In each case the strata are arranged in order as though they represented a complete vertical section of horizontal strata irrespective of their actual inclination. When measuring strata in outcrop their dip must be considered in correcting for the true thickness of the formation when placed in a theoretically horizontal attitude. Cor-

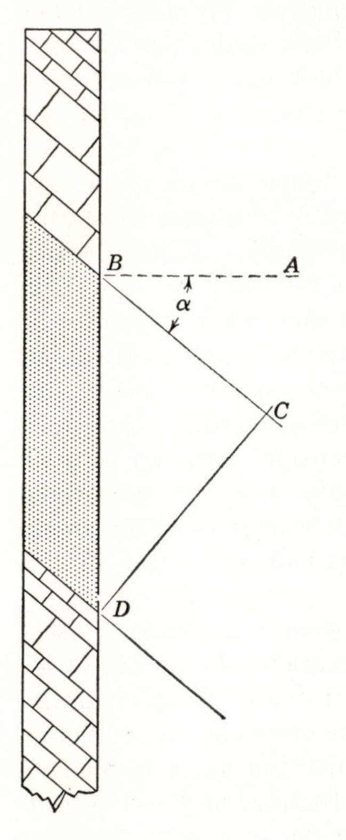

Fig. 12.13 Sketch of inclined strata in vertical bore hole in which *CD* rather than *BD* represents true thickness of strata. $\cos \alpha \times BD = CD$, hence the actual thickness of strata can be calculated from bore-hole information.

rection for dip is important when dips exceed a few degrees, particularly if the section measured is spread out over a horizontal distance and the section as drawn is a composite of several partially complete sections. Records of borings are indicated in the same manner, and if the strata are known to dip correction must be made on the observed (drilled) thicknesses to represent the true value perpendicular to the bedding. (See Fig. 12.13.)

Rock-unit charts constructed for an extensive area provide one of the most satisfactory devices for showing the relationship between groups of strata as they are traced from shelf into basinal conditions of deposition. Strata can be observed to thicken and change their lithologic quality when localities of thin shelf accumulation are compared with the thick deposits which characterize basins. Local formerly uplifted, or stable, areas are indicated, as for example the vicinty of Cedar Creek (Fig. 12.12) which can be interpreted to have undergone relative elevation sometime between deposition of the Meagher and Hasmark formations.

Frequently the rock-unit chart is called a *cross section* which in a restricted sense is somewhat true. In this connection it presents a distorted picture of the actual conditions by establishing a datum of a stratigraphic horizon rather than referring all strata to their actual position in space. All dips are eliminated, and, hence, in complexly folded regions the actual attitude of strata is not properly displayed. The true cross section is a factual representation of the conditions in a slice through a local part of the earth's crust, and the only permitted distortion is in vertical scale. This type of representation is designed to indicate the actual attitude of the strata, whereas the rock-unit chart is designed to indicate the relative relationship of strata by connecting a series of isolated measured sections.

Selected Supplementary Readings

Garrels, R. M., *A Textbook of Geology,* Harper and Bros., 1955, Chapter 14. An illustration of geologic reasoning in reconstruction of earth history.

Grout, F. F., *Kemp's Handbook of Rocks,* 6th ed., D. Van Nostrand Co., 1952, Chapters VII and VIII. Concerns the processes of development of sedimentary rocks as well as their description.

Krumbein, W. C., and Sloss, L. L., *Stratigraphy and Sedimentation,* W. H. Freeman and Co., 1951, Chapter 10. Describes the principles of stratigraphic correlation.

Pirsson, L. V., and Knopf, Adolph, *Rocks and Rock Minerals,* 3rd ed., John Wiley and Sons, 1947, Chapters IX, X, and XI. Describes and classifies sedimentary rocks.

Shrock, R. R., *Sequence in Layered Rocks,* McGraw-Hill Book Co., 1948, Chapters IV and V. A description of primary and secondary structures in sedimentary rocks.

Spock, L. E., *Guide to the Study of Rocks,* Harper and Bros., 1953, Chapter 8. Describes sedimentary rocks and processes of their accumulation.

13 Crustal deformation

STRUCTURE DEFINED

To a greater or lesser degree all rocks undergo some deformation as diastrophism causes changes in the level of the earth's crust. In some localities folding and faulting have been severe. These are sites of great geosynclines or closely associated linear depositional troughs. Certain other regions, for example the central part of the United States, since Precambrian time have been affected by mild diastrophic activity only. Here, rock failure is restricted to jointing and gentle undulations measured in tens of feet per mile. Localities of ancient or present-day mountain chains characteristically are regions of pronounced deformation where sharp folds are observed at closely spaced intervals. Also, these are sites where faulting has been intense in contrast to the regions of essentially horizontal rocks. Considerable importance is attached, therefore, to the degree to which an area has been folded,

---

Folded strata, East Greenland. (Photograph by Boyd, courtesy American Geographical Society.)

and in any geologic report particular reference is made to the structural attitude of the rocks.

Structure is considered to be a fundamental attribute of all rocks inasmuch as it supplies certain characterizing aspects which are independent of the rock's composition and texture. A complete description

Table 13.1. Primary Structures of Rocks

Igneous	Sedimentary	Metamorphic*
1. Lineation of crystals	1. Bedding	1. Foliation and gneissic banding
2. Foliation of crystals	2. Fossil replacement	2. Relict structures of sedimentary or igneous rocks
3. Flow lines	3. Concretion	
4. Ropelike flow structure in lava	4. Ripple mark	
5. Pillow structures in lavas	5. Mud crack	
6. Cooling joints	6. Raindrop imprint	
7. Vesicles	7. Gradation of grain size	
	8. Flow cast, load cast, and associated structures	
	9. Worm borings and other traces of organisms	
	10. Swash mark and associated marks produced on the beach	
	11. Intraformational conglomerate	

* Although foliation and gneissic banding are typical of the metamorphic environment it may be argued that all such structures are secondary and belong in the same category as folds.

of the rock, therefore, requires consideration of its structure in addition to its composition and texture. Of these basic properties structure involves large masses of rock and is not recognized in thin sections or very small specimens except under very exceptional circumstances.

The term "structure" is used in more than one sense and is applied to relatively large-scale features which are produced during the development of the rock as well as to the deformed attitude of strata. There are igneous structures and metamorphic structures as well as those which are typical of sediments. In such a connection they are characterizing attributes identifying some aspect of the environment which produced the rock and are not to be associated with those developed through later deformation. Structures which are hallmarks of the conditions under which the rock developed are called *primary* in contrast to *secondary* structures which result from the application of stress at some later time.

Primary structures consist of features known to be produced under special conditions during the initial development of the rock or sediment. As such they are indicators of the former existence of certain environmental conditions, and their identification is important in indicating the nature of the rock body. They are particularly valuable in reconstruction of the rock's environmental history. (See Table 13.1.)

PRIMARY STRUCTURES

In Igneous Rocks

Flow Lines

Among igneous rocks primary structures are useful in indicating characteristics of magma or lava flow. Oftentimes coarsely crystalline igneous masses show strongly pronounced parallelism, particularly among the tabular crystals of feldspar. The mean orientation of such crystals, determined by statistical procedures, has been demonstrated to yield important information on the nature and direction of emplacement of the igneous body.

Among important primary structures in igneous rocks are flow lines left by lava moving slowly under a gravity drive just prior to final solidification. Some flow lines are to be observed on a small scale, i.e., in terms of a matter of inches. The small scale makes recognition difficult except on weathered surfaces upon which the flow lines have been etched, but small plates of solidified crust are broken and oriented much as their large counterparts. (See Fig. 13.1.)

Pillow Structure

Pillow structures are useful in identifying ancient lavas which have been altered by partial metamorphism such that they no longer reveal ordinary aspects of igneous rocks. This is particularly true of certain

Fig. 13.1 Flow banding in a rhyolite produced by movement of lava during cooling. Scale approximately one-half size. (From Stose and Bascom, *U. S. Geological Survey Folio* 225.)

flows of Precambrian age which are now called greenstones. These former basalts are largely altered to a mixture of the minerals chlorite and epidote and except for the pillow structures are not readily identified as former flows. Pillow structures range in size from a few inches to several feet across. In cross section they are distinguished one from the other by a "rind" of chilled (very finely crystalline) lava which separates individual pillows. (See Figs. 13.2 and 13.3.) The internal structure is more coarsely crystalline although also fine grained and sometimes vesicular. As pillows build one upon the other their lower surfaces are flat or concave and digitated as the liquid lava pours over the surface and fills depressions between already solidified pillows. Their upper surfaces tend to be statistically convex upward, and this aspect is used to indicate the direction of the top of a major flow in a series of highly contorted greenstones.

Tubular Vesicles

Tubular vesicles formed by escaping gas near the top of a lava flow not only are useful in identifying the nature of the igneous rock body,

Fig. 13.2 Single "pillow" in greenstone, Precambrian, Marquette, Michigan. Coarse material to left of pencil and along right margin is filling of ejecta between individual pillows.

but also are helpful in indicating the direction of former movement. Frequently the tubes are bent in the direction of flow, particularly in their upper portions. (See Fig. 13.4.)

Columnar Jointing

Basaltic lavas in particular often develop a fracture system upon solidification. Ideally the system of fractures is arranged in an hexagonal outline producing an aspect like a giant honeycomb. Fractures are arranged with their long surfaces in planes perpendicular to the cooling surface and break the rock into a series of individual prisms. (See Fig. 13.5.) This structure called *columnar jointing* often is well developed, and great areas of individual columns produce a landscape of unusual features.[1] From a geologic viewpoint they are important because the joint system establishes important channelways for water which may be carried considerable distances. For example, when lavas

[1] Such localities become sites of tourist interest because of the unusual appearance, and names such as "Devil's Postpile" are typical.

possessing such structures border a lake imponded by a dam considerable leakage from the reservoir may occur.

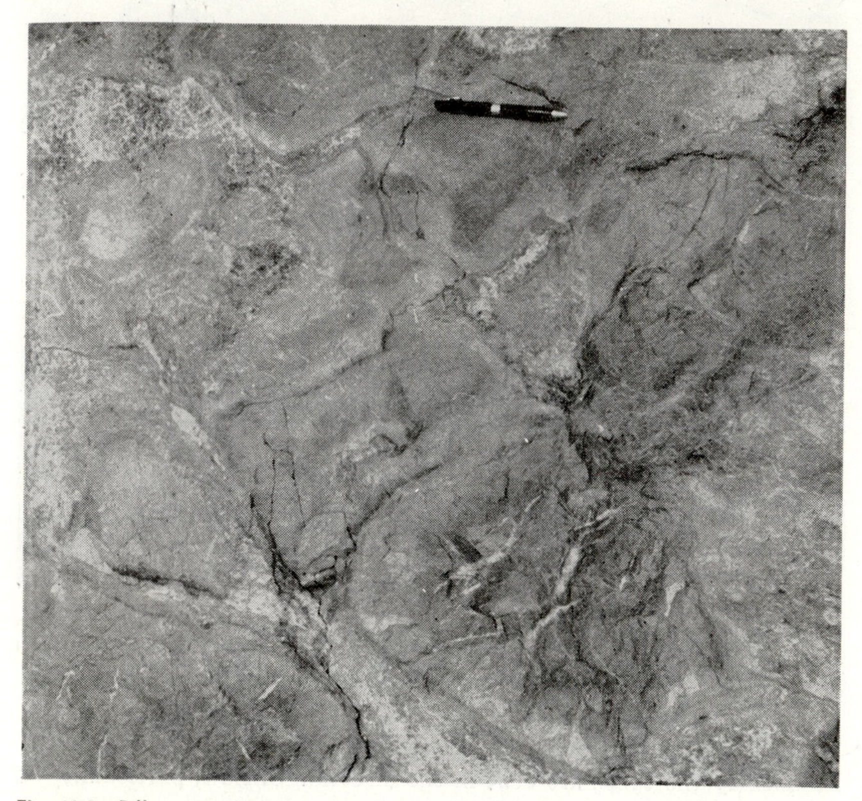

Fig. 13.3 Pillow structure in greenstone, Precambrian, Marquette, Michigan. Note the flat base of the pillow near lower margin of the photograph. Note also the tendency for small pillows to fill spaces between larger ones. The chilled borders of glassy texture appear slightly darker than the asphanitic texture of the interior of each pillow.

In Metamorphic Rocks

Identification of primary structures in metamorphic rocks is difficult because of the influence which metamorphism has had upon changing the textural aspect and mineral composition of the original rock. The processes involved in metamorphism tend to erase pre-existing textures; hence, only relict structures remain. In some cases faint indication of the former existence of bedding, fossils, ripple marks, and mud cracks are preserved, but as the intensity of metamorphism increases even these are obliterated entirely. In the sense that all metamorphic

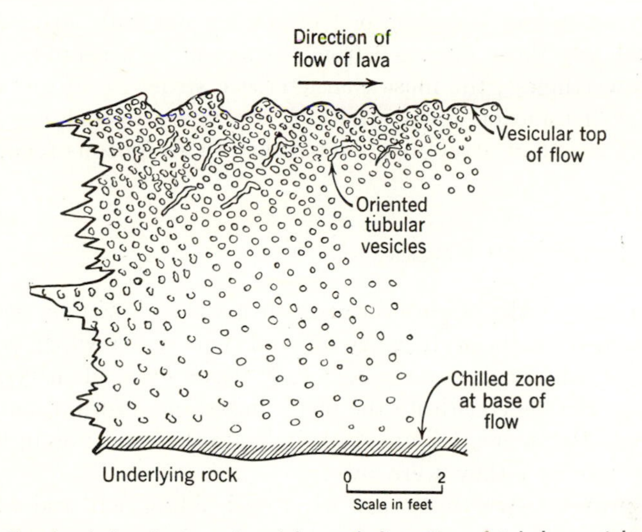

Fig. 13.4 Sketch of distribution of vesicles and distortion of tubular vesicles resulting from movement of lava during cooling.

Fig. 13.5 Columnar structure in basalt lava near Tower Falls, Yellowstone National Park. The flow lies upon, and is overlain by, Tertiary and Pleistocene river gravels now exposed in the walls of the recent canyon. Such a sequence of rocks produces an excellent aquifer. (Photograph courtesy L. H. Nobles.)

rocks are secondary, foliation or banding are not truly primary structures, and only those inherited from the parent rock should be considered. Nevertheless, the most characteristic structure of metamorphic rocks is their foliation, a feature which is primary to such a rock in the same way that flow structures are primary to an igneous rock.

In Sedimentary Rocks

Sedimentary rocks display the greatest variety of primary structures. This results from the wide range of conditions under which sediments accumulate, and each environment may leave some identifying characteristic. Fossils constitute the most important information-yielding structures. By their attributes they indicate the condition of the environment in which they were dwellers.

Some primary structures such as cross-bedding, cut and fill, ripple mark, and mud crack are extremely common in occurrence and are useful in the solution of several types of problems. Cross-bedding is an important indicator of the strength and direction of currents. Cut and fill indicates stream channels or active bar movement along shorelines. Ripple mark is useful in identifying bedding planes in massive strata and by special types (oscillatory) in identifying water-laid sediments (see Fig. 12.1). Mud crack, raindrop imprint, and swash mark indicate a former position of the earth's surface which was exposed to sunlight and atmospheric agencies. In the case of swash mark a specific environment of a beach is indicated, whereas the other two indicate desiccation of the sediments following wetting.

Concretions

Concretions appear to grow as the result of the influence of the environment of deposition. They occur as spheroids or nodules of various sizes ranging between tiny oolites and ball-like masses several feet in diameter. With the exception of chert which shows no internal structure other concretions display some indication of growth in concentric layers or radiating from some center. (See Fig. 12.4*d*.)

Certain concretions are more common than others, for example those composed of chert, siderite, and calcite. Chert concretions almost exclusively are limited to limestones and dolomites and are rare in shale or sandstone. (See Fig. 13.6.) Siderite and calcite concretions are typically developed in shales. The common habit of concretions is as irregularly shaped nodules with rounded surfaces scattered throughout the body of the rock. As a general rule such occurrences

Fig. 13.6 Zones of individual nodules of chert (very light gray) parallel to stratification of Galena dolomite, Ordovician, Dixon, Illinois. (Photograph courtesy L. L. Sloss.)

are in zones with distribution parallel to bedding planes, but the irregularity of the nodule frequently transgresses the bedding, or small veins may cut transversely across the rock. Some carbonates appear to be chert-free, whereas other formations contain so much chert as to characterize the unit and become part of the formation name.

Chert. The common color of chert is white or gray, but black, green, and red is found. Dark-gray or black colors are attributed to finely divided organic matter, red colors to ferric oxide, and green colors generally to ferrous iron. On weathered surfaces the nodules stand out in relief due to their superior hardness. Moreover, resistance to weathering causes the nodules to accumulate on and within the soil. In some localities the soil is principally fragmental chert which even under the best circumstances constitutes very poor agricultural soil.

Boundaries of the nodules are sharply defined against the enclosing carbonate, and any zone of intergradation which does exist must be of the order of a few millimeters in thickness. The boundary between chert and the enclosing rock is a zone of weakness, and on exposed surfaces the nodules tend to loosen and dislodge. Also, in quarried rock the chert tends to fracture on weathering and break loose from the parent material. This habit of easy fracture upon weathering is not clearly understood, but the chert becomes extremely brittle upon exposure and tends to shatter easily when struck. Whatever the cause the effect of weathering is a pronounced reduction in hardness and tensile strength and a tendency to break into irregular fragments.

Chert-bearing rocks are unsatisfactory engineering materials in general. When used as road metal the tendency of chert to soften and break upon weathering is a rather undesirable attribute. In this connection the freshly quarried chert is likely to indicate suitable strength and resistance to shattering, but upon short-term weathering a very pronounced decline in the values of such physical properties is to be observed. Perhaps the most undesirable characteristic is to be found when chert is used as part of the mix of concrete aggregate. In such a mixture chert enters into chemical reaction with the cement. This reaction concerns the addition of excess alkalis within the cement to the silica of the chert. The nature of the reaction appears to be generation of silica gel which absorbs water and swells with large expansive pressures. The concrete cracks, spalls in large masses, and becomes generally unsafe. Reaction of high-alkali cement is not limited to chert but is observed also with glassy volcanic rocks (rhyolites through andesites) and some mica schists.

Calcareous concretions. Certain calcareous concretions consisting principally of calcite and siderite ($FeCO_3$) occur typically in gray to dark gray shales as ovaloids, lenticles, or spheroids, weathering to rusty brown color (see Fig. 13.7). Less commonly such sideritic deposits are in thin 1- to 2-inch layers in the same shales. The concretions are typical of subgraywacke shales of the coal-bearing beds and frequently occur in zones with distinct stratigraphic continuity. This is particularly true of the small masses a few inches in length, and less true of large spheroids a foot or more in diameter. Shale-concretion boundaries are sharply defined, and thin laminas of shale can be observed to follow the contour of the concretion indicating that the shale layers were forced apart as the concretion grew. Frequently, such forces result in differential slipping between the shale and concretion such that the surfaces are striated and polished by this movement.[2] The effect is to loosen the concretion in its matrix of shale. Wherever such concretions are exposed through tunneling, quarrying, grading, or mining operations they tend to dislodge suddenly. This feature is

[2] A polished and striated surface which results from differential movement between sliding masses of rock is called a surface of *slickensides.* Such surfaces are ideally developed along faults.

troublesome in tunneling or mining particularly and often results in property damage or loss of life as large concretions fall from the roof or walls.

Fig. 13.7 Large concretions of siderite (approximately 2 feet in maximum diameter) in Pennsylvanian shales near Rolla, Missouri. Note contortion of bedding around the concretions and the tendency to separate from the enclosing strata.

Flow and Load Casts

Flow casts, or flow structures, and load casts are developed in muds and silts which are rendered plastic by virtue of their high water content. When buried under a load of overlying sandy sediments the water-saturated layer begins to shift, and the result is a strong deformation of laminas of the silt or mud. Siltstones and shales associated with coals frequently display such features when overlain by massive sandstone beds. The crenulations are to be interpreted as having developed when the sediments were still soft and in a thoroughly water-saturated condition. (See Fig. 12.1.) Flow structures in sediments are not to be interpreted as having any relationship to post-depositional folding.

SECONDARY STRUCTURES

Certain very commonly occurring structures observed in rocks are recognized as secondary in origin, i.e., they have developed under circumstances which were brought about after the rock had become a solid mass. Principal among such secondary structures are those which result from the application of stress against the rock mass, but others appear to be produced through chemical activity or erosion (see Table 13.2). Among the latter are veins, stylolites, and geodes.

Table 13.2. Secondary Structures in Rocks

Producing Agent	Structure	Frequency of Occurrence		
		Igneous	Sedimentary	Metamorphic
Stress	Joint	Common	Common	Common
	Major fold	Recognized in lava flows, sills, or dikes	Common	Recognized in slates, marbles, and quartzites
	Minor or drag fold	Recognized in lava flows, sills, or dikes	Common	Recognized in slates, marbles, and quartzites
	Fault	Common	Common	Common
	Schistosity	Common in large intrusions	Absent	Common
	Fracture cleavage	Recognized in lava flows, sills, or dikes	Uncommon	Common
Chemical reaction	Soil profile	Present	Present	Present
	Veins	Common	Not common	Very common
	Solution caves	Not present	Common in carbonates	Common in marbles
	Stylolites	Not present	Common in carbonates	Common in marbles
	Geodes	Rare in lavas	Common in carbonates	Rare in marbles
Erosion	Unconformity	Frequent	Common	Difficult to recognize

Veins

Veins are enlarged joints, fissures, or fault surfaces along which mineral matter has been precipitated. They are extremely important

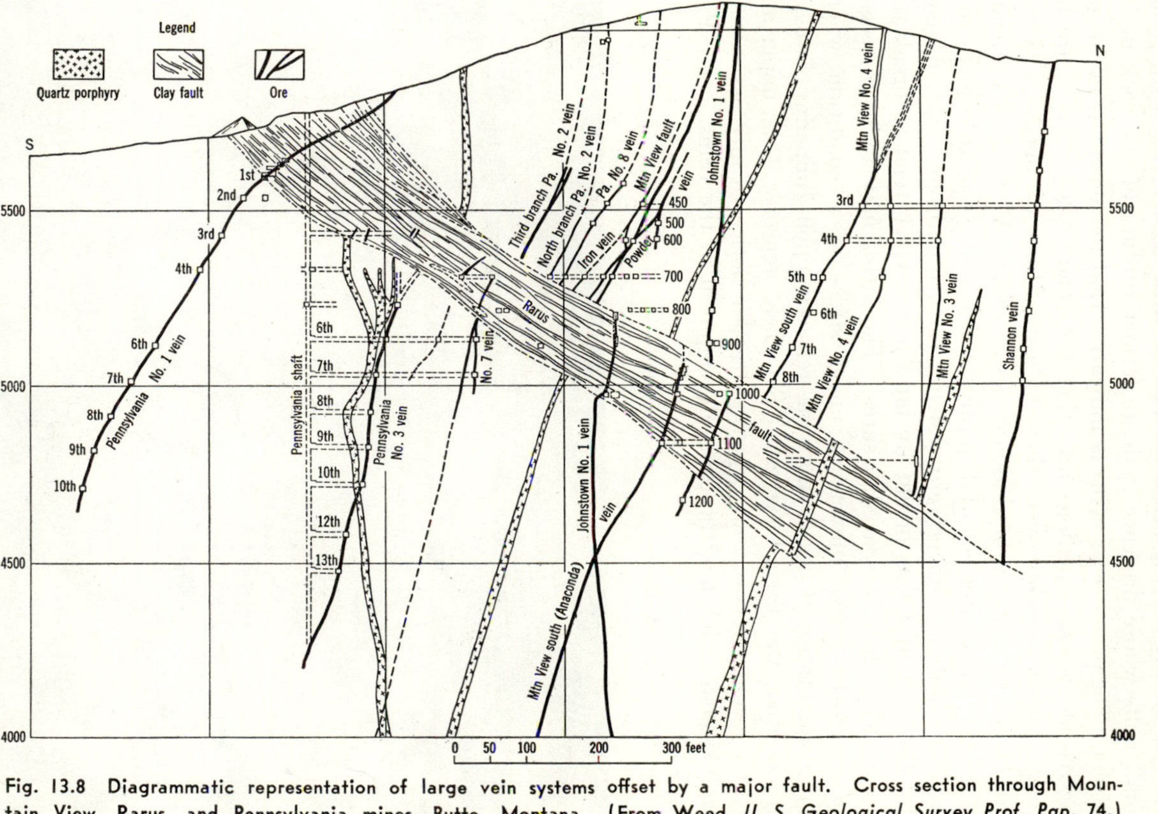

Fig. 13.8 Diagrammatic representation of large vein systems offset by a major fault. Cross section through Mountain View, Rarus, and Pennsylvania mines, Butte, Montana. (From Weed, *U. S. Geological Survey Prof. Pap. 74.*)

from a mining viewpoint and contain such a complex mineralogy as to require attention far beyond the scope of this text. They are important in an engineering sense because of their association with joints and faults whose presence indicate former rupture of the rock under stress. The presence of veins is indicative that movement is no longer active along such openings and that the conditions of stress are no longer present. Moreover, they indicate filling of openings which might otherwise carry large quantities of water. Very often veins are displaced by a later set of fractures which may become mineralized and constitute a younger vein system. In turn both vein systems may be displaced by faults or cut by later joints which tend to break the rock with regularly spaced fissures. The latter may indicate recent or currently active movement under stress or a condition where large quantities of water may be released upon tunneling into or intersecting the fissure system. For these reasons it is important to map vein systems and to pay particular notice to any open fracture system which may postdate development of the veins. (See Fig. 13.8.)

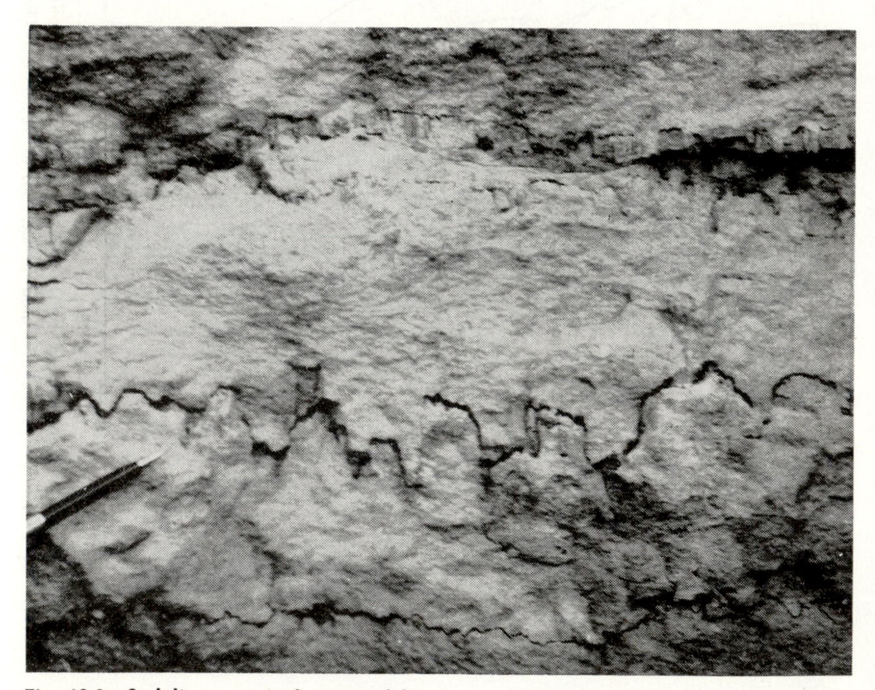

Fig. 13.9 Stylolite seam in fragmental limestone. Note interpenetration of the columns and the thin separation of clay (dark) capping the interpenetrations.

Stylolites and Geodes

Stylolites and geodes are of minor importance. The former are found in carbonate rocks almost exclusively and consist of a complicated system of small serrated separations between rock units, principally arranged parallel to the bedding. (See Fig. 13.9.)

Geodes are concretionarylike masses whose centers are partially open and filled with well-terminated crystals of such minerals as quartz and calcite.

Structures Produced by Stress

Joints

Table 13.2 contains a list of the principal structures produced when stress is applied to rock masses. This stress is either tensional or compressional, but in either case rock failure may occur. Most rocks are brittle and tend to fail by fracture, but under slow and intense compressional stress and particularly under conditions of high confining pressures failure occurs by contortion of rocks into folds. Scarcely any rocks exist in natural exposure or in quarries which are not cut by at least one set of fractures with parallel orientation. Such planes of parting separating a once continuous block of rock are called *joints*. (See Fig. 13.10.) Ideally no differential movement has occurred between the blocks separated by a joint, and rupture has not been followed by continued application of stress.

Some joint systems are tensional and represent openings with a wide separation between the walls. Tensional fractures frequently are the site of vein development, and precipitation of mineral matter takes place from the rock walls toward the center of the opening. Oftentimes filling of the veins has been incomplete, and the orientation of the crystals is clearly visible growing outward from the rock walls.[3]

Other joint systems indicate rupture under the application of compressional stresses. Generally these joints are tightly closed and consist of two sets of diagonally inclined fractures intersecting at acute or obtuse angles. In sedimentary rocks one of the compressional joint planes is oriented commonly along the bedding, and such strata appear to be broken by only one set of fractures. Similarly, in metamorphic rocks one set of joints may parallel the planes of mineral foliation.

[3] Incompletely filled veins of this type are called *crusted* veins.

Tensional joints are rarely as perfect in orientation as compressional joints. The former tend to have a jagged outline which clearly demonstrates that the walls of rock on either side of the joint have been pulled apart. In igneous rocks cooling cracks are of this variety, and columnar jointing represents the most typical of all the tensional fractures. (See Fig. 13.5.) Homogeneous igneous masses typical of some granitic bodies are cut by arcuate tensional joints (sheeting),

Fig. 13.10 Joint system in homogeneous, massive granite of the Sierra Nevada, California. The open character of the joints suggests they might have been developed by tensional stresses, the horizontal joints (sheeting) by unloading, and vertical joints by lateral stress. See also Fig. 2.15*a*. (Photograph by Gilbert, from Mathes, *U. S. Geological Survey Prof. Pap.* 160.)

generally attributed to development by unloading through removal of overlying rocks by erosion. (See p. 143 and Fig. 6.3.)

Theoretical concepts of stress and strain. All stresses impressed upon a mass of rock may be resolved into three mutually perpendicular directions. These can be regarded as axes of stress, and their magnitude may be expressed as greatest, intermediate, and least. When the magnitude of the stresses are equal for each axis the stress is considered to be hydrostatic, and the changes which occur in the rock are in terms only of volume and not of shape. When the stresses are unbalanced a deformational stress exists, and the shape of the rock is altered. (See Fig. 6.2.)

Application of stress produces strain. Distribution of strain may be regarded in terms of principal axes whose orientations coincide with those of the stress axes. The axis of the greatest application of stress is the axis of greatest compression. This is regarded as the axis of

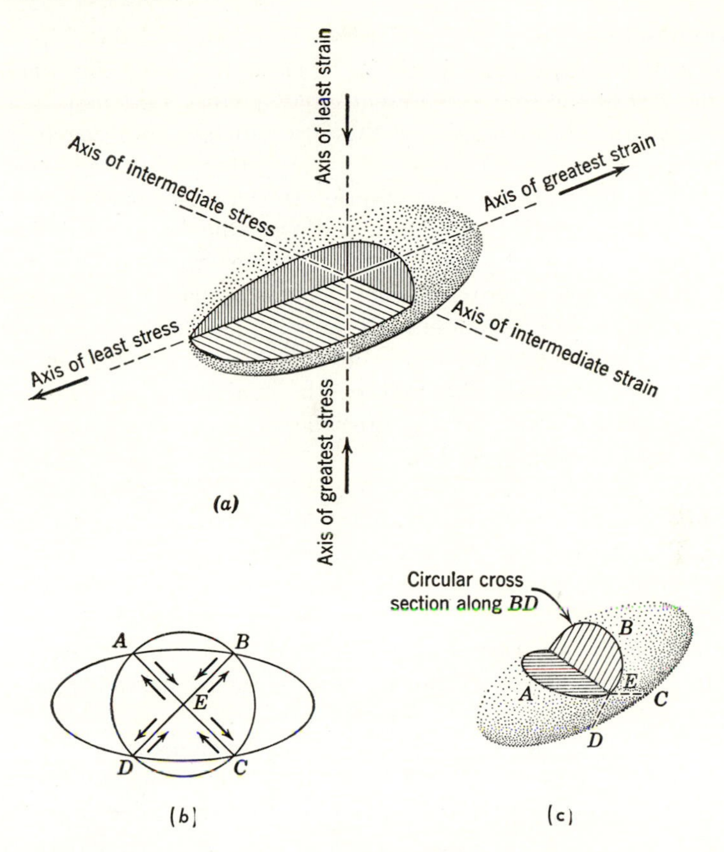

Fig. 13.11 Sketches of the theoretical strain ellipsoid. (*a*) Ellipsoid cut so as to show planes of reference and orientation of axes of stress and strain. The axis of greatest stress is the axis of least strain; the axis of least stress is the axis of greatest strain; the third axis is intermediate in stress and strain. (*b*) Longitudinal slice through ellipsoid demonstrating relative shear along planes as the originally circular cross section was distorted into an ellipse. (c) Ellipsoid cut to show the position of planes of shear of circular cross section. Note that the lines *AC* and *BD* are traces of such planes and are lines of shear.

least strain. The axis of least compression (i.e., greatest elongation) is the axis of greatest strain, whereas the axis of intermediate stress is also the axis of intermediate strain. In rocks these concepts are best illustrated in terms of the strain ellipsoid.

The strain ellipsoid is a conceptual figure which is considered to be produced when a sphere of homogeneous rock is subjected to the application of deforming stress. The ellipsoid formed from the sphere can be oriented in such a fashion that the three mutually perpendicular axes of strain can be indicated (see Fig. 13.11). In deforming the sphere into the ellipsoid the surface of the sphere and the ellipsoid are coincident at four points A, B, C, D (Fig. 13.11*b*). The diameters AC and BD of the sphere represent circular cross sections in both the sphere and the ellipsoid. Along these two surfaces of circular outline no change has occurred in shape, but any other section has been distorted by being changed to an ellipse. In the segment of the ellipsoid BEC the strain has been elongation, whereas in the segment AEB the strain has been shortening. Between these two segments lies the plane, BD, of circular cross section. This is a surface of shear because the distribution of stress on either side is applied in opposing direction. Two such planes of shear ideally arranged approximately in perpendicular position to each other transect the strain ellipsoid. These shear planes become surfaces of rupture when the compressional strength of the rock has been exceeded.

In the strain ellipsoid the axis of greatest strain indicates the direction of elongation. This is also the direction of internal tensional stress. When stress is applied sufficient to rupture the rock failure tends to occur in a plane whose surface ranges in orientation between a position perpendicular to the axis of elongation and that of a plane approaching the position of the surface of maximum shear. Failure does not occur, however, along both planes of maximum shear as happens when the rock is under compression.

The concept of the strain ellipsoid was developed from the theory of behavior of materials within their elastic limit. Orientation of the axes of stress and strain and the planes of shear are in accordance with Hookes' law that strain is proportional to stress. When actual rupture takes place the theory no longer holds, but the assumption is made that immediately prior to failure the planes of maximum shear are oriented in the position in which shear actually occurs. Following rupture continued application of stress results in rotation of the planes of shear, and their position in space may be modified to a large extent.

Orientation of joints. Mapping joint systems requires a means of orienting such surfaces in space for purposes of recording their positions. This has led to certain definitions of direction and inclination based upon a line and a plane of reference. By custom the line of reference is true north, and the plane of reference is the horizontal. The trend of a joint can be indicated with respect to north, whereas the inclination of the joint may be indicated by its angular departure from the horizontal. Figure 13.12 is a diagrammatic representation of an idealized rock mass cut by a single joint whose surface has been exposed for purposes of illustration. The surface of this joint intersects an imaginary horizontal plane ($ABCD$) along the line EF. This line of intersection is defined as the *strike* of the joint, and its direction is measured from north (HG) such as angle GHF. The maximum angular inclination of the joint as measured from the horizontal plane ($ABCD$) is called the *dip* and is the angle CHJ. Each joint has one strike and one angle of dip, but several joints may have the same strike and different angles of dip. Also, the joint is not completely placed in space by these two measurements inasmuch as the direction in which

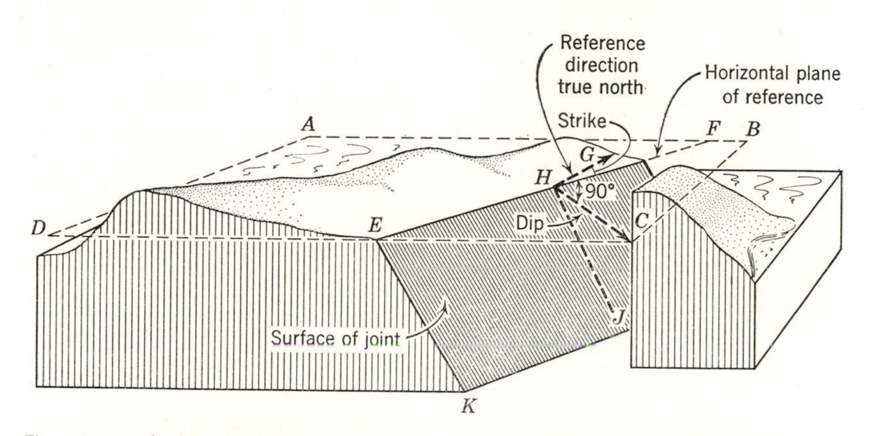

Fig. 13.12 Idealized sketch of a single block of homogeneous rock along a sea coast and cut by a single joint. The diagram is "exploded" to show the surface of the joint. A parallel to the water surface *ABCD* is the theoretical horizontal surface. The plane *EHJK* is the surface of the joint. The angle *CHJ* is its *dip*. The *strike*, angle *GHF*, is the direction of the trace of the joint, line *EF*, measured from north in the horizontal plane. A perpendicular, *HC*, to this direction in the horizontal plane is the *direction of dip* and indicates the direction in which the joint is inclined or dipping.

the joint is inclined must be known. A direction, HC, defined as the *direction of dip* completes the orientation of a joint plane. The direction of dip is described as the direction of maximum inclination of the joint and is measured customarily from north.[4]

Certain relationships exist between the three measurements orienting the joint in space. The directions of dip and strike are at right angles to one another; and the angle of dip must be measured in the plane of the direction of dip. Using Fig. 13.12 as an illustration, the procedure employed in the field is to measure by means of a clinometer the maximum angle of the inclination of the joint from the horizontal (i.e., CHJ). Line HJ or any line along the joint surface parallel to HJ is the limit of this angle. The angular reading (CHJ) is recorded in degrees as the angle of dip. This angle has been measured in a vertical

[4] The direction of the dip is the trace of the angle of dip in the horizontal plane.

plane whose trace on the horizontal plane is the line HC (or some parallel direction). The direction of the line HC is the direction of dip of the joint and is measured from north by means of a compass. From the value of the direction of dip the strike (HF) is computed. Sometimes it is convenient to measure the strike and angle of dip and to merely note the general direction of dip of the joint. From these observations the true direction of dip may be computed.[5]

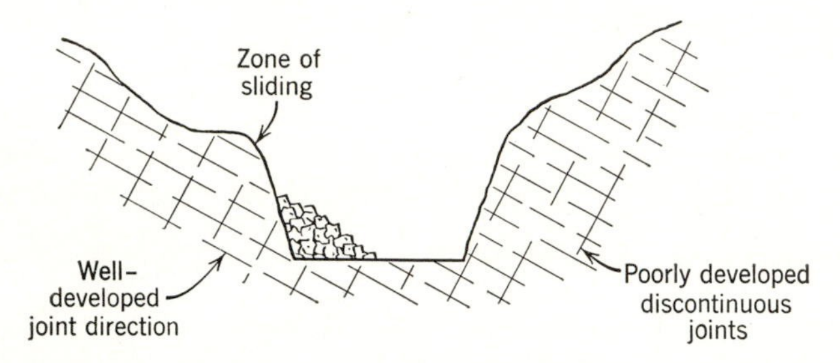

Fig. 13.13 Schematic cross section of a road cut in jointed rocks, illustrating potential zones of rock falls and landsliding along a well-developed joint system oriented properly to direct movement into cut.

An analysis of the orientation of hundreds of joints employing the concepts of the strain ellipsoid oftentimes is useful in prediction of the probable alignment of larger structures such as folds or faults. Upon other occasions knowledge of the joint orientation will reveal the position of expected rock slides into cuts. (See Figs. 13.13 and 13.14.) A quarry properly laid out with respect to a joint system is likely to produce rock at far lower costs than an adjacent quarry improperly located. An understanding of local joint systems, therefore, is important in all cases requiring removal of rock.

Faults

A fracture along which displacement has occurred between formerly adjacent points is called a *fault*. The fault differs from the joint in the continued application of stress following rock rupture. Some circumstances of faulting has forced rock masses to override others for

[5] Following are examples of dip and strike determinations:

1. A joint has a dip of 30° in a direction N 40° E. What is its orientation? *Ans:* The angle of dip is 30°. The direction of dip is N 40° E., and the strike is computed as N 50° W.

2. A joint strikes N 70° E. and dips 42° in a northwesterly direction. What is its orientation? *Ans:* The angle of dip is 42°. The strike is N 70° E., and the direction of dip is computed as N 20° W. inasmuch as the dip is inclined in a northwesterly direction.

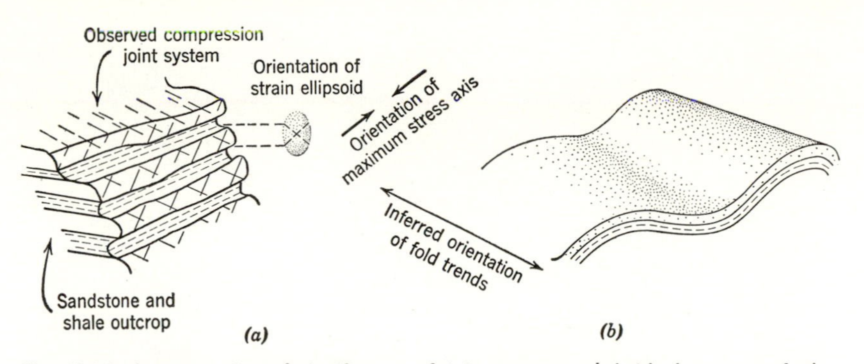

Fig. 13.14 Interpretation of significance of joint pattern. (*a*) Ideal outcrop of alternating beds of sandstone and shale cut by two systems of compressional joints. Orientation of strain ellipsoid is shown in accordance with position of joint planes. (*b*) Inferred orientation of folding which must be present in the strata in order to agree with interpreted position of strain ellipsoid. Note that for this orientation of the ellipsoid the axis of greatest strain must be vertical, whereas the axis of greatest stress, or least strain, must be in a horizontal position. Also, the direction of orientation of the axis of greatest stress must be perpendicular to the trace of the joints on a horizontal surface.

miles, whereas in other cases the displacement can be measured in feet or even inches. Joints are found in large numbers in parallel sets or systems, whereas faults generally occur as single fractures. Occasionally faults develop in parallel sets or systems, and normally these are widely spaced in terms of miles. However, infrequent occurrences are known where several faults may be spaced a few feet apart and a network may occur within 1 sq. mile.[6]

Faults result from tensional as well as compressional stress; hence, use is made of the concept of the strain ellipsoid to analyze the direction of application of the forces involved in their development. In this connection rotation of the fault surfaces by continued movement frequently causes difficulty in proper orientation of the ellipsoid, and uncertainty exists in the precise direction from which stress was applied to produce the faulting. However, a rough estimate of the distribution of forces nearly always is possible. The analysis is based largely upon the knowledge that tensional faulting tends to cause elongation of a part of the earth's crust and faulting due to compressional stress shortens a segment of the crustal layers.

Fault terminology. Oftentimes there is difficulty in identifying the nature of the stresses involved, and as a result fault terminology has been based upon factual relationships between displaced segments of rock. For example, the surface of rupture, generally called the *fault*

[6] Closely spaced faults may produce a mass of broken rock fragments in chaotic orientation and the entire broken area is identified as a *shear zone*.

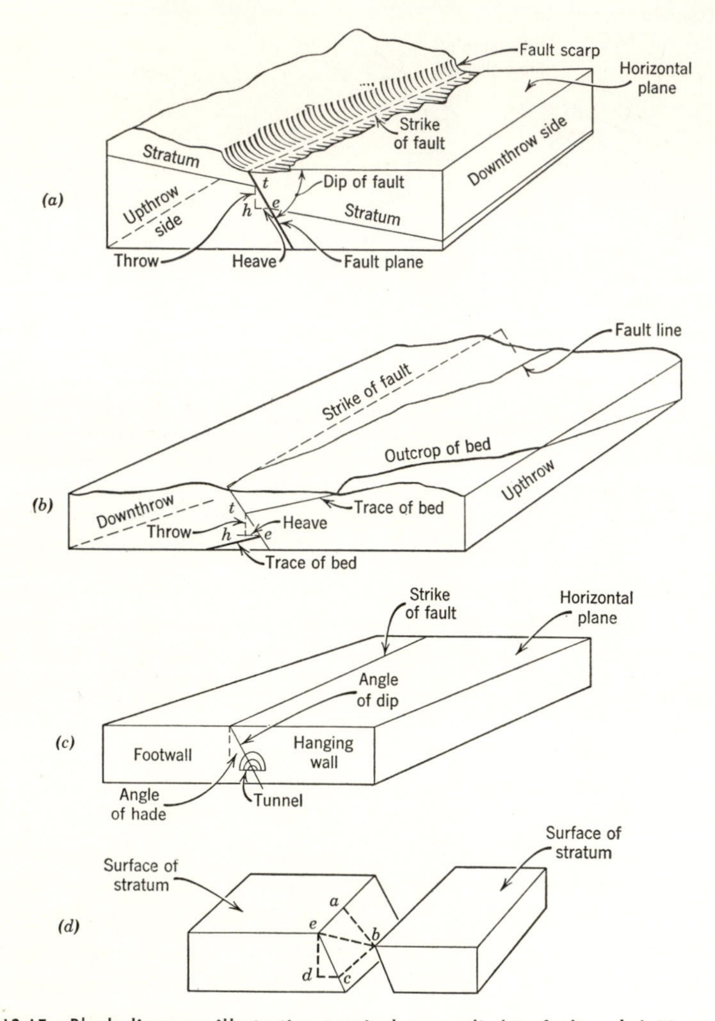

Fig. 13.15 Block diagrams illustrating terminology applied to faults. (*a*) Homogeneous mass of rock cut by a single fault. Note that the position of the stratum indicates that the right-hand block has moved down with respect to the left-hand block. The fault plane dips toward the *downthrow side*, a condition which defines a *normal fault*. (*b*) Block similar to (*a*) but where erosion has removed all evidence of the *scarp*. The direction of the *fault line* is controlled by topography and dip of fault plane and does not necessarily coincide with the *strike*. The fault dips toward the *upthrow side*, a condition which defines a *thrust fault; th* is *throw, he* is *heave* of fault. Note that these are vertical and horizontal displacements, respectively. (*c*) Other common terms used in mining or tunneling. (*d*) Common terms of reference between points once in contact but now on opposite sides of the fault plane; *eb* is *net slip, ab* is *dip slip, cb* is *strike slip, ed* is *throw* and *dc* is *heave.*

plane, separates masses of rock. On one side of the fault plane a point has moved relatively up with respect to its counterpart on the opposite side of the fault plane. The side which appears to have moved up is called the *upthrow side,* and the other is called the *downthrow side.* Faults in which the fault plane dips toward the downthrow side are called *normal* (see Fig. 13.15*a*). *Thrust or reverse faults* are defined as those in which the fault plane dips toward the upthrow side. (See Fig. 13.15*b*.)

Other terms in common use are strike, dip, and direction of dip which are defined as for the joint. In Fig. 13.15*b* the strike of the fault plane is represented by a dashed line, whereas the *fault line* or actual trace of emergence of the fault plane at the earth's surface departs from the strike direction. This difference is due to the effect of topographic relief which in departing from the horizontal plane causes the trace of the fault line to change direction. Many special terms are employed in the nomenclature of faulting. Among these is *fault scarp* which refers to the cliff which initially is formed along the upthrow side (Fig. 13.15*a*). Scarps retreat under the impress of erosion, and their base is not necessarily coincidental with the fault plane. *Throw* is a term applied to indicate the displacement in a vertical component between formerly coinciding points (*th* in Fig. 13.15*a* and *b*), and *heave* refers to the amount of displacement in a horizontal component (*he* in Figs. 13.15*a* and *b*). *Net slip* (Fig. 13.15*d*) is defined as the total amount of displacement, *eb*, measured along the fault plane between two points formerly coinciding. *Dip slip* is the amount of displacement, *ab*, measured on the fault surface in the plane of the direction of dip, whereas strike slip, *cb*, is the displacement measured on the fault surface in the plane of the strike. Other terms used to describe the geometry of faults are indicated in Fig. 13.15*c*.[7]

Overthrusts. With very few exceptions all faults can be classified as normal or thrust, but a special variety called an *overthrust* is recognized where strata are thrust for long distances along fault planes of low angle. Overthrusts are typically observed in mountain chains and especially within the site of the former geosynclines. An example is illustrated by Fig. 13.16 which schematically represents a cross section across the Appalachian Highlands in eastern Tennessee. Strata of gentle inclination in the Cumberland Plateau increase gradually in structural complexity eastward toward the Great Smoky Mountains. Although not indicated by the scale of the cross section some of this

[7] For additional terminology, explanation, and solution of fault problems the reader is referred to M. P. Billings, *Structural Geology,* 2nd ed., Prentice-Hall, Inc., 1954.

complexity is reflected in local folding, but great thrust faults are the principal structures. In the locality of the Cumberland Plateau the fault planes dip steeply, and the displacement is not large. Eastward the fault planes are much more closely spaced and in general tend to be more gently inclined. Repetition of identical sequences of strata by closely spaced thrusts is typical and is termed *imbricate faulting*.

Downdip, fault planes are known to flatten as determined in a few cases where they have been crossed during drilling operations. This

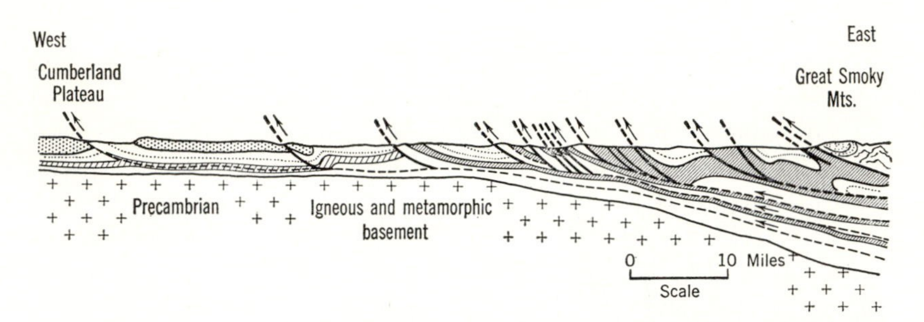

Fig. 13.16 Diagrammatic cross section from west to east through the Appalachian Plateau and into Great Smoky Mountains (Tennessee–North Carolina) showing *imbricate* thrust slices. Movement of thrust occurs along the major *sole faults* which are confined to shale beds. Note that the sediments have been pushed over the rigid basement which is not involved in the overthrusting. (Simplified from King, Princeton University Press, 1951. Originally drawn by Rodgers.)

change in dip is interpreted as representing the gradual junction of a network system into a *sole* fault along which extensive horizontal movement has occurred. This is the great overthrust fault. In most localities of imbricate faulting major sole thrusts cannot be observed because by the nature of the structure they may be several thousand feet underground. Their presence is difficult to prove, and their position is projected primarily from anomalous relationships between rock layers and large masses of rock isolated from their main outcrops by miles of distance and from theoretical calculations based upon model behavior. Considered generally, the location of important sole faults appears to be controlled by important layers of shale. Steep thrust faults transgress strata until they intersect a shale layer at the proper angle, whereupon a sole fault develops. Such faults restrict their position to the shale within which the great differential movement occurs. Shale layers are believed to act as plastic media and during the movement to generally perform as lubricants. This concept has been employed in the cross section Fig. 13.16 in which the position of the shales in the Appalachian area are shown as zones within which overthrust faulting has been restricted.

Recognition of faults. Exposures of great cliffs, road cuts, or quarries frequently reveal the presence of a fault plane which marks the boundary between relatively displaced strata on either side. Where strata are thin and change in type vertically such beds can be seen to be offset as indicated in diagrams such as Fig. 13.15*a*. This obvious relationship is rarely to be observed, and indirect criteria of recognition must be used. (See Table 13.3.) In outcrop the actual

Table 13.3. Criteria for the Recognition of Faults

A. In outcrop.
1. Displacement of strata.
2. Evidence of shearing along walls of a fissure.
3. Slickensides along walls of a fissure.
4. Prominent addition of silica (silicification) to beds which nearby are normally cemented sandstones, shales, or limestones
5. Veins of calcite or other mineralized zones particularly ore.
6. Abrupt change in rock type in lateral direction.
7. Scarp offsetting valleys or with hanging valleys.
8. Scarp showing triangular facets.

B. In subsurface drilling.
1. Repetition or omission of beds.
2. Abrupt change in expected position of strata (expected position is known from other borings).
3. Abrupt appearances of metamorphic strata in a sedimentary sequence.

C. From geologic maps (i.e., including area larger than a single outcrop).
1. In areas of dipping strata, repetition of the same sequence of beds.
2. In dipping strata, omission of a bed known from drilling to be present in the area.
3. Offset in a sequence of dipping beds.
4. Abrupt change of strike in a set of prominent joints.
5. Abrupt change in strike of a sequence of beds.
6. Abrupt displacement of parts of a single fold. (See folded strata p. 545.)
7. Junction, along a common boundary, of rocks known to be developed under markedly different environments (e.g., strongly metamorphosed strata in contact with sedimentary rocks).
8. Offset of dikes.

fault plane may be revealed. Often this is a clearly defined zone of shearing and crushing in which rock has been ground to clay dimensions. This material is called *gouge* and contains breccia of rock torn from the walls of the fault plane. In natural exposures the soft clay gouge is eroded and leaves a well-defined indentation in the rock face. Walls of the fault plane generally show physical evidence of shearing. Local areas are smoothed and polished or streaked by slickensides. Slickensides are useful in identifying the direction of net slip and sometimes the upthrow and downthrow sides may be indicated by rubbing the hand on the corrugated surfaces to indicate the "smooth" or "rough" direction.[8]

The fault may be a zone through which water moves freely, particularly when the gouge is poorly developed or permeable, and intersection of a fault plane in tunneling operations may suddenly release enormous amounts of water. Under natural conditions the zone is a channelway for solutions from which silica and calcite are often precipitated. Silicification of the fault zone and of the rock walls bordering the fault is commonplace. On weathered exposures such silicified zones stand in bold relief, forming prominent outcrops.

In outcrop, quarry, or tunnel operations a stratigraphic sequence may suddenly terminate in a lateral direction against a wall of different rock type. For example, alternating beds of shale and sandstone when traced laterally may terminate abruptly against a limestone and shale sequence or against metamorphic rock. This is not the normal condition to be expected and almost always indicates a fault.

Where mountain fronts rise abruptly from a plain's surface such as is characteristic of the Basin and Range section of the United States fault scarps are often a prominent topographic feature. Sometimes the scarp is only a few feet high, whereas in other cases it may exceed several thousand feet. In either case its angle of steepness must exceed that of the normal land slope. The new scarp is cut by stream erosion, and large alluvial fans often develop at its base where youthful valleys emerge on the flat surface of the basin. One of the products of such erosion is the development of triangular facets on the face of the scarp. (See Fig. 13.17*a*.)

Intersection of a fault plane by a boring is identified by comparison of bore records. For example, in Fig. 13.17*b* a cross section illustrates the conditions imposed by a normal fault. Bore hole *a* represents the normal stratigraphic section with strata arranged in their proper order and thickness. The record of bore hole *c* is anomalous inasmuch as the conglomerate bed is abnormally thin and the underlying limestone is missing entirely. Normal faulting is recognized on the basis of abnormally thin stratigraphic units and the omission of others. Thrust faults are indicated by repetition of strata within the same boring or abnormal excess thickness of a bed. The record of bore hole

[8] The smooth direction is the direction of movement of the overriding rock mass.

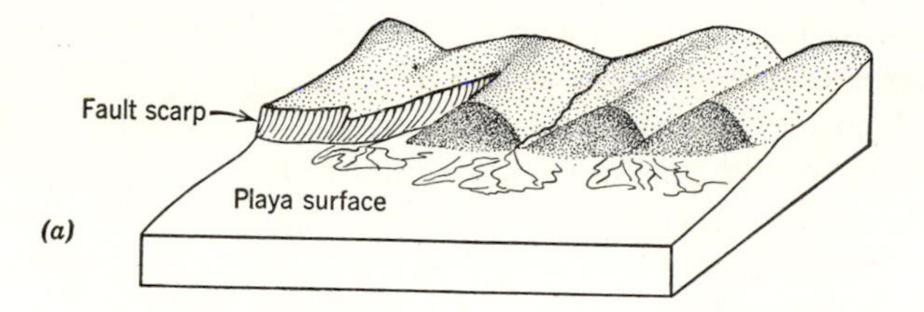

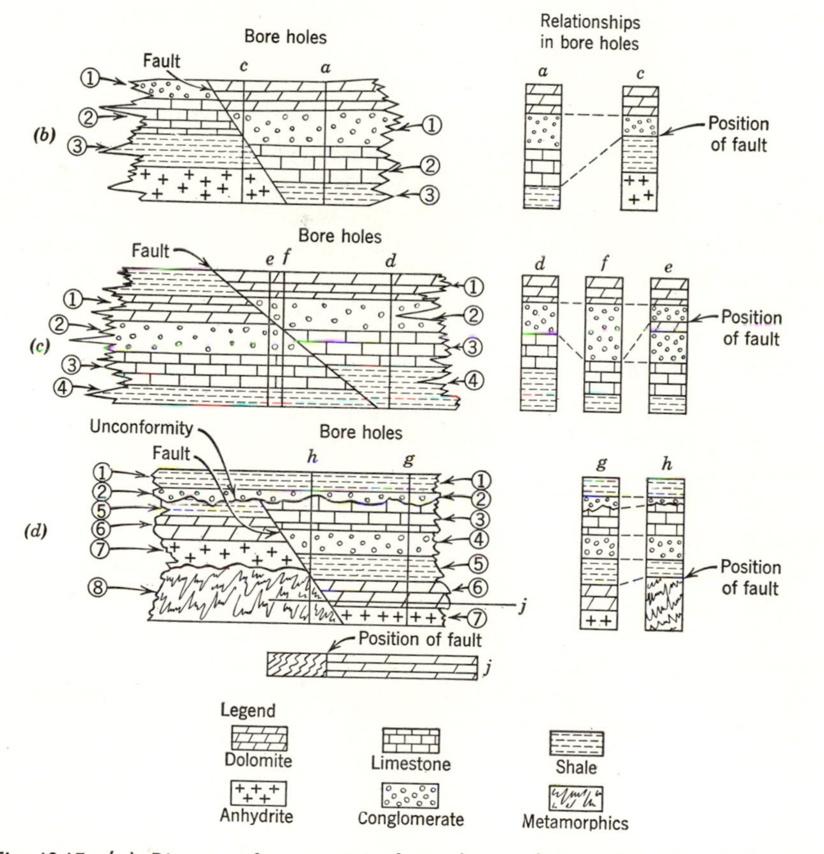

Fig. 13.17 (*a*) Diagram of a mountain front showing how erosion of a fault scarp produces triangular facets. (*b*) Cross section of a normal-fault relationship interpreted from records of bore holes *a* and c. (c) Cross section of thrust fault interpreted from bore holes *d*, e, and *f*. Note that recognition of a normal fault is indicated by strata which are abnormally thin or absent, whereas in thrust faults strata are repeated in the bore hole. (*d*) Strata cut by a normal fault overlain by a surface of *unconformity* above which strata are not offset, as interpreted from bore holes *g*, *h*, and *j*. Note abnormal position of schist in *h* and *j* due to faulting.

e in Fig. 13.17*c* indicates repetition of conglomerate and dolomite units. Note that boring *f* does not show a repetition but an abnormal thickness of conglomerate. Borings *h* and *j* (Fig. 13.17*d*) illustrate two conditions of normal faulting. Beds known to be present in the area from the record of bore hole *g* are not found in bore hole *h*, but in the example metamorphic rock is encountered instead.

Frequently the presence of faults traversing an area can be recognized from the geologic map of the area. A geologic map is a record of the rock lying immediately below the cover of soil, alluvium, glacial till, loess, or other unconsolidated material. The map indicates the kind of rock, its formation name, and the geologic age to which the rock unit has been assigned. In areas of dipping or folded strata individual formations appear on the map in belts whose pattern indicates the nature and the type of folds involved. Faults are to be interpreted where such patterns are offset abruptly or where their direction of strike is changed sharply. Offset in the outcrop position of a dike or abrupt displacement of the axial crest of a single fold are among the most reliable criteria of recognition of faulting.

Folding of Rock Layers

When compressional stress is applied slowly under high confining pressure even the most brittle rock tends to yield by "plastic" flow rather than rupture. As a result rock layers are contorted into gentle undulations or very sharp folds depending upon the magnitude of the stresses involved. Each fold can be classified either as an upward arch structure or a downfolded trough, irrespective of its size or symmetry. A fold arched upward is known as an *anticline,* whereas a downwarp is called a *syncline.* A single anticline or syncline may range in size between a few inches across and several miles, but in either case the bulk characteristics are similar. Folds are called open and symmetrical such as shown in Fig. 13.18*a* or asymmetric as in cross section *b*. The degree of asymmetry is indicated by the relative dip of each flank of the anticline or syncline. The measure of asymmetry is also indicated by the inclination of the axial plane. This is defined as that plane passing through the fold in such a manner as to divide it as nearly as possible into mirror images. When the fold is symmetrical the axial plane passes through the crest, or trough, and is oriented vertically, but in asymmetric folds the axial plane is inclined and not necessarily coincidental with the crest line. Inclination of the axial plane is variable and ranges between vertical and horizontal. As a general rule axial planes are not strongly inclined, but in very intense folding associated with great overthrusts the axial planes may be approximately horizontal. Folds of such orientation are called *recumbent.* (See Fig. 13.16 near Great Smoky Mountains.)

Folding may be open in which case the limbs or flanks of the fold dip in opposing directions. In complexly folded areas the folding is tight, whereupon under extreme conditions all strata dip in the same

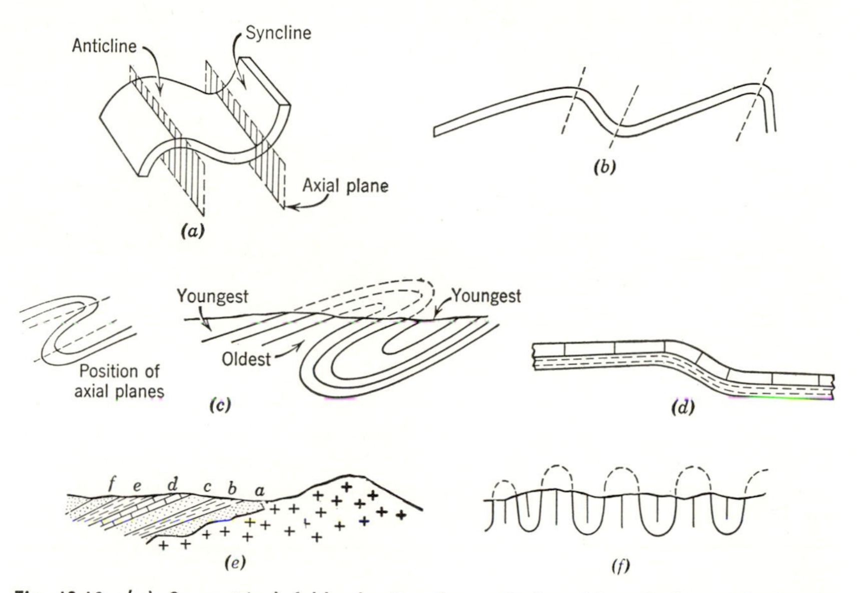

Fig. 13.18 (*a*) Symmetrical folds showing the vertical position of the *axial plane* in the *anticline* to the left and in the *syncline* to the right. (*b*) Asymmetric folds as indicated by differences in dip of each limb on either side of the axial plane. (*c*) Cross section of an overturned fold in which on one *limb* younger strata dip beneath older strata and the strata of both limbs dip in the same direction. (*d*) *Monocline* a type of asymmetric anticline in otherwise horizontal strata. (*e*) *Homocline* in which successively younger strata dip in the same direction from a central uplift. (*f*) Isoclinal folds in which the limbs of alternate anticlines and synclines are compressed into very steeply dipping or overturned strata.

direction. The overturned fold is a case in point, and in one limb strata are reversed in order of age superposition. Note in Fig. 13.18*c* that strata of the overturned limb are arranged with older resting upon younger.[9] Isoclinal folding (Fig. 13.18*f*) is indicated when strata have been so closely compressed that the same bed is in repeated contact with itself. Isoclinal folding is difficult to recognize but is

[9] This exception to the law of superposition of strata is used to demonstrate the existence of an overturned fold in cases where the fold is obscured by lack of outcrop.

associated with vertical or overturned strata. Note the similarity between Figs. 13.18*e* and *f*. In drawing *e* the strata obey the law of superposition and decrease in age from *a* to *f*, whereas in drawing *f* there is no indication as to which are younger and older strata. Note also that the thickness of strata in drawing *e* can be measured from bed *a* to bed *f*, whereas in drawing *f* the presumed thickness of strata across the diagram is entirely incorrect inasmuch as only one bed is involved and is repeated over and over again by the isoclinal folding. Except in connection with metamorphic strata such as marbles, schists,

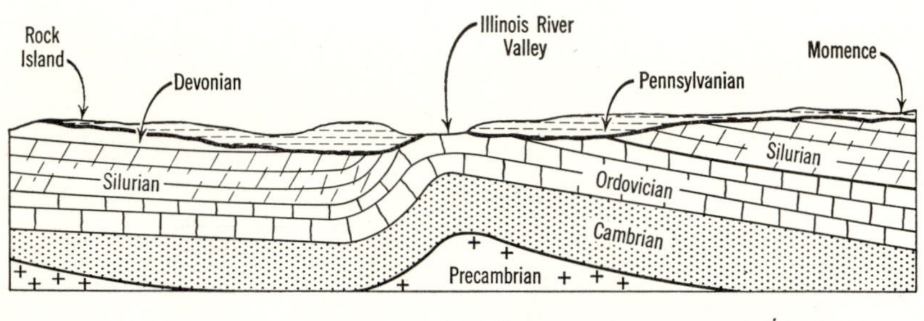

Fig. 13.19 Cross section from west to east across northern Illinois illustrating the monoclinal flexure known as the LaSalle Anticline. Note that strata of Pennsylvanian age lie unconformably upon the fold. Locally these have been folded by renewed movement. (Simplified from Weller, *Geologic Map of Illinois,* Illinois Geol. Survey, 1942.)

and gneisses isoclinal folding is rarely observed. Generally the latter rocks are associated with the axial position of a great geosyncline and are so strongly altered that bedding planes are unrecognizable. In slate belts where the intensity of metamorphism is less than in the region of the gneiss the bedding can be deciphered, and the presence of isoclinal folding established. Based upon this analogy belts of schists, gneisses, and marbles which appear to represent former deposits as much as 50 miles thick are considered to be actually only a fraction of this apparent thickness and are due to stratigraphic duplication brought about by isoclinal folding.

Isoclinal folding need not be oriented with beds in vertical dip, but this is usually the case, and without supporting evidence to the contrary distinction is made between isoclinal folding and strata more gently inclined as in diagram *e*. Beds dipping uniformly in the same direction for a distance of the order of a mile or more are designated as dipping *homoclinally,* and the structure is identified as a *homocline.*

Strata of the homocline obey the law of superposition of strata, and progressively younger beds are to be found in the direction of dip.

A special type of anticline associated with very gently inclined strata is a flexure designated a *monocline*. (See Fig. 13.18*d*.) Monoclines are abrupt departures from the regional extremely gentle dips of adjacent areas. Oftentimes they are prominent folds several miles in width, extend for miles in length, and are appropriately named after some geographic locality where the structure is well exposed. (See Fig. 13.19.) Some monoclines not only are miles in length but maintain a uniform directional trend. Deep drilling and geophysical information are revealing that such monoclinal folds are strata draped over important normal faults within the Precambrian basement. Strata overlying the basement have been deformed by folding, whereas the more massive and homogeneous basement rocks (igneous and metamorphic) have yielded by rupture.

THE GEOLOGIC MAP

Effect of Erosion on Folded Strata

In its ideal development a simple fold consists of an anticline and its coupled syncline. Such simple upwarp and downwarp can be considered to have topographic expression such as indicated in Fig. 13.20*a*. The anticline stands as a ridge, and the syncline is the adjacent lowland. Except for the soil cover the formerly horizontal strata indicated by layers 3,2, and 1 are contorted as shown. This topographic condition is unstable under conditions of normal erosion, and the anticline is beveled to approach the level of the syncline. For purposes of simplicity assume that the surface of erosion eventually attains a horizontal plane as shown in Fig. 13.20*b*. Erosion has disclosed (beneath the soil cover) an exposure pattern of bedrock as indicated. Note that formations 3 and 2 crop out in parallel strips, whereas, formation 1 remains buried below the surface. By the law of superposition we know the relative ages of the formations to be 3,2,1 in order of increasing age, but except for formation 1 they now outcrop along the same horizontal plane. Diagram *b* indicates the positions of the respective axial planes, and the cross section shows that for the anticline the strata are inclined in opposite directions away from the axial plane. In the syncline the strata are inclined in opposite directions on either side of the axial plane but toward the trough of the fold.

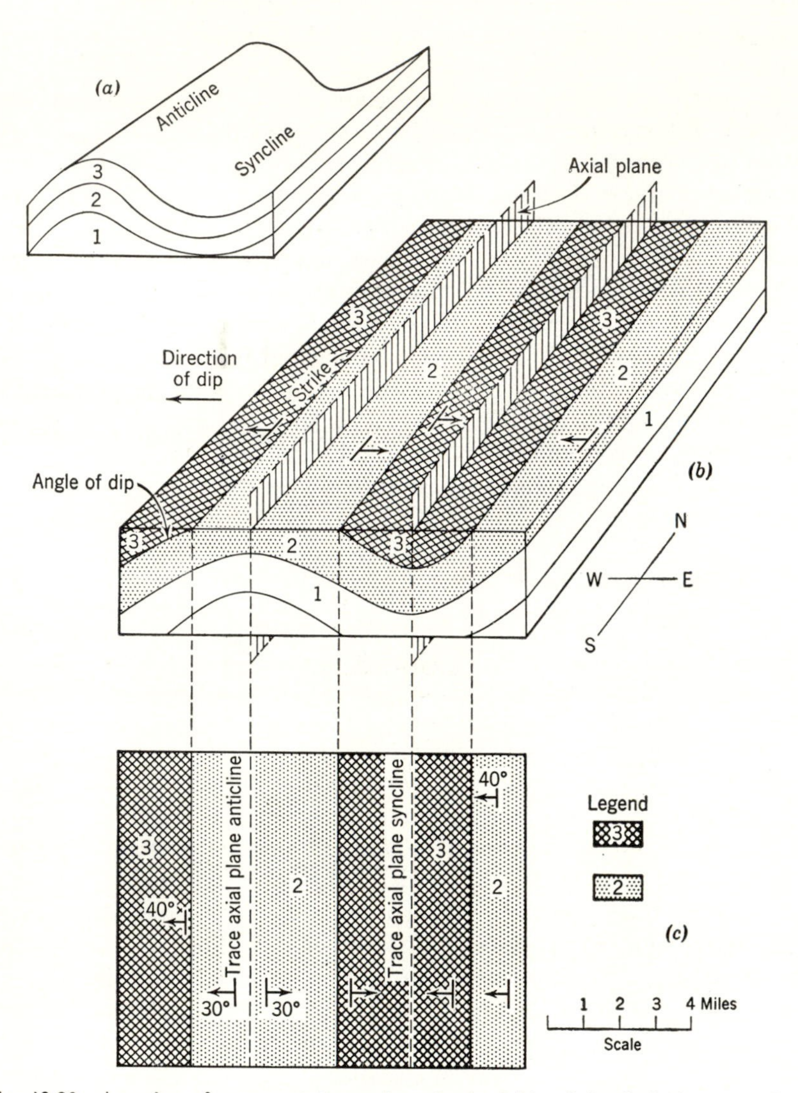

Fig. 13.20 A series of representations of a simple fold. (*a*) Block diagram of an anticline and syncline showing strata 1, 2, 3, in order of decreasing age. (*b*) Block diagram of (*a*) eroded to an horizontal surface, showing by the pattern on the surface the outcrop of strata illustrated in the cross section. Note the positions of the axial planes, the symbols marking dip and strike, and the order of arrangement of outcropping strata. (*c*) Geologic map of eroded folds (*b*) above. Legend blocks of strata appearing on the map are arranged in order of decreasing age upward; angles of dip are expressed in degrees.

Dip and Strike

In order to show the position of a fold on a geologic map it must be oriented in the same relative position as on the earth's surface. The trend and inclination of strata must be measured. This is done in the same manner as for the joint, namely in terms of dip and strike. The strike of an inclined layer is defined as the direction of the line of intersection between the horizontal plane and the bedding plane. The angle of dip is the angle of maximum inclination of the stratum measured from the horizontal plane, and the direction of dip is the direction in which the stratum is inclined in the plane of the dip (this direction is at right angles to the strike).[10] These relationships are indicated in diagram *b* along with the appropriate symbol for indicating dip and strike.

Characteristics of the Map

Figure 13.20*c* is a geologic map. This is a representation, to scale, of the observed and inferred relationships of the structural attitudes of strata as shown in the block diagram *b*. The distribution of outcrop of each formation is shown by a pattern, and the relative age of strata exposed is shown by a legend. By custom the legend always is arranged in order of decreasing age from the bottom upward. Oldest formations are indicated at the bottom of the column, and youngest in their appropriate order at the top. The geologic map also displays positions of observed dip and strike of a rock layer. A conventional symbol of a line and arrow is used to designate the direction of dip, and the strike is shown by a straight line perpendicular to that direction. Angles of dip are indicated by values in numbers placed above the dip and strike symbol.

Much information can be obtained from the geologic map, principally in connection with the structural attitude of strata. For example, the legend indicates that formation 2 is older than formation 3; hence, by the law of superposition formation 2 must pass from its outcrop under formation 3. Along the left-hand margin (west side) of map

[10] In the field the angle of dip and direction of dip are measured, and the strike can be calculated in the same manner as for the joint. Sometimes the strike is observed directly, and the data are recorded as follows: Strike N 87° E. dip 11° to the northwest. In this case the direction of dip can be calculated as N 3° W.

formation 2 must be present at depth, and, hence, formation 2 must dip toward the west from its area of outcrop. Also, formation 2 must dip eastward from a part of its outcrop in the west half of the map in order to pass beneath the outcrop of formation 3 (in the east half) and emerge again to crop out along the map's eastern margin. The anticline and syncline represented, therefore, are indicated without reference to the dip and strike symbols. In a general way, also, the axial positions of the anticline and syncline are indicated. Note from the outcrop relationships that the anticline is indicated by older strata surrounded by younger (i.e., formation 2 bordered by formation 3). The opposite relationships indicate the syncline (younger strata, formation 3, are bordered by older strata, formation 2). Therefore, an anticlinal axis is known to pass through formation 2 and a synclinal axis to pass through formation 3.

The Plunging Fold

The anticline and syncline illustrated in Fig. 13.20 are simple structures whose crest, or trough, is horizontal in position. Such folds are said to be nonplunging. A *plunging fold* is inclined along its crest line as illustrated in Fig. 13.21*a*, and the direction and degree of this inclination are measured from lines of reference. One of these reference lines lies within the axial plane in a horizontal position such as *AB*, Fig. 13.21*b*. The other line of reference is the *axial line,* or *axis of the fold,* and is defined as the line of intersection between the axial plane and a bedding plane surface. The intersection of the axis and a horizontal line in the axial plane is the *angle of plunge or rake of the fold,* and the direction in which this line is inclined (i.e., trace of angle of plunge or rake in the horizontal plane) is called the *direction of plunge* or *strike of the fold.*[11] Wherever the plunging fold has been eroded to a horizontal surface such as Fig. 13.21*b*, the axis of the fold is determined by relationships of the outcropping beds in the axial plane. The upper surface of formation 3 extends from *C* to *D*, and the line *CD* is the axis. The angle *BCD* is the plunge or rake, and the direction of the line *CB* is the strike of the fold. Similar relationships hold for the adjacent syncline. The axial line between formation 4 and 5 is inclined northward (toward the background of the block diagram). Note that a bore hole drilled along the north line of the

[11] The strike of the fold should not be confused with the strike of an individual bed. The strike of the fold refers to the fold as a whole, whereas the strike of a bed refers to a local measurement made along a part of the fold.

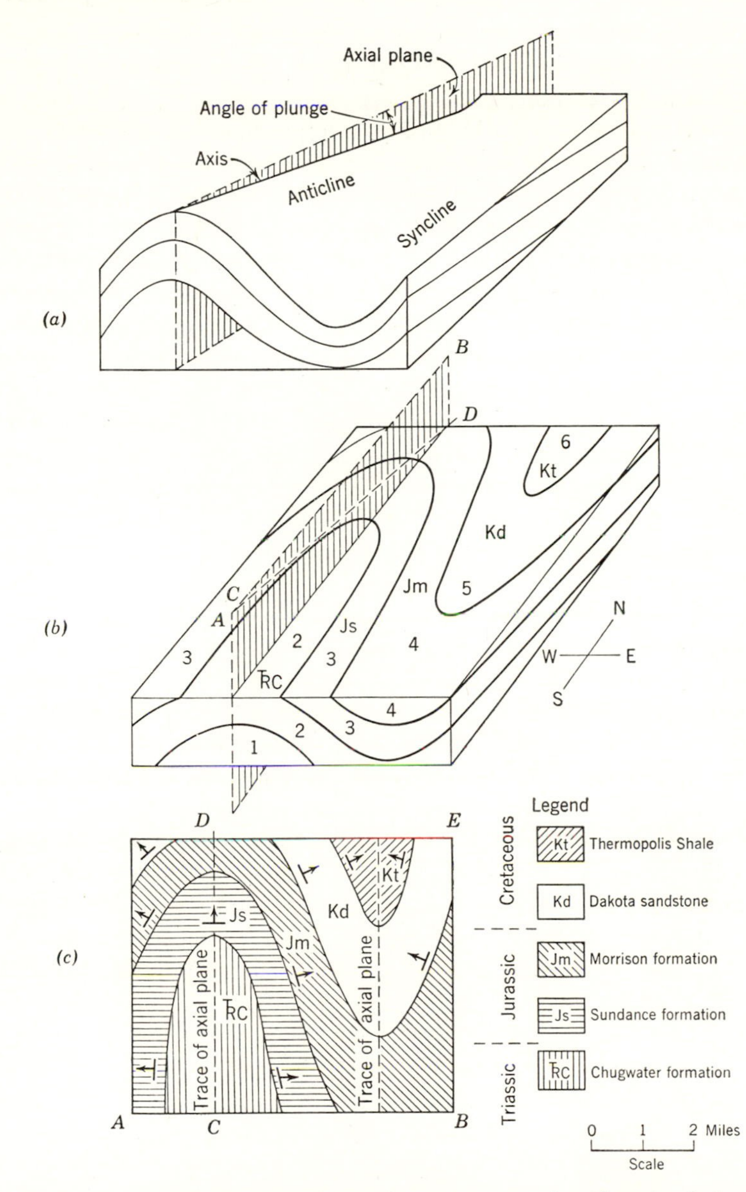

Fig. 13.21 (*a*) Anticline and syncline with *plunge* to north. The axis of the fold is inclined from the horizontal line of reference. (*b*) Block diagram of (*a*) eroded to an horizontal surface and exposing beds of decreasing age from 2 to 6. Note the position of the axis, *CD*, in the axial plane. The value of the *plunge* of the fold is the angle *BCD*. (c) Geologic map of (*b*). Symbols indicate geologic age and the names of formations and are used instead of numbers; relative age is indicated in the legend also.

block diagram in formation 6 will reach the contact between formation 4 and 5 (i.e., along the fold axis) at depth, whereas to the south this contact emerges at the surface.

The block diagram of Fig. 13.21*b* is reproduced as a geologic map in diagram *c*. This map shows by conventional symbols the identifying mark of the formation. The first letter, or symbol, designates the geologic age, and the second letter represents the officially recognized name of the formation. These are indicated in their appropriate age relationships in the legend. Distribution of strata on the geologic map shows the oldest formation (Chugwater) is bordered by progressively younger formations outward. This is the structure of an anticline, and the trace of the axial plane must pass through the fold (line CD). In the eastern half of the map the youngest formation (Thermopolis) is bordered by progressively older strata. This relationship is that of a syncline, and the axial plane cuts the fold as indicated. Both anticline and syncline are known to plunge toward the north, a condition which causes the outcrop to turn and double back producing a "nose." The nose of the fold is an important indicator of the direction of plunge or strike. In the case of the anticline the plunge is always in the direction of the nose, whereas in the syncline the plunge is always away from the nose, or toward the open end.

Dip and strike symbols indicating observed relationships at the designated localities show that the strike of strata changes with position on the plunging fold. For any selected point on a perfectly horizontal surface a tangent to the bounding line between formations is the directon of strike of the bed. At all points along the trace of the axial plane on a horizontal surface the strike is perpendicular to the line, and the angle of dip is coincidental with the angle of plunge or rake. Elsewhere, values of strike and dip change with the position on the fold. (See Fig. 13.21*c*.)

THE CROSS SECTION

Although the geologic map is in part an interpretation of the geology of an area it is intended to be as factual as possible. Indeed, it is based upon as many observations of rock type, occurrence, and structural attitude as is deemed necessary to construct a highly accurate representation commensurate with the scale. The geologic map is the basis for much less precise interpretation of the structural attitude which prevails several thousands of feet below the surface. In some areas much subsurface information is available through the records of many borings in search of oil or water. Elsewhere, tunnels and mines provide access to investigation of the subsurface, but in most areas construction of the geologic map is a prerequisite to drilling, and much geologic interpretation must be done in advance of the drill. In this connection it becomes necessary to construct cross sections showing the conditions postulated at depth based upon the geology observed at the surface. As an example of the procedure Figs. 13.22*a*, *b*, and *c* have been drawn from the geologic map of the plunging anticline and syncline (Fig. 13.21*c*). The cross sections constructed are along the lines *AB*, *CD*, and *AE*. Such cross sections are based upon the width of outcrop of each formation along the line of the cross section, the

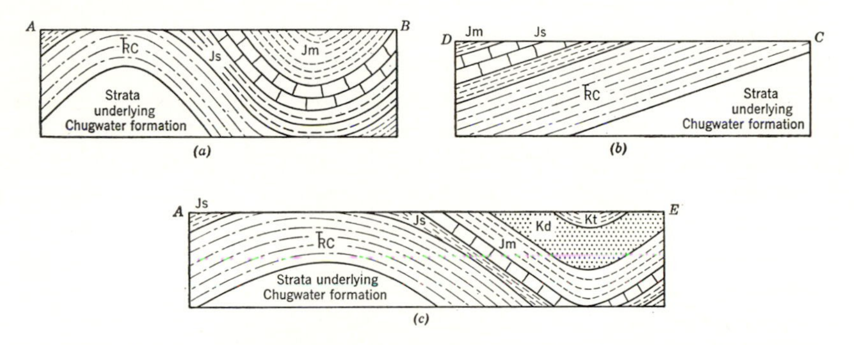

Fig. 13.22 Cross sections of geologic map Fig. 13.21c along lines indicated on the map. Cross section (*a*) along line *AB*, (*b*) along line *DC*, and (*c*) along line *AE*. Compare with block diagram Fig. 13.21*b*.

structural attitude as indicated by the computed component of dip, and the thickness of each formation. The latter may be described in some columnar section of the strata (see Fig. 12.12) or computed from the scale of the map, the width of outcrop, and the dip. From such data and any available from the subsurface is constructed the interpretative cross section. Rocks which lie beneath the oldest stratum cropping out in the map area are known only from scattered information which may be available. The presence of deeply buried igneous rocks and unconformities which separate the structural attitude of the underlying strata from those above the unconformity must remain in the realm of speculation unless some information is available from deep drilling or interpretation of geophysical observations.

INFLUENCE OF TOPOGRAPHY ON OUTCROP PATTERNS

The geologic map as illustrated in Fig. 13.21*c* is a hypothetical representation of an anticline and a syncline eroded to a perfectly horizontal surface. This condition is approached as a limiting surface, but as a general rule the effect exerted by topography upon the position of outcrop is particularly noticeable in the field. In map representation the influence of topography decreases in importance in direct proportion to the map scale. For example, the geologic map of the United States prepared by the U. S. Geological Survey is printed on a base whose scale is 1:2,500,000 or 1 inch equals 207,500 feet, approximately. In the field the maximum relief at any one locality does not exceed 15,000 feet. On the scale of the geologic map of the United States a distance or height of 15,000 feet represents a length of .072 inch. The effect of topography of a great mountain range, therefore, is to cause a maximum deviation in a line of outcrop of less than one tenth of an inch from its position on a perfectly horizontal surface. Hence, on a map of such a scale all deviations in formational boundaries greater than $\frac{1}{10}$ inch must be attributed to geologic structure and not to the effects of topography. In short, for a map of such scale the surface can be considered to be a horizontal plane. The opposite is true on maps of large scale where the influence of topography on the outcrop pattern is especially noticeable. For example Fig. 13.23*a* shows, in a coastal area of rough topography, the outcrop pattern of a dike dipping approximately 45 degrees. The outcrop pattern migrates in the direction of dip as the elevation is lowered and climbs the mountainside with increased height. Figure 13.23*b* illustrates the detail of a part of diagram *a* with topography shown in contours and on a scale five times as large. In this diagram the dip of the dike is about 27 degrees. Note the intricate pattern of outcrop produced. Diagram *b* is constructed on a base with topographic contours. The dike, known to dip at a rate of 500 feet of vertical drop in a horizontal distance of 1000 feet (approximately 27 degrees), has been drawn in the plane of the direction of dip to appropriate scale in the cross section at the lower end of diagram *b*. Intersections of the dike with sea level elevations of 0, 500, 1000, 1500, and 2000 feet are drawn on the base map and represent lines of strike of the dike at actual or theoretical elevations assuming the dike continues to dip at 27 degrees. A point of coincidence between contours of surface elevation and those of the dike elevation is a position of outcrop. This is the location of an intersection between the line of strike of the dike and a topographic contour line of the same elevation. Interpolation of respective elevations of the dike and topography indicate the positions of other points of outcrop, and a line showing the trace of continuous outcrop may be drawn as has been done in diagram *b*. Comparison of the position of outcrop between diagram *a* and *b* indicates that as the dip decreases

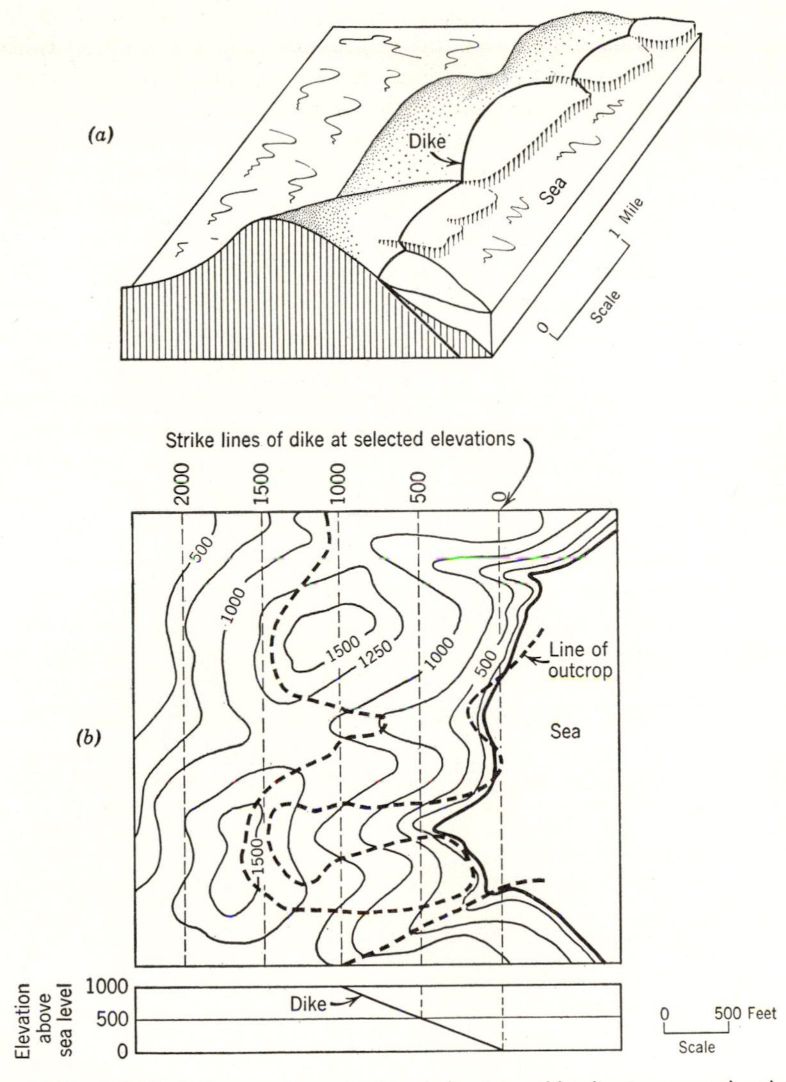

Fig. 13.23 (*a*) Block diagram of coastal island showing a dike dipping seaward and the line of outcrop changing direction with change in topographic elevation. (*b*) Topographic map of part of (*a*), on an enlarged scale, illustrating the position of the outcrop of the dike as determined from the elevation in the cross section below the map. Parallel north-south dashed lines on the map are elevations of the dike projected from the cross section onto the map. Coincidence of an elevation of the dike surface and the topography is a position of outcrop. If all such points are connected the line of outcrop is as shown on the map.

the sinuosity in outcrop increases. A vertical dike, or bed, continues across country as a straight line irrespective of topography. The outer extreme is the case of horizontal strata in which the outcrop belt is controlled entirely by the nature of the topography. In a district of pronounced stream dissection the stream system is dendritic and so also is the pattern of outcrop.

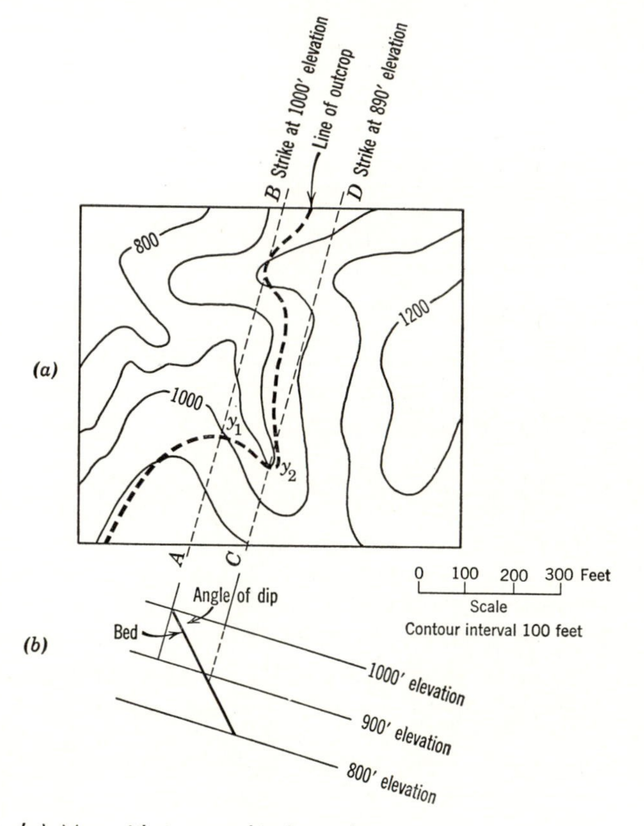

Fig. 13.24 (*a*) Map with topographic base, showing the line of outcrop of a bed of uniform dip. The line of strike is drawn where the bed crosses the same elevation at two points. Projection onto a surface in the plane of the dip as indicated in (*b*) permits graphic solution of the dip when plotted to the scale of the map.

Computing the Dip and Strike

Common use is made of the outcrop pattern on a geologic map with topographic base to compute the dip and strike of a rock unit.[12] An example is illustrated by Fig. 13.24. The outcrop of the upper surface

[12] The method described assumes the dip of the bed to be a plane surface between the positions of outcrop measured.

of a bed is shown in diagram *a*. This line intersects the 1000-foot contour at three points which can be connected by a line such as *AB*. By definition this is the strike of the bed inasmuch as it represents the line of intersection of the stratum with a horizontal plane (in this case the horizontal surface of 1000 feet of elevation). Its direction may be measured with a protractor. A parallel line *CD* is drawn representing the strike at another elevation (in this case 890 feet). The graphical solution to the problem consists in extending the strike lines of the map and constructing a cross section to the same scale in the plane of the dip (diagram *b*). Intersections of the bed at the 1000- and 890-foot planes are indicated, and the line of the trace of the bed is drawn in the vertical cross section. The angle of dip is measured with the protractor. The dip may also be computed using the following equation:

$$\tan \alpha = \frac{y}{x} \tag{1}$$

where α = angle of dip.

y = vertical distance between two positions of outcrop, y_1 and y_2 as determined by difference in elevation contours.

x = horizontal distance along the line of trace of the direction of dip between the two points y_1 and y_2.

On many geologic maps lacking a contour base a qualitative appraisal of the direction and steepness of dip can be determined from the outcrop pattern developed by stream valleys. A diagrammatic illustration is presented by Fig. 13.25 which represents two anticlines and an intervening syncline all plunging in a general northeasterly direction. The outcrop pattern at locality *A* indicates a general dip toward the northwest at a relatively low angle, whereas the eastern limb of the same fold at *B* is vertical as indicated by the failure of the formation boundary to shift as it passes in and out of a river valley. East of the trace of the axial plane of the syncline at locality *C* the dip is indicated as westward by a shift in the formation boundary. The syncline is recognized as being asymmetric with a vertical western limb and a gentle eastern limb. To the east the adjacent anticline appears to be only slightly asymmetric as indicated by the relative indentation in formation boundaries between *C* and *D*.

Effect of Faulting

The geologic map also indicates the effect which faulting has upon the outcrop pattern of strata. An example is illustrated by Fig. 13.26

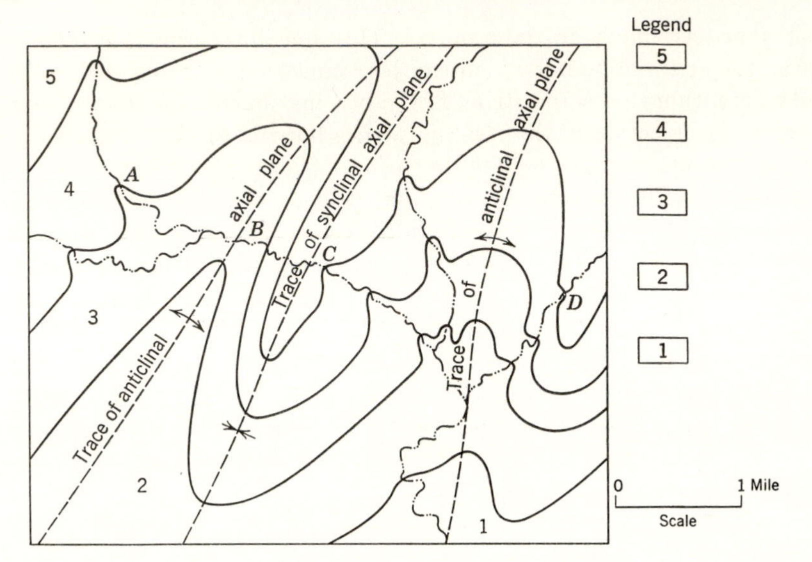

Fig. 13.25 Geologic map of two anticlines and an intervening syncline, all plunging northeast, showing effects of topography upon the outcrop boundary between formations. Note the migration of the boundary downdip is least where beds are steeply dipping (*B*) and greatest where gently dipping (*A*).

which is a geologic map of a symmetrical anticline and syncline plunging to the north. The folds have been cut by a fault striking generally north of west, and the south side is the upthrow. A deep stream valley transects the area from north to south. This valley reveals the dip along the western limb of the anticline by causing the boundary between formations to shift in the direction of dip. Southward the same valley crosses the fault plane. As it does so the trace of the fault plane swings northward and loses elevation crossing the valley.

A shift in the fault trace indicates the dip of the fault plane in the same manner as does the shift in the trace of a formation boundary. On the basis of this relationship the fault is known to dip northward, i.e., in the direction of the downthrow side, and can be described as a normal fault. For purposes of illustration, this normal fault is considered to have moved with a vertical component only, i.e., the net slip and dip slip are one and the same, and the strike slip is zero. On the map this is indicated by absence of any horizontal shift in the axial plane of the folds. Nevertheless, despite the absence of any lateral shift there exists a clearly defined offset in the boundaries between formations. Without knowing that horizontal motion had not occurred an observer would consider that a strong strike slip was present. The apparent paradox is the result of relative migration of formation boundaries downdip. As erosion lowers the upthrow side each formation boundary migrates downdip. For example, the boundary between formations 2 and 3 on the west limb of the anticline must migrate in a westerly direction. This situation causes the gap to widen between two equivalent points on either side of the fault as erosion lowers the upthrow side to that of the downthrow. The ultimate width of the gap

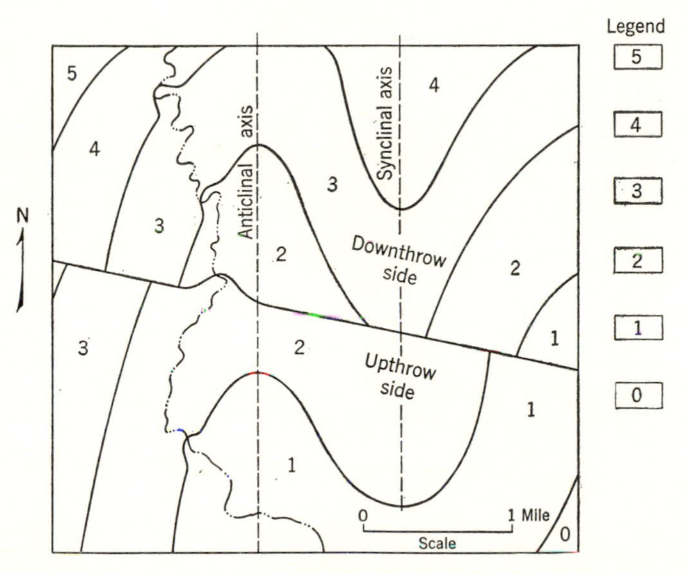

Fig. 13.26 Geologic map of an anticline and a syncline, plunging north, cut by a normal fault with no horizontal displacement. Note the migration of the boundary of the fault plane downdip in the stream valley, the apparent horizontal shift in formation boundaries and that older strata on upthrow side are in contact with younger ones on the downthrow side.

between the two boundaries is a function of the amount of throw on the fault and the dip of the strata.

Migration in gently dipping beds is greater than in steeply dipping ones. In the case of the anticline migration of outcrop boundaries is outward from the axial plane, and on the upthrow side one formation (in this case formation 2) displays a much wider outcrop belt than its counterpart on the downthrow side. Formation boundary migration is downdip, also, in the case of the syncline. This means that equivalent boundaries will migrate toward the axial plane, and on the upthrow side of a fault the migration will be nearer the axial position than the counterpart on the downthrow side (note boundaries between formations 1 and 2 on the east limb of the syncline). Also, along the synclinal axis it is commonplace for older beds on the upthrow side to be in contact with younger ones on the downthrow side. (Note formation 2 in contact with formation 3.)

INTERPRETATION OF FOLDS IN THE FIELD

Some folds are sufficiently small to be readily observed in their entirety in the field; the majority, however, are large and are recognized only by means of assembling a series of isolated observations onto a map which reveals the structure. In their most simple form the data may consist of isolated observations of dip and strike of strata plotted in their respective positions on a map. By appropriate connection of the lines of strike a form line of the fold can be indicated. In other cases a complete geologic map is prepared to indicate the folds. Under still other circumstances a structure contour map is prepared which shows the position of some datum surface (generally a well-defined bed) by means of contours. The outline and number of contours reveal the slope, disposition, and magnitude of the anticlines and synclines.

Drag Fold

Gathering the necessary data in the field is not always a simple task. In some areas information is unavailable because it is covered by a lake or hidden by a thick cover of glacial drift. In some cases certain information which may be obtained at the available outcrops proves extremely useful in construction of the fold pattern. An example is the information provided by drag folds. A *drag fold* is a small fold produced in a thin bed of shale which separates massive beds of limestone or sandstone. Wherever such a sequence is folded differential shear takes place along the bedding planes, and the weak and plastic shale layer is contorted into folds as the massive beds shear over one another without deformation. (See Fig. 13.27.)

Drag folds are important because their shape is indicative of relative movement between two massive layers and because they generally approximate the shape of the large fold of which they are a part. Relative movement between beds producing the drag fold is always the same: younger strata move over older toward an anticlinal crest and outward from a synclinal trough. The greatest amount of differential movement is to be noted along the limbs of the major fold, and in

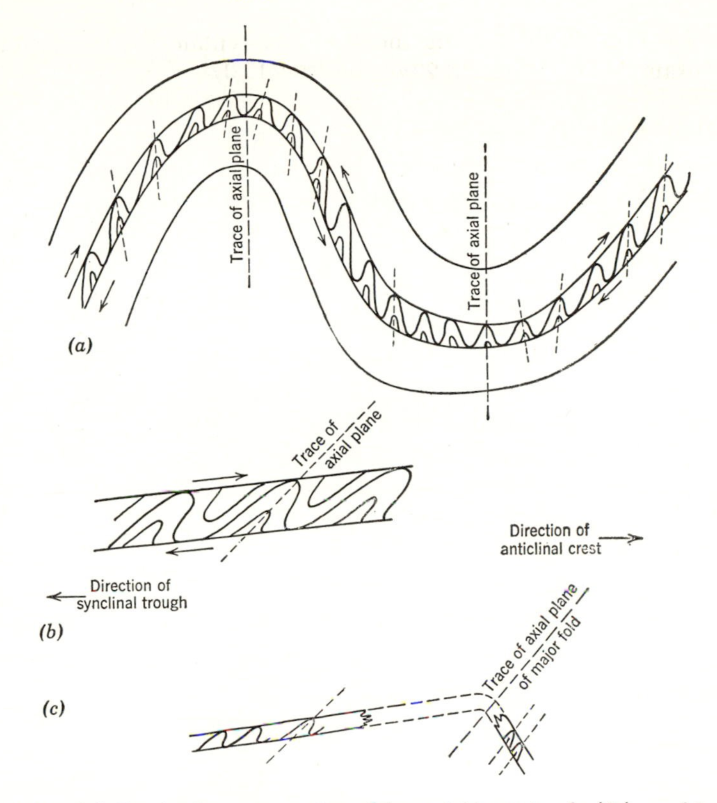

Fig. 13.27 (*a*) Sketch of a cross section of large folds, parts of which consist of three beds, a central one of shaly rock enclosed between two massive layers. As folding occurs differential movement (note arrows) between the massive beds contorts the soft unit into *drag* folds. Note general parallelism of axial planes to one another and to those of the major folds. (*b*) Part of a unit similar to (*a*) in which at the local outcrop the directions of differential movement of the massive layers may be determined by the curvatures of the drag folds. The axial plane of the drag fold approximates the general attitude of the axial plane of the major fold. Note that the interpretation indicates an anticlinal crest in the right-hand direction and that the limb opposite to the one here observed must be steeply inclined in order to preserve the general parallelism of the axial plane of the major fold with that of the drag fold. (*c*) Interpretation of structure of major fold as indicated by (*b*).

such positions the axial planes of the drag folds are inclined so as to indicate the relative movement of the strong, massive beds. Figure 13.27*b* is a representation of a part of an outcrop in which the orientation of the drag folds indicates the direction of the major anticlinal crest and synclinal trough.

As a rule the axial plane of the drag fold is roughly parallel to that of the major fold of which it is a part. This characteristic provides a very useful means of approximating the general symmetry of the major fold. For example, in Fig. 13.27*a* the axial planes of the drag folds are oriented approximately vertically, indicating that the axial plane of the major fold is similarly disposed. If the axial plane of the major fold is vertical the fold must be symmetrical, and the dip of one limb is approximately the mirror image of the other. Based upon this line of reasoning the dip at the locality of observation must be representative of the dip of a corresponding position on the opposite limb of the fold. Thus, the shape of the fold has been constructed upon information gathered at a single outcrop. One such isolated locality of observation is shown in Figure 13.27*b*. Here, strata are inclined gently but the axes of drag folds are not vertically oriented. From this relationship the limb on the other side of the anticlinal axis is inferred to dip somewhat more steeply than at the position of observation in order to preserve the symmetry required by general parallelism of the axial planes between the drag fold and the major fold. This analysis is carried out in diagram *c* in which the conditions of the west limb of the fold are shown as known, and dip of the east limb is postulated on the basis of the angular relationship between the dips of the axial plane and the bed at the point of observation.

Drag Folds and the Plunging Fold

Relative movement between two massive layers separated by a weaker shaly layer is indicated by shear distributed throughout the fold and is to be noted in sections selected with random orientation across the plunging fold. This fortunate circumstance permits reconstruction of the major fold from any suitable exposure of the drag-folded section irrespective of its position on the major fold. In Fig. 13.28 is shown the distribution of drag folds in an asymmetric anticline. Note the position of arrows pointing in the directions of relative shear of massive beds. As previously described such relative motion is revealed in cross section by drag folds. A similar relationship is shown for plunging anticlines or synclines when the exposed surface is horizontal. Indeed, the orientation of the surface of exposure has no influence upon the interpretation of the directions of relative shear. Observed relationships at any position along the fold permit prediction of the nature and symmetry of the major fold. For example, at locality *A* the drag fold relationships indicate that the westernmost bed has moved north with respect to underlying units. This situation requires that an anticlinal nose must exist to the north of the point of observation and, also, that the position of observation must be on the western limb of the major fold. If the uppermost bed has moved north with respect to underlying ones as indicated in plan view a vertical section must show the relationship as drawn in the cross section of the block diagram as shown at the south, and the

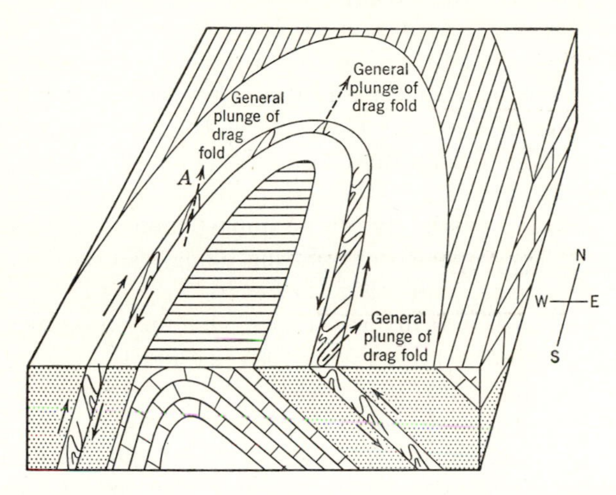

Fig. 13.28 Block diagram of an anticline plunging north showing orientation of drag folds. Note that in plan view or cross section the general relationship of the curvature of drag folds to differential movement between the massive layers is the same and also that a rough parallelism of axial planes between major and minor folds is maintained.

position of the western limb of an anticline whose axis lies to the east is substantiated. At locality *A* a statistical mean of the plunge of the drag folds will also provide an approximate measure of the general plunge of the major fold.

CLEAVAGE

Definition

Among generally shaly rocks there is developed a structure called cleavage which is a parting along smooth-faced planes. This tendency for shaly strata to split is doubtless controlled in part by the orientation of clay minerals which dominate rocks of this type. Cleavage parallel to bedding is one of the characterizing aspects of shales and in some cases is so well developed that laminas of nearly paper thinness can be separated. These shales are called *fissile*. *Fissile* shales commonly consist of alternating layers of silt- and clay-sized particles, and to some degree parting takes place along these planes of natural separation.

Upon application of intense stress to such rocks the clay minerals tend to alter to micas and to rotate in position with their longer dimension parallel to the axis of greatest strain in the strain ellipsoid. When this mineral orientation becomes pronounced the cleavage imparted by the parallelism of micaceous minerals is no longer parallel to bedding and in fact is oriented in some position independent of the original bedding separation. Cleavage of this variety is much more strikingly developed than fissility; individual crystals are welded together by intergrowth, and the rock emits a clear tonal sound when struck. The rock is distinctly more compact than shale and is identified as slate. The capacity of splitting along well-marked planes is called *slaty cleavage*. (See Fig. 14.1.)

If an area continues to experience the metamorphic processes described above, mica crystals grow in size, and there is produced a coarsely crystalline micaceous rock known as a schist. (See Fig. 14.2*b*.) Its appearance is no longer that of slate, but orientation of the mica crystals remains the same, i.e., in the plane of the axis of greatest strain of the strain ellipsoid. The cleavage is distinguished by being called *schistosity*, but there is no difference in its orientation from that in slaty cleavage.

A much less intense variety of cleavage is developed within a thin bed of shaly sandstone folded with massive strata of sandstone or limestone. As such interlayered strata are folded relative movement between the massive layers tends to cause rupture within the shaly sandstone, and there is developed a series of closely spaced joints. The jointing is identified as *fracture cleavage*, and although always associated with folding it need not be found with metamorphic rocks.

Fracture Cleavage

Addition of sand to a clay sediment tends to reduce its plasticity and increase its brittleness. For this reason differential movement between massive layers above and below a thin, shaly sandstone leads to its rupture rather than folding. Repetitive failure of the thin sandstone produces the structure called fracture cleavage, and there is a general absence of yielding by plastic flow as in development of drag folds. The cleavage consists of a very closely spaced set of joints, oftentimes less than one inch apart, which imparts to the rock a distinct platy aspect. Fracture cleavage is the product of stress along a shear couple induced as the strata are compressed into folds. The immediate cause of rupture is relative movement in the overlying and underlying layers of massive rock when under compressional stress. Failure of the shaly sandstone is along a plane identified as a com-

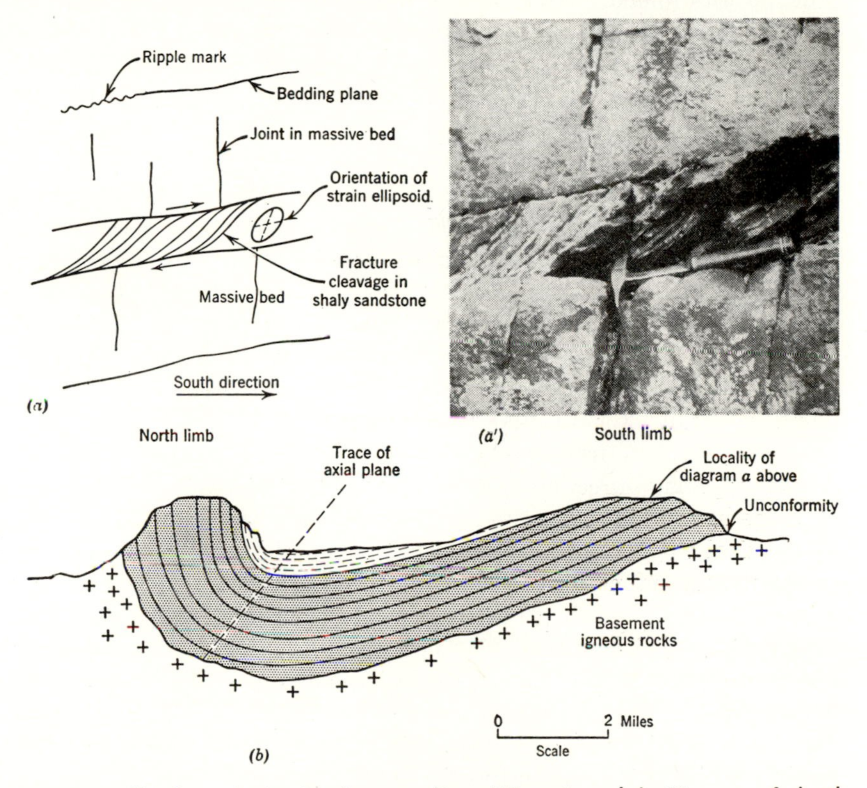

Fig. 13.29 Sketches of the Baraboo syncline, Wisconsin. (*a*) Diagram of local exposure (*a*′) along the south limb of the syncline showing fracture cleavage developed in a shaly bed, interpretation of differential movement, and orientation of strain ellipsoid. An anticlinal axis must exist to the south and a synclinal axis to the north. The vertical trace of the axial plane of the major fold must parallel approximately that of the axis of greatest strain of the strain ellipsoid. (*b*) Cross section of the major fold as known from field mapping and drilling.

pressional joint, and, therefore, the general orientation of this joint set must be in accordance with shear planes of the strain ellipsoid. Commonly, one of the positions of the shear plane is parallel to bedding; hence, the cleavage is dominated by one set of joints. An illustration of this condition is shown in Fig. 13.29*a* which is a representation of an outcrop at the south limb of a large syncline.

Outcrops in the south limb of the fold show beds of massive quartzite separated by beds of shaly quartzite highly jointed by fracture cleavage. Close inspection of the surfaces of fracture reveal the directions of relative movements of the massive beds as the great syncline was folded. Note how bending of the fracture cleavage so as to produce the aspect of a reverse curve demonstrates movement of the quartzite layers. No doubt exists that the uppermost bed moved to the right relative to the lowermost bed. Ripple marks on the bedding surfaces indicate that the strata are in a normal attitude and the law of superposition holds. We know, therefore, that the youngest bed sheared over the oldest bed, and from the position of observation an anticlinal axis must lie to the south and a synclinal axis to the north.

Proper orientation of the strain ellipsoid requires that fracture cleavage is coincidental with one of the planes of shear. When this is done the other plane of shear falls with only slight inclination to the bedding. Inasmuch as bedding already is a surface of separation no new fractures will develop, and the bedding becomes one of the planes of shear. For this reason the fracture cleavage consists of only one parallel set of fractures, the bedding having absorbed the shear in the opposing direction. Although all of the planes of fracture cleavage have been rotated to some extent some are less inclined from the vertical than others and are considered to be more representative of the original orientation at the time of initial rupture. On this basis orientation of the strain ellipsoid is as shown in Fig. 13.29*a*.

When the strain ellipsoid is oriented on the basis of drag folds the position of the axis of greatest strain is placed in a direction parallel to the greatest elongation of the small fold. The axial plane of the major fold must lie also in a position parallel to the axis of greatest strain. From this relationship the axial plane of the major fold is inclined to the south indicating that the major fold is asymmetric. At the locality of observation, namely the south limb of the fold, the strata dip to the north at an angle between 10 and 15 degrees; hence, in order to preserve the asymmetry of the fold strata of the north limb must dip steeply. This can be verified by field observation of dip of the north limb.

The foregoing has shown that fracture cleavage may be used to provide interpretations in the same manner as drag folds, the inclination of the fracture cleavage indicating the direction of relative movement of strata. In like manner the trace of fracture cleavage on a horizontal surface of outcrop of a plunging fold indicates the nature of the major fold, and the technique employed in interpretation is the same as that used with drag folds. (See Fig. 13.30.)

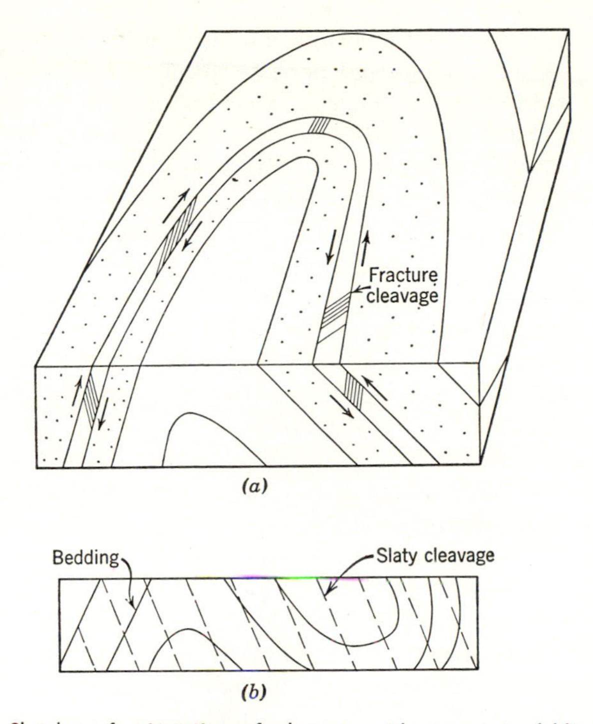

Fig. 13.30 Sketches of orientation of cleavage with respect to folding. (*a*) Block diagram of *fracture cleavage* showing the relationship to differential movement between massive beds both in cross section and plan view. Note that the interpretation of the relationships is the same as in the case of drag folds. (*b*) Cross section of folds showing *slaty cleavage* cutting across the bedding of both a syncline and an anticline. The plane of slaty cleavage is parallel to the axial planes of the major folds.

Slaty Cleavage or Schistosity

When rocks are subjected to strong compression the structural attitude is characterized by tight folds, thrust faults, and overthrusts. As intensity of the stresses increases so also does the tendency to develop schistose cleavage. This is best developed in argillaceous rocks, but lava flows also readily display this cleavage. The process of development involves growth and orientation of platy minerals, particularly micas or those of rodlike shape such as hornblende. The latter shift in position until their long axes lie parallel to the theoretical axis of greatest strain, whereas micas become oriented with their large tabular surface in this plane. The effect of this mineral orientation is to impart to the rock a very strong cleavage typically identified with slate or schist.

Slaty cleavage is unlike the localized occurrence of fracture cleavage but extends for many miles and persists throughout an entire belt of argillaceous rocks or lavas. Thus, in a general way slaty cleavage has parallelism over the entire area of its occurrence, and this regional parallelism imparts a characteristic aspect of monotony to the rocks. Close inspection of the cleavage surface reveals the trace of individual thin beds recognized by changes in color. The bedding can be seen to be tightly folded, and as the individual anticlines and synclines are indicated the slaty cleavage is uniformly oriented, cutting across the bedding and folds alike. This condition provides the regional parallelism mentioned above.

The position of slaty cleavage has a direct bearing upon the orientation of the fold and can be demonstrated to closely parallel axial planes of anticlines and synclines. For this reason when the rocks become schistose and folding can no longer be recognized because of the absence of recognizable bedding the cleavage is called axial-plane schistosity. The latter proves to be virtually the sole means of interpreting the folds of an area. (See Figs. 13.30*b* and 14.1.) Because of the structural importance of this cleavage an understanding of the well-defined relationship of slaty cleavage to the individual fold is the key to unraveling the nature of the fold. In this connection some general rules have been enunciated of which the following are considered of primary importance:

1. If the cleavage is vertical the fold is known to be symmetrical and normal in position, i.e., the law of superposition of strata holds.
2. Wherever the cleavage dips more steeply than the bedding and in the same general direction the strata are known to be normal in attitude (i.e., younger beds lie on older), and a synclinal axis lies in the direction in which the beds dip.
3. At localities where the cleavage dips more gently than the bedding the strata are overturned, and older strata lie upon younger ones, and a synclinal axis lies in a direction opposite to that in which the strata are inclined.
4. The trace of the bedding on the cleavage face makes an angle with respect to a horizontal line. This is the angle of plunge, and the direction of its inclination is the strike of the fold.

Much importance should be placed upon the geologic structure of an area, and as its complexity is increased use must be made of the orientation of slaty cleavage to unravel the pattern of folding. Many of the problems of interpretation which arise are beyond the scope of this text, but the general principles which have been discussed are those most commonly employed. Where isoclinal folding dominates an area the region becomes characterized by metamorphic rocks. These are the sites of the most difficult of all structural interpretations, and only through knowledge of the conditions of development of these rocks can some understanding of the structure be achieved. The following chapter is designed to provide an insight into the complex conditions which produce metamorphic rocks.

Selected Supplementary Readings

Billings, M. P., *Structural Geology,* Prentice-Hall, Inc., 2nd Edition, 1954, Chapters 8 and 9. An excellent description, classification, and criteria for recognition of faults.

DeSitter, L. U., *Structural Geology,* McGraw-Hill Book Co., 1956, Chapters 2, 3, and 4. The theoretical and observational behavior of rocks under stress in the laboratory.

Emmons, W. H., Thiel, G. A., Stauffer, C. R., and Allison, I. S., *Geology Principles and Processes,* 4th ed., McGraw-Hill Book Co., 1955, Chapter 18. Describes the processes of diastrophism in the earth's crust.

Leith, C. K., *Structural Geology,* Henry Holt and Co., 1923, Chapter II. One of the earliest presentations of the concept of the strain ellipsoid in geology.

Low, Julian W., *Geologic Field Methods,* Harper and Bros., 1957, Chapter VI. A comprehensive treatment of the methods employed to decipher geologic structure in the field.

Willis, Bailey, and Willis, Robin, *Geologic Structures,* McGraw-Hill Book Co., 1934, Chapter IV. A classification of folds.

14

Metamorphism

GENERAL CONCEPTS

Among the important concepts which have been developed in preceding chapters is that of geologic processes, operating in opposition, to attain steady-state conditions. The intent has been to impress upon the reader the importance of balance in all phases of geologic processes. Also, much evidence is on hand to indicate that once steady-state conditions are attained they do not remain indefinitely but are in time displaced by new releases of energy. In so far as surficial phenomena are concerned the driving force is principally in the same direction, i.e., toward lowering of potential energy by removal of irregularities on the earth's surface. But intensive study of the geologic history of the earth has failed to disclose any evidence of exhaustion of the earth's internal energy. Rather, the evidence available suggests more or less constant release of this energy in the form of volcanism or mountain building.

About one hundred years ago certain geologists became convinced that release of internal energy is somewhat periodic and may simul-

Gneiss, shore of Lake Superior.

taneously display its effects in more than one continent as indicated by repeated uplift of great mountain chains at approximately the same position in the geologic time scale. Following the uplift of one of these mountain chains the trend of equilibrium for erosion and deposition once more is reversed and directed toward denudation and its associated processes. Through these processes the great mountain core of metamorphic rocks is exposed. Are these rocks the products of equilibrium conditions within the crustal segment of the earth? Do they represent the products of steady-state conditions when sedimentary rocks are subjected to high temperature and high pressure? Such are the questions of concern in the last chapter of this book.

The Geosyncline

Except for certain forms of direct information such as lava emerging from volcanoes and local geothermal gradients the remainder of the evidence bearing upon the earth's internal energy is indirect and must be assembled from scattered data gathered in the field. An example is the current understanding concerning the origin and behavior of geosynclinal belts. Our present concept is dependent upon interpretation of data which in themselves are not directly concerned with the geosyncline. The cause of geosynclinal development lies within the realm of speculation, and even the general appearance of the geosynclinal site is not recognizable with absolute assurance. Reconstruction of the site and understanding of the behavior of the geosyncline are based upon observed characteristics of associated sediments, igneous rocks, folding, faulting, and intense metamorphism. Following is a list of the principal observed characteristics and interpretations:

1. The geosyncline must be a site of very extensive and intensive deposition of clastic debris (graywacke), much of it coarse in size, which has been dumped by rivers draining areas of complex geology. The evidence suggests that the transporting rivers are not of great length because many of the particles are angular; rounding is known to be a property which when plotted against distance of transport increases very rapidly initially and then approaches the distance axis asymptotically.
2. The geosyncline must be a site in which crustal downwarp must keep pace with the accumulation of sediments. For this reason the sedimentary column must be abnormlly thick when contrasted with the extra-geosynclinal positions. Measurement of the acceleration of gravity in such sites should be abnormally low when compared to

adjacent areas of little or no downwarp. This condition is to be expected inasmuch as the downwarp has depressed heavy subcrustal rocks and their space is occupied by sediments of lower density. The over-all gravitational value as measured at the earth's surface correspondingly is reduced at the site of the geosyncline.

3. The site must be one in which basic-lava flows sporadically are deposited with the sediments, and small intrusions of ultrabasic rocks are to be expected as evidences of instability within the subcrust.

4. The geosyncline must be a region which periodically is subject to compressional stress which produces folds and thrust faults contemporaneously with deposition of the sediments.

5. An episode of intense folding and overthrust faulting must terminate the depositional history of the geosyncline. Folding and overthrusting are minor along the peripheral boundaries but increase in intensity toward the axial position of the geosyncline. (See Fig. 13.16.) Here, rocks once deeply buried are brought to the surface by the magnitude of the folding and faulting.

6. With approach toward the geosynclinal axis and as the intensity of folding increases a corresponding change is to be observed in the rocks. In the outer borders the alteration is restricted to fracture or slaty cleavage and the development of great belts of slate from rocks which were formerly mostly shales. Deeper within the geosyncline metamorphism has been more intense, and special rocks are produced under conditions which promote reassembling of elements into new minerals stable under high stress and high temperature. In the same locality and transitional into the highly metamorphic rocks are the great batholiths, or plutonic injections. At some places the granitic *plutons* (batholiths) transgress metamorphic rocks. In others transition between the two is so gradual that the development of the igneous rocks is believed to have occurred in place; that is to say, no intrusion ever occurred, nor did crystallization of minerals at any time ever take place from a liquid mass (see Fig. 5.16). Rather, the typical igneous minerals are considered to have grown by ionic diffusion through the solid state.

Among such characterizing features of the geosynclinal belts little can be tested or compared with currently existing depositional conditions. Regions of abnormally low values of gravity are known along the chains of islands such as the Japanese, Philippine, East Indies, and West Indies which form arcuate outlines bordering and locally connected to the continental mass. These island arcs also are localities of active volcanism, much erosion by streams of short length, and intense local diastrophism, as indicated by the concentration of earthquake activity. They are sites of complex geology, close folding, and occurrence of metamorphic rocks. These are the principal features which are suggestive that geosynclines are in current progress and that localities of island arcs are such sites.

Great belts of metamorphic rocks are known to occupy the central part of geosynclinal troughs which existed in Paleozoic and Mesozoic times. In fact the relationship between geosynclines and metamorphism is so unique that certain metamorphic rocks are not found outside the former geosynclinal sites.

Small zones of metamorphosed strata occur localized around intrusions such as stocks and frequently are associated with the occurrence of ore deposits. (See Figs. 5.14 and 5.16.) Here, there is clear proof that metamorphism can be produced by igneous injection and is in no way connected with the distribution of geosynclinal belts. Indeed, under these circumstances the conditions which produce metamorphism are high temperatures and movement of aqueous solutions capable of establishing an environment of chemical activity in which new minerals are produced through the destruction of others.

From the foregoing the reader must be aware that high temperature and high pressure acting in concert or high temperature alone can establish an environment of chemical mobility. Both conditions must lead to the development of minerals and rocks which are the equilibrium products of the special environment of metamorphism. Locally, this environment approaches that of the igneous rock, but temperatures are lower than those necessary for complete melting. Instead, minerals typical of igneous rocks may grow by ionic diffusion in the solid state when elevated temperatures dictate the existence of such mobility. In the extreme case the mineralogy of the igneous rock appears, but textures and structures of the parent sedimentary rock are preserved.[1] It would appear that with increased depth within the earth pressure, temperature, and chemical zones overlap in such a manner as to establish a general locality of development of extensive belts of metamorphic rocks. If such is the case only the great geosynclinal downwarps can attain this position, and only in such localities will be found certain metamorphic rocks. There is reason to believe, also, that the lower part of the geosynclinal downwarp may penetrate the zone where igneous rocks can be generated either through migmatization or by actual melting of sediments.

[1] Igneous rocks developed without passing through the stage of complete melting are known as *migmatites*. They contain structures typical of the host (sedimentary) rock with which they are gradational. Migmatites are common among plutons such as batholiths. They are not observed in stocks which show clearly marked instrusive relationships.

Kinds of Metamorphism

All metamorphism can be defined as a mineralogic response to change in temperature, pressure, and chemical environment which permits minerals to alter without passing through a fluid condition. The definition excludes effects in the zone of weathering or cementation of buried sediments. Generally, the processes involved bring about recrystallization of the parent-rock material, but simple crushing of grains within a rock accompanied by minor recrystallization should be included. There are to be recognized, therefore, four kinds of metamorphism which are described below.

Predominantly Thermal

Heat dominates the processes, and this type of metamorphism is associated with igneous intrusions, principally stocks. Liquid and gaseous emanations from the magma accelerate mineral transformations. As a general rule there is no change in the bulk chemical composition of the rock, i.e., only small quantities of new material are introduced from the magma. The effect is principally recrystallization, or the construction of new stable minerals from others which were unstable under high temperatures. Such effects are localized to a "rind" of variable thickness surrounding the intrusion, and the type of metamorphism is called *contact.*

Predominantly Directed Stress

Directional, or unbalanced, stress is typical in the upper parts of the crustal zone of the earth. Structures such as joints, faults, and folds particularly reveal this condition. At the immediate surface application of these stresses does not appear to be as great as somewhat deeper in the crust especially in the axial portion of the geosyncline. This appears to be the site of initiation of very intense horizontally directed stress which is manifested by isoclinal folding and overthrusting. Inasmuch as the intensity of this stress condition increases from the geosynclinal margins toward the axial portion the visible effects of stress upon the rocks increase in the same direction. Attention has been directed to such features as fracture cleavage and slaty cleavage which are developed in rocks as stress is applied. Except for simple folding without any corresponding change in the properties of the rock fracture cleavage can be considered as the first manifestation of metamorphism due to stress. In its most extreme development this cleavage consists of a series of such closely spaced joints as to impart to the rock a very crude slatelike appearance. As a general rule, however, fracture cleavage is associated with unmetamorphosed strata, and suitable argument could be proffered to show that fracture cleavage is not a feature of metamorphism. However slaty or axial-plane cleavage is recognized as clearly within the category in question. Slaty cleavage or schistosity displays an orientation of mica minerals with respect to the stress axis which indicates that stress has causal relationship to the development of the mica

Fig. 14.1 Bedding in slate as revealed by large light gray band and by faint alternating light and dark bands inclined from left to right. Slaty cleavage is the fracture system which is nearly vertical and cuts bedding. (See Fig. 13.30.) Siamo slate, Precambrian, near Negaunee, Michigan.

crystals. (See Fig. 14.1.) The term *dynamic* metamorphism is applied to the condition which produces such rocks.

Sandstones pass through a stage of intense penetration of quartz grains into one another to develop the texture recognized as quartzite. This texture lacks the typical relationship of grains being held together by cement; rather, individual grains are welded together. For this reason the rock breaks with irregular or conchoidal fracture through individual mineral grains. Under the influence of intense stress internal crushing of the rock occurs. Individual mineral grains are crushed into smaller dimensions and generally are rewelded together. Along planes of intense shearing micaceous minerals are produced, but otherwise the mineral composition of the rock remains unchanged. This process is identified as *cataclastic* metamorphism.

Predominantly Directed Stress at Elevated Temperatures

One of the most powerful environments leading to generally complete mineral recrystallization and development of new structures is directed stress at high temperature. High pressures tend to raise melting points of individual minerals; hence, some minerals remain unchanged in the new environment, but others change easily under the conditions of ionic mobility which are established. Near the axial position of the geosyncline horizontally applied stress and elevated temperatures prevail. This is the type locality of metamorphism which has been called *dynamo-thermal,* or *regional,* and typical rocks are gneisses and schists.[2] (See Fig. 14.2.)

Predominantly Hydrostatic Stress at Elevated Temperature

In many respects the conditions imposed upon rocks by this environment are not dissimilar to those in which directed stress is applied. However, hydrostatic stress is dominant only in the deeper zones of the crust. An attempt to show the distribution of these conditions appears in Fig. 3.2 in which the change from shear stress to dominant hydrostatic stress is indicated as occurring principally at depths below 100 km. Although the trough of the geosyncline is not considered to be depressed to this depth the condition must be approached during generation of the great batholithic plutons. In rocks of the plutons are some portions which contain minerals of high density. Typical among these is garnet in which the concentration of mass is high per unit volume and which has a crystal habit approaching a sphere or cube.

The environment of hydrostatic stress is considered ideally suited to plastic flow in which rocks although not heated to their melting point behave much like liquids. Impressive evidence of plastic flow is to be observed in the great plutons. Principal among the items of evidence is the highly contorted mineral banding and tight folding of thin "sill-like" masses of quartz and feldspar which lie separated by thin zones of schistose rock. (See Fig. 14.2*b*.) This form of gneiss is so characteristic of the environment that the type of metamorphism is designated *plutonic.*

In this zone of approach to magmatic conditions distinction virtually is impossible between a true igneous rock in the sense of having been crystallized from a melt and crystallization in the solid state by

[2] Gneiss and schist are characteristic of central parts of folded mountain chains. The close association of the two has led to the application of the term *orogenic* (meaning mountain building) to designate such rocks.

(*a*)

(*b*)

Fig. 14.2 (*a*) Outcrop of Precambrian gneiss, Central Mineral Region, Texas. (From Paige, *U. S. Geological Survey Folio* 183.) (*b*) Outcrop of Precambrian schist with thin sill-like intrusions of gneiss (light colored bands), Marquette, Michigan.

movement of ions to loci of crystallization. Some rocks subjected to these conditions are former sandstones particularly high in silica. Recrystallization of such rocks produces phaneritic and porphyritic textures not unlike those of igneous rocks, but the mineral composition lacks the requisite amount of feldspar. However, recrystallization of graywackes whose bulk chemical analysis is similar to that of an igneous rock leads to a final product whose mineralogy is indistinguishable from an igneous rock. For such reasons confusion concerning the origin of certain igneous rocks must prevail until more knowledge is forthcoming.

Zones, Facies, and Grades of Metamorphism

During the first decade of the twentieth century a concept of metamorphism was developed by which certain zones were recognized as indicating conditions to be expected with increased depth below the surface. Rocks of the *epizone,* or near-surface zone, are those clearly associated with shear stress and generally low temperature conditions. The *mesozone* is defined as a position where considerable temperature and pronounced directed pressure prevail. Its position is considered intermediate between the epizone and the underlying *katazone.* The katazone is recognized as the zone of high temperature and generally hydrostatic stress. Each zone can be identified approximately in Fig. 3.2, but the maximum depth of the katazone cannot be fixed except as ranging between 20 and 100 km. (See Table 14.1.) Despite objections to vagueness in boundaries and characteristics the thesis of depth zones has remained in continued use.

The concept of a depth zone as a dominant factor in the development of metamorphic rocks is based upon the influence of low temperature and application of shear stress in the epizone, a condition of moderate temperature and gradual transition from shear to hydrostatic stress in the mesozone, and an environment of high temperature and hydrostatic stress in the katazone. Nevertheless original difference in rock type is recognized as having a very marked control upon the type of metamorphic rock produced within the same depth zone. Rocks of the epizone identified as quartzites, slates, and marbles are recognized with complete certainty as former sandstones, shales, and carbonates. However, some rocks of the katazone have been altered so thoroughly that no currently known criteria can be employed to indicate their original nature. There was introduced, therefore, a concept identified as the *metamorphic facies.* This is defined as a

Table 14.1. Characteristic Aspects of Zones of Metamorphism

Zone	Temperature	Dominant Pressure Condition	Nature of Minerals Formed	Rock Types Formed	Kind of Metamorphism
Epizone	Less than 300°C	Strong shear stress, near-surface conditions.	Stable under shear-stress conditions. Muscovite, chlorite, epidote, generally flaky and containing hydroxyl in their lattice.	Slates, chlorite and mica schists.	Chiefly dynamic
Mesozone	300 to 500°C	Stress shear and static, condition of intermediate depth.	Stable under shear stress. Biotite, hornblende, garnet.	Schists, biotite and hornblende schist, garnet schist.	Dynamothermal
Katazone	500 to 800°C	Strong static stress, deep-seated crustal condition.	Stable under static stress. Minerals generally equidimensional; orthoclase, augite, olivine.	Gneisses, biotite-orthoclase gneiss, augite gneiss.	Deep-seated metamorphism associated with plutonic intrusions

group of rocks of varying chemical composition characterized by a definite set of minerals which have arrived approximately at equilibrium under a given combination of pressure and temperature conditions. Facies distinguished by mineral associations are named without necessarily specifying the nature of the original rock or the precise condition of development. Examples are *greenschist facies* and *amphibolite facies*. Of late this useful concept has been largely supplanted by the idea of a degree, or grade, of metamorphism.

By *metamorphic grade* is meant a stage, degree, or rank of metamorphism which has been attained by a varied assemblage of rocks. The intent is to indicate that rocks of similar, but not necessarily identical, mineral facies belong to the same grade of metamorphism. This should not be interpreted as signifying that all rocks belonging to the same grade necessarily contain the same minerals. Rather, they contain the minerals which characterize the facies, but, more important, rocks of different chemical constitution contain different minerals each of which has been produced under identical conditions of temperature and pressure. Some facies are more sensitive indicators of metamorphic grade than others and, hence, have been made the basis for indicating the progression of metamorphism from lower to higher grades.

The grade of metamorphism which the rocks have attained is defined by key minerals developed in shaly rocks. Chlorite, biotite, garnet, and aluminum silicates of which *kyanite* (Al_2SiO_5) is one appear in order of increasing rank. Rocks are described as in the chlorite grade or biotite grade as the index minerals are employed to connote the stage of metamorphism.

METAMORPHIC MINERALS

Influence of Original Rock Composition

As minerals appear in the metamorphic environment the particular ones which develop are influenced not only by the kind of metamorphism, but also by the composition of the rock which is undergoing alteration. Unless additional material is supplied from some outside source no important change in the bulk composition of the rock is to be anticipated. A shale consisting of clay minerals, quartz, muscovite, and iron oxides will form a rock called a *hornfels* under contact metamorphism. This hornfels may consist of quartz, biotite, kyanite, and feldspar. With the exception of quartz new minerals have been developed by reassembling the available elements into new crystal structures. The bulk composition of the original shale and the secondary hornfels thus remain the same. Likewise, if the same shale is subjected to dynamic metamorphism a schist consisting of quartz, muscovite, biotite, and garnet can result. Each end product of metamorphism can be considerably different in texture and mineralogy, but the chemical analyses will be identical. The hornfels has a finely crystalline texture in which individual minerals are not recognized by the unaided eye, the entire appearance of the rock being dense and dark. By way of contrast the schist will consist of large well-oriented flakes of muscovite and biotite more or less wrapped around small spheroidlike crystals of garnet.

Alteration during metamorphism affects certain igneous and sedimentary rocks in more or less the same manner. These are rocks whose basic chemical composition is much the same. Thus the grouping of rocks into those which will develop similar metamorphic rock types is not dependent upon whether they were originally sedimentary or igneous but upon the similarity of their original bulk composition. Four basic rock types can be distinguished as the parent rocks from which the metamorphic rocks are derived, namely:

1. Shales and dominantly shaly rocks.
2. Sandstones, high-silica igneous rocks, high-silica tuffs.

3. Limestones, dolomites, and dominantly carbonate rocks.
4. Lavas, intrusive bodies, and tuffs of low to intermediate silica content.

Argillaceous (clay-rich) rocks are the most sensitive to changes brought about by metamorphic conditions and are capable of producing a well-graduated series of changes which are suitable to the erection of metamorphic grades. Sandstones are dominated by quartz which is stable under all geologic conditions. Quartzose sandstones tend to alter very slightly, and except for intersuturing and welding of quartz grains one to another scarcely any change occurs even when the conditions are those of high metamorphic rank. On the other hand graywackes contain mineral diluents to the quartz which tend to make such a rock increasingly sensitive to metamorphic change as the percentage of quartz declines. Limestones and dolomites tend to be stable under rather high temperatures and pressures, and the pure carbonate rocks recrystallize to marble and fail by plastic flow rather than to undergo very marked change. However, addition of clay or sand to the carbonate establishes a rock composition which is quite unstable in the metamorphic environment, and new minerals are developed with ease. Igneous rocks high in silica tend to resist change. This is the result of the stability of quartz, mica, and feldspar in the metamorphic environment. Low-silica igneous rocks, however, contain such minerals as pyroxene and olivine which are subject to change in the epizone and mesozone, and these rocks are well suited as indicators of progressive metamorphism.

Influence of Heat and Pressure

Considered broadly the environment of high temperature is also the environment in which stress tends to be uniformly applied from all sides. The environment of low temperatures is that of strong application of shear stress. Contact metamorphism and that typical of the katazone characteristically display the effects of heat and hydrostatic stress. The effect of the latter is the tendency to maintain conditions in a state of constant or reduced volume and permit Van't Hoff's law to operate; namely, with rise in temperature the equilibrium is displaced toward the absorption of heat. Reactions will move in the direction of developing minerals which absorb heat while producing the lattice structure. A rise in temperature also causes an increase in the volatile constituents which, in a "closed" system established by the condition of hydrostatic pressure, tends to cause a rise in vapor pressure.

The effect of stress is predicted by the law of Le Chatelier which dictates that a chemical system at constant temperature will favor those reactions in which a mineral is produced of smaller volume or changed shape. Hydrostatic stress favors the formation of minerals

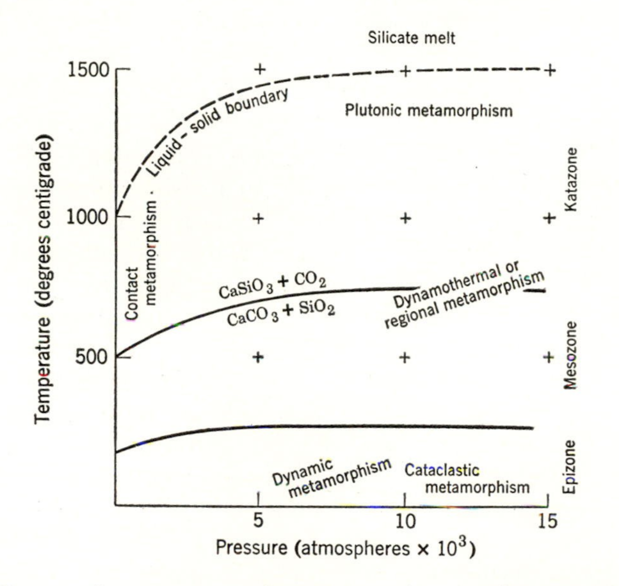

Fig. 14.3 Diagram of temperature and pressure conditions to be expected in various types of metamorphism, and boundaries of epi-, meso-, and katazones. The upper limit of the katazone is set by the solid-liquid boundary; the lower limit is the equilibrium between wollastonite and calcite. The lower limit of the mesozone is set by an indefinite temperature boundary marking the limits below which chlorite is stable.

which are more dense and, hence, occupy less volume than those undergoing decomposition. Shear stress favors the development of minerals of flake, sheet, rodlike, and tabular outline.

The influence of heat and pressure as controlling the type of metamorphism is indicated in Fig. 14.3 in which are shown the general fields of metamorphism. Along the pressure axis at very low temperatures is the condition of cataclastic metamorphism in which the change in rock is brought about by shear stress alone. Contact metamorphism is typical of low pressure but high temperature. Between these two extremes are conditions which produce rocks of the epizone, mesozone, and katazone. Rocks of the epizone show virtually no influence of temperature, and the effects of shear stress are observed in the development of slate and marble. This condition encourages the appearance of minerals which tend to contain OH ion and a lattice which is typical of sheet-structure silicates such as chlorite and muscovite. As temperatures increase in the mesozone chlorite becomes unstable and is replaced by coarse crystals of biotite which tend to form schists. Carbonate rocks containing quartz sand provide the most important indicator of temperature by a reversible reaction between calcite and quartz. The reaction expressed as

$$\underset{\text{(calcite)}}{CaCO_3} + \underset{\text{(quartz)}}{SiO_2} \overset{500°}{\rightleftharpoons} \underset{\text{(wollastonite)}}{CaSiO_3} + CO_2 \tag{1}$$

indicates that calcite and quartz combine at elevated temperatures to produce a silicate mineral *wollastonite* and carbon dioxide. At atmospheric pressure the reaction is driven toward the right at 500°C, and wollastonite and carbon dioxide are formed. The effect of pressure is to cause the reaction to proceed toward the left. If the temperature is kept constant and the pressure is raised the stable condition is the one in which calcite and quartz are the components. However, as the temperature is increased it exerts greater influence than does pressure upon the change from calcite and quartz to wollastonite, and even at very high pressure wollastonite will form at less than 1000°C. The reaction, therefore, is an important indicator of temperature condition, and the presence of wollastonite in marbles is significant in establishing a geologic thermometer.

Wollastonite forms only in carbonates and in such rocks is used to indicate conditions of the katazone. Other minerals which are typical of high-temperature conditions, particularly in shales, are *andalusite* (Al_2SiO_5), *sillimanite* (Al_2SiO_5), pyroxene, olivine, and under extreme conditions orthoclase feldspar.

Minerals such as micas, chlorite, talc, and hornblende have crystal habits which exhibit pronounced difference between their thickness and length. These are minerals whose growth is favored by shear stress, and as has been previously mentioned they tend to develop with their long axes in the plane or direction of the axis of greatest strain of the strain ellipsoid (i.e., the axial plane of folds). Chlorite in particular tends to develop under condition of low temperature, and as shales are subjected to intense shear stress chlorite forms by alteration of clay minerals. Each chlorite crystal is systematically oriented during its growth, and the axial-plane cleavage typical of slate appears in the rocks. Sandstones contain small quantities of clay or none at all, and under the same conditions no axial-plane cleavage can be developed; rather, the welding and intersuturing of quartz grains produce the rock identified as quartzite. Epizone metamorphism of alternating units of shale and sandstone results in alternating slates and quartzites. The slates will reveal a clearly defined axial-plane cleavage, whereas the quartzite will fail by rupture and will show fracture cleavage or a joint pattern oriented in the positions dictated by the theoretical strain ellipsoid. (See Fig. 14.1.)

Textures and Structures

Under conditions of high temperatures certain minerals such as garnet exhibit tendencies to crystallize in more or less equidimensional crystals despite the existence of strong shear stress. As a result rather distinct textures appear which characterize metamorphic rocks. Some textures show large crystals surrounded by a more finely crystalline matrix, and the aspect is not unlike that of a porphyry. In order to distinguish the metamorphic textures from igneous ones and to imply development of minerals by ionic diffusion through solid material the suffix *"blastic"* is employed. Thus the term *porphyroblastic* is used to indicate a texture resembling an igneous porphyry but without any connotation of crystallization from a melt. Textural terms such as *granoblastic* and *crystalloblastic* are used to indicate presence of crystals with grainlike and crystal-outline form respectively. Other important textural terms are defined in Table 14.2.

Table 14.2. Principal Textures of Metamorphic Rocks

Name	Definition
Granoblastic	Intergrowth of minerals whose crystal boundaries are irregular. General appearance of a granular rock.
Porphyroblastic	Large crystals surrounded by a matrix of smaller crystal size.
Xenoblastic	Poorly developed crystal outline.
Crystalloblastic	Well-developed crystal outline of recrystallized minerals.
Idioblastic	Grains of well-developed crystal outline. Frequently such grains are associated with others of poorly formed outline.

Some minerals form better crystal outlines than others and tend to be *idioblastic* (i.e., well-developed crystal outline), whereas the surrounding ones show poor crystal outline (*xenoblastic*). An order of tendency toward good crystal outline has been recognized in minerals and is useful in understanding the textures developed and in identification of mineral species. (See Table 14.3.)

Table 14.3. Development of Geometric Shape among Minerals of Metamorphic Rocks

Crystal Outline	Minerals
Well developed	Magnetite, garnet, andalusite, staurolite, and kyanite.
Moderately well developed	Epidote, amphiboles, pyroxenes, and wollastonite.
Fairly well developed	Mica, chlorite, talc, calcite, and dolomite.
Poorly developed	Feldspar and quartz.

Recognition of structures in metamorphic rocks is not a simple process of cataloguing into primary and secondary such as can be done with sedimentary rocks. Rather, there are relic structures which are inherited from the parent rock, and these frequently influence the structure of the metamorphosed product. The impress of metamorphism, however, develops a structure of its own such as schistosity. This is a structure exclusively developed in metamorphic rocks and is also diagnostic of certain environments of metamorphism. Other structures, some of which are not so clearly recognizable, are those which are only textural distinctions and involve the association of several textures to produce a structure in the same hand specimen. Thus, in some metamorphic rocks the distinction between structure and texture is not clear. For this reason the term *fabric* is used to generally refer to combinations of textures. By employing such an inclusive definition five different types of fabric, or structure, can be recognized in metamorphic rocks. These are *maculose, cataclastic, foliate, granulose,* and *gneissic.* (See Tables 14.4 and 14.5.)

Maculose

Maculose fabrics are typically developed in contact metamorphism where high temperatures have been attained within a border of the host rock as it was invaded by a large intrusion, particularly a stock. In this connection sandstones show very little change other than silicification to quartzite, but shales and limestones are altered extensively and intensively. Among shales the principal effect is the appearance of centers of crystallization of minerals such as andalusite (Al_2SiO_5), chlorite, and biotite which produce a spotted aspect to the rock. Under increasingly severe conditions the spots develop into large porphyroblasts which are defined clearly against the matrix. The matrix is baked and silicified but consists of an intimate intergrowth of chlorite, quartz, biotite, and feldspar in fine crystalloblastic texture. To the unaided eye the appearance is often homogeneous, and

Table 14.4. Structure of Metamorphic Rocks

Name of Structure	Processes Involved	General Texture	General Rock Types
Maculose	Thermal metamorphism of shaly rocks. Typical of contact metamorphism.	Well-developed porphyroblasts, textures of "knots" or "spots" of minerals such as biotite, andalusite, or graphite concentrated at random.	Spotted hornfels, or spotted slates, or hornfels.
Cataclastic	Fragmentation of rocks under shear stress. Typical of the epizone.	Irregular grains showing crushing, large crystals tend to form a porphyroblastic texture. Tendency toward orientation of grain.	Sheared quartzites and slates, crush breccia, mylonites.
Foliate, slaty, or schistose	Shear stress dominates. Temperatures range between low in slates and moderately high in schists. Typical epizone and mesozone rocks.	Well-oriented micas, chlorite, talc, and amphiboles to form laminas in parallel bands. Foliation ranges between moderate in slates and high in schists. Folia may wrap around porphyroblasts.	Slates, phyllites, schists.
Granulose	High temperature and hydrostatic stress typical of the katazone or plutonic metamorphism. In quartzites and marbles pressures and temperatures also may be individually high or low.	Equidimensional minerals prominent, micas and rod-like minerals minor. Typical texture granoblastic, some lenticular streaking and parallel banding.	Quartzites, marbles, granulites.
Gneissose	High temperatures and hydrostatic stress typical of the katazone or plutonic metamorphism.	Alternation of schistose and granulose bands and small lenses of distinctly different composition. Splitting occurs in the planes of schistosity. Micas and hornblendes are segregated into schist bands, feldspar and quartz into granulose bands.	Gneiss.

the rock is called *hornfels* appropriately. Variation in the appearance of hornfels depends upon the extent to which porphyroblasts have developed, and they may grow to large well-formed crystals. (See Fig. 14.4*a*.)

Table 14.5. Hand-Specimen Classification of Metamorphic Rocks

Structure	Texture	Principal Mineral Composition	Minor Mineral Composition	Parent Rock	Metamorphic Rock Name
Cata-clastic	Strongly sheared, crushed grains, grains drawn to lenticles.	Similar to parent rock	Similar to parent rock	Quartzite, conglomer-ate, granite	Sheared quartz-ite, sheared conglomerate, crush breccia, mylonite
	With large por-phyroblasts.	Porphyroblasts of feldspar in rock of granitic composi-tion		Granite	Augen gneiss
Maculose	Generally fine grained, silicified or baked appear-ance, porphyro-blastic.	Quartz, andalusite, sillimanite, pyrox-ene, plagioclase	Biotite, mus-covite	Shale	Hornfels
	In marbles coarsely crystalline.	Calcite, dolomite, wollastonite, am-phibole, pyroxene, garnet	Olivine, serpentine	Limestone or dolomite	Marble
	Intersutured grains of quartz.	Quartz	Muscovite, orthoclase	Sandstone	Quartzite
Foliate	Individual crystals not recognized by unaided eye.	Chlorite, muscovite, quartz	Feldspar	Shale	Slate
	Individual crystals not recognized by unaided eye, but cleavage surface has a silky sheen due to larger flakes of muscovite than in slate.	Chlorite, muscovite, quartz	Feldspar, biotite	Shale	Phyllite
	Individual mica crystals recognized and well oriented.	Muscovite, biotite, chlorite, quartz	Hornblende, feldspar, graphite, pyrite	Shale or sandy shale	Schist
	Bands of coarse calcite or dolomite crystals separating schist folia.	Muscovite, calcite, dolomite	Amphibole	Calcareous shale	Calc-schist
	Schistose bands of calcite or dolomite.	Calcite, dolomite	Amphibole	Limestone or dolomite	Schistose marble
	Individual crystals oriented and recog-nized.	Chlorite	Epidote, amphibole	Basic igneous	Chlorite schist
	Crystals oriented but not always individually recog-nized.	Hornblende, plagioclase	Biotite	Basic igneous	Amphibolite
		Talc, serpentine	Chlorite, dolomite, quartz	Ultrabasic igneous	Talc schist, soapstone
		Chlorite	Epidote	Basic igneous	Greenstone
Granulose	Individual crystals interlocked, grano-blastic.	Calcite, dolomite	None	Limestone or dolomite	Marble
	Slightly foliated.	Calcite, dolomite	Pyroxene, wollastonite	Limestone or dolomite	Schistose marble
Gneissose	Alternation of schistose and gran-ulose bands and lenses (foliation interrupted).	Quartz, plagioclase, orthoclase, biotite, hornblende	Garnet, kyanite	Sandstone, conglom-erate, granitic rocks	Gneiss

Cataclastic

Cataclastic structures in rocks result from the application of intense shear stress. In some cases simple crushing of rock and individual minerals occurs producing microbreccia and granulated structureless aggregates. Brittle rocks tend to brecciate to a far greater degree than

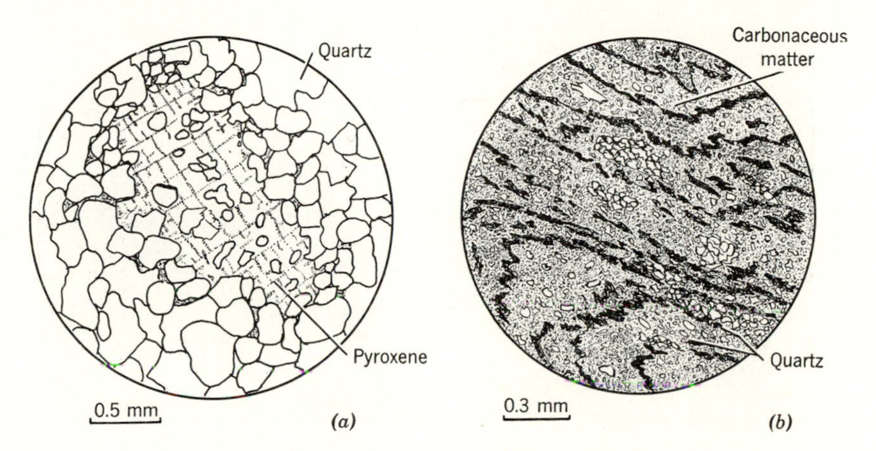

Fig. 14.4 (*a*) Sketch of thin section of hornfels, Stillwater Valley, Montana; a large porphyroblast of pyroxene has formed under the high temperatures involved. (*b*) Drawing of a thin section of mylonite, San Andreas fault zone, California, developed from a quartzite. All grains of quartz have been crushed to small dimensions, and shearing in the rock is indicated by the contorted lines of shaly streaks and carbonaceous matter originally present.

do shales. Nevertheless, cataclastic structures in the latter appear in the form of sheared slates.

Crush breccia and crush conglomerate develop through mechanical fragmentation of former sandstones and conglomerates. Pebbles in the latter are stretched and broken, and the sandstone becomes a thoroughly fractured quartzite (see Fig. 14.5). The extreme condition is called *mylonite* in which the grains are pulverized and the texture is made very finely granular. (See Fig. 14.4*b*.) Mylonites form along the sole surfaces of great overthrust sheets, and recognition of the mylonitic fabric is a satisfactory criterion for indicating the near presence of such faulting. Rocks consisting of quartz and feldspar crush without reconstitution into new minerals, but similar thrusts in basalts produce chlorite schists, and the cataclastic fabric is not so well developed. Locally, frictional heat generated during thrusting may cause fusion of mineral particles, and a glassy texture appears.

Fig. 14.5 Elongated, rolled, and crushed pebbles whose shapes are developed by intense folding. Gypsy quartzite near Metaline, Washington. (From Park and Cannon, *U. S. Geological Survey Prof. Pap.* 202.)

Foliate

Foliate structures are characterized by well-oriented flakes, or "leaves," of micas. As pressures and temperatures rise foliation in slates becomes increasingly pronounced, and the lamination produced by micaceous minerals is enhanced by the increase in size of individual crystals. The transition is from slate to *phyllite* into mica schist.[3] Phyllites and schists are typical rocks of the mesozone, and in those portions which attained the highest temperature garnets are found separating the folia of mica. (See Fig. 14.6*a*.) The tendency for minerals to increase in size continues into the katazone, and foliates become gradational with coarsely crystalline rocks of gneissic fabric. Foliates also merge with rocks of cataclastic structure, particularly brittle quartzites as zones are encountered which are typical of low temperature and intense shear.

[3] Phyllite is somewhat higher than slate in metamorphic grade and contains some visible mica crystals. Often the chief identifying feature is a silky sheen which is produced by mica crystals on the cleavage surfaces, otherwise the outward appearance is that of slate,

Granulose

The fabric described as *granulose* is developed principally in rocks which are characterized by granoblastic textures. Individual crystals tend to be irregular in outline and well intersutured with their neighbors. This aspect of xenoblastic shape imparts an appearance of grains as in a sedimentary rock. "Grains" appear to be generally

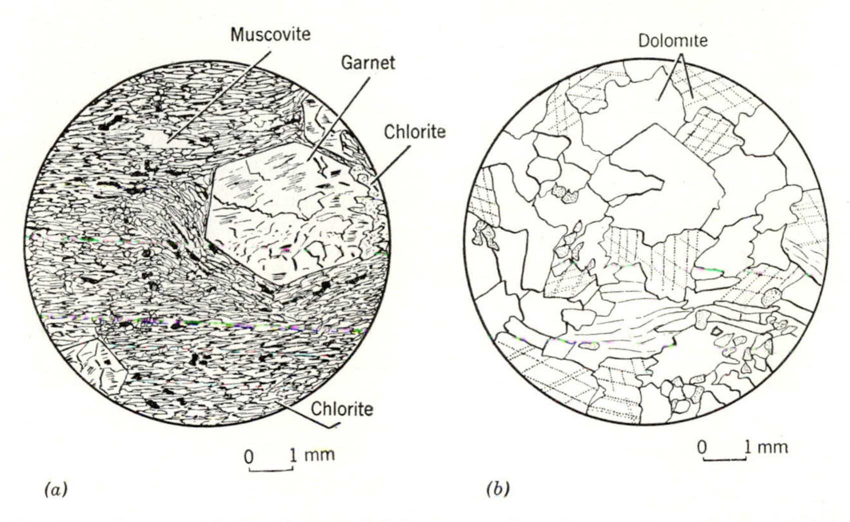

Fig. 14.6 Diagram of thin section of (*a*) garnet schist, Zermatt, Switzerland, and (*b*) granulose texture in marble, Randville, Michigan.

uniform in size, and there is no tendency toward cleavage or schistosity. Quartzite, marble, and certain nonfoliated quartz-feldspar rocks belong in this structural category. (See Fig. 14.6*b*.) The latter are called *granulites* and are distinguished from gneiss on the basis of the absence of bands of mica which the gneisses display. Granulites are considered to develop under conditions of exceptionally high temperature and hydrostatic stress and are not nearly so abundant as gneiss.

Gneissose

As the name implies *gneissose* fabric is characterized by banded gneiss. The structure consists of alternations of schistose and granulose bands and lenses of dissimilar mineral composition. Generally, the schistose bands are layers of biotite arranged in parallel orientation. These are separated or interrupted by zones, lenses, or layers of quartz and feldspar of granoblastic texture. (See Fig. 14.2*a*.) The schistose layers impart a cleavage to the rock, and fragments will break with rudely plane surfaces within the zones of concentrated micas. Fundamentally, the aspect of gneissose structure is to be attributed to the parallel orientation of the minerals in the schistose layers. As this fabric becomes less and less well defined gneissose structure becomes gradational with that described as granulose. Also, with increase in the amount of schistose bands the structure merges into rocks identified as schists.

Gneisses are extremely common rock types, and extensive belts are to be found in the axial positions of the geosynclines. Also, they are variable in mineralogy, and complex varieties are the rule. Where they are gradational with plutons banding in the gneiss is accentuated by the presence of small lenses of quartz-feldspar rock of granitelike aspect. The result is to produce an appearance of a hybrid rock being partly granite and partly schist or gneiss. This is the typical fabric of the migmatite, and the rock may be called a granite gneiss or gneissic granite depending upon which is the dominating type.

INTERPRETATION OF METAMORPHIC ROCKS

Structure

From the viewpoint of the engineer the importance of metamorphic rocks is primarily in connection with their contained ores, their variable strength, and their use in recognition of geologic structure. With this viewpoint in mind the engineer must be quick to recognize axial plane positions on the basis of slaty cleavage and schistosity. Also, the presence of schistosity must automatically indicate very pronounced weakness in the tensile strength of rocks. Orientation of the schistosity becomes a very important item with reference to conditions of quarrying, tunneling, stability of large road cuts, and directional strength of foundations. Each cleavage plane is a potential position of failure under stress, and when slates and schists are loaded differentially the tendency is great to slide along the cleavage. In contrast are rocks of enormous strength. Quartzite, marble, and hornfels by virtue of their crystalloblastic or granoblastic fabric and absence of cleavage display tensile strength uniform in all directions. Except for jointing which may cut such rocks and thus establish planes of weakness they are to be considered extremely desirable foundation materials. Thus, foliate and gneissose structures are known indicators of strong distinction in tensile strength in different directions, whereas maculose and granulose fabrics imply the rocks have uniformity of strength. For rocks of cataclastic structures strength is dependent entirely upon the degree of crushing and the extent to which intergranular welding has occurred. Some highly crushed and stretched pebble conglomerates have been thoroughly welded during the recrystallization following crushing. Such rocks are strong, and the stretched character of the pebbles is not necessarily an indicator of planes or directions of rock weakness. In contrast, certain brecciated zones associated with overthrusting are localities of rocks of virtually no tensile strength. Recognition of the nature of cataclastic fabric, therefore, is extremely important in prediction of the expected behavior of the rock under stress.

Distribution

Examination of outcrops in the field and large specimens in the laboratory has enabled the geologist to arrive at the following generalizations:

1. Rocks showing fabrics which are foliate, gneissose, or granulose are known to be widespread and distributed in uniform belts which may be miles in width and length. Selected specimens taken from geographically identified localities can be arranged on the basis of fabric and mineralogy in a crude order of metamorphic grade. Thus belts of slates, schists, marbles, quartzites, gneisses, and others can be roughly outlined on maps. Moreover, a first approximation can be made regarding the direction of expected increase or decrease in metamorphic grade, and the expected nature of rocks in the adjoining areas.

2. Recognition of textures and rocks described as hornfels provides the basis for an entirely different interpretation of the rock character of an area. In this case metamorphism is indicated as contact, and the presence of an intrusive "plug" of some considerable size is to be expected. Moreover, the width of the contact zone can be anticipated as less than a mile beyond which unmetamorphosed sediments are to be encountered. This interpretation has provided the basis for expecting rapid local changes in rock type, an important body of igneous rock nearby, and absence of any belts of metamorphic rocks.

3. Contact metamorphism is a frequent indicator of the presence of ore deposits, or conversely presence of ore deposits associated with a granitic stock or similar intrusion should provide the basis to expect a contact zone. The association is not to point out the probable existence of a belt of hornfelsic rocks but rather a frequent condition of rocks in ore districts. The rock condition to be expected is intense

alteration of the igneous rocks to clays. Typical igneous minerals have been transformed into clay minerals by the release of large volumes of hot waters which carried the ores from the cooling silicic magma. Hydrothermal alteration of the igneous mass is deep seated and must not be confused with surface weathering. Thus, the crumbling clayey quality of the altered igneous rock will not change with depth, and the undesirable nature of foundations can be predicted in advance of additional explorations.

Significance of Change in Rock Type

Routine drilling in an area of sediments may reveal the presence of metamorphic rock in sharp contact with unmetamorphosed beds (see Fig. 13.17*d*). Recognition of this condition provides the basis for prediction of a fault or unconformity. Representation of the geologic structure based upon an interpretation of the existence of a fault differs radically from a representation based upon an interpretation of an unconformity. The ability to recognize the possibility of such distinctly different interpretations provides the basis for directing drilling operations to provide the information necessary to make the correct interpretation. There have been occasions when the amount of metamorphic rock cut by the initial drill hole has been of the order of magnitude of less than a foot, and recognition of the metamorphic character of the rock has been of utmost importance.

Localities of rocks displaying regional or dynamic metamorphism often are very extensive as for example much of New England. Engineering structures within such a region are perforce erected upon metamorphosed rocks, and understanding of local differences in grade and rock type frequently is important in selection of alternative sites for construction. Here, the engineer is dependent upon the geologist to produce the proper interpretation, but the engineer must be sufficiently versed in the significance of the geology in order to offset the natural difficulties by engineering design.

Soils and Sediments

Lastly, under the conditions imposed by weathering and erosion soils are developed upon metamorphic rocks. Generally considered, the nature of the soils will be responses in part to the prevailing climatic conditions. But the influence of the parent rock on soil type is extremely important. For example, marbles will produce loamy soils similar to those developed on limestones, and slates tend to form clay soils but not exactly like those of shales. Quartzites resist soil formation, and only alluvial soils are to be expected to blanket such rocks. Schists provide a condition of marked contrast with quartzites, and in some cases exceptionally thick soils (as much as 50 to 75 feet) are developed in the same locality as the soil-free quartzite areas. Gneisses are variable soil producers. Those gneisses which contain much mica and feldspar usually yield a well-developed soil profile; others tend to resist weathering but never to the extent of slates and quartzites.

Glacial tills found in areas of noncarbonate metamorphic rocks generally lack the large quantity of clay matrix which is so characteristic of the tills from the limestone and shale bedrocks of Illinois, Iowa, Indiana, and Michigan. Till developed from gneiss is much more sandy than that from slate and schist, and the engineering properties vary accordingly. Much the same can be said for properties of gravels in the outwash deposits. Gravels of schist and slate are extremely undesirable engineering materials, those of gneiss are somewhat better but not to be compared with gravels of marble or quartzite.

The preceding examples of predicable soil and sediment conditions are representative of many other possible ones using the rapid techniques introduced in Chapter 2 (Soil Materials). Approximations of expected conditions can be made upon the basis of relatively meager information, but the accuracy of the approximation depends almost entirely upon the understanding of the interpreter. In turn the skill of the interpreter is reckoned to be dependent upon the extent of his geologic knowledge.

Such was the thesis which we proceeded to investigate in the first chapters of the book. Thus, in a figurative manner the reader has been returned to the subject of initial consideration. Each chapter has revealed a branch of geologic specialization in which much remains to be investigated. Yet, as the field of specialization is extended the student becomes increasingly aware of the interdependence of these arbitrarily bounded limits of interest. With this concept in mind the writer has moved through a series of chapters in which there is no firmly established order of procedure. Outwardly, the subject of metamorphism appears to demand the greatest amount of geologic background. Generally considered this appears to be true, but the intent has been directed principally to the creation of a desire on the part of the reader to seek further into some aspect of what has been presented herein as a sequence of individual chapters.

Selected Supplementary Readings

Emmons, W. H., Thiel, G. A., Stauffer, C. R., and Allison, I. S., *Geology Principles and Processes,* 4th ed., McGraw-Hill Book Co., 1955, Chapter 21. An elementary description of the processes of metamorphism.

Grout, F. F., *Kemp's Handbook of Rocks,* 6th ed., D. Van Nostrand Co., 1952, Chapters IX, X, XI, and XII. Concerns the processes of rock metamorphism and a description of the common rocks produced.

Pirsson, L. V., and Knopf, Adolph, *Rocks and Rock Minerals,* 3rd ed., John Wiley and Sons, 1947, Chapters XII and XIII. Describes and classifies metamorphic rocks.

Spock, L. E., *Guide to the Study of Rocks,* Harper and Bros., 1953, Chapter 9. Summarizes metamorphic processes and describes metamorphic rocks.

Appendix

Tables for the determination of common minerals*

GENERAL CLASSIFICATION OF THE TABLES

Basis of Table Organization

In the following tables the general organization consists of subdividing all minerals on the basis of luster. This can be done by recognition of two fundamental luster types namely *metallic* or *submetallic*, and *nonmetallic*. Minerals with metallic luster often display a well-defined streak, whereas among minerals of nonmetallic luster the streak is generally colorless. Streak is used, therefore, as a means of classification principally among the minerals with recognized metallic or submetallic luster.

Hardness is an extremely important property as a means of recognition of minerals. With few exceptions the hardness of individual mineral species remains surprisingly uniform despite the change in

* These tables are much condensed and slightly modified from C. S. Hurlbut, *Dana's Manual of Mineralogy,* 16th ed., John Wiley and Sons, 1952. The reader is referred to the text for introductory mineralogy.

habit. For this reason it is common practice to arrange mineral identification tables on the basis of hardness.

Cleavage, specific gravity, and color are used to subdivide mineral species within the large groups selected on the basis of luster and hardness. In this connection some minerals display characterizing cleavage, weight, or color which are important in their recognition.

Special properties such as magnetism are used very effectively as a means of identification and oftentimes are sufficiently diagnostic to permit recognition of the mineral without further testing. A case in point is the elasticity displayed by cleavage laminas of mica. The student of minerals should become familiar with such diagnostic properties but not at the expense of a systematic procedure of mineral identification. For this reason the following table of organization is presented.

Organization of Tables

Luster—Metallic or Submetallic

Hardness

- I. <2½ (will leave mark on paper), p. 585.
- II. >2½, <5½ (can be scratched by knife; will not readily leave mark on paper), p. 585.
- III. >5½ (cannot be scratched by knife), p. 586.

Luster—Nonmetallic

Streak

- I. Definitely colored, p. 587.
- II. Colorless—Hardness.
  - *A*. <2½ (can be scratched by fingernail), p. 587.
  - *B*. >2½ to 5 (cannot be scratched by fingernail; can be scratched by cent).
    1. Cleavage prominent, p. 588.
    2. Cleavage not prominent.
       Splinter infusible in candle flame, p. 589.
  - *C*. >3, <5½ (cannot be scratched by cent; can be scratched by knife).
    1. Cleavage prominent, p. 589.
    2. Cleavage not prominent, p. 590.
  - *D*. >5½, <7 (cannot be scratched by knife; can be scratched by quartz).
    1. Cleavage prominent, p. 591.
    2. Cleavage not prominent, p. 592.
  - *E*. >7 (cannot be scratched by quartz).
    1. Cleavage prominent, p. 593.
    2. Cleavage not prominent, p. 593.

Luster: Metallic or Submetallic

I. Hardness: <$2\frac{1}{2}$. (Will leave a mark on paper.)

Streak	Color	G	H	Remarks	Name, Composition, Crystal System, Group
Black	Iron black	4.7	1–2	Usually splintery or in radiating fibrous aggregates.	PYROLUSITE MnO_2 Tetragonal Sedimentary (ore)
	Steel gray to iron black	2.3	$1–1\frac{1}{2}$	Cleavage perfect basal. May be in hexagonal-shaped plates. Greasy feel.	GRAPHITE C Hexagonal Metamorphic
Gray black	Blue black to lead gray	7.6	$2\frac{1}{2}$	Cleavage perfect cubic. In cubic crystals or massive granular.	GALENA PbS Isometric (Cubic) Ore
Red brown	Red to vermilion	5.2	1+	Earthy. Frequently as pigment in rocks. Crystalline hematite is harder.	HEMATITE Fe_2O_3 Hexagonal (Rhombohedral) Sedimentary (ore)
Yellow brown	Yellow brown	3.6 to 4.0	1+	Earthy. Limonite is usually harder.	LIMONITE $FeO(OH)\cdot nH_2O + Fe_2O_3\cdot nH_2O$ Amorphous Sedimentary (ore)

II. Hardness: >$2\frac{1}{2}$, <$5\frac{1}{2}$.

(Can be scratched by a knife; will not readily leave a mark on paper.)

Streak	Color	G	H	Remarks	Name, Composition, Crystal System, Group
Black. May leave a mark on paper	Iron black	4.7	1–2	Usually splintery or in radiating fibrous aggregates.	PYROLUSITE MnO_2 Tetragonal Sedimentary (ore)
Gray black. Will mark paper	Lead gray	7.5	$2\frac{1}{2}$	Cleavage perfect cubic. In cubic crystals and granular masses. If a fragment is held in the candle flame, it does not fuse but is slowly reduced and small globules of metallic lead collect on the surface.	GALENA PbS Isometric (Cubic) Ore
Black	Brass yellow	4.1 to 4.3	$3\frac{1}{2}–4$	Usually massive, but may be in crystals resembling tetrahedrons. Associated with other copper minerals and pyrite.	CHALCOPYRITE $CuFeS_2$ Tetragonal Ore
Light to dark brown	Dark brown to coal black. More rarely yellow or red	3.9 to 4.1	$3\frac{1}{2}–4$	Cleavage perfect. Usually cleavable granular; may be in tetrahedral crystals. The darker the specimen the higher the percentage of iron. The streak is always of a lighter color than the specimen.	SPHALERITE ZnS Isometric (Cubic) Ore

Luster: Metallic or Submetallic (*Continued*)

Streak	Color	G	H	Remarks	Name, Composition, Crystal System, Group
Red brown to Indian red	Dark brown to steel gray to black	4.8 to 5.3	$5\frac{1}{2}$–$6\frac{1}{2}$	Usually harder than knife. Massive, radiating, reniform, micaceous.	HEMATITE Fe_2O_3 Hexagonal (Rhombohedral) Sedimentary (ore)
Yellow brown. Yellow ocher	Dark brown to black	3.6 to 4.0	5–$5\frac{1}{2}$	Vitreous luster. Usually contains about 15 per cent water, whereas goethite contains 10 per cent.	LIMONITE $FeO(OH){\cdot}nH_2O + Fe_2O_3{\cdot}nH_2O$ Amorphous Sedimentary (ore)
Yellow brown. Yellow ocher	Dark brown to black	4.37	5–$5\frac{1}{2}$	Cleavage fair to good. In radiating fibers, mammillary and stalactitic forms. Rarely in crystals. Definitely distinguished from limonite by presence of cleavage or crystal form.	GOETHITE $FeO(OH)$ Orthorhombic Sedimentary (ore)
				III. Hardness: $>5\frac{1}{2}$. (Cannot be scratched by a knife.)	
	Pale brass-yellow	5.0	6–$6\frac{1}{2}$	Often in pyritohedrons or striated cubes. Massive granular. Most common sulfide.	PYRITE FeS_2 Isometric (Cubic) In all rocks
	Pale yellow to almost white	4.9	6–$6\frac{1}{2}$	Frequently in "cock's comb" crystal groups and radiating fibrous masses.	MARCASITE FeS_2 Orthorhombic In all rocks
	Black	5.18	6	Strongly magnetic. Crystals octahedral. Many show octahedral parting.	MAGNETITE Fe_3O_4 Isometric (Cubic) Chiefly metamorphic (ore)
Red brown. Indian red	Dark brown to steel gray to black	4.8 to 5.3	$5\frac{1}{2}$–$6\frac{1}{2}$	Radiating, reniform, massive, micaceous (specularite). Rarely in steel-black rhombohedral crystals. Some varieties softer than 3.	HEMATITE Fe_2O_3 Hexagonal (Rhombohedral) Metamorphic when well crystallized (ore)
Yellow brown to yellow ocher	Dark brown to black	3.6 to 4.0	5–$5\frac{1}{2}$	Vitreous luster. Usually contains about 15 per cent water, whereas goethite contains 10 per cent.	LIMONITE $FeO(OH){\cdot}nH_2O + Fe_2O_3{\cdot}nH_2O$ Amorphous Sedimentary (ore)
Yellow brown to yellow ocher	Dark brown to black	4.37	5–$5\frac{1}{2}$	Cleavage. Radiating, colloform, stalactitic. Distinguished from limonite by presence of cleavage or crystal form.	GOETHITE $FeO(OH)$ Orthorhombic Sedimentary (ore)

Luster: Nonmetallic

I. Streak definitely colored

Streak	Color	G	H	Remarks	Name, Composition, Crystal System, Group
Red brown. Indian brown	Dark brown to steel gray to black	4.8 to 5.3	$5\frac{1}{2}$–$6\frac{1}{2}$	Radiating, reniform, massive, micaceous. Rarely in steel-black rhombohedral crystals. Some varieties softer.	HEMATITE Fe_2O_3 Hexagonal (Rhombohedral) Metamorphic when well crystallized (ore)
Yellow brown to yellow ocher	Dark brown to black	3.6 to 4.0	5–$5\frac{1}{2}$	Usually hard, with vitreous luster. Usually contains about 15 per cent water, whereas goethite contains 10 per cent.	LIMONITE $FeO(OH)\cdot nH_2O + Fe_2O_3\cdot nH_2O$ Amorphous Sedimentary (ore)
Yellow brown to yellow ocher	Dark brown to black	4.4	5–$5\frac{1}{2}$	Cleavage fair to good. In radiating fibers, mammillary and stalactitic forms. Rarely in crystals. Definitely distinguished from limonite only by presence of cleavage or crystal form. Usually metallic.	GOETHITE $FeO(OH)$ Orthorhombic Sedimentary (ore)
Brown	Light to dark brown	3.83 to 3.88	$3\frac{1}{2}$–4	In cleavable masses or in small curved rhombohedral crystals. Becomes magnetic after heating in candle flame.	SIDERITE $FeCO_3$ Hexagonal (Rhombohedral) Sedimentary
Light brown	Light to dark brown	3.9 to 4.1	$3\frac{1}{2}$–4	Cleavage perfect. Usually cleavable granular; may be in tetrahedral crystals. The darker the specimen the higher the percentage of iron. The streak is always of a lighter color than the specimen.	SPHALERITE ZnS Isometric (Cubic) Ore
Light green	Bright green	3.9 to 4.03	$3\frac{1}{2}$–4	Radiating fibrous, mammillary. Associated with azurite and may alter to it. Effervesces in cold acid.	MALACHITE $Cu_2CO_3(OH)_2$ Monoclinic Sedimentary (ore)

II. Streak colorless

A. Hardness: $<2\frac{1}{2}$. (Can be scratched by fingernail.)

Cleavage, Fracture	Color	G	H	Remarks	Name, Composition, Crystal System, Group
Perfect cleavage in one direction	Pale brown, green, yellow, white	2.76 to 3.0	2–$2\frac{1}{2}$	In foliated masses and scales. Crystals tabular with hexagonal or diamond-shaped outline. Cleavage flakes elastic. Common mica.	MUSCOVITE $KAl_3Si_3O_{10}(OH)_2$ Monoclinic Igneous and metamorphic
	Usually dark brown, green to black; may be yellow	2.95 to 3	$2\frac{1}{2}$–3	Usually in irregular foliated masses. Crystals have hexagonal outline but rare. Cleavage flakes elastic. Dark mica.	BIOTITE $K(Mg,Fe)_3AlSi_3O_{10}(OH)_2$ Monoclinic Igneous and metamorphic

Luster: Nonmetallic (*Continued*)

Cleavage, Fracture	Color	G	H	Remarks	Name, Composition, Crystal System, Group
	Green of various shades	2.6 to 2.9	2–2½	Usually in irregular foliated masses. May be in compact masses of minute scales. Thin sheets flexible but not elastic.	CHLORITE $(Mg,Fe)_5(Al,Fe''')_2Si_3O_{10}(OH)_8$ Monoclinic Chiefly metamorphic
Perfect cleavage in one direction	White, apple green, gray. When impure, as in soapstone, dark gray, dark green to almost black	2.7 to 2.8	1	Greasy feel. Frequently distinctly foliated or micaceous. Cannot be positively identified by physical tests.	TALC $Mg_3Si_4O_{10}(OH)_2$ Monoclinic Metamorphic
Perfect good	Colorless, white, gray. May be colored by impurities	2.32	2	Occurs in crystals, broad cleavage flakes. May be compact massive without cleavage or fibrous with silky luster.	GYPSUM $CaSO_4 \cdot 2H_2O$ Monoclinic Sedimentary
Perfect, seldom seen, earthy fracture	White; may be darker	2.6 to 2.63	2–2½	Generally claylike and compact. When breathed upon gives argillaceous odor. Will adhere to the dry tongue. Does not swell on wetting.	KAOLINITE $Al_4Si_4O_{10}(OH)_8$ Monoclinic Sedimentary
Uneven fracture	Yellow brown, gray white	2.0 to 2.55	1–3	In rounded grains also pisolitic, often earthy and claylike. Usually harder than 2½.	BAUXITE $Al_2O_3 \cdot nH_2O$ Amorphous Sedimentary

B. Hardness: >2½, <5. (Cannot be scratched by fingernail; can be scratched by cent.)

1. Cleavage prominent

Cleavage, Fracture	Color	G	H	Remarks	Name, Composition, Crystal System, Group
Cubic	Colorless white, red, blue	2.1 to 2.3	2½	Common salt, soluble in water, taste salty, fusible in candle flame. In granular cleavable masses or in cubic crystals.	HALITE NaCl Isometric (Cubic) Sedimentary
	Colorless, white, blue, gray, red	2.89 to 2.98	3–3½	Commonly in massive fine aggregates, not showing cleavage; then can be distinguished only by chemical tests.	ANHYDRITE $CaSO_4$ Orthorhombic Sedimentary
Cleavage in three directions not at right angles. Rhombohedral	Colorless, white, and variously tinted	2.72	3	Effervesces in cold acid. Crystals show many forms. Occurs in large masses as limestone and marble. Clear varieties show strong double refraction.	CALCITE $CaCO_3$ Hexagonal (Rhombohedral) Sedimentary and metamorphic

Luster: Nonmetallic (*Continued*)

Color	G	H	Remarks	Name, Composition, Crystal System, Group
Colorless, white, pink	2.85	$3\frac{1}{2}$–4	Usually harder than copper coin. Often in curved rhombohedral crystals with pearly luster. In coarse masses as dolomitic limestone and marble. Powdered mineral will effervesce in cold acid.	DOLOMITE $CaMg(CO_3)_2$ Hexagonal (Rhombohedral) Sedimentary and metamorphic
			2. Cleavage not prominent Infusible in candle flame	
Colorless or white	2.6 to 2.63	2–$2\frac{1}{2}$	Usually compact earthy. When breathed upon gives an argillaceous odor. Does not swell on wetting.	KAOLINITE $Al_2Si_2O_5(OH)_4$ Monoclinic Sedimentary
Colorless, white, blue, gray, red	2.89 to 2.98	3–$3\frac{1}{2}$	Commonly in massive fine aggregates not showing cleavage, and can be distinguished only by chemical tests.	ANHYDRITE $CaSO_4$ Orthorhombic Sedimentary
Yellow, brown, gray white	2.0 to 2.55	1–3	Usually pisolitic; in rounded grains and earthy masses. Often impure.	BAUXITE A mixture of aluminum hydroxides Sedimentary (ore)
Olive to blackish green, yellow green, white	2.2	2–5	Massive. Fibrous in the asbestos variety, chrysotile. Frequently mottled green in the massive variety.	SERPENTINE $Mg_3Si_2O_5(OH)_4$ Monoclinic Metamorphic

II. Streak colorless

C. Hardness: >3, $<5\frac{1}{2}$. (Cannot be scratched by cent; can be scratched by knife.)

1 Cleavage prominent

Cleavage	Color	G	H	Remarks	Name, Composition, Crystal System, Group
Two cleavage directions Prismatic at angles of 55° & 125°	White, green, black	3.0 to 3.3	5–6	Crystals usually slender, fibrous, asbestiform. Tremolite (white, gray, violet), actinolite (green) common in metamorphic rocks. Hornblende (dark green to black) common in igneous and metamorphic rocks. Cleavage angle characteristic.	AMPHIBOLE GROUP Essentially calcium magnesium silicates (see p. 79) Monoclinic Igneous and metamorphic
Two cleavage directions Prismatic at 90° angles	White, green, black	3.1 to 3.5	5–6	In stout prisms with rectangular cross section. Often in granular crystalline masses. Diopside (colorless, white, green), augite (dark green to black) are rock-forming minerals. Characterized by rectangular cross section and cleavage.	PYROXENE GROUP Essentially calcium magnesium silicates (see p. 77) Monoclinic Igneous and metamorphic

Luster: Nonmetallic (*Continued*)

Cleavage, Fracture	Color	G	H	Remarks	Name, Composition, Crystal System, Group
Three cleavage directions	Colorless, white and	2.72	3	Effervesces in cold acid. Crystals show many forms.	CALCITE $CaCO_3$
Three directions not at right angles. Rhombohedral	variously tinted			Occurs in large masses as limestone and marble. Clear varieties show strong double refraction.	Hexagonal (Rhombohedral) Sedimentary and metamorphic
	Colorless, white, pink	2.85	$3\frac{1}{2}$–4	Often in curved rhombohedral crystals with pearly luster. In coarse masses as dolomitic limestone and marble. Powdered mineral will effervesce in cold acid.	DOLOMITE $CaMg(CO_3)_2$ Hexagonal (Rhombohedral) Sedimentary and metamorphic
	Light to dark brown	3.83 to 3.88	$3\frac{1}{2}$–4	In cleavable masses, or in small curved rhombohedral crystals. Becomes magnetic after heating in the candle flame.	SIDERITE $FeCO_3$ Hexagonal (Rhombohedral) Sedimentary
Three cleavage directions	Colorless, white, blue, gray, red	2.89 to 2.98	3–$3\frac{1}{2}$	Commonly in massive fine aggregates not showing cleavage, and can be distinguished only by chemical tests.	ANHYDRITE $CaSO_4$ Orthorhombic Sedimentary
Four cleavage directions	Colorless, violet, green, yellow, pink. Usually has a fine color	3.18	4	In cubic crystals often in penetration twins. Characterized by excellent cleavage.	FLUORITE CaF_2 Sedimentary and igneous
Six cleavage directions	Yellow, brown, white	3.9 to 4.1	$3\frac{1}{2}$–4	Luster resinous. Small tetrahedral crystals rare. Usually in cleavable masses. If massive difficult to determine.	SPHALERITE ZnS Isometric (Cubic) Ore

2. Cleavage not prominent

Color	G	H	Remarks	Name, Composition, Crystal System, Group
Colorless, white, yellow, red, brown	2.72	3	May be fibrous or fine granular, banded Effervesces in cold hydrochloric acid. Mexican onyx variety of calcite.	CALCITE $CaCO_3$ Hexagonal (Rhombohedral) Sedimentary and metamorphic
Light to dark brown	3.83 to 3.88	$3\frac{1}{2}$–4	Usually cleavable but may be in compact concretions in clay or shale—clay ironstone variety. Becomes magnetic on heating.	SIDERITE $FeCO_3$ Hexagonal (Rhombohedral) Sedimentary

Luster: Nonmetallic *(Continued)*

Color	G	H	Remarks	Name, Composition, Crystal System, Group
Yellow. brown, gray, white	2.0 to 2.55	1–3	Usually pisolitic; in rounded grains and earthy masses. Often impure.	BAUXITE A mixture of aluminum hydroxides Sedimentary (ore)
Olive to blackish green, yellow green, white	2.2	2–5	Massive. Fibrous in the asbestos variety, chrysotile. Frequently mottled green in the massive variety.	SERPENTINE $Mg_3Si_2O_5(OH)_4$ Monoclinic Metamorphic

D. Hardness: $>5\frac{1}{2}$, <7. (Cannot be scratched by knife; can be scratched by quartz.)

1. Cleavage prominent

Cleavage	Color	G	H	Remarks	Name, Composition, Crystal System, Group
One cleavage direction, perfect	Hair brown, grayish green	3.23	6–7	Commonly in long, slender, prismatic crystals. May be in parallel groups—columnar or fibrous. Found in schistose rocks.	SILLIMANITE Al_2SiO_5 Orthorhombic Metamorphic
	Yellowish to blackish green	3.35 to 3.45	6–7	In prismatic crystals striated parallel to length. Found in metamorphic rocks and crystalline limestones.	EPIDOTE $Ca_2(Al,Fe)_3(SiO_4)_3(OH)$ Monoclinic Chiefly metamorphic
	Blue, usually darker at center of crystal. May be gray or green	3.56 to 3.66	5–7	In bladed aggregates with cleavage parallel to length. Can be scratched by knife parallel to length of crystal but not in a direction at right angles to length. Found in schists.	KYANITE Al_2SiO_5 Triclinic Metamorphic
Two cleavage directions, good	Colorless, white, gray	2.8 to 2.9	$5–5\frac{1}{2}$	Usually cleavable massive to fibrous. Also compact. Associated with highly metamorphic limestone.	WOLLASTONITE $CaSiO_3$ Triclinic Metamorphic
Two cleavage directions at or nearly 90° angles	Colorless, white, gray, cream, red, green	2.54 to 2.50	6	In cleavable masses or in irregular grains as rock constituents. May be in crystals in pegmatite. Distinguished with certainty only under polarizing microscope. Green amazonstone is microcline.	ORTHOCLASE (Monoclinic) MICROCLINE (Triclinic) $KAlSi_3O_8$ Chiefly igneous and metamorphic
	Colorless, white, gray, bluish. Often shows a beautiful play of colors	2.62 (albite) to 2.76 (anorthite)	6	In cleavable masses or in irregular grains as a rock constituent. On the better cleavage can be seen a series of fine parallel striations due to albite twinning; these distinguish it from orthoclase.	PLAGIOCLASE Various combinations of albite, $NaAlSi_3O_8$, and anorthite, $CaAl_2Si_2O_8$ Chiefly igneous and metamorphic

Luster: Nonmetallic (*Continued*)

Cleavage	Color	G	H	Remarks	Name, Composition, Crystal System, Group
	White, green, black	3.1 to 3.5	5–6	In stout prisms with rectangular cross section. Often in granular crystalline masses. Diopside (colorless, white, green), augite (dark green to black) are rock-forming minerals. Characterized by rectangular cross section and cleavage.	PYROXENE GROUP Essentially calcium magnesium silicates Monoclinic Igneous and metamorphic
At 55° and 125° angles	White, green, black	3.0 to 3.3	5–6	Crystals usually slender, fibrous asbestiform. Tremolite (white, gray, violet) and actinolite (green) are common in metamorphic rocks. Hornblende (dark green to black) is common in igneous rocks. The group is characterized by its broad cleavage angle.	AMPHIBOLE GROUP Essentially calcium magnesium silicates (see p. 79) Monoclinic Igneous and metamorphic

2. Cleavage not prominent

Color	G	H	Remarks	Name, Composition, Crystal System, Group
Colorless, white, smoky, amethyst. Variously colored when impure	2.65	7	Crystals usually show horizontally striated prism with pyramid.	QUARTZ SiO_2 Rhombohedral Igneous, sedimentary, and metamorphic
Light brown, yellow, red, green	2.65	7	Luster waxy to dull. Commonly colloform. May be banded or lining cavities. Also as nodules of chert.	CHALCEDONY or CHERT SiO_2 Cryptocrystalline quartz Chiefly sedimentary
Olive to grayish green, brown	3.27 to 3.37	$6\frac{1}{2}$–7	Usually in disseminated grains in basic igneous rocks. May be massive granular.	OLIVINE $(Mg,Fe)_2SiO_4$ Orthorhombic Igneous
Black, green, brown, blue, red, pink, white	3.0 to 3.25	7–$7\frac{1}{2}$	In slender prismatic crystals with triangular cross section. Crystals may be in radiating groups. Found usually in pegmatites. Black most common, other colors associated with lithium minerals.	TOURMALINE $WX_3B_3Al_3(AlSiO_9)_3(O,OH,F)_4$ W = Na,Ca X = Al,Fe,Li,Mg Hexagonal (Rhombohedral) Igneous and metamorphic
Red brown to brownish black	3.65 to 3.75	7–$7\frac{1}{2}$	In prismatic crystals; commonly in cruciform penetration twins. Frequently altered on the surface and then soft. Found in schists.	STAUROLITE $(Fe''OH)Al_4(AlSi_2)O_{12}$ Orthorhombic Metamorphic

Luster: Nonmetallic (*Continued*)

Color	G	H	Remarks	Name, Composition, Crystal System, Group
Reddish brown, flesh red, olive green	3.16 to 3.20	$7\frac{1}{2}$	Prismatic crystals with nearly square cross section. Cross section may show black cross (chiastolite). May be altered to mica and then soft. Found in schists.	ANDALUSITE Al_2SiO_5 Orthorhombic Metamorphic

E. Hardness: >7. (Cannot be scratched by quartz.)

1. Cleavage prominent

Cleavage	Color	G	H	Remarks	Name, Composition, Crystal System, Group
One cleavage direction	Brown, gray, greenish gray	3.23	6–7	Commonly in long slender prismatic crystals. May be in parallel groups, columnar or fibrous. Found in schistose rocks.	SILLIMANITE Al_2SiO_5 Orthorhombic Metamorphic

2. Cleavage not prominent

Color	G	H	Remarks	Name, Composition, Crystal System, Group
Colorless, white, smoky, amethyst. Variously colored when impure	2.65	7	Crystals usually show horizontally striated prism with pyramid.	QUARTZ SiO_2 Hexagonal (Rhombohedral) Igneous, sedimentary, metamorphic
Green, brown, blue, red, pink, white, black	3.0 to 3.25	$7–7\frac{1}{2}$	In slender prismatic crystals with triangular cross section. Crystals may be in radiating groups. Found usually in pegmatites. Black most common, other colors associated with lithium minerals.	TOURMALINE $WX_3B_3Al_3(AlSi_2O_9)_3(O,OH,F)_4$ W = Na,Ca X = Al,Fe,Li,Mg, Hexagonal (Rhombohedral) Igneous and metamorphic
Olive to grayish green, brown	3.27 to 3.37	$6\frac{1}{2}–7$	Usually in disseminated grains in mafic or ultramafic igneous rocks. May be massive granular.	OLIVINE $(Mg,Fe)_2SiO_4$ Orthorhombic Igneous
Reddish brown, flesh red, olive green	3.16 to 3.20	$7\frac{1}{2}$	Prismatic crystals with nearly square cross section. Cross section may show black cross (chiastolite). May be altered to mica and then soft. Found in schists.	ANDALUSITE Al_2SiO_5 Orthorhombic Metamorphic
Red brown to brownish black	3.65 to 3.75	$7–7\frac{1}{2}$	In prismatic crystals; commonly in cruciform penetration twins. Frequently altered on the surface and then soft. Found in schists.	STAUROLITE $Fe''Al_5Si_2O_{12}(OH)$ Orthorhombic Metamorphic

Luster: Nonmetallic (*Continued*)

Color	G	H	Remarks	Name, Composition, Crystal System, Group
Usually brown to red. Also yellow, green, pink	3.5 to 4.3	$6\frac{1}{2}$–$7\frac{1}{2}$	A minor mineral in igneous rocks and pegmatites. Commonly in metamorphic rocks. Also occurs as a concentrate in some sand.	GARNET $R_3''R_2'''(SiO_4)_3$ R'' = Ca,Mg,Fe,Mn R''' = Al,Fe,Ti,Cr Isometric (Cubic) Metamorphic

Index

7042 BH 921
7-30-96 161192 MS
ICI
INFORMATION CONSERVATION, INC